AF412053

Synaptic Plasticity in Pain

Marzia Malcangio
Editor

Synaptic Plasticity in Pain

Editor
Marzia Malcangio
Wolfson Centre for Age Related Diseases
King's College London
London SE1 1UL
United Kingdom
marzia.malcangio@kcl.ac.uk

ISBN 978-1-4419-0225-2   e-ISBN 978-1-4419-0226-9
DOI 10.1007/978-1-4419-0226-9
Springer Dordrecht Heidelberg London New York

Library of Congress Control Number: 2009928134

Printed on acid-free paper

Springer is part of Springer Science+Business Media (www.springer.com)

Contents

Contributors

Mark L. Baccei MED-Anaesthesiology, University of Cincinnati, OH, USA, bacceiml@ucmail.uc.edu

Rita Bardoni Department of Biomedical Sciences, University of Modena and Reggio Emilia, Modena, Italy, bardoni@unimo.it

Lucy Bee Neuroscience, Physiology and Pharmacology, University College London, London, UK, l.bee@ucl.ac.uk

Simon Beggs Program in Neurosciences and Mental Health, Hospital for Sick Children, University of Toronto Centre for the Study of Pain, Toronto, Ontario, Canada, simon.beggs@utoronto.ca

Fernando Cervero Anaesthesia Research Unit, McGill University, Montreal, Quebec, Canada, fernando.cervero@mcgill.ca

Victoria Chapman School of Biomedical Sciences, University of Nottingham, Nottingham, UK, victoria.chapman@nottingham.ac.uk

Anna K. Clark Wolfson Centre for Age Related Diseases, King's College London, London, UK, anna.clark@kcl.ac.uk

Joyce A. DeLeo Department of Pharmacology and Toxicology, Dartmouth College, Hanover, NH, USA; Neuroscience Center at Dartmouth, Dartmouth College, Hanover, NH, USA; Department of Anesthesiology, Dartmouth-Hitchcock Medical Center, Lebanon, NH, USA, joyce.a.deleo@dartmouth.edu

Anthony Dickenson Neuroscience, Physiology and Pharmacology, University College London, London, UK, anthony.dickenson@ucl.ac.uk

Ruth Drdla Department of Neurophysiology, Center for Brain Research, Medical University of Vienna, Vienna, Austria, ruth.drdla@meduniwien.ac.at

Andrea Ebersberger Department of Physiology, Friedrich-Schiller-University of Jena, Jena, Germany, andrea.ebersberger@mti.uni-jena.de

Maria Fitzgerald Department of Anatomy and Developmental Biology, University College London, London, UK, m.fitzgerald@ucl.ac.uk

Xiao Ying Hua Moores Cancer Center, University of California, San Diego, CA, USA, xyhua@ucsd.edu

Rachel A. Ingram Department of Anatomy and Developmental Biology, University College London, London, UK, rachel.ingram@ucl.ac.uk

Ru-Rong Ji Department of Anesthesiology, Brigham and Women's Hospital and Harvard Medical School, Pain Research Center, Boston, MA, USA, rrji@zeus.bwh.harvard.edu

Torbjörn Johansson Institute of Pharmacology and Toxicology, University of Zurich, Zurich, Switzerland; Institute of Pharmaceutical Sciences, ETH Zurich, Switzerland, johansson@pharma.uzh.ch

David Kendall School of Biomedical Sciences, University of Nottingham, Nottingham, UK, david.kendall@nottingham.ac.uk

Josephine Lai Department of Pharmacology, University of Arizona Health Sciences Center, Tucson, AZ, USA, lai@email.arizona.edu

Jennifer M.A. Laird Department of Pharmacology and Therapeutics and Alan Edwards Centre for Research on Pain, McGill University, Montréal, Quebec, Canada; AstraZeneca R&D Montréal, Montréal, Quebec, Canada, jennifer.laird@astrazeneca.com

Melissa J. Mahoney Department of Chemical and Biological Engineering, University of Colorado at Boulder, Boulder, CO, USA, melissa.mahoney@ colorado.edu

Marzia Malcangio Wolfson Centre for Age Related Diseases, King's College London, London, UK, marzia.malcangio@kcl.ac.uk

Juan Carlos Marvizon Center for Neurobiology of Stress, Division of Digestive Diseases, Department of Medicine, David Geffen School of Medicine at UCLA, Los Angeles, CA, USA, marvizon@ucla.edu

Adalberto Merighi Department of Veterinary Morphophysiology, University of Turin, Turin, Italy; Istituto Nazionale di Neuroscienze, Turin, Italy

Erin D. Milligan Department of Neurosciences, University of New Mexico Health Sciences Center, Albuquerque, NM, USA, emilligan@salud.unm.edu

Volker Neugebauer Department of Neuroscience and Cell Biology, The University of Texas Medical Branch, Galveston, TX, USA, voneugeb@utmb.edu

Frank Porreca Department of Pharmacology, University of Arizona Health Sciences Center, Tucson, AZ, USA, frankp@email.arizona.edu

John V. Priestley Barts and The London School of Medicine and Dentistry, Institute of Cell and Molecular Science, Neuroscience Centre, Queen Mary University of London, London, UK, j.v.priestley@qmul.ac.uk

Mirjana Randić Department of Biomedical Sciences, Iowa State University, Ames, IA, USA, mrandic@iastate.edu

Devi Rani Sagar School of Biomedical Sciences, University of Nottingham, Nottingham, UK, devi.sagar@nottingham.ac.uk

Jürgen Sandkühler Department of Neurophysiology, Center for Brain Research, Medical University of Vienna, Vienna, Austria, juergen.sandkuehler@meduniwien.ac.at

Hans-Georg Schaible Department of Physiology, Friedrich-Schiller-University of Jena, Jena, Germany, hans-georg.schaible@mti.uni-jena.de

Ryan G. Soderquist Department of Chemical and Biological Engineering, University of Colorado at Boulder, Boulder, CO, USA, ryan.soderquist@colodado.edu

Camilla I. Svensson Department of Physiology and Pharmacology, Karolinska Institutet, Stockholm, Sweden, camilla.svensson@ki.se

Vivianne L. Tawfik Department of Pharmacology and Toxicology, Dartmouth College, Hanover, NH, USA; Dartmouth College, Neuroscience Center at Dartmouth, Hanover, NH, USA, vivianne.tawfix@dartmouth.edu

Stephen W.N. Thompson Biomedical Science, University of Plymouth, Plymouth, PL4 8AA, UK, stephen.thompson@plymouth.ac.uk

Andrew J. Todd Institute of Biomedical and Life Sciences, University of Glasgow, Glasgow, UK, a.todd@bio.gla.ac.uk

Ruizhong Wang Department of Pharmacology, University of Arizona Health Sciences Center, Tucson, AZ, USA, wangrz@email.arizona.edu

Robert Witschi Institute of Pharmacology and Toxicology, University of Zurich, **Zurich**, Switzerland; Institute of Pharmaceutical Sciences, ETH Zurich, Switzerland, witschi@pharma.unizh.ch

Tony L. Yaksh Molecular Biology, University of California, San Diego, CA, USA, tyaksh@ucsd.edu

Hanns Ulrich Zeilhofer Institute of Pharmacology and Toxicology, University of Zurich, Winterthurerstrasse 190, CH-8057 Zurich, Switzerland; Institute of Pharmaceutical Sciences, ETH Zurich, Wolfgang-Pauli Strasse 10, 8093 Zurich, Switzerland, zeilhofer@pharma.uzh.ch

Introduction

The chapters in this book have evolved around the concept that the first sensory synapse between the central terminals of primary sensory neurons and dorsal horn neurons in the spinal cord is plastic and modifiable. Thus, the book title reflects the effort of the several authors to address this idea of plasticity in pain from their own perspective. I am grateful to colleagues who have contributed with enthusiasm and competence to this task and particularly Dr. Sandkuhler for his advice and suggestions.

As extensively stressed throughout the book chapters, the detection and perception of pain have multi-dimensional nature and pain is defined by the International Association for the Study of Pain (IASP) as 'an unpleasant sensory and emotional experience associated with actual or potential tissue damage, or described in terms of such damage'.

Primary sensory neurons respond to peripheral stimulation and project to the spinal cord. Specifically, the population of neurons which respond to damaging stimuli terminate in the superficial layers of the dorsal horn. The passage of sensory inputs from the dorsal horn of the spinal cord to higher centres in the brain is modulated by both descending facilitatory and inhibitory neurones. Therefore, the dorsal horns constitute the first relay site for nociceptive fibre terminals which make synaptic contacts with second order neurons. It has recently become clear that the strength of this first sensory synapse is plastic and modifiable by several modulators, including neuronal and non-neuronal regulators. Undoubtedly, the studies on the fundamental processes regulating the plasticity of the first pain synapse have resulted in the identification of new targets for the treatment of chronic pain.

This book includes six sections which start from the delineation of some anatomical circuits for pain in the dorsal horn.

The next two sections are focussed on the main players of the fast and slow transmissions at the pain synapse including GABA and the opioids as well as substance P, calcitonin gene-related peptide and brain derived neurotrophic factor.

The fourth section is concerned with synaptic plasticity and the application of sensory information in the dorsal horn of the spinal cord.

The final section consists of several chapters on mechanisms and targets for chronic pain in the dorsal horn, including the arthritic pain, visceral pain and neuropathic pain. Specifically, a number of contributors have expressed their views on the role played in the modulation of pain mechanisms by non neuronal cells, astrocytes and microglia which have recently become the focus of intensive research.

This book will be of interest to a wide readership in the pain field including PhD students, post-doc scientists and academics. Drug discovery teams in the private sector will find in this book some solid scientific support to their research.

Furthermore this book will arouse scientists interested in synaptic plasticity associated with other CNS functions such as hippocampal plasticity in learning processes.

Finally, I wish to thank Ann Avouris from Springer who has ideated this project and has supported this initiative with optimism.

Part I
Anatomical Plasticity of Dorsal Horn Circuits

Chapter 1
Changes in NK1 and Glutamate Receptors in Pain

Andrew J. Todd

Abstract The amino acid glutamate and the neuropeptide substance P are contained in many nociceptive primary afferents that terminate mainly in the superficial part of the dorsal horn. Both glutamate and substance P are released from the central terminals of nociceptive afferents following noxious stimulation. Glutamate acts on a variety of ionotropic and metabotropic receptors, while substance P acts on the neurokinin 1 receptor (NK1r), and both transmitters contribute to the processing of nociceptive information at the spinal level. Noxious stimulation of the hindpaw causes rapid (within minutes) internalisation of the NK1r, phosphorylation of the GluR1 subunit of the AMPA-type glutamate receptor and phosphorylation of the NR1 subunit of NMDA-type glutamate receptors. These plastic changes of SP and glutamate receptors that occur in acute and chronic pain states presumably contribute to sensitisation of dorsal horn neurons (central sensitization).

Abbreviations

AMPA	α-amino-2,3-dihydro-5-methyl-3-oxo-4-isoxazolepropanic acid
CGRP	calcitonin gene related peptide
CFA	complete Freund's adjuvant
CVLM	caudal ventrolateral medulla
DRG	dorsal root ganglion
LPb	lateral parabrachial area
LTP	long term potentiation
NK1r	neurokinin 1 receptor
NMDA	N-methyl-D-aspartate
PAG	periaqueductal grey

A.J. Todd (✉)
Institute of Biomedical and Life Sciences, University of Glasgow, Glasgow, UK
e-mail: a.todd@bio.gla.ac.uk

M. Malcangio (ed.), *Synaptic Plasticity in Pain*,
DOI 10.1007/978-1-4419-0226-9_1, © Springer Science+Business Media, LLC 2009

1.1 Introduction

Substance P, and the neurokinin 1 receptor (NK1r) on which it acts, have long been thought to play an important role in pain mechanisms. Substance P is contained in many nociceptive primary afferents, released following noxious stimulation and activates NK1rs on certain dorsal horn neurons. Glutamate is also released by central terminals of nociceptive afferents and acts on a variety of ionotropic and metabotropic receptors in the dorsal horn. The receptors that form the subject of this chapter are therefore activated following noxious stimulation and contribute to the processing of nociceptive information at the spinal level. The chapter will focus on aspects of dorsal horn anatomy that are relevant to substance P and the NK1r and to glutamatergic transmission, and will discuss evidence for plasticity involving these receptors. Only a brief description of other aspects of dorsal horn anatomy will be given here, and for more detailed descriptions the reader is referred to other recent reviews (Todd and Koerber, 2005; Ribeiro-da-Silva and De Koninck, 2008). This account is based on findings in the rat (unless stated otherwise), since most anatomical data have been obtained in this species.

1.2 Anatomical Components of the Dorsal Horn

The dorsal horn receives its major input from primary afferent axons, which arborise in a modality-specific pattern. It contains a diverse collection of neurons that can be divided into two main classes: (1) those with axons that remain in the spinal cord (interneurons), and (2) projection neurons, with axons that ascend through the white matter and terminate in the brain, forming a major output from the dorsal horn. The interneurons include both excitatory (glutamatergic) and inhibitory (mainly GABAergic) cells, while most projection neurons are glutamatergic (Broman, 1994; Todd and Koerber, 2005). The dorsal horn also receives inputs from descending axons that originate in several brain regions and modulate the transmission of sensory information (see Chapter 19).

Rexed (1954) divided the dorsal horn of the cat spinal cord into 6 laminae and this scheme, which has been applied to other species, is widely used for descriptive purposes. Laminae I and II are often referred to as the superficial dorsal horn, and form the main termination zone for nociceptive primary afferents. Laminae III–VI (the deep dorsal horn) receive their major primary afferent input from low-threshold mechanoreceptive afferents. However, this region is also important in pain mechanisms, since it contains projection neurons that convey nociceptive information, and because the low-threshold afferents that terminate within it can give rise to tactile allodynia (touch-evoked pain) in certain pain states.

1.3 Substance P and the NK1r

1.3.1 Sources of Substance P in the Dorsal Horn

Lamina I and the outer part of lamina II (IIo) contain a dense plexus of substance P-containing axons, most of which are of primary afferent origin. In addition, scattered substance P axons (including primary afferents) terminate in deeper laminae. Information about the functions of substance P-containing primary afferents has come from studies in which immunostaining has been carried out after electrophysiological characterisation of individual afferents recorded in the guinea pig dorsal root ganglion (DRG) (Lawson et al., 1997). Substance P was found most commonly in cells that gave rise to unmyelinated (C) fibres, but also in some with small (Aδ) or large (Aβ) myelinated axons. All substance P-containing afferents were nociceptors, but not all nociceptive afferents contained the peptide. Substance P was particularly associated with nociceptive afferents that had deep cutaneous receptive fields, although it was also seen in some polymodal C nociceptors that innervated glabrous skin. In the rat, all peptidergic primary afferents contain calcitonin gene-related peptide (CGRP) (Ju et al., 1987), which is only found in primary afferent axons in the dorsal horn. Therefore the presence of CGRP can be used in double-labelling immunocytochemical studies to distinguish between substance P-containing axons that are primary afferents and those that are not (Sakamoto et al., 1999). Cell bodies that contain substance P or the mRNA for its precursor protein (preprotachykinin 1) are present in the dorsal horn (Hökfelt et al., 1977; Warden and Young, 1988) and give rise to the non-primary substance P-containing axons.

1.3.2 Anatomical Distribution of NK1r

Several immunocytochemical studies have described the distribution of the NK1r in the spinal dorsal horn (Bleazard et al., 1994; Liu et al., 1994; Nakaya et al., 1994; Brown et al., 1995; Littlewood et al., 1995; Mantyh et al., 1995; Todd et al., 1998). NK1r-immunoreactivity is present on the cell bodies and dendrites of certain dorsal horn neurons, but not on axons in the spinal cord. Immunostaining for the receptor is particularly dense in lamina I and is scattered throughout the deeper laminae (III–VI), but is present on very few neurons in lamina II. Within the dorsal horn, we have estimated that ~45% of neurons in lamina I and 10–30% of those in laminae III–VI are NK1r-immunoreactive. Most of the dendrites of the NK1r-immunoreactive lamina I cells are restricted to this lamina, where they make up a dense plexus. Cheunsuang and Morris (2000) have demonstrated that there is a bimodal size distribution of NK1r-positive neurons in lamina I, with a population of small weakly stained cells, and a group of large cells that generally show strong immunoreactivity. Among the NK1r-expressing neurons in deeper laminae, there is a population of large neurons with dendrites that travel dorsally to enter lamina I (Liu et al., 1994; Brown et al., 1995; Littlewood et al., 1995; Mantyh et al., 1995; Naim et al., 1997). Although these cells are very distinctive, there are only ~20–25 of them on either side in each mid-lumbar segment and slightly fewer per segment in the cervical enlargement (Todd et al., 2000; Al-Khater et al., 2008).

1.3.3 Projection Neurons and the NK1r

Cell bodies of projection neurons can be identified by injection of retrograde tracers into brain regions where their axons terminate. Studies of this type have shown that in rat lumbar enlargement projection neurons are concentrated in lamina I and scattered throughout laminae III–VI and the lateral spinal nucleus. Many of these cells have axons that cross the midline and ascend in the contralateral white matter to terminate in various regions of the brainstem and thalamus (Todd, 2002). Brainstem regions that receive inputs from lamina I projection neurons include the caudal ventrolateral medulla (CVLM), lateral parabrachial area (LPb) and periaqueductal grey matter (PAG).

It has been estimated that lamina I contains approximately 400 projection neurons on each side in the L4 segment in the rat (Todd et al., 2000; Spike et al., 2003). Most of these (~85%) project to contralateral LPb, with around 30% sending collaterals to the PAG. Only ~15 lamina I neurons/segment project to the thalamus from the midlumbar cord, although the number of lamina I spinothalamic neurons is much higher (~90 cells/segment) in the cervical enlargement (Al-Khater et al., 2008). Most lamina I projection neurons send their axons only to contralateral brain targets, but some have bilateral projections (Spike et al., 2003).

Since the NK1r is present on many lamina I neurons, several studies have investigated the extent to which the receptor is expressed by projection neurons in this lamina (Ding et al., 1995; Marshall et al., 1996; Li et al., 1998; Todd et al., 2000; Spike et al., 2003; Al-Khater et al., 2008). We have estimated that ~80% of lamina I neurons that project to thalamus, LPb, PAG or the medulla are NK1r-immunoreactive (Marshall et al., 1996; Todd et al., 2000; Spike et al., 2003; Al-Khater et al., 2008), and these correspond to the large NK1r-positive cells identified by Cheunsuang and Morris (2000) (Polgár et al., 2002). All of the large NK1r-positive cells in laminae III and IV with long dorsal dendrites that enter the superficial dorsal horn are projection neurons, since virtually all of them can be labelled with tracer injected into the CVLM, while two-thirds project to LPb (Todd et al., 2000). We have recently shown that approximately 20% of these cells in the lumbar enlargement, and about 85% of those at cervical levels, belong to the spinothalamic tract (Al-Khater et al., 2008).

The NK1r-immunoreactive projection neurons in lamina I, as well as those located in laminae III and IV, receive a dense synaptic input from substance

P-containing primary afferents (Naim et al., 1997; Todd et al., 2002). These afferents not only innervate dendrites of these cells that lie within the dense plexus of substance P axons in laminae I–IIo, but also make numerous synapses on dendrites of the lamina III/IV cells that lie below the plexus. For the lamina III/IV cells, it has been shown that primary afferent inputs are organised in a selective manner, since these cells receive very few contacts from C fibres that do not contain substance P (Sakamoto et al., 1999).

1.3.4 Plasticity of NK1rs in the Dorsal Horn

Mantyh et al. (1995) demonstrated that acute noxious stimulation of the rat hindpaw caused internalisation of NK1rs on many lamina I neurons and on dorsal dendrites of the large lamina III/IV cells. Internalisation causes loss of NK1r-immunoreactivity from the plasma membrane and the appearance of immunoreactive endosomes (Fig. 1.1). There is also a structural alteration, with thin dendrites showing a marked beading. These changes develop rapidly

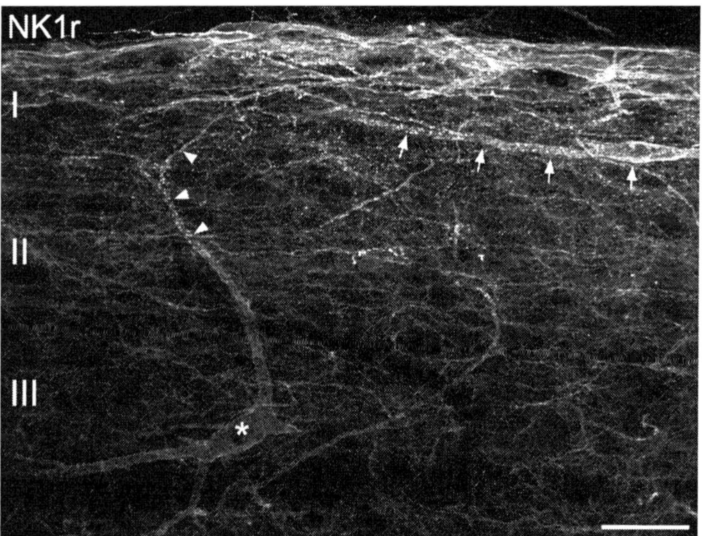

Fig. 1.1 Internalisation of NK1 receptor following acute noxious stimulation. A confocal image showing part of a parasagittal section through the dorsal horn of the L4 segment of a rat stained for the NK1r. The rat had received noxious mechanical stimulation of the ipsilateral hindpaw (pinching of the skin) under terminal general anaesthesia 5 minutes before perfusion fixation. Internalisation of the receptor is seen on a dendrite that belongs to a lamina I neuron (*arrows*) and on the distal part of a dorsal dendrite of a large lamina III cell (*arrowheads*). The soma (*asterisk*) and proximal dendrites of the lamina III cell show the normal distribution of NK1r on the surface. Scale bar = 50 μm. Modified from Polgár et al. (2007) with permission from BioMed Central

(within 1 minute) and last less than an hour. On lamina I cells, the entire
somatodendritic membrane is affected, and since substantial parts of these
cells are not in contact with substance P-containing axons (Todd et al., 2002),
this is consistent with the view that substance P diffuses from its release sites and
acts through volume transmission.

It was subsequently shown that following inflammation of the hindpaw with
complete Freund's adjuvant (CFA), internalisation of the NK1r in the ipsilateral dorsal horn was substantially increased (Abbadie et al., 1997). This change
was reflected in an increase in the number of lamina I neurons that showed
internalisation after noxious stimulation, as well as significant internalisation
in NK1r-positive cell bodies in deeper laminae, which was not seen in normal
rats. In addition, previously innocuous mechanical stimuli could cause internalisation of the receptor. Allen et al. (1999) reported that during CFA-induced
inflammation, NK1r internalisation was evoked by electrical stimulation of
sciatic nerve only at Aδ or C fibre strength. This indicates that the internalisation seen during inflammation is caused by substance P released from Aδ and C
fibres. Increased internalisation of the receptor may also play a role in visceral
pain and hyperalgesia, since Honoré et al. (2002) reported that colonic inflammation led to increased internalisation in lamina I neurons following noxious
colo-rectal distension, as well as internalisation following non-noxious visceral
stimuli (see Chapter 13).

There is also evidence for increased NK1r expression in the ipsilateral dorsal
horn following inflammation or nerve injury. There is a higher level of NK1r
mRNA after injection of CFA (Schäfer et al., 1993; McCarson and Krause,
1994), while NK1r-immunoreactivity increases after inflammatory stimuli or
various types of nerve injury (Abbadie et al., 1996, 1997; Goff et al., 1998;
Honoré et al., 1999) (Fig. 1.2). Up-regulation of NK1r was seen in both lamina I

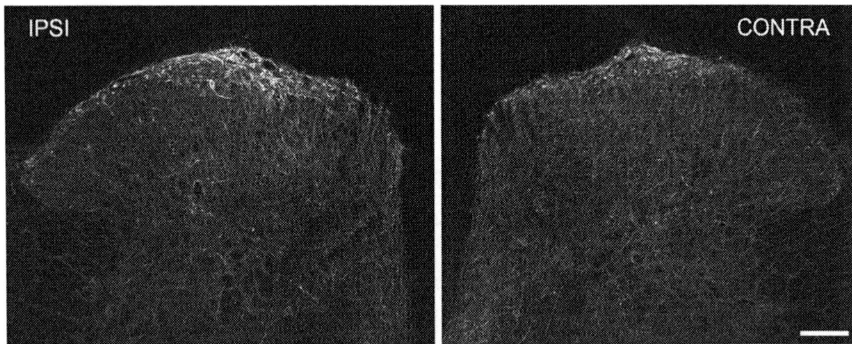

Fig. 1.2 Upregulation of NK1r in the L4 dorsal horn of a rat two weeks after chronic
constriction injury of the left sciatic nerve. NK1r-immunoreactivity is densest in lamina I
on both sides, but the staining is considerably stronger on the ipsilateral side (ipsi), compared
to that on the contralateral side (contra). Scale bar = 100 μm. Reproduced from Todd and
Ribeiro-da-Silva (2007) with permission

and laminae III–IV, and extended into regions outside those innervated by the affected nerves or skin territories. Abbadie et al. (1997) reported that the number of NK1r-positive lamina I neurons was not altered following inflammation with CFA, although their staining intensity increased. However, palecek et al. (2003) reported that inflammation of the colon resulted in *de novo* expression of NK1r by a small number of lamina III/IV projection neurons belonging to the post-synaptic dorsal column pathway, which are not normally NK1r-immunoreactive (Polgár et al., 1999).

1.4 Sources of Glutamatergic Input to the Dorsal Horn

All primary afferents use glutamate as a neurotransmitter (De Biasi and Rustioni, 1988; Broman et al., 1993), and their central terminations are arranged in a highly ordered way that depends on fibre diameter and sensory modality (Ribeiro-da-Silva and De Koninck, 2008). Unmyelinated afferents, most of which function as nociceptors or thermoreceptors, can be divided into two main groups, those with neuropeptides and those without. Most peptidergic afferents project to laminae I–IIo, with some arborising further ventrally. In contrast, the majority of non-peptidergic C fibres terminate in a narrow band that occupies the central part of lamina II. Aδ afferents project to two different regions in the dorsal horn: Aδ nociceptors terminate mainly in lamina I (with additional branches to laminae V and X), while those that innervate down hairs (D-hair afferents) arborise in inner lamina II (IIi) and lamina III. Low-threshold mechanoreceptive Aβ afferents end in a region extending ventrally from lamina IIi.

Excitatory interneurons and projection cells in the dorsal horn provide another major source of glutamatergic axons. Between 25 and 40% of neurons in laminae I–III are GABA-immunoreactive (Polgár et al., 2003) and the remainder are thought to be excitatory, glutamatergic cells. Until the discovery of the vesicular glutamate transporters it was difficult to identify the axons of these cells. However, it is now known that most (if not all) glutamatergic neurons in the dorsal horn express VGLUT2, and these cells are likely to give rise to the great majority of VGLUT2-immunoreactive boutons that are present in large numbers throughout the dorsal horn (Oliveira et al., 2003; Todd et al., 2003; Alvarez et al., 2004).

There are also descending glutamatergic axons (e.g. corticospinal tract), although little is known about the synaptic arrangements that these form.

1.5 Glutamate Receptors

1.5.1 Ionotropic Receptors at Glutamatergic Synapses

In situ hybridization studies have shown that all 3 types of ionotropic glutamate receptor (NMDA, AMPA and kainate) are present in the dorsal horn

(Furuyama et al., 1993; Tölle et al., 1993; Watanabe et al., 1994). However, although immunocytochemistry can be used to reveal the subunits of these receptors at non-synaptic sites such as the perikaryal cytoplasm (Tachibana et al., 1994; Jakowec et al., 1995; Popratiloff et al., 1996), it is difficult to detect receptors at synapses because of the extensive cross-linking of synaptic proteins that results from fixation. We have therefore used an antigen retrieval method based on pepsin treatment to reveal the synaptic distribution of AMPA (Nagy et al., 2004a; Polgar et al., 2008) and NMDA (Nagy et al., 2004b) receptor subunits in the rat spinal cord.

AMPA receptors (AMPArs) are tetramers made up from 4 subunits (GluR1-4, or GluRA-D), and both heteromeric and homomeric arrangements can form functional receptors. Most AMPArs contain the GluR2 subunit, which renders them impermeable to Ca^{2+}. Following antigen retrieval, AMPArs can be detected at most if not all, glutamatergic synapses throughout the spinal grey matter (Nagy et al., 2004a; Polgar et al., 2008). Virtually all of these synapses contain GluR2, while the other 3 subunits have distinct laminar distributions. In laminae I–II, GluR1 and GluR3 are each present in ~60–65% of AMPAr-containing synaptic puncta, with only 10% of puncta lacking both of these subunits, whereas GluR4 is present in ~25% of puncta in lamina I and <10% of those in lamina II. Further ventrally, expression of GluR4 increases, while that of GluR1 decreases, and in lamina IV–IX most glutamatergic synapses contain GluR2, GluR3 and GluR4. Ca^{2+}-permeable (GluR2-lacking) AMPArs are present in the dorsal horn (Engelman et al., 1999) and since virtually all glutamatergic synapses appear to contain GluR2, this suggests that there is heterogeneity of subunit composition, with Ca^{2+}-permeable and - impermeable receptors being intermingled at synapses (Tong and MacDermott, 2006).

The NMDA receptor (NMDAr) is also a tetramer, but unlike the AMPAr only heteromeric arrangements form functional receptors. These normally contain two NR1 and two NR2 subunits. There are four different NR2 subunits (A–D) and 8 splice variants of the NR1 subunit, which result from inclusion or exclusion of two cassettes (N1 and C1) and a further modification that can lead to replacement of another cassette (C2 or C2'). In situ hybridisation data has suggested that all spinal neurons express the NR1 subunit (Tölle et al., 1993), and we have found with immunocytochemistry after antigen retrieval that synaptic staining for this subunit is widespread throughout the spinal grey matter (Nagy et al., 2004b). In contrast, NR2A puncta were most numerous in the deep dorsal horn (particularly laminae III–IV) and ventral horn, while NR2B puncta were concentrated in laminae I–II. As expected, both of these NR2 subunits showed substantial co-localisation with NR1. The laminar distribution of synaptic NR1, NR2A and NR2B subunits that we observed with immunocytochemistry closely matches the distributions of their mRNAs reported by Watanabe et al. (1994).

At present, little is known about the synaptic distribution of kainate receptors in the spinal cord.

1.5.2 Metabotropic Glutamate Receptors

Eight different metabotropic glutamate receptors (mGluR1-8) have been identified, and several of these are present in the spinal cord (Vidnyánszky et al., 1994; Ohishi et al., 1995; Boxall et al., 1998; Jia et al., 1999; Alvarez et al., 2000; Azkue et al., 2001; Walker et al., 2001). Staining for mGluR5 is very dense in the superficial laminae, where it is thought to be present in both local neurons and primary afferent terminals. mGluR1 is expressed by neurons in deeper laminae of the dorsal horn and in the ventral horn. Immunostaining with an mGluR2/3 antibody (probably representing mGluR3) is concentrated in laminae IIi and III, and is located in both intrinsic neurons and primary afferent terminals. mGluR4 and mGluR7 are both concentrated in the superficial dorsal horn, with much of the receptor being associated with primary afferent terminals and some with local neurons.

1.5.3 Plasticity Involving Glutamate Receptors

1.5.3.1 AMPA Receptors

Central sensitisation of dorsal horn neurons, which contributes to both inflammatory and neuropathic pain states, shows certain similarities to long-term potentiation (LTP) (see Chapter 9). LTP in the hippocampus is thought to involve insertion of AMPArs into the affected synapses, leading to an increase in synaptic strength. Phosphorylation of the GluR1 subunit seems to be required for this form of activity-dependent insertion during hippocampal LTP, and this also results in increased current flow through the receptor. Several studies have therefore examined whether there is upregulation or phosphorylation of AMPAr subunits following noxious stimulation, induction of inflammation, or nerve injury, since such changes might be at least partially responsible for central sensitisation in the dorsal horn.

There is evidence for an increase in the mRNA or protein level of all four AMPAr subunits in the ipsilateral dorsal horn after various types of nerve injury (Harris et al., 1996; Popratiloff et al., 1998; Garry et al., 2003; Yang et al., 2004). Popratiloff et al. (1998) carried out post-embedding immunogold labelling with an antibody that recognises both GluR2 and GluR3 subunits on spinal cords following sciatic nerve transection and reported that the density of receptors was increased at synapses formed by Aδ D-hair afferents. However, the functional significance of this observation is hard to interpret, as these afferents will have lost their peripheral receptive fields due to axotomy.

Plasticity involving AMPArs has also been investigated following acute noxious stimuli or during inflammation. Zhou et al. (2001) examined AMPAr subunit mRNA levels after injection of CFA into one hindpaw, and reported that there were changes in the ipsilateral dorsal horn. mRNA for GluR1 was upregulated within the first few hours, while mRNAs for GluR2 and GluR3 showed a slower increase (5–24 hours after CFA injection). Fang et al. (2003) performed Western blots on tissue from both dorsal horns in rats that had received injections of capsaicin into the hindpaw. They observed an increase in the level of phosphorylation of the GluR1 subunit at two serine residues (S831 and S845) on the ipsilateral side, which developed within 5 minutes and lasted for 1 hour after the stimulus. They also reported an increase in staining in the ipsilateral dorsal horn with antibodies that recognised the two phosphorylated forms of GluR1. However, it is unlikely that this staining represents synaptic receptors, since these are generally masked by tissue fixation (see above). We have used antigen retrieval to investigate phosphorylation of GluR1 at synapses in the dorsal horn (Nagy et al., 2004a). We found that the basal level of S845 phosphorylation was very low, but that 10 minutes after injection of capsaicin into the hindpaw there was a significant increase in the level of GluR1-S845 at synapses in laminae I and II in the somatotopically appropriate part of the ipsilateral dorsal horn (Fig. 1.3a, b). In contrast, there was a significant basal

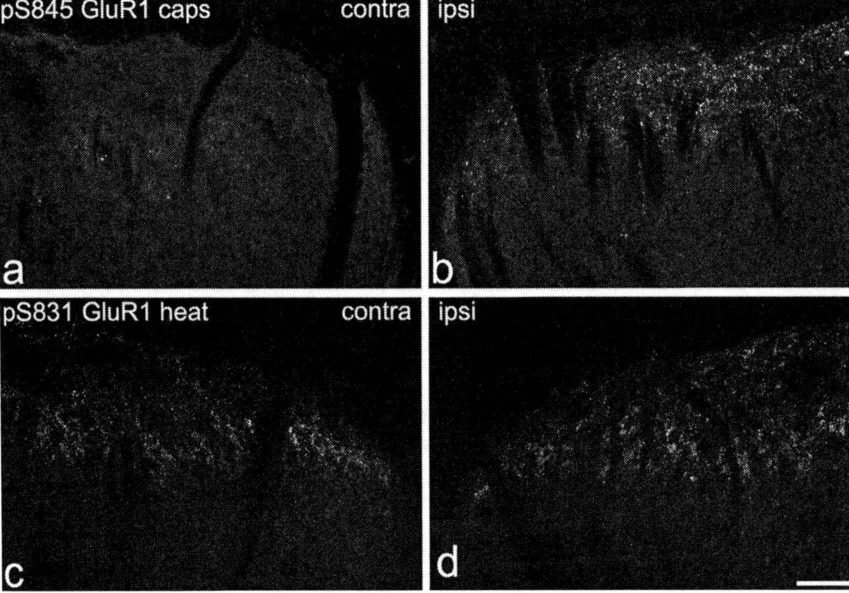

Fig. 1.3 Immunostaining for GluR1 phosphorylated at S845 **a,b** or S831 **c,d** residues following antigen retrieval. The confocal images show transverse sections from the L4 segment of rats that had received a capsaicin injection into one hindpaw **a,b** or had one hindpaw immersed in water at 52°C for 45 seconds **c,d**. In each case the stimulus was applied under terminal general anaesthesia 10 minutes prior to perfusion fixation, and the sections were reacted by an antigen-retrieval method that reveals synaptic receptors. **a,b**: Note that there is little phospho-S845-GluR1 on the contralateral (contra) side, but a relatively high level in laminae I and II on the ipsilateral (ipsi) side. **c,d**: there is significant basal level of phospho-S831-GluR1 (as seen on the contralateral side), and this is not altered following the stimulus. Scale bar = 50 μm. Reproduced from Todd (2008) with permission

level of GluR1 phosphorylated at S831, and this did not change after noxious stimulation (Nagy and Todd, unpublished observations) (Fig. 1.3c, d). We have also found that phosphorylation of GluR1 on S845 at synapses in laminae I–II can be induced by noxious thermal stimulation and that it begins within 5 minutes of the stimulus and lasts for at least 2 hours (E. Polgár and A.J. Todd, unpublished observations). This provides direct anatomical evidence for a type of plasticity involving AMPArs at glutamatergic synapses in the dorsal horn. Galan et al. (2004) observed a rapid increase in the level of GluR1 (but not of GluR2/3) in the plasma membrane fraction of dorsal horns from rats in a visceral pain model, and a similar finding was reported by Pezet et al. (2008) after formalin injection into the hindpaw. These results suggest that insertion of GluR1 subunits into dorsal horn synapses contributes to hyperalgesia.

1.5.3.2 NMDA Receptors

NMDA receptor subunits have numerous serine and tyrosine phosphorylation sites in their C terminal regions, and several studies have investigated phosphorylation of these in inflammatory or neuropathic pain states. NR1 can be phosphorylated at 3 sites, S890, S896 and S897, which appear in the C1 cassette (see above). Phosphorylation of these sites may result in increased or decreased clustering of the receptor at synapses (Chen and Roche, 2007). NR1 subunits in the Golgi apparatus and endoplasmic reticulum are highly phosphorylated, but they appear to be rapidly dephosphorylated following exit from the endoplasmic reticulum, suggesting a role for these phosporylation sites in intracellular trafficking (Scott et al., 2003). Zou et al. (2000) found that the number of spinothalamic tract neurons showing cytoplasmic staining with an antibody that recognises NR1 phosphorylated at the S897 site was increased in the ipsilateral dorsal horn following injection of capsaicin into one hindpaw, while Brenner et al. (2004) reported an increase in the number of neurons staining with antibody against NR1 phosphorylated at S896 after a noxious heat stimulus. There is also evidence that NR1 phosphorylation is increased following nerve injury, since Gao et al. (2005) showed that spinal nerve ligation led to an increase in the number of phospho-S897-NR1-immunoreactive neurons in the ipsilateral dorsal horn that lasted from 3 days to 4 weeks after the operation. These results provide evidence that trafficking of the NR1 subunit is altered in both acute and neuropathic pain states, although it is not yet clear whether this is reflected in an alteration in the number of NMDA receptors at synapses, or in their functional properties. An additional issue that needs to be considered is that of splice variants. Prybylowski et al. (2001) reported that only ~5% of the NR1 protein in the spinal cord contained the C1 cassette, which includes these phosphorylation sites. Although there may be significant regional differences within the grey matter, this suggests that the changes in phophorylation reported above may affect only a minority of NMDA receptors in the dorsal horn.

There is also evidence that NR2B subunits in the spinal cord are phosphorylated during inflammatory and neuropathic pain states (Guo et al., 2002; Abe et al., 2005). NR2B has several potential phosphorylation sites and specific antibodies have been used to demonstrate phosphorylation of serine 1303 and tyrosine 1472 (Abe et al., 2005; Zhang et al., 2005; Pezet et al., 2008). Phosphorylation at these sites may result in retention of the receptors at the synapse and increased current flow. Abe et al. (2005) reported that after transection of the L5 spinal nerve, there was a substantial increase in the level of NR2B phosphorylated at tyrosine 1472 in the superficial dorsal horn, and showed with electron microscopy that some of this was located at synapses. Caudle et al. (2005) reported a progressive down-regulation of NR2B protein from the spinal cord during the five days following injection of carrageenan into the hindpaw, although Guo et al. (2002) found no change in NR2B protein level following inflammation induced with CFA.

1.5.3.3 Metabotropic Glutamate Receptors

There have been relatively few studies of plasticity involving mGluRs. Boxall et al. (1998) reported that ultraviolet-irradiation of one hindpaw sufficient to cause hyperalgesia was associated with a significant bilateral increase in the level of mRNA for mGluR3 in the dorsal horn, while Dolan et al. (2003, 2004) demonstrated an increase in mRNA and protein for both mGluR3 and mGluR5 in the dorsal horn of the sheep in cases of persistent inflammatory pain due to foot infection, and in a post-surgical pain model. Evidence for plasticity of mGluR expression following nerve injury has been provided by Hudson et al. (2002) who reported an increase in mGluR5 protein in A fibre somata in L4 and L5 DRG, the proximal sciatic nerve stump and the ipsilateral superficial dorsal horn after sciatic nerve transection. Following L5 spinal nerve ligation, they observed upregulation of mGluR5 not only in the affected ganglion, but also in the undamaged L4 DRG, and these increases occurred mainly in ganglion cells that gave rise to myelinated fibres. A recent study by Pitcher et al. (2007) has investigated possible redistribution of mGluR1 and mGluR5 in the ipsilateral dorsal horn following injection of CFA into one hindpaw. They found evidence for up-regulation of mGluR5 and a significant increase in the amount associated with the plasma membrane. Although there was no difference in the amount of mGluR1 labelling associated with the membrane, immunoparticles coding for mGluR1 were located significantly closer to synapses. These results therefore suggest that there is trafficking of the group I mGluRs towards the plasma membrane and towards synaptic active zones, presumably resulting in increased efficiency of synaptic transmission involving these receptors.

1.6 Concluding Remarks

The ionotropic and metabotropic receptors for glutamate and the NK1r for SP undergo plastic changes in the dorsal horn of the spinal cord in models of acute as well as chronic pain. Given the important roles of both glutamate and substance P in nociceptive signalling in the dorsal horn, these changes are likely to contribute to the central sensitisation of dorsal horn neurons that underlies hyperalgesia.

References

Abbadie C, Brown JL, Mantyh PW et al (1996) Spinal cord substance P receptor immunoreactivity increases in both inflammatory and nerve injury models of persistent pain. Neuroscience 70:201–209

Abbadie C, Trafton J, Liu H et al (1997) Inflammation increases the distribution of dorsal horn neurons that internalize the neurokinin-1 receptor in response to noxious and non-noxious stimulation. J Neurosci 17:8049–8060

Abe T, Matsumura S, Katano T et al (2005) Fyn kinase-mediated phosphorylation of NMDA receptor NR2B subunit at Tyr1472 is essential for maintenance of neuropathic pain. Eur J Neurosci 22:1445–1454

Al-Khater KM, Kerr R, Todd AJ (2008) A quantitative study of spinothalamic neurons in laminae I, III and IV in lumbar and cervical segments of the rat spinal cord. J Comp Neurol 511:1–18

Allen BJ, Li J, Menning PM, Rogers SD et al (1999) Primary afferent fibers that contribute to increased substance P receptor internalization in the spinal cord after injury. J Neurophysiol 81:1379–1390

Alvarez FJ, Villalba RM, Carr PA et al (2000) Differential distribution of metabotropic glutamate receptors 1a, 1b, and 5 in the rat spinal cord. J Comp Neurol 422:464–487

Alvarez FJ, Villalba RM, Zerda R et al (2004) Vesicular glutamate transporters in the spinal cord, with special reference to sensory primary afferent synapses. J Comp Neurol 472: 257–280

Azkue JJ, Murga M, Fernandez-Capetillo O et al (2001) Immunoreactivity for the group III metabotropic glutamate receptor subtype mGluR4a in the superficial laminae of the rat spinal dorsal horn. J Comp Neurol 430:448–457

Bleazard L, Hill RG, Morris R (1994) The correlation between the distribution of the NK1 receptor and the actions of tachykinin agonists in the dorsal horn of the rat indicates that substance P does not have a functional role on substantia gelatinosa (lamina II) neurons. J Neurosci 14:7655–7664

Boxall SJ, Berthele A, Laurie DJ et al (1998) Enhanced expression of metabotropic glutamate receptor 3 messenger RNA in the rat spinal cord during ultraviolet irradiation induced peripheral inflammation. Neuroscience 82:591–602

Brenner GJ, Ji RR, Shaffer S, Woolf CJ (2004) Peripheral noxious stimulation induces phosphorylation of the NMDA receptor NR1 subunit at the PKC-dependent site, serine-896, in spinal cord dorsal horn neurons. Eur J Neurosci 20:375–384

Broman J, Anderson S, Ottersen OP (1993) Enrichment of glutamate-like immunoreactivity in primary afferent terminals throughout the spinal cord dorsal horn. Eur J Neurosci 5:1050–1061

Broman J (1994) Neurotransmitters in subcortical somatosensory pathways. Anat Embryol (Berl) 189:181–214

Brown JL, Liu H, Maggio JE et al (1995) Morphological characterization of substance P receptor-immunoreactive neurons in the rat spinal cord and trigeminal nucleus caudalis. J Comp Neurol 356:327–344

Caudle RM, Perez FM, Valle-Pinero AY et al (2005) Spinal cord NR1 serine phosphorylation and NR2B subunit suppression following peripheral inflammation. Mol Pain 1:25

Chen BS, Roche KW (2007) Regulation of NMDA receptors by phosphorylation. Neuropharmacology 53:362–368

Cheunsuang O, Morris R (2000) Spinal lamina I neurons that express neurokinin 1 receptors: morphological analysis. Neuroscience 97:335–345

De Biasi S, Rustioni A (1988) Glutamate and substance P coexist in primary afferent terminals in the superficial laminae of spinal cord. Proc Natl Acad Sci USA 85:7820–7824

Ding YQ, Takada M, Shigemoto R et al (1995) Spinoparabrachial tract neurons showing substance P receptor-like immunoreactivity in the lumbar spinal cord of the rat. Brain Res 674:336–340

Dolan S, Kelly JG, Monteiro AM et al (2003) Up-regulation of metabotropic glutamate receptor subtypes 3 and 5 in spinal cord in a clinical model of persistent inflammation and hyperalgesia. Pain 106:501–512

Dolan S, Kelly JG, Monteiro AM et al (2004) Differential expression of central metabotropic glutamate receptor (mGluR) subtypes in a clinical model of post-surgical pain. Pain 110: 369–377

Engelman HS, Allen TB, MacDermott AB (1999) The distribution of neurons expressing calcium-permeable AMPA receptors in the superficial laminae of the spinal cord dorsal horn. J Neurosci 19:2081–2089

Fang L, Wu J, Zhang X, Lin Q, et al (2003) Increased phosphorylation of the GluR1 subunit of spinal cord alpha-amino-3-hydroxy-5-methyl-4-isoxazole propionate receptor in rats following intradermal injection of capsaicin. Neuroscience 122:237–245

Furuyama T, Kiyama H, Sato K et al (1993) Region-specific expression of subunits of ionotropic glutamate receptors (AMPA-type, KA-type and NMDA receptors) in the rat spinal cord with special reference to nociception. Brain Res Mol Brain Res 18: 141–151

Galan A, Laird JM, Cervero F (2004) In vivo recruitment by painful stimuli of AMPA receptor subunits to the plasma membrane of spinal cord neurons. Pain 112:315–323

Gao X, Kim HK, Chung JM et al (2005) Enhancement of NMDA receptor phosphorylation of the spinal dorsal horn and nucleus gracilis neurons in neuropathic rats. Pain 116:62–72

Garry EM, Moss A, Rosie R et al (2003) Specific involvement in neuropathic pain of AMPA receptors and adapter proteins for the GluR2 subunit. Mol Cell Neurosci 24:10–22

Goff JR, Burkey AR, Goff DJ et al (1998) Reorganization of the spinal dorsal horn in models of chronic pain: correlation with behaviour. Neuroscience 82:559–574

Guo W, Zou S, Guan Y et al (2002) Tyrosine phosphorylation of the NR2B subunit of the NMDA receptor in the spinal cord during the development and maintenance of inflammatory hyperalgesia. J Neurosci 22:6208–6217

Harris JA, Corsi M, Quartaroli M et al (1996) Upregulation of spinal glutamate receptors in chronic pain. Neuroscience 74:7–12

Hokfelt T, Ljungdahl A, Terenius L et al (1977) Immunohistochemical analysis of peptide pathways possibly related to pain and analgesia: enkephalin and substance P. Proc Natl Acad Sci USA 74:3081–3085

Honore P, Menning PM, Rogers SD et al (1999) Spinal substance P receptor expression and internalization in acute, short-term, and long-term inflammatory pain states. J Neurosci 19:7670–7678

Honoré P, Kamp EH, Rogers SD et al (2002) Activation of lamina I spinal cord neurons that express the substance P receptor in visceral nociception and hyperalgesia. J Pain 3:3–11.

Hudson LJ, Bevan S, McNair K et al (2002) Metabotropic glutamate receptor 5 upregulation in A-fibers after spinal nerve injury: 2-methyl-6-(phenylethynyl)-pyridine (MPEP) reverses the induced thermal hyperalgesia. J Neurosci 22:2660–2668

Jakowec MW, Fox AJ, Martin LJ et al (1995) Quantitative and qualitative changes in AMPA receptor expression during spinal cord development. Neuroscience 67:893–907

Jia H, Rustioni A, Valtschanoff JG (1999) Metabotropic glutamate receptors in superficial laminae of the rat dorsal horn. J Comp Neurol 410:627–642

Ju G, Hokfelt T, Brodin E et al (1987) Primary sensory neurons of the rat showing calcitonin gene-related peptide immunoreactivity and their relation to substance P-, somatostatin-, galanin-, vasoactive intestinal polypeptide- and cholecystokinin-immunoreactive ganglion cells. Cell Tissue Res 247:417–431

Lawson SN, Crepps BA, Perl ER (1997) Relationship of substance P to afferent characteristics of dorsal root ganglion neurones in guinea-pig. J Physiol 505 (Pt 1):177–191

Li JL, Ding YQ, Xiong KH et al (1998) Substance P receptor (NK1)-immunoreactive neurons projecting to the periaqueductal gray: distribution in the spinal trigeminal nucleus and the spinal cord of the rat. Neurosci Res 30:219–225

Littlewood NK, Todd AJ, Spike RC et al (1995) The types of neuron in spinal dorsal horn which possess neurokinin-1 receptors. Neuroscience 66:597–608

Liu H, Brown JL, Jasmin L et al (1994) Synaptic relationship between substance P and the substance P receptor: light and electron microscopic characterization of the mismatch between neuropeptides and their receptors. Proc Natl Acad Sci USA 91:1009–1013

Mantyh PW, DeMaster E, Malhotra A et al (1995) Receptor endocytosis and dendrite reshaping in spinal neurons after somatosensory stimulation. Science 268:1629–1632

Marshall GE, Shehab SA, Spike RC et al (1996) Neurokinin-1 receptors on lumbar spinothalamic neurons in the rat. Neuroscience 72:255–263

McCarson KE, Krause JE (1994) NK-1 and NK-3 type tachykinin receptor mRNA expression in the rat spinal cord dorsal horn is increased during adjuvant or formalin-induced nociception. J Neurosci 14:712–720

Nagy GG, Al Ayyan M, Andrew D et al (2004a) Widespread expression of the AMPA receptor GluR2 subunit at glutamatergic synapses in the rat spinal cord and phosphorylation of GluR1 in response to noxious stimulation revealed with an antigen-unmasking method. J Neurosci 24:5766–5777

Nagy GG, Watanabe M, Fukaya M et al (2004b) Synaptic distribution of the NR1, NR2A and NR2B subunits of the N-methyl-d-aspartate receptor in the rat lumbar spinal cord revealed with an antigen-unmasking technique. Eur J Neurosci 20:3301–3312

Naim M, Spike RC, Watt C et al (1997) Cells in laminae III and IV of the rat spinal cord that possess the neurokinin-1 receptor and have dorsally directed dendrites receive a major synaptic input from tachykinin-containing primary afferents. J Neurosci 17: 5536–5548

Nakaya Y, Kaneko T, Shigemoto R et al (1994) Immunohistochemical localization of substance P receptor in the central nervous system of the adult rat. J Comp Neurol 347: 249–274

Ohishi H, Nomura S, Ding YQ et al (1995) Presynaptic localization of a metabotropic glutamate receptor, mGluR7, in the primary afferent neurons: an immunohistochemical study in the rat. Neurosci Lett 202:85–88

Oliveira AL, Hydling F, Olsson E et al (2003) Cellular localization of three vesicular glutamate transporter mRNAs and proteins in rat spinal cord and dorsal root ganglia. Synapse 50:117–129

Palecek J, Paleckova V, Willis WD (2003) Postsynaptic dorsal column neurons express NK1 receptors following colon inflammation. Neuroscience 116:565–572

Pezet S, Marchand F, D'Mello R et al (2008) Phosphatidylinositol 3-kinase is a key mediator of central sensitization in painful inflammatory conditions. J Neurosci 28:4261–4270

Pitcher MH, Ribeiro-da-Silva A, Coderre TJ (2007) Effects of inflammation on the ultrastructural localization of spinal cord dorsal horn group I metabotropic glutamate receptors. J Comp Neurol 505:412–423

Polgar E, Shehab SA, Watt C et al (1999) GABAergic neurons that contain neuropeptide Y selectively target cells with the neurokinin 1 receptor in laminae III and IV of the rat spinal cord. J Neurosci 19:2637–2646

Polgar E, Puskar Z, Watt C et al (2002) Selective innervation of lamina I projection neurones that possess the neurokinin 1 receptor by serotonin-containing axons in the rat spinal cord. Neuroscience 109:799–809

Polgar E, Hughes DI, Riddell JS et al (2003) Selective loss of spinal GABAergic or glycinergic neurons is not necessary for development of thermal hyperalgesia in the chronic constriction injury model of neuropathic pain. Pain 104:229–239

Polgár E, Campbell A, MacIntyre LM et al (2007) Phosphorylation of ERK in neurokinin 1 receptor-expressing neurons in laminae III and IV of the rat spinal dorsal horn following noxious stimulation. Mol Pain 3:4

Polgar E, Watanabe M, Hartmann B et al (2008) Expression of AMPA receptor subunits at synapses in laminae I–III of the rodent spinal dorsal horn. Mol Pain 4:5

Popratiloff A, Weinberg RJ, Rustioni A (1996) AMPA receptor subunits underlying terminals of fine-caliber primary afferent fibers. J Neurosci 16:3363–3372

Popratiloff A, Weinberg RJ, Rustioni A (1998) AMPA receptors at primary afferent synapses in substantia gelatinosa after sciatic nerve section. Eur J Neurosci 10:3220–3230

Prybylowski KL, Grossman SD, Wrathall JR et al (2001) Expression of splice variants of the NR1 subunit of the N-methyl-D-aspartate receptor in the normal and injured rat spinal cord. J Neurochem 76:797–805

Rexed B (1954) A cytoarchitectonic atlas of the spinal cord in the cat. J Comp Neurol 100:297–379

Ribeiro-da-Silva A, de Koninck Y (2008) Morphological and neurochemical organization of the spinal dorsal horn. In: Bushnell MC, Basbaum AI (eds). Pain, Academic Press, San Diego

Sakamoto H, Spike RC, Todd AJ (1999) Neurons in laminae III and IV of the rat spinal cord with the neurokinin-1 receptor receive few contacts from unmyelinated primary afferents which do not contain substance P. Neuroscience 94:903–908

Schafer MK, Nohr D, Krause JE et al (1993) Inflammation-induced upregulation of NK1 receptor mRNA in dorsal horn neurones. Neuroreport 4:1007–1010

Scott DB, Blanpied TA, Ehlers MD (2003) Coordinated PKA and PKC phosphorylation suppresses RXR-mediated ER retention and regulates the surface delivery of NMDA receptors. Neuropharmacology 45:755–767

Spike RC, Puskar Z, Andrew D et al (2003) A quantitative and morphological study of projection neurons in lamina I of the rat lumbar spinal cord. Eur J Neurosci 18:2433–2448

Tachibana M, Wenthold RJ, Morioka H et al (1994) Light and electron microscopic immunocytochemical localization of AMPA-selective glutamate receptors in the rat spinal cord. J Comp Neurol 344:431–454

Todd AJ, Spike RC, Polgar E (1998) A quantitative study of neurons which express neurokinin-1 or somatostatin sst2a receptor in rat spinal dorsal horn. Neuroscience 85:459–473

Todd AJ, McGill MM, Shehab SA (2000) Neurokinin 1 receptor expression by neurons in laminae I, III and IV of the rat spinal dorsal horn that project to the brainstem. Eur J Neurosci 12:689–700

Todd AJ (2002) Anatomy of primary afferents and projection neurones in the rat spinal dorsal horn with particular emphasis on substance P and the neurokinin 1 receptor. Exp Physiol 87:245–249

Todd AJ, Puskar Z, Spike RC et al (2002) Projection neurons in lamina I of rat spinal cord with the neurokinin 1 receptor are selectively innervated by substance p-containing afferents and respond to noxious stimulation. J Neurosci 22:4103–4113

Todd AJ, Hughes DI, Polgar E et al (2003) The expression of vesicular glutamate transporters VGLUT1 and VGLUT2 in neurochemically defined axonal populations in the rat spinal cord with emphasis on the dorsal horn. Eur J Neurosci 17:13–27

Todd AJ, Koerber HR (2005) Neuroanatomical substrates of spinal nociception. In: McMahon S, Koltzenburg M (eds) Melzack and Wall's textbook of pain, Churchill Linvingstone, Edinburgh

Todd AJ, Ribeiro-da-Silva A (2007) Anatomical changes in the spinal dorsal horn after peripheral nerve injury. In Zhuo M (ed) Molecular Pain, Beijing Higher Education Press/Springer, Beijing

Todd AJ (2008) Neuronal circuits and receptors involved in spinal cord pain processing. In Castro-Lopes M (ed) Current topics in pain: 12th World Congress on Pain, IASP Press, Seattle

Tölle TR, Berthele A, Zieglgansberger W et al (1993) The differential expression of 16 NMDA and non-NMDA receptor subunits in the rat spinal cord and in periaqueductal gray. J Neurosci 13:5009–5028

Tong CK, MacDermott AB (2006) Both Ca2 + -permeable and -impermeable AMPA receptors contribute to primary synaptic drive onto rat dorsal horn neurons. J Physiol 575:133–144

Vidnyánszky Z, Hámori J, Negyessy L (1994) Cellular and subcellular localization of the mGluR5a metabotropic glutamate receptor in rat spinal cord. Neuroreport 6:209–213

Walker K, Reeve A, Bowes M (2001) mGlu5 receptors and nociceptive function II. mGlu5 receptors functionally expressed on peripheral sensory neurones mediate inflammatory hyperalgesia. Neuropharmacology 40:10–19

Warden MK, Young WS III (1988) Distribution of cells containing mRNAs encoding substance P and neurokinin B in the rat central nervous system. J Comp Neurol 272:90–113

Watanabe M, Mishina M, Inoue Y (1994) Distinct spatiotemporal distributions of the N-methyl-D-aspartate receptor channel subunit mRNAs in the mouse cervical cord. J Comp Neurol 345:314–319

Yang L, Zhang FX, Huang F et al (2004) Peripheral nerve injury induces trans-synaptic modification of channels, receptors and signal pathways in rat dorsal spinal cord. Eur J Neurosci 19:871–883

Zhang X, Wu J, Lei Y et al (2005) Protein phosphatase modulates the phosphorylation of spinal cord NMDA receptors in rats following intradermal injection of capsaicin. Brain Res Mol Brain Res 138:264–272

Zhou QQ, Imbe H, Zou S et al (2001) Selective upregulation of the flip-flop splice variants of AMPA receptor subunits in the rat spinal cord after hindpaw inflammation. Brain Res Mol Brain Res 88:186–193

Zou X, Lin Q, Willis WD (2000) Enhanced phosphorylation of NMDA receptor 1 subunits in spinal cord dorsal horn and spinothalamic tract neurons after intradermal injection of capsaicin in rats. J Neurosci 20:6989–6997

Chapter 2
Trophic Factors and Their Receptors in Pain Pathways

John V. Priestley

Abstract Trophic factors play a key role in the plasticity of pain pathways. They shape the circuitry and neurochemistry of acute pain pathways and also contribute to changes that occur in chronic inflammatory and neuropathic pain. Adult dorsal root ganglion (DRG) neurons express neurotrophin receptors and localization studies have shown that different DRG subtypes express different receptors. Thus large diameter neurons (low threshold mechanoreceptors) express either trkB (the receptor for BDNF) or trkC (the receptor for neurotrophin-3, NT-3) and small diameter neurons express trkA (the receptor for nerve growth factor, NGF). However the trkA expression is confined to the population of nociceptors that constitutively express neuropeptides (peptidergic nociceptors). Another population of nociceptors exists (non-peptidergic nociceptors) which normally do not express neuropeptides, which can be identified using the lectin *Griffonia simplifolia* IB4, and which express receptor components for GDNF. After nerve injury or inflammation major changes take place that contribute to the development of chronic pain and many of these changes appear to be driven by changes in the availability of growth factors. The role of NGF in such changes has been well documented but less is known about the role of GDNF. However there is growing evidence that endogenous GDNF contributes to inflammatory pain, and that exogenous GDNF can be used to treat neuropathic pain. In each case the GDNF effects are mediated primarily by the non-peptidergic (IB4) population of nociceptors. There is also evidence that neuropoetic cytokines act on non-peptidergic nociceptors.

J.V. Priestley (✉)
Barts and The London School of Medicine and Dentistry, Institute of Cell and Molecular Science, Neuroscience Centre, Queen Mary University of London, 4 Newark Street, Whitechapel, London E1 2AT, UK
e-mail: j.v.priestley@qmul.ac.uk

M. Malcangio (ed.), *Synaptic Plasticity in Pain*,
DOI 10.1007/978-1-4419-0226-9_2, © Springer Science+Business Media, LLC 2009

Abbreviations

BDNF   brain derived neurotrophic factor
CCI   chronic constriction injury
CGRP   calcitonin gene-related peptide
CFA   complete Freund's adjuvant
CNTF   ciliary neurotrophic factor
DRG   dorsal root ganglion
GDNF   glial cell-derived neurotrophic factor
IB4   isolectin B4
IL6   inteleukin 6
LIF   leukaemia inhibitory factor
MCP-1   monocyte chemoattractant protein 1
NGF   nerve growth factor
OM   oncostatin M
NPY   neuropeptide Y
SNL   spinal nerve ligation
Trk   tropomyosin kinase
TRP   transient receptor potential (ion channels)
VIP   vasoactive intestinal peptide

2.1 Introduction

Trophic factors play a key role not only in development and regeneration, but also in regulating the phenotype of cells in the adult nervous system. In the context of pain, they shape the circuitry and neurochemistry of acute pain pathways and also contribute to changes that occur in chronic inflammatory and neuropathic pain. This chapter will review the organization of adult dorsal root ganglion (DRG) neurons and the expression of trophic factors and trophic factor receptors belonging to the neurotrophin, neuropoetic cytokine and glial cell line-derived neurotrophic factor (GDNF) families. Expression in spinal pain pathways, and changes that take place in inflammation and nerve injury, will also be briefly reviewed. Excellent recent reviews on neurotrophins (Pezet and McMahon, 2006) and on immune factors (Marchand et al., 2005) in chronic pain states are already available and so this review will focus particularly on non-peptidergic nociceptive DRG neurons and their regulation by the GDNF and neuropoetic cytokine families. Specific aspects of the role of brain-derived neurotrophic factor (BDNF) in pain pathways are covered in Chapters 5 and 22. Many viscera receive innervation not only from DRG neurons but also from specialised primary sensory neurons located in the nodose and jugular ganglia. Those neurons have distinct properties and will not be discussed, but are covered in several recent reviews (Belvisi, 2002; Blackshaw et al., 2007; Jia and Lee, 2007).

2.2 Expression of Trophic Factors and Their Receptors by DRG Neurones

2.2.1 Subtypes of DRG Neurons

DRG neurons can be divided into various subpopulations based on their neurochemistry, anatomy and physiology (see Fig. 2.1 and reviews by Lawson, 1992; Snider and McMahon, 1998; Lawson, 2002; Priestley, 2008). About 40% of (lumbar) DRG neurons are low threshold mechanoreceptors (Aα/β conduction velocity) which have large or medium sized cell bodies and heavily myelinated axons and which form specialised peripheral endings. These include Pacinian corpuscles, hair follicle afferents, and Merkel touch domes in skin, and spindle afferents in muscle. However they have in common that they can be histochemically distinguished by their high content of phosphorylated heavy chain (200 kDa) neurofilament (Lawson and Waddell, 1991). Aβ high threshold

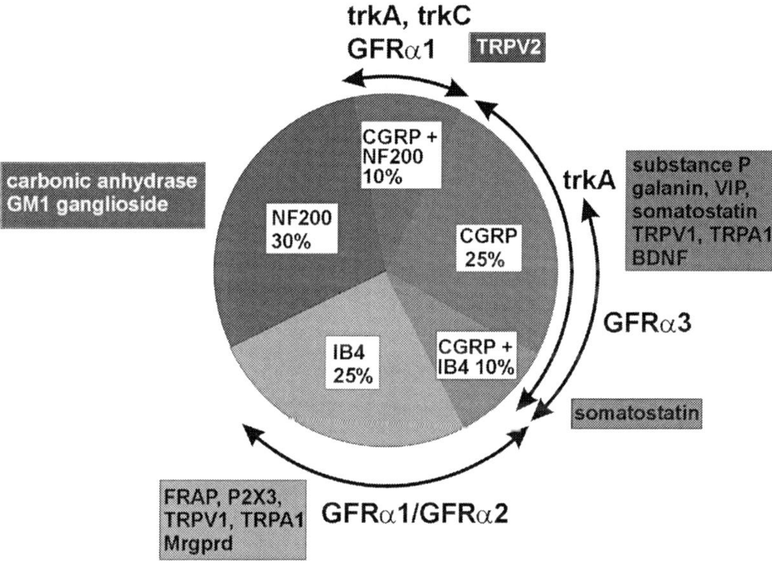

Fig. 2.1 Pie-chart summarising the main neurochemical populations of DRG neurons. Modified from Priestley (2008). One population comprises large neurons which give rise to myelinated axons and have high levels of neurofilament (NF200). Small neurons, which have mainly unmyelinated axons and are predominantly nociceptors, comprise two populations. One population consists of neurons that constitutively synthesise the neuropeptide CGRP. The other population is characterised by neurons that bind the lectin *Griffonia simplicifolia* IB4. An overlap between the NF200 and CGRP populations corresponds to Aδ nociceptors. There is also some overlap between the IB4 and CGRP populations. Neurotrophin receptors and GDNF receptors expressed by small and medium sized neurons (predominantly nociceptors) are indicated around the periphery of the chart, together with molecules found in each major population. For further details see text

mechanoreceptors (nociceptors) also exist (Djouhri and Lawson, 2004), but are less common than Aδ and C fibre nociceptors (discussed below).

Small diameter (10–30 μm) DRG neurons express the intermediate filament peripherin, have unmyelinated axons (C fibres, <1.3 m/s conduction velocity), and have unencapsulated nerve endings that function as nociceptors and/or thermoreceptors. Several subtypes exist but most C-fibres (70%) are polymodal nociceptors, and respond to noxious mechanical, thermal and chemical stimuli. This property is largely due to the fact that their peripheral terminals express TRPV1 (formerly VR1) and/or TRPA1 (formerly ANKTM1), nonselective cation channels that are members of the vanilloid and ankyrin subfamilies of transient receptor potential (TRP) ion channels (Moran et al., 2004). TRPV1 is activated by temperatures above 43°C, by molecules that possess a vanilloid moiety (e.g. capsaicin, the pungent component of capsicum peppers), and by a range of factors associated with tissue damage including the endocannabinoid anandamide, the leukotriene HPETE and low pH (<5.9) (Szallasi et al., 2007). TRPA1 is activated by a variety of pungent compounds including cinnamaldehyde, allyl isothiocianate (mustard oil), formaldehyde, acrolein, and 4-hydroxynoneal (an aldehyde that accumulates in membranes during inflammatory or oxidative stress) (Bautista et al., 2006; Trevisani et al., 2007; Bandell et al., 2007; Macpherson et al., 2007). TRPA1 was originally reported to also sense noxious cold (<15°C) (Story et al., 2003) although this has not been replicated by more recent studies (Jordt et al., 2004). Most (67%) TRPV1 expressing neurons also express TRPA1, and TRPA1 is not expressed in the absence of TRPV1 (Kobayashi et al., 2005). Another TRPV member, TRPV2, responds to noxious heat (>52°C) and is an important receptor expressed by finely myelinated Aδ high threshold mechanoreceptors (Tamura et al., 2005; Lawson et al., 2008).

2.2.2 *Peptidergic and Non-Peptidergic Nociceptors*

In addition to the broad distinction between large and small DRG neurons outlined above, histochemical studies have revealed that small sized DRG neurons can be further divided into two main subgroups depending on their expression of trophic factor receptors (see Sections 2.2.3 and 2.2.4) and whether or not they constitutively express neuropeptides. Just over half of all small DRG neurons (50–60%) constitutively express the neuropeptide calcitonin gene-related peptide (CGRP) and may also express other neuropeptides, including substance P, somatostatin, vasoactive intestinal polypeptide (VIP) and galanin (Ju et al., 1987). Histochemical (Michael and Priestley, 1999) and functional studies (Lawson et al., 1997, 2002) have shown that the majority of CGRP neurons (59–65%) and virtually all substance P neurons possess TRPV1 and are likely to be polymodal nociceptors. Neuropeptides are generally not found in the large diameter mechanoreceptive neurons, but CGRP is present in many medium sized 200 kDa neurofilament-rich neurons which are also mainly nociceptors (high threshold mechanoreceptors). The role of the peptide-expressing (peptidergic) population of C-fibres, especially in inflammatory pain, is reasonably well understood. Thus many peptidergic neurons express TRPV1 and TRPA1 and respond to inflammatory factors such as NGF (see Section 2.2.3), bradykinin, eicosanoids and cytokines (Woolf and Costigan, 1999; Kidd and Urban, 2001; Marchand et al., 2005; Pezet and McMahon, 2006; Hensellek et al., 2007). Inflammation leads to both sensitization and upregulation of TRPV1 and TRPA1, together with upregulation of substance P and CGRP. These neuropeptides are released peripherally to cause vasodilation and extravasation, and are released centrally to contribute to sensitization in the spinal cord (see Hua and Yaksh, Chapter 6).

The other half of the small DRG neurons either do not contain neuropeptides or only contain low levels (Kashiba et al., 2001) and their role is much less well understood. However they will be discussed now in some detail because they selectively express GDNF receptors (see Section 2.2.4) and respond to neuro-poietic cytokines (Section 2.2.5). The most widely used method of identifying this population of neurons involves staining with the lectin *Griffonia simplicifolia* IB4 (Silverman and Kruger, 1990) and so they are normally referred to as "IB4-binding", or "IB4-labelled" or simply as "IB4" neurons. However they can also be identified based on a number of other features including staining with the monoclonal antibody "LA4" (Alvarez et al., 1991) and selective expression of the following molecules: a non-lysosomal fluoride-resistant acid phosphatase (FRAP) (Silverman and Kruger, 1990), the purinoceptor $P2X_3$ (Bradbury et al., 1998a), the bradykinin B1 receptor (Wotherspoon and Winter, 2000), the canonical TRP channel TRPC3 (Elg et al., 2007), a subtype of sensory neuron specific G protein coupled receptor (GPR) referred to as Mas-related G protein coupled receptor subtype "d" (Mrgprd) (Dong et al., 2001; Dussor et al., 2008). The function of Mrgprd is unknown but activation of Mrgprd by β-alanine has recently been shown to inhibit KCNQ2/3 potassium channels and hence enhance DRG excitability (Shinohara et al., 2004; Crozier et al., 2007). The majority of IB4 neurons express TRPV1 (75% of IB4 cells, see Michael and Priestley, 1999) and/or respond to ATP (Mrgprd cells, Dussor et al., 2008). IB4 neurons in mouse have been reported not to express TRPV1 immunoreactivity (Zwick et al., 2002; Woodbury et al., 2004), but in culture they do respond to capsaicin (Dirajlal et al., 2003; Hjerling-Leffler et al., 2007). Thus they do appear to express TRPV1 but it may be a different form to that expressed in rat.

Nociceptors can therefore be divided neurochemically into three main subtypes (Fig. 2.2), namely (1) medium diameter TRPV2 expressing peptidergic (CGRP) neurons that have finely myelinated (Aδ) axons, (2) small diameter TRPV1/TRPA1 expressing peptidergic (CGRP, substance P) neurons that give rise to unmyelinated axons (C fibres), and (3) small diameter TRPV1/TRPA1, $P2X_3$ expressing IB4 non-peptidergic neurons that give rise to unmyelinated axons (C fibres). Note, however, that not all C-fibres are nociceptive. A rare subpopulation of unmyelinated, nonpeptidergic, DRG neurons that express the

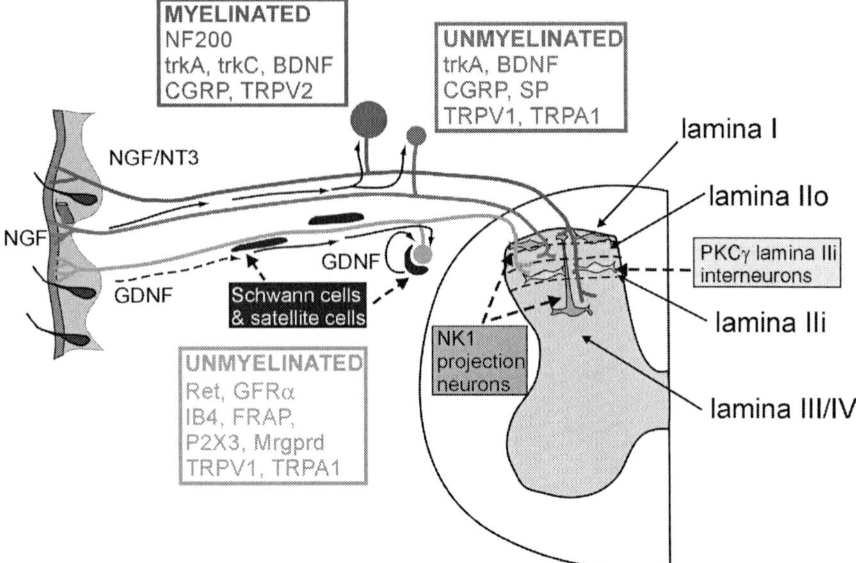

Fig. 2.2 Diagram showing the three main populations of nociceptors, namely: (1) medium diameter peptidergic (CGRP) neurons (*red*) that have finely myelinated axons, function as high threshold mechanoreceptors (express TRPV2), and respond to NGF and NT3. Their axons innervate predominantly laminae I and III/IV. (2) small diameter peptidergic (CGRP, substance P) neurons (*blue*) that function as polymodal nociceptors (express TRPV1 and TRPA1), and respond to NGF. Their axons innervate predominantly laminae I and II outer, and their targets include NK1 expressing projection neurons. (3) small diameter IB4-binding non-peptidergic neurons (*green*) that function as polymodal nociceptors (express TRPV1and TRPA1), and respond to GDNF. Their axons innervate a region of mid lamina II that is sandwiched between the termination zone of peptidergic C-fibres in lamina II outer, and the zone of PKCγ interneurons in lamina II inner. Adapted from Priestley (2008). For further details see text

Mrgpr subtype B4 (MrgprdB4) may mediate gentle touch (Liu et al., 2007). In addition, cold-sensitive (TRPM8 expressing) DRG neurons include (peptidergic) C-fibres but are distinct from the TRPV1 population (Kobayashi et al., 2005). However thermoreceptors have not been characterised as thoroughly as nociceptors, and it is not known whether they also comprise peptidergic and non peptidergic populations. It should be noted also that major changes in neuropeptide expression take place after nerve injury or inflammation and under such conditions non-peptidergic neurons can begin to express neuropeptides (reviewed by McMahon and Priestley, 2005). The most dramatic example of this involves galanin, which is expressed in the majority of DRG neurons after axotomy (see Section 2.2.5, Villar et al., 1989; Shortland et al., 2006) and under such circumstances is thought to reduce excitability in the spinal cord (Wynick et al., 2001; Liu and Hokfelt, 2002).

The significance of having two populations of C-fibre nociceptors (peptidergic and non-peptidergic) is not known, but one possibility is that they signal different types of pain because of their different peripheral targets and receptor properties. For example, features that characterize the non-peptidergic (IB4) population include:

- IB4 neurons innervate dermis and epidermis, including hair follicles, but generally do not form perivascular endings. More IB4 DRG neurons project to the epidermis (Lu et al., 2001), and their axons terminate more superficially (Zylka et al., 2005) than peptidergic neurons.
- Targeted deletion of IB4 neurons using saporin indicates that IB4 neurons mediate acute thermal and mechanical nociception (Vulchanova et al., 2001).
- IB4 neurons express the P2X3 purinoceptor so may signal ATP-mediated pain and/or certain types of mechanosensation. P2X3 antisense studies suggest functions in both inflammatory & neuropathic pain (Barclay et al., 2002).
- IB4 neurons show subtly different responses to capsaicin and to low pH compared to peptidergic neurons (Stucky and Lewin, 1999; Petruska et al., 2002; Dirajlal et al., 2003; Liu et al., 2004). There are species differences between mouse and rat, but IB4 neurons in rat may only be activated when a more severe acidosis occurs and then respond with a sustained response but without associated neuropeptide release (Liu et al., 2004).
- IB4 neurons have high Nav1.9 and Kv1.4 expression which gives them distinct membrane properties (Vydyanathan et al., 2005; Fang et al., 2006).
- IB4 cells utilise the cAMP-activated guanine exchange factor, Epac, in mediating cAMP-to-PKC signalling in β-adrenoreceptor mediated inflammatory pain (adrenaline-induced mechanical hyperalgesia) (Hucho et al., 2005).

It is also possible that each population targets distinct central pathways. For example it is well established that IB4-binding axons terminate more deeply in lamina II than peptidergic ones (e.g. Lorenzo et al., 2008), although there are disagreements in the literature regarding their exact depth within lamina II (reviewed by Woodbury et al., 2000). Recent studies indicate that peptidergic C-fibres innervate dorsal horn neurons that project to traditional CNS pain relay sites such as the parabrachial nucleus and the ventrolateral thalamus, whereas non-peptidergic C-fibres may activate neurons that project to limbic and striatal sites including the bed nucleus of the stria terminalis and the globus pallidus (Braz et al., 2005). Thus the non-peptidergic neurons may innervate pathways that convey primarily the affective component of pain rather than the sensory-discriminatory component, and may also activate brain centres that control motor responses to pain.

2.2.3 Neurotrophins and Neurotrophin Receptors

Adult DRG neurons express neurotrophin receptors and localization studies have shown that the different DRG subtypes reviewed above express different receptors (Fig. 2.1). Thus large diameter neurons (low threshold mechanoreceptors) express either trkB (the receptor for BDNF) (see Bardoni and Merighi Chapter 5) or trkC (the receptor for neurotrophin-3, NT-3) and small diameter neurons express trkA (the receptor for nerve growth factor, NGF) (McMahon et al., 1994; Karchewski et al., 1999; Kashiba et al., 2003). However the trkA expression is confined to the peptidergic population of nociceptors with little expression in non-peptidergic IB4 neurons (Averill et al., 1995). All DRG neurons that express trk receptors also express p75 (Kashiba et al., 2003), a non-selective neurotrophin receptor that is a member of the TNF receptor superfamily. p75 acts as a trk coreceptor but can also signal directly by recruitment of various adaptor proteins. Consistent with the lack of expression of trks in non-peptidergic neurons, IB4 neurons also lack p75 (Bennett et al., 1996).

The source of the endogenous neurotrophins that act on these trk receptors is generally the peripheral target tissue and supporting glial cells (reviewed by Del Fiacco and Priestley, 2001). In the case of trkA nociceptors, for example, NGF is produced by keratinocytes and fibroblasts in the skin. Schwann cells and various types of immune cells (lymphocytes, macrophages and mast cells) also synthesize NGF and are important sources after nerve injury and/or inflammation. DRG neurons are generally considered not to synthesise NGF, although NGF mRNA has been reported in DRG neurons after spinal cord injury (Brown et al., 2007). In contrast numerous studies have reported synthesis by DRG neurons of BDNF. BDNF is constitutively expressed by a small percentage of DRG neurons and expression increases selectively in the trkA peptidergic neurons in response to exogenous NGF (Apfel et al., 1996; Michael et al., 1997a) or peripheral inflammation (Cho et al., 1997b, a). In response to nerve injury, synthesis declines in peptidergic neurons but is increased in large diameter trkB/trkC expressing neurons (Michael et al., 1999). In each cell type in which it is synthesized, the BDNF is packaged into dense-cored vesicles (Salio et al., 2007) and axonally transported into the central terminals of primary afferents within the spinal cord and brainstem. Here the BDNF is released (Lever et al., 2001) to affect local neurons in a variety of different ways, including activation of mitogen-activated protein kinase (MAPkinase) pathways (Pezet et al., 2002) and transcription factors (Kerr et al., 1999; Miletic et al., 2004; Jongen et al., 2005), modulation of NMDA receptors (Kerr et al., 1999; Slack et al., 2004), and induction of a depolarizing shift in the chloride reversal potential (Coull et al., 2005) (see Bardoni and Merighi Chapter 5). Note however that this last effect is thought to be due to BDNF released from microglial cells rather than from primary afferent terminals (Coull et al., 2005) (see Beggs Chapter 22).

2.2.4 GDNF Receptors

Low threshold mechanoreceptors, and the non-peptidergic population of nociceptors, respond to GDNF and express GDNF receptor components such as the tyrosine kinase RET and the accessory subunits GFRα1, GFRα2 and/or GFRα3 (Bennett et al., 1998). The peptidergic C-fibre nociceptors do not express RET but some do express GFRα3 (Orozco et al., 2001) and this expression increases after nerve injury (Bennett et al., 2000). GDNF is thought to regulate non-peptidergic neurons as a target-derived factor but a more important role is probably as a local injury-derived factor. GDNF levels are higher in peripheral nerve than they are in targets such as muscle or skin. GDNF is synthesised by Schwann cells and by DRG satellite glial cells and synthesis is increased further by axotomy (Hammarberg et al., 1996). GDNF immunoreactivity has also been reported in DRG neurons (Holstege et al., 1998) but it is not clear whether this is locally synthesised or derived from uptake. Studies using radiolabelled exogenous GDNF suggest that Schwann cell- or satellite cell-derived GDNF may be taken up by DRG neurons and transported anterogradely along the axons for release from terminals in the spinal cord, resulting in neuronal transcytosis (the vesicular transport of macromolecules from one side of a cell to the other) (Rind and von Bartheld, 2002). Interestingly, the selective expression of GDNF receptors by IB4 neurons is established only postnatally. During development the survival of DRG neurons depends on NGF and NT-3 and virtually all DRG neurons express trkA or trkC (Farinas et al., 2002). In the adult, DRG neurons no longer require neurotrophins for survival, and there are postnatal changes that establish a new pattern of receptor expression (Fig. 2.3). Under the control of the

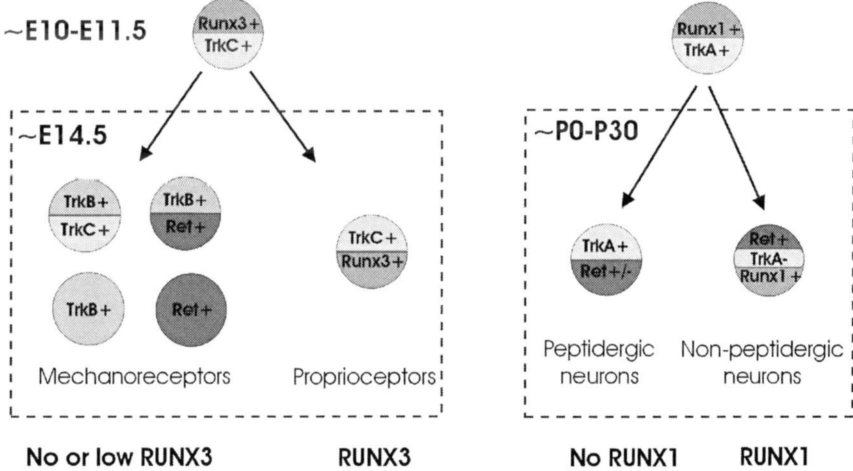

Fig. 2.3 Trophic factors and sensory neuron development. The level of RUNX proteins helps to control the diversification of sensory neurons into different cell types. Large diameter neurons are derived from an early wave of neurogenesis and at E10–E11.5 express both *Runx3* and trkC. Neurons from this group that maintain *Runx3* expression keep trkC and become proprioceptors, whereas neurons that lose or reduce *Runx3* expression make trkB and trkC, trkB alone, RET alone or RET and trkB mechanoreceptive neurons. Small diameter

transcription factor RUNX1 (Marmigere and Ernfors, 2007), IB4 neurons downregulate trkA and upregulate RET and GFRα1 (Bennett et al., 1996; Molliver et al., 1997).

2.2.5 Neuropoietic Cytokines

The neuropoietic cytokines include ciliary neurotrophic factor (CNTF), oncostatin M (OM), leukaemia inhibitory factor (LIF) and interleukin-6 (IL-6) (for review, see Bauer et al., 2007). They are produced mainly by resident and infiltrating inflammatory cells (Bauer et al., 2007) and by Schwann cells (Matsuoka et al., 1997; Rutkowski et al., 1999), but in some circumstances have been reported to be produced by DRG neurons themselves (Murphy et al., 1995; Scott et al., 2000). Expression increases markedly after nerve injury (Kurek et al., 1996; Ito et al., 2000) and cytokine signalling between macrophages, Schwann cells and DRG neurons is thought to play a key role in both Wallerian degeneration and regeneration. For example, Schwann cells respond to nerve injury by synthesizing monocyte chemoattractant protein-1 (MCP-1, CCL2) and the resulting infiltrating macrophages produce IL-6 themselves and also promote its production by Schwann cells (see Subang and Richardson, 2001). The IL-6, in turn, can affect the axons and cell bodies of DRG neurons. However such is the complexity of the various interactions, that it has proved very difficult to unravel the role of individual neuropoietic cytokines. For example some authors (Zhong et al., 1999), but not others (Inserra et al., 2000), report that regeneration is reduced in IL-6 knockout mice. Similarly, IL-6, LIF and CNTF have all been shown to mimic aspects of the conditioning response (a peripheral nerve injury that promotes the subsequent regeneration of central axons) but their exact contribution remains unclear (Cafferty et al., 2001, 2004; Cao et al., 2006; Wu et al., 2007).

In addition to effects on regeneration, neuropoietic cytokines have been implicated in some of the changes in DRG phenotype and consequent pain signalling that occur after nerve injury. Probably the most dramatic of these is the upregulation of galanin that occurs in most DRG neurons after axotomy (Villar et al., 1989). Galanin expression is negatively regulated by NGF (Verge et al., 1995) and positively regulated by LIF and IL-6 (Corness et al., 1996; Sun and Zigmond, 1996; Thompson et al., 1998; Murphy et al., 1999). IL-6 also contributes to the upregulation of BDNF and modulates the downregulation of substance P that is seen after axotomy (Murphy et al., 2000). These effects of

◄——

Fig. 2.3 (continued) neurons are derived from two later waves of neurogenesis and at E10–E11.5 express *Runx1* and trkA. Maintained *Runx1* expression drives a trkA–, non-peptidergic phenotype (*TrkA–/Ret+/Runx1+* neurons), whereas downregulation of *Runx1* allows the neurons to acquire a peptidergic phenotype (*TrkA+/Ret+/–*). E, embryonic day; P, postnatal day. Modified from Marmigere and Ernfors (2007)

IL-6 may underlie the reductions in neuropathic pain (cutaneous hypersensitivity and mechanical allodynia) that have been reported in the IL-6 knockout (Ramer et al., 1998; Murphy et al., 1999).

Relatively few studies have examined the particular DRG cell types affected by the neuropoietic cytokines, or the expression of their receptors by DRG neurons. The gp130 signalling component is expressed by all DRG neurons (Mizuno et al., 1997; Gardiner et al., 2002) and LIFRβ mRNA expression and LIF binding by DRG neurons has been reported (Qiu et al., 1997; Scott et al., 2000). LIF retrograde transport and LIFRβ immunoreactivity labels a subgroup of both CGRP and IB4 neurons (Thompson et al., 1997; Gardiner et al., 2002), suggesting some selectivity for nociceptors. However studies on galanin expression in the LIF knockout mouse (Corness et al., 1996) suggest that LIF may also affect large diameter (non-nociceptive) neurons. In contrast, expression of the oncostatin M receptor beta has been reported to be confined mainly to P2X3 (IB4) neurons (Tamura et al., 2003), where it appears to play a role not only in the adult but also in development. Oncostatin M knockout mice show decreased numbers of P2X3 neurons and reduced noxious responses in models of acute thermal, mechanical, chemical, and visceral pain (Morikawa et al., 2004).

In addition to nerve injury, cytokines are of course though to play an important role in inflammatory pain, and IL-6 is amongst the pro-inflammatory cytokines reported to increase with inflammation (De Jongh et al., 2003; Summer et al., 2008). IL-6, injected into the rat hind paw, has been reported to cause mechanical allodynia, whereas local pretreatment with IL-6 antibody reduces carrageenin-evoked hyperalgesia (Cunha et al., 1992). In contrast to IL-6, LIF has been reported to increase in peripheral inflammation but to have an anti-inflammatory role (Zhu et al., 2001). However many cytokine effects on sensory neurons appear to be mediated via NGF (Woolf et al., 1994) and most changes that occur in DRG neurons following inflammation are associated with the trkA expressing peptidergic population (see Pezet and McMahon, 2006). In Section 2.4.1, the possibility that inflammation-associated GDNF leads to an increase of TRPV1 in IB4 cells will be discussed, and recently we have reported changes in IB4 cells following intraplantar complete Freunds adjuvant (CFA) that may be mediated by a neuropoietic cytokine. Reg2, a Schwann cell mitogen which is known to be regulated by neuropoietic cytokines, is upregulated selectively in IB4 cells as early as one day after CFA treatment (Averill et al., 2008). Intrathecal delivery of LIF, but not of GDNF or NGF treatment, upregulates Reg2 expression in DRG neurons (Fig. 2.4) but with an expression pattern that is slightly less selective for IB4 cells than is seen after CFA. The reason for this is not known, but may relate to changes in LIF receptor expression after CFA, and/or targeting by inflammatory cells of particular DRG subpopulations. The IB4 population of DRG neurons have been reported to produce MCP-1 in experimental inflammation (Jeon et al., 2008) and following nerve injury (Thacker et al., 2009). The activity-dependent release of MCP-1 from the central terminals of sensory neurons may modulate nociceptive mechanisms via activation of CCR2 receptors (Abbadie et al., 2003) (see Abbadie and Sullivan Chapter 15).

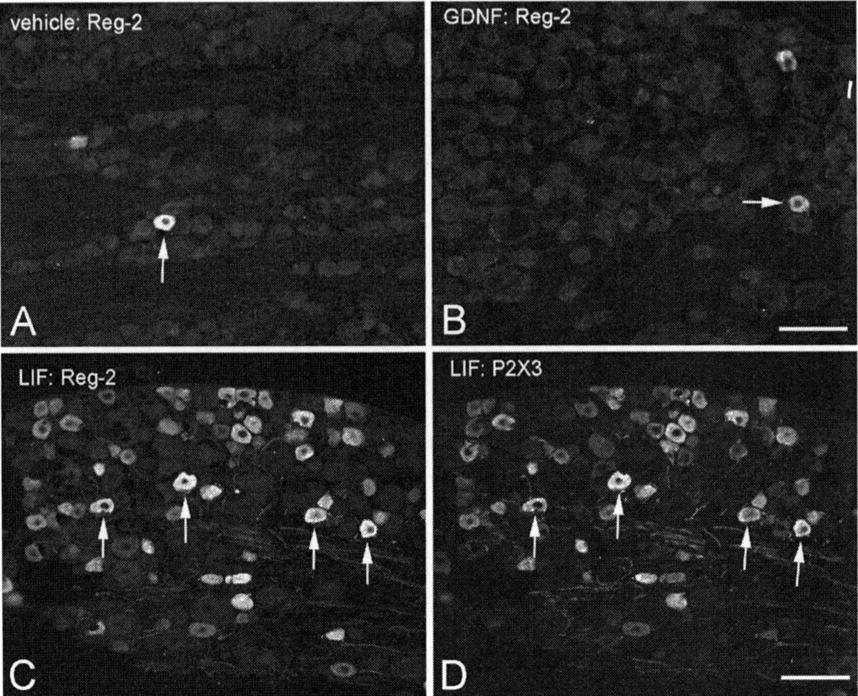

Fig. 2.4 LIF and Reg-2. After LIF treatment, Reg-2 is expressed in many DRG neurons and these are predominantly of the IB4 type. (**A, B**) Low magnification micrographs showing Reg-2 immunoreactivity after 2 weeks' intrathecal vehicle (**A**) or GDNF (**B**) treatment. Only one or two Reg-2 immunoreactive neurons are visible (*arrows*). (**C, D**) Low magnification micrographs showing Reg-2 (**C**) and P2X3 (**D**) double labelling after 2 weeks intrathecal LIF treatment. Many Reg-2 immunoreactive neurons are present and are double-labelled for P2X3 (*arrows*), indicating that they belong to the IB4 population. Scale bars = 100 μm. Reproduced from Averill et al. (2008)

2.3 Expression of Trophic Factors and Their Receptors by CNS Spinal Pain Pathways

Most neurons in the CNS express both trkB and trkC and, although not widely studied, this appears also to apply to CNS pain pathways (Fig. 2.5). Thus lumbar projection neurons (Bradbury et al., 1998b) and raphe-spinal neurons (King et al., 1999) have been shown to express trkB and trkC, and exogenous BDNF has been shown to activate dorsal horn neurons (Jongen et al., 2005), including spinothalamic projection neurons (Slack et al., 2005) (see Bardoni and Merighi Chapter 5). TrkA, in contrast, is confined to a small population of propriospinal neurons (Michael et al., 1997b). Synthesis of neurotrophins by spinal neurons is limited, but BDNF expression has been reported in adult human dorsal horn neurons (Josephson et al., 2001). Truncated trkB receptors

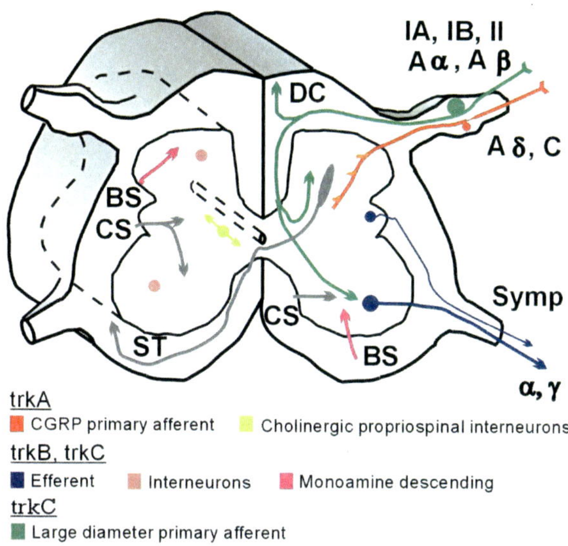

Fig. 2.5 Expression of neurotrophin receptors in spinal cord pathways. TrkA, trkB and trkC are expressed in distinct types of primary afferent (DRG) neuron. In contrast trkB and trkC are generally coexpressed within spinal cord neurons, and trkA is confined to a small population of cholinergic interneurons

are expressed by astrocytes, which are able to bind, endocytose, and subsequently release, BDNF (Rubio, 1997). Microglia also release BDNF and this can modulate neuronal excitability by causing a depolarizing shift in the anion reversal potential (see also Section 2.2.3, Coull et al., 2005) (see Beggs Chapter 22).

GDNF receptors (Ret, GFRα1) and neuropoietic cytokine receptors (LIFRβ, CNTFRα, gp130) are expressed in motoneurons and regulated by axotomy (Hammarberg et al., 2000; Josephson et al., 2001). However fewer studies have examined their expression in pain pathways. Jongen et al. (2007) report that Ret immunoreactivity is present in many dorsal horn neurons, including NK1 expressing ones, and that Ret immunoreactive neurons in the superficial dorsal horn are in close proximity to (primary afferent) GDNF immunoreactive terminals. Consistent with this, intrathecal delivery of GDNF induces c-Fos and phosporylation of ERK in dorsal horn neurons (Jongen et al., 2005). IL-6 expression by spinal cord neurons has been reported following spinal or sciatic nerve injury (DeLeo et al., 1996; Arruda et al., 1998). CFA-induced peripheral inflammation evokes glial activation and IL-6 expression (together with IL-1β and TNF-α) in microglia and astrocytes (Bao et al., 2001; Raghavendra et al., 2004).

2.4 GDNF in Inflammation and Nerve Injury

The neurochemical features of nociceptive DRG neurons reviewed in Section 2.2, and their expression of trophic factor receptors, applies to the intact nervous system and the signaling of acute pain. However after nerve injury or inflammation major changes take place that contribute to the development of chronic pain and many of these changes appear to be driven by changes in the availability of growth factors. The role of NGF in such changes has been well documented (Pezet and McMahon, 2006) but less is known about the role of GDNF. However there is growing evidence that endogenous GDNF contributes to inflammatory pain, and that exogenous GDNF can be used to treat neuropathic pain (Sah et al., 2005), and that in each case the GDNF effects are mediated primarily by the non-peptidergic (IB4) population of nociceptors.

2.4.1 *Inflammation*

The strongest evidence of a role for GDNF in inflammation comes from studies by Amaya et al. (2004) who have shown that intraplantar CFA leads to increased GDNF protein in dorsal root ganglia, and increased TRPV1 in IB4 neurons. The increase in TRPV1 and the associated thermal hyperalgesia is reduced by intrathecal delivery of anti-GDNF (Amaya et al., 2004). Consistent with this data, peripheral inflammation has been reported to selectively increase TRPV1 function in IB4 neurons cultured from adult mouse (Breese et al., 2005). Fang et al. (2003) also report that intrathecal anti-GDNF reduces CFA-induced pain (mechanical allodynia) although observed that GDNF protein in DRGs and immunoreactivity in IB4 neurons decreases with CFA. Acute intraplantar administration of GDNF sensitizes nociceptors and produces mechanical hyperalgesia in the rat, and intrathecal IB4-saporin attenuates the GDNF but not the NGF hyperalgesia (Bogen et al., 2008). In addition to increasing TRPV1 expression, GDNF modulates several other key molecules in IB4 neurons. GDNF and neurturin increase the B1 bradykinin receptor in IB4 neurons in culture (Vellani et al., 2004). Intrathecal GDNF increases somatostatin expression in small DRG neurons and its electrically-evoked release (Charbel-Issa et al., 2001; Malcangio et al., 2002), and increases CGRP expression in non-trkA neurons (Ramer et al., 2003). Intrathecal delivery can, of course, also affect spinal cord neurons and so changes in pain thresholds in response to intrathecal agents may not be exclusively due to effects on DRG neurons. GDNF induces c-fos in Ret-immunoreactive dorsal horn neurons (Jongen et al., 2005). GDNF mRNA has also been reported to decrease in dorsal horn of CFA rats, although it is not known whether this is neuronal or glial (Fang et al., 2000).

In addition to GDNF, there is good evidence for the involvement of another member of the GDNF family (artemin) in inflammation. Malin et al. (2006) report no increase in GDNF or neurturin in inflamed skin, but a large increase in artemin in skin and of its cognate receptor (GFRα3) in dorsal root ganglia. Hindpaw injection of GDNF, neurturin or artemin (0.2 μg/20 μl) produces acute thermal hyperalgesia (Malin et al., 2006). Artemin overexpression in skin increases expression of TRPV1 and TRPA1 in DRGs (Elitt et al., 2006). However since GFRα3 is expressed predominantly by peptidergic (CGRP) DRG neurons, it is likely that artemin acts on those neurons rather than the IB4 ones.

2.4.2 Nerve Injury

In parallel with the neuroprotective effects described for NGF on trkA expressing peptidergic DRG neurons (see McMahon and Priestley, 2005), GDNF has been reported to prevent a wide range of axotomy-induced changes in IB4 neurons. Thus GDNF reduces the axotomy-evoked upregulation of ATF3 (Averill et al., 2004) and NaV1.3 (Leffler et al., 2002) in IB4 neurons, and prevents the axotomy-evoked down regulation of the following molecules in IB4 neurons: P2X3 (Bradbury et al., 1998a), NaV1.8 (SNS) and NaV1.9 (NaN) (Cummins et al., 2000), IB4, TMP and somatostatin (Bennett et al., 1998), and TRPV1 (Priestley et al., 2002). GDNF also prevents the axotomy-induced upregulation of GFRα1 and GFRα3 and downregulation of GFRα2 (Bennett et al., 2000). These effects may underlie the dramatic analgesic effects reported for GDNF in neuropathic pain models. Spinal nerve ligation (SNL) produces a downregulation in IB4 and P2X3, and an upregulation of ATF3, galanin, and NPY. All these changes are prevented by intrathecal GDNF, which also prevents development of mechanical and thermal hyperalgesia (Wang et al., 2003). Most strikingly, delayed administration of GDNF in SNL injury reversed the mechanical and thermal hyperalgesia for the duration of the infusion (Boucher et al., 2000). Mechanical and thermal hyperalgesia after partial sciatic ligation is also blocked, as is the upregulation of NaV1.3 and downregulation of NaV1.8 and NaV1.9 (Boucher et al., 2000). The chronic administration of GDNF to normal animals led to no change in responses to noxious thermal or mechanical stimuli. Acute peripheral treatment with GDNF (plantar injections of 0.1, 1, and 10 µg) did not affect pain-related behavior (Boucher et al., 2000). However a 30 day infusion of GDNF in the lateral ventricles has been reported by other authors to produce weight loss and allodynia (Hoane et al., 1999). The role of endogenous GDNF in neuropathic pain is, however, less clear. Chronic constriction injury (CCI) increases GDNF and GFRα1 in DRGs, and the accompanying thermal hyperalgesia is reduced by intrathecal GFRα1 antisense (Dong et al., 2006). CCI and SNL leads to decreased levels of GDNF within dorsal root ganglia and decreased numbers of Ret immunoreactive DRG neurons, and GDNF delivery reduces mechanical and thermal hyperalgesia (Nagano et al., 2003).

Aertemin has also been shown to have potent neuroprotective effects in nerve injury and neuropathic pain models. Systemic (subcutaneous) artemin prevents SNL-induced downregulation of substance P, CGRP, IB4, P2X3, NaV1.8, and upregulation of galanin and NPY for the duration of the treatment (Gardell et al., 2003). Artemin did not prevent the SNL-induced increase in GFRα3 (Gardell et al., 2003). Artemin prevents, and reverses, SNL-induced thermal and mechanical hyperalgesia (Gardell et al., 2003). Artemin reverses axotomy-evoked changes in SP, somatostatin, CGRP, TRPV1, IB4, galanin, and ATF3 but not NPY and maintains C-fibre conduction velocity, and C-fibre evoked substance P release within the dorsal horn (Bennett et al., 2006).

2.5 Concluding Remarks

The studies reviewed in this chapter have given new insights into the organization of primary afferent neurons and have identified growth factors as both central players in, and potential treatments for, neuropathic and inflammatory pain states. The development of therapeutic agents that target growth factors in pain pathways remains and exciting prospect, but also a major challenge.

Acknowledgments Many thanks to Dr Marzia Malcangio and Dr Bared Safieh-Garabedian for helpful suggestions and comments on the draft manuscript.

References

Abbadie C, Lindia JA, Cumiskey AM, Peterson LB, Mudgett JS, Bayne EK, DeMartino JA, MacIntyre DE, Forrest MJ (2003) Impaired neuropathic pain responses in mice lacking the chemokine receptor CCR2. Proc Natl Acad Sci USA 100: 7947–7952.

Alvarez FJ, Morris HR, Priestley JV (1991) Sub-populations of smaller diameter trigeminal primary afferent neurons defined by expression of calcitonin gene-related peptide and the cell surface oligosaccharide recognized by monoclonal antibody LA4. J Neurocytol 20: 716–731.

Amaya F, Shimosato G, Nagano M, Ueda M, Hashimoto S, Tanaka Y, Suzuki H, Tanaka M (2004) NGF and GDNF differentially regulate TRPV1 expression that contributes to development of inflammatory thermal hyperalgesia. Eur J Neurosci 20: 2303–2310.

Apfel SC, Wright DE, Wiideman AM, Dormia C, Snider WD, Kessler JA (1996) Nerve growth-factor regulates the expression of brain-derived neurotrophic factor messenger-RNA in the peripheral nervous-system. Mol Cell Neurosci 7: 134–142.

Arruda JL, Colburn RW, Rickman AJ, Rutkowski MD, DeLeo JA (1998) Increase of interleukin-6 mRNA in the spinal cord following peripheral nerve injury in the rat: potential role of IL-6 in neuropathic pain. Mol Brain Res 62: 228–235.

Averill S, Inglis JJ, King VR, Thompson SW, Cafferty WB, Shortland PJ, Hunt SP, Kidd BL, Priestley JV (2008) Reg-2 expression in dorsal root ganglion neurons after adjuvant-induced monoarthritis. Neurosci 155: 1227–1236.

Averill S, McMahon SB, Clary DO, Reichardt LF, Priestley JV (1995) Immunocytochemical localization of trkA receptors in chemically identified subgroups of adult rat sensory neurons. Eur J Neurosci 7: 1484–1494.

Averill S, Michael GJ, Shortland PJ, Leavesley RC, King VR, Bradbury EJ, McMahon SB, Priestley JV (2004) NGF and GDNF ameliorate the increase in ATF3 expression which occurs in dorsal root ganglion cells in response to peripheral nerve injury. Eur J Neurosci 19: 1437–1445.

Bandell M, Macpherson LJ, Patapoutian A (2007) From chills to chilis: mechanisms for thermosensation and chemesthesis via thermoTRPs. Curr Opin Neurobiol 17: 490–497.

Bao L, Zhu Y, Elhassan AM, Wu Q, Xiao B, Zhu J, Lindgren JU (2001) Adjuvant-induced arthritis: IL-1 beta, IL-6 and TNF-alpha are up-regulated in the spinal cord. Neuroreport 12: 3905–3908.

Barclay J, Patel S, Dorn G, Wotherspoon G, Moffatt S, Eunson L, Abdel'al S, Natt F, Hall J, Winter J, Bevan S, Wishart W, Fox A, Ganju P (2002) Functional downregulation of P2X3 receptor subunit in rat sensory neurons reveals a significant role in chronic neuropathic and inflammatory pain. J Neurosci 22: 8139–8147.

Bauer S, Kerr BJ, Patterson PH (2007) The neuropoietic cytokine family in development, plasticity, disease and injury. Nat Rev Neurosci 8: 221–232.

Bautista DM, Jordt SE, Nikai T, Tsuruda PR, Read AJ, Poblete J, Yamoah EN, Basbaum AI, Julius D (2006) TRPA1 mediates the inflammatory actions of environmental irritants and proalgesic agents. Cell 124: 1269–1282.

Belvisi MG (2002) Overview of the innervation of the lung. Curr Opin Pharmacol 2: 211–215.

Bennett DLH, Averill S, Clary DO, Priestley JV, McMahon SB (1996) Postnatal changes in the expression of the trkA high-affinity NGF receptor in primary sensory neurons. Eur J Neurosci 8: 2204–2208.

Bennett DL, Boucher TJ, Armanini MP, Poulsen KT, Michael GJ, Priestley JV, Phillips HS, McMahon SB, Shelton DL (2000) The glial cell line-derived neurotrophic factor family receptor components are differentially regulated within sensory neurons after nerve injury. J Neurosci 20: 427–437.

Bennett DL, Boucher TJ, Michael GJ, Popat RJ, Malcangio M, Averill SA, Poulsen KT, Priestley JV, Shelton DL, McMahon SB (2006) Artemin has potent neurotrophic actions on injured C-fibres. J Peripher Nerv Syst 11: 330–345.

Bennett DLH, Michael GJ, Ramachandran N, Munson JB, Averill S, Yan Q, McMahon SB, Priestley JV (1998) A distinct subgroup of small DRG cells express GDNF receptor components and GDNF is protective for these neurons after nerve injury. J Neurosci 18: 3059–3072.

Blackshaw LA, Brookes SJ, Grundy D, Schemann M (2007) Sensory transmission in the gastrointestinal tract. Neurogastroenterol Motil 19: 1–19.

Bogen O, Joseph EK, Chen X, Levine JD (2008) GDNF hyperalgesia is mediated by PLCgamma, MAPK/ERK, PI3K, CDK5 and Src family kinase signaling and dependent on the IB4-binding protein versican. Eur J Neurosci 28: 12–19.

Boucher TJ, Okuse K, Bennett DL, Munson JB, Wood JN, McMahon SB (2000) Potent analgesic effects of GDNF in neuropathic pain states. Science 290: 124–127.

Bradbury EJ, Burnstock G, McMahon SB (1998a) The expression of P2X3 purinoreceptors in sensory neurons: Effects of axotomy and glial-derived neurotrophic factor. Mol Cell Neurosci 12: 256–268.

Bradbury EJ, King VR, Simmons LJ, Priestley JV, McMahon SB (1998b) NT-3, but not BDNF, prevents atrophy and death of axotomized spinal cord projection neurons. Eur J Neurosci 10: 3058–3068.

Braz JM, Nassar MA, Wood JN, Basbaum AI (2005) Parallel "pain" pathways arise from subpopulations of primary afferent nociceptor. Neuron 47: 787–793.

Breese NM, George AC, Pauers LE, Stucky CL (2005) Peripheral inflammation selectively increases TRPV1 function in IB4-positive sensory neurons from adult mouse. Pain 115: 37–49.

Brown A, Ricci MJ, Weaver LC (2007) NGF mRNA is expressed in the dorsal root ganglia after spinal cord injury in the rat. Exp Neurol 205: 283–286.

Cafferty WB, Gardiner NJ, Das P, Qiu J, McMahon SB, Thompson SW (2004) Conditioning injury-induced spinal axon regeneration fails in interleukin-6 knock-out mice. J Neurosci 24: 4432–4443.

Cafferty WB, Gardiner NJ, Gavazzi I, Powell J, McMahon SB, Heath JK, Munson J, Cohen J, Thompson SW (2001) Leukemia inhibitory factor determines the growth status of injured adult sensory neurons. J Neurosci 21: 7161–7170.

Cao Z, Gao Y, Bryson JB, Hou J, Chaudhry N, Siddiq M, Martinez J, Spencer T, Carmel J, Hart RB, Filbin MT (2006) The cytokine interleukin-6 is sufficient but not necessary to mimic the peripheral conditioning lesion effect on axonal growth. J Neurosci 26: 5565–5573.

Charbel-Issa P, Lever IJ, Michael GJ, Bradbury EJ, Malcangio M (2001) Intrathecally delivered glial cell line-derived neurotrophic factor produces electrically evoked release of somatostatin in the dorsal horn of the spinal cord. J Neurochem 78: 221–229.

Cho HJ, Kim SY, Park MJ, Kim DS, Kim JK, Chu MY (1997b) Expression of mRNA for brain-derived neurotrophic factor in the dorsal root ganglion following peripheral inflammation. Brain Res 749: 358–362.

Cho HJ, Kim JK, Zhou XF, Rush RA (1997a) Increased brain-derived neurotrophic factor immunoreactivity in rat dorsal root ganglia and spinal cord following peripheral inflammation. Brain Res 764: 269–272.

Corness J, Shi TJ, Xu ZQ, Brulet P, Hokfelt T (1996) Influence of leukemia inhibitory factor on galanin/GMAP and neuropeptide Y expression in mouse primary sensory neurons after axotomy. Exp Brain Res 112: 79–88.

Coull JA, Beggs S, Boudreau D, Boivin D, Tsuda M, Inoue K, Gravel C, Salter MW, De Koninck Y (2005) BDNF from microglia causes the shift in neuronal anion gradient underlying neuropathic pain. Nature 438: 1017–1021.

Crozier RA, Ajit SK, Kaftan EJ, Pausch MH (2007) MrgD activation inhibits KCNQ/M-currents and contributes to enhanced neuronal excitability. J Neurosci 27: 4492–4496.

Cummins TR, Black JA, Dib-Hajj SD, Waxman SG (2000) Glial-derived neurotrophic factor upregulates expression of functional SNS and NaN sodium channels and their currents in axotomized dorsal root ganglion neurons. J Neurosci 20: 8754–8761.

Cunha FQ, Poole S, Lorenzetti BB, Ferreira SH (1992) The pivotal role of tumour necrosis factor alpha in the development of inflammatory hyperalgesia. Br J Pharmacol 107: 660–664.

De Jongh RF, Vissers KC, Meert TF, Booij LH, De Deyne CS, Heylen RJ (2003) The role of interleukin-6 in nociception and pain. Anesth Analg 96: 1096–1103, table.

Del Fiacco M, Priestley JV (2001) Neurotrophins and adult somatosensory neurons. In: Neurobiology of the neurotrophins (Mocchetti I, ed), pp 205–236. Johnson City, TN: F. P.Graham Publishing.

DeLeo JA, Colburn RW, Nichols M, Malhotra A (1996) Interleukin-6-mediated hyperalgesia/allodynia and increased spinal IL-6 expression in a rat mononeuropathy model. J Interferon Cytokine Res 16: 695–700.

Dirajlal S, Pauers LE, Stucky CL (2003) Differential response properties of IB4 -positive and -negative unmyelinated sensory neurons to protons and capsaicin. J Neurophysiol 89: 513–524.

Djouhri L, Lawson SN (2004) Abeta-fiber nociceptive primary afferent neurons: a review of incidence and properties in relation to other afferent A-fiber neurons in mammals. Brain Res Brain Res Rev 46: 131–145.

Dong X, Han S, Zylka MJ, Simon MI, Anderson DJ (2001) A diverse family of GPCRs expressed in specific subsets of nociceptive sensory neurons. Cell 106: 619–632.

Dong ZQ, Wang YQ, Ma F, Xie H, Wu GC (2006) Down-regulation of GFRalpha-1 expression by antisense oligodeoxynucleotide aggravates thermal hyperalgesia in a rat model of neuropathic pain. Neuropharmacology 50: 393–403.

Dussor G, Zylka MJ, Anderson DJ, McCleskey EW (2008) Cutaneous sensory neurons expressing the mrgprd receptor sense extracellular ATP and are putative nociceptors. J Neurophysiol 99: 1581–1589.

Elg S, Marmigere F, Mattsson JP, Ernfors P (2007) Cellular subtype distribution and developmental regulation of TRPC channel members in the mouse dorsal root ganglion. J Comp Neurol 503: 35–46.

Elitt CM, McIlwrath SL, Lawson JJ, Malin SA, Molliver DC, Cornuet PK, Koerber HR, Davis BM, Albers KM (2006) Artemin overexpression in skin enhances expression of TRPV1 and TRPA1 in cutaneous sensory neurons and leads to behavioral sensitivity to heat and cold. J Neurosci 26: 8578–8587.

Fang X, Djouhri L, McMullan S, Berry C, Waxman SG, Okuse K, Lawson SN (2006) Intense isolectin-B4 binding in rat dorsal root ganglion neurons distinguishes C-fiber nociceptors with broad action potentials and high Nav1.9 expression. J Neurosci 26: 7281–7292.

Fang M, Wang Y, He QH, Sun YX, Deng LB, Wang XM, Han JS (2003) Glial cell line-derived neurotrophic factor contributes to delayed inflammatory hyperalgesia in adjuvant rat pain model. Neurosci 117: 503–512.

Fang M, Wang Y, Liu HX, Liu XS, Han JS (2000) Decreased GDNF mRNA expression in dorsal spinal cord of unilateral arthritic rat. Neuroreport 11: 737–741.

Farinas I, Cano-Jaimez M, Bellmunt E, Soriano M (2002) Regulation of neurogenesis by neurotrophins in developing spinal sensory ganglia. Brain Res Bull 57: 809–816.

Gardell LR, Wang R, Ehrenfels C, Ossipov MH, Rossomando AJ, Miller S, Buckley C, Cai AK, Tse A, Foley SF, Gong B, Walus L, Carmillo P, Worley D, Huang C, Engber T, Pepinsky B, Cate RL, Vanderah TW, Lai J, Sah DW, Porreca F (2003) Multiple actions of systemic artemin in experimental neuropathy. Nat Med 9: 1383–1389.

Gardiner NJ, Cafferty WB, Slack SE, Thompson SW (2002) Expression of gp130 and leukaemia inhibitory factor receptor subunits in adult rat sensory neurones: regulation by nerve injury. J Neurochem 83: 100–109.

Hammarberg H, Piehl F, Cullheim S, Fjell J, Hokfelt T, Fried K (1996) GDNF mRNA in Schwann cells and DRG satellite cells after chronic sciatic nerve injury. Neuroreport 7: 857–860.

Hammarberg H, Piehl F, Risling M, Cullheim S (2000) Differential regulation of trophic factor receptor mRNAs in spinal motoneurons after sciatic nerve transection and ventral root avulsion in the rat. J Comp Neurol 426: 587–601.

Hensellek S, Brell P, Schaible HG, Brauer R, Segond VB (2007) The cytokine TNFalpha increases the proportion of DRG neurones expressing the TRPV1 receptor via the TNFR1 receptor and ERK activation. Mol Cell Neurosci 36: 381–391.

Hjerling-Leffler J, Alqatari M, Ernfors P, Koltzenburg M (2007) Emergence of functional sensory subtypes as defined by transient receptor potential channel expression. J Neurosci 27: 2435–2443.

Hoane MR, Gulwadi AG, Morrison S, Hovanesian G, Lindner MD, Tao W (1999) Differential in vivo effects of neurturin and glial cell-line-derived neurotrophic factor. Exp Neurol 160: 235–243.

Holstege JC, Jongen JLM, Kennis JHH, van Rooyen-Boot AAMA, Vecht CJ (1998) Immunocytochemical localization of GDNF in primary afferents of the lumbar dorsal horn. Neuroreport 9: 2893–2897.

Hucho TB, Dina OA, Levine JD (2005) Epac mediates a cAMP-to-PKC signaling in inflammatory pain: an isolectin B4(+) neuron-specific mechanism. J Neurosci 25: 6119–6126.

Inserra MM, Yao M, Murray R, Terris DJ (2000) Peripheral nerve regeneration in interleukin 6-deficient mice. Arch Otolaryngol Head Neck Surg 126: 1112–1116.

Ito Y, Yamamoto M, Li M, Mitsuma N, Tanaka F, Doyu M, Suzumura A, Mitsuma T, Sobue G (2000) Temporal expression of mRNAs for neuropoietic cytokines, interleukin-11 (IL-11), oncostatin M (OSM), cardiotrophin-1 (CT-1) and their receptors (IL-11Ralpha and OSMRbeta) in peripheral nerve injury [In Process Citation]. Neurochem Res 25: 1113–1118.

Jeon SM, Lee KM, Park ES, Jeon YH, Cho HJ (2008) Monocyte chemoattractant protein-1 immunoreactivity in sensory ganglia and hindpaw after adjuvant injection. Neuroreport 19: 183–186.

Jia Y, Lee LY (2007) Role of TRPV receptors in respiratory diseases. Biochim Biophys Acta 1772: 915–927.

Jongen JL, Haasdijk ED, Sabel-Goedknegt H, van der BJ, Vecht C, Holstege JC (2005) Intrathecal injection of GDNF and BDNF induces immediate early gene expression in rat spinal dorsal horn. Exp Neurol 194: 255–266.

Jongen JL, Jaarsma D, Hossaini M, Natarajan D, Haasdijk ED, Holstege JC (2007) Distribution of RET immunoreactivity in the rodent spinal cord and changes after nerve injury. J Comp Neurol 500: 1136–1153.

Jordt SE, Bautista DM, Chuang HH, McKemy DD, Zygmunt PM, Hogestatt ED, Meng ID, Julius D (2004) Mustard oils and cannabinoids excite sensory nerve fibres through the TRP channel ANKTM1. Nature 427: 260–265.

Josephson A, Widenfalk J, Trifunovski A, Widmer HR, Olson L, Spenger C (2001) GDNF and NGF family members and receptors in human fetal and adult spinal cord and dorsal root ganglia. J Comp Neurol 440: 204–217.

Ju G, Hökfelt T, Brodin E, Fahrenkrug J, Fischer JA, Frey P, Elde RP, Brown JC (1987) Primary sensory neurons of the rat showing calcitonin gene- related peptide immunoreactivity and their relation to substance P-, somatostatin-, galanin-, vasoactive intestinal polypeptide- and cholecystokinin-immunoreactive ganglion cells. Cell Tissue Res 247: 417–431.

Karchewski LA, Kim FA, Johnston J, McKnight RM, Verge VM (1999) Anatomical evidence supporting the potential for modulation by multiple neurotrophins in the majority of adult lumbar sensory neurons. J Comp Neurol 413: 327–341.

Kashiba H, Uchida Y, Senba E (2001) Difference in binding by isolectin B4 to trkA and c-ret mRNA-expressing neurons in rat sensory ganglia. Mol Brain Res 95: 18–26.

Kashiba H, Uchida Y, Senba E (2003) Distribution and colocalization of NGF and GDNF family ligand receptor mRNAs in dorsal root and nodose ganglion neurons of adult rats. Brain Res Mol Brain Res 110: 52–62.

Kerr BJ, Bradbury EJ, Bennett DLH, Trivedi PM, Dassan P, French J, Shelton DB, McMahon SB, Thompson SWN (1999) Brain-derived neurotrophic factor modulates nociceptive sensory inputs and NMDA-evoked responses in the rat spinal cord. J Neurosci 19: 5138–5148.

Kidd BL, Urban LA (2001) Mechanisms of inflammatory pain. Br J Anaesth 87: 3–11.

King VR, Michael GJ, Joshi RK, Priestley JV (1999) TrkA, trkB, and trkC messenger RNA expression by bulbospinal cells of the rat. Neurosci 92: 935–944.

Kobayashi K, Fukuoka T, Obata K, Yamanaka H, Dai Y, Tokunaga A, Noguchi K (2005) Distinct expression of TRPM8, TRPA1, and TRPV1 mRNAs in rat primary afferent neurons with adelta/c-fibers and colocalization with trk receptors. J Comp Neurol 493: 596–606.

Kurek JB, Austin L, Cheema SS, Bartlett PF, Murphy M (1996) Up-regulation of leukaemia inhibitory factor and interleukin-6 in transected sciatic nerve and muscle following denervation. Neuromuscul Disord 6: 105–114.

Lawson SN (1992) Morphological and biochemical cell types of sensory neurons. In: Sensory neurons. Diversity, Development, and Plasticity (Scott SA, ed), pp 27–59. New York: OUP.

Lawson SN (2002) Phenotype and function of somatic primary afferent nociceptive neurones with C-, Adelta- or Aalpha/beta-fibres. Exp Physiol 87: 239–244.

Lawson SN, Crepps BA, Perl ER (1997) Relationship of substance P to afferent characteristics of dorsal root ganglion neurones in guinea-pig. J Physiol 505: 177–191.

Lawson SN, Crepps B, Perl ER (2002) Calcitonin gene-related peptide immunoreactivity and afferent receptive properties of dorsal root ganglion neurones in guinea-pigs. J Physiol 540: 989–1002.

Lawson JJ, McIlwrath SL, Woodbury CJ, Davis BM, Koerber HR (2008) TRPV1 unlike TRPV2 is restricted to a subset of mechanically insensitive cutaneous nociceptors responding to heat. J Pain 9: 298–308.

Lawson SN, Waddell PJ (1991) Soma neurofilament immunoreactivity is related to cell size and fibre conduction velocity in rat primary sensory neurons. J Physiol 435: 41–63.

Leffler A, Cummins TR, Dib-Hajj SD, Hormuzdiar WN, Black JA, Waxman SG (2002) GDNF and NGF reverse changes in repriming of TTX-sensitive Na(+) currents following axotomy of dorsal root ganglion neurons. J Neurophysiol 88: 650–658.

Lever IJ, Bradbury EJ, Cunningham JR, Adelson DW, Jones MG, McMahon SB, Marvizon JC, Malcangio M (2001) Brain-derived neurotrophic factor is released in the dorsal horn by distinctive patterns of afferent fiber stimulation. J Neurosci 21: 4469–4477.

Liu HX, Hokfelt T (2002) The participation of galanin in pain processing at the spinal level. Trends Pharmacol Sci 23: 468–474.

Liu Q, Vrontou S, Rice FL, Zylka MJ, Dong X, Anderson DJ (2007) Molecular genetic visualization of a rare subset of unmyelinated sensory neurons that may detect gentle touch. Nat Neurosci 10: 946–948.

Liu M, Willmott NJ, Michael GJ, Priestley JV (2004) Differential pH and capsaicin responses of Griffonia simplicifolia IB4 (IB4)-positive and IB4-negative small sensory neurons. Neurosci 127: 659–672.

Lorenzo LE, Ramien M, St Louis M, De Koninck Y, Ribeiro-da-Silva A (2008) Postnatal changes in the Rexed lamination and markers of nociceptive afferents in the superficial dorsal horn of the rat. J Comp Neurol 508: 592–604.

Lu J, Zhou XF, Rush RA (2001) Small primary sensory neurons innervating epidermis and viscera display differential phenotype in the adult rat. Neurosci Res 41: 355–363.

Macpherson LJ, Xiao B, Kwan KY, Petrus MJ, Dubin AE, Hwang S, Cravatt B, Corey DP, Patapoutian A (2007) An ion channel essential for sensing chemical damage. J Neurosci 27: 11412–11415.

Malcangio M, Getting SJ, Grist J, Cunningham JR, Bradbury EJ, Charbel-Issa P, Lever IJ, Pezet S, Perretti M (2002) A novel control mechanism based on GDNF modulation of somatostatin release from sensory neurones. FASEB J 16: 730–732.

Malin SA, Molliver DC, Koerber HR, Cornuet P, Frye R, Albers KM, Davis BM (2006) Glial cell line-derived neurotrophic factor family members sensitize nociceptors in vitro and produce thermal hyperalgesia in vivo. J Neurosci 26: 8588–8599.

Marchand F, Perretti M, McMahon SB (2005) Role of the immune system in chronic pain. Nat Rev Neurosci 6: 521–532.

Marmigere F, Ernfors P (2007) Specification and connectivity of neuronal subtypes in the sensory lineage. Nat Rev Neurosci 8: 114–127.

Matsuoka I, Nakane A, Kurihara K (1997) Induction of LIF-mRNA by TGF-beta 1 in Schwann cells. Brain Res 776: 170–180.

McMahon SB, Armanini MP, Ling LH, Phillips HS (1994) Expression and coexpression of trk receptors in subpopulations of adult primary sensory neurons projecting to identified peripheral targets. Neuron 12: 1161–1171.

McMahon SB, Priestley JV (2005) Nociceptor plasticity. In: The Neurobiology of Pain (Hunt SP, Koltzenburg M, eds), pp 35–64. Oxford: Oxford University Press.

Michael GJ, Averill S, Nitkunan A, Rattray M, Bennett DLH, Yan Q, Priestley JV (1997a) Nerve growth factor treatment increases brain-derived neurotrophic factor selectively in trkA-expressing dorsal root ganglion cells and in their central terminations within the spinal cord. J Neurosci 17: 8476–8490.

Michael GJ, Averill S, Shortland PJ, Yan Q, Priestley JV (1999) Axotomy results in major changes in BDNF expression by dorsal root ganglion cells: BDNF expression in large trkB and trkC cells, in pericellular baskets, and in projections to deep dorsal horn and dorsal column nuclei. Eur J Neurosci 11: 3539–3551.

Michael GJ, Kaya E, Averill S, Rattray M, Clary DO, Priestley JV (1997b) TrkA immunoreactive neurones in the rat spinal cord. J Comp Neurol 385: 441–455.

Michael GJ, Priestley JV (1999) Differential expression of the mRNA for the vanilloid receptor subtype 1 in cells of the adult rat dorsal root and nodose ganglia and its down-regulation by axotomy. J Neurosci 19: 1844–1854.

Miletic G, Hanson EN, Miletic V (2004) Brain-derived neurotrophic factor-elicited or sciatic ligation-associated phosphorylation of cyclic AMP response element binding protein in the rat spinal dorsal horn is reduced by block of tyrosine kinase receptors. Neurosci Lett 361: 269–271.

Mizuno M, Kondo E, Nishimura M, Ueda Y, Yoshiya I, Tohyama M, Kiyama H (1997) Localization of molecules involved in cytokine receptor signaling in the rat trigeminal ganglion. Mol Brain Res 44: 163–166.
Molliver DC, Wright DE, Leitner ML, Parsadanian AS, Doster K, Wen D, Yan Q, Snider WD (1997) IB4-binding DRG neurons switch from NGF to GDNF dependence in early postnatal life. Neuron 19: 849–861.
Moran MM, Xu H, Clapham DE (2004) TRP ion channels in the nervous system. Curr Opin Neurobiol 14: 362–369.
Morikawa Y, Tamura S, Minehata K, Donovan PJ, Miyajima A, Senba E (2004) Essential function of oncostatin m in nociceptive neurons of dorsal root ganglia. J Neurosci 24: 1941–1947.
Murphy PG, Borthwick LA, Altares M, Gauldie J, Kaplan D, Richardson PM (2000) Reciprocal actions of interleukin-6 and brain-derived neurotrophic factor on rat and mouse primary sensory neurons. Eur J Neurosci 12: 1891–1899.
Murphy PG, Grondin J, Altares M, Richardson PM (1995) Induction of interleukin-6 in axotomized sensory neurons. J Neurosci 15: 5130–5138.
Murphy PG, Ramer MS, Borthwick L, Gauldie J, Richardson PM, Bisby MA (1999) Endogenous interleukin-6 contributes to hypersensitivity to cutaneous stimuli and changes in neuropeptides associated with chronic nerve constriction in mice. Eur J Neurosci 11: 2243–2253.
Nagano M, Sakai A, Takahashi N, Umino M, Yoshioka K, Suzuki H (2003) Decreased expression of glial cell line-derived neurotrophic factor signaling in rat models of neuropathic pain. Br J Pharmacol 140: 1252–1260.
Orozco OE, Walus L, Sah DW, Pepinsky RB, Sanicola M (2001) GFRalpha3 is expressed predominantly in nociceptive sensory neurons. Eur J Neurosci 13: 2177–2182.
Petruska JC, Napaporn J, Johnson RD, Cooper BY (2002) Chemical responsiveness and histochemical phenotype of electrophysiologically classified cells of the adult rat dorsal root ganglion. Neurosci 115: 15–30.
Pezet S, Malcangio M, Lever IJ, Perkinton MS, Thompson SW, Williams RJ, McMahon SB (2002) Noxious stimulation induces Trk receptor and downstream ERK phosphorylation in spinal dorsal horn. Mol Cell Neurosci 21: 684–695.
Pezet S, McMahon SB (2006) Neurotrophins: mediators and modulators of pain. Annu Rev Neurosci 29: 507–538.
Priestley JV (2008) Neuropeptides – sensory systems. In: The New Encyclopedia of Neuroscience (Squire LR, ed), Amsterdam: Elsevier.
Priestley JV, Michael GJ, Averill S, Liu M, Willmott N (2002) Regulation of nociceptive neurons by nerve growth factor and glial cell line derived neurotrophic factor. Can J Physiol Pharmacol 80: 495–505.
Qiu L, Towle MF, Bernd P, Fukada K (1997) Distribution of cholinergic neuronal differentiation factor/leukemia inhibitory factor binding sites in the developing and adult rat nervous system in vivo. J Neurobiol 32: 163–192.
Raghavendra V, Tanga FY, DeLeo JA (2004) Complete Freunds adjuvant-induced peripheral inflammation evokes glial activation and proinflammatory cytokine expression in the CNS. Eur J Neurosci 20: 467–473.
Ramer MS, Bradbury EJ, Michael GJ, Lever IJ, McMahon SB (2003) Glial cell line-derived neurotrophic factor increases calcitonin gene-related peptide immunoreactivity in sensory and motoneurons in vivo. Eur J Neurosci 18: 2713–2721.
Ramer MS, Murphy PG, Richardson PM, Bisby MA (1998) Spinal nerve lesion-induced mechanoallodynia and adrenergic sprouting in sensory ganglia are attenuated in interleukin-6 knockout mice. Pain 78: 115–121.
Rind HB, von Bartheld CS (2002) Anterograde axonal transport of internalized GDNF in sensory and motor neurons. Neuroreport 13: 659–664.

Rubio N (1997) Mouse astrocytes store and deliver brain-derived neurotrophic factor using the non-catalytic gp95(trkb) receptor. Eur J Neurosci 9: 1847–1853.

Rutkowski JL, Tuite GF, Lincoln PM, Boyer PJ, Tennekoon GI, Kunkel SL (1999) Signals for proinflammatory cytokine secretion by human Schwann cells. J Neuroimmunol 101: 47–60.

Sah DW, Ossipov MH, Rossomando A, Silvian L, Porreca F (2005) New approaches for the treatment of pain: the GDNF family of neurotrophic growth factors. Curr Top Med Chem 5: 577–583.

Salio C, Averill S, Priestley JV, Merighi A (2007) Costorage of BDNF and neuropeptides within individual dense-core vesicles in central and peripheral neurons. Dev Neurobiol 67: 326–338.

Scott RL, Gurusinghe AD, Rudvosky AA, Kozlakivsky V, Murray SS, Satoh M, Cheema SS (2000) Expression of leukemia inhibitory factor receptor mRNA in sensory dorsal root ganglion and spinal motor neurons of the neonatal rat. Neurosci Lett 295: 49–53.

Shinohara T, Harada M, Ogi K, Maruyama M, Fujii R, Tanaka H, Fukusumi S, Komatsu H, Hosoya M, Noguchi Y, Watanabe T, Moriya T, Itoh Y, Hinuma S (2004) Identification of a G protein-coupled receptor specifically responsive to beta-alanine. J Biol Chem 279: 23559–23564.

Shortland PJ, Baytug B, Krzyzanowska A, McMahon SB, Priestley JV, Averill S (2006) ATF3 expression in L4 dorsal root ganglion neurons after L5 spinal nerve transection. Eur J Neurosci 23: 365–373.

Silverman JD, Kruger L (1990) Selective neuronal glycoconjugate expression in sensory and autonomic ganglia: relation of lectin reactivity to peptide and enzyme markers. J Neurocytol 19: 789–801.

Slack SE, Grist J, Mac Q, McMahon SB, Pezet S (2005) TrkB expression and phospho-ERK activation by brain-derived neurotrophic factor in rat spinothalamic tract neurons. J Comp Neurol 489: 59–68.

Slack SE, Pezet S, McMahon SB, Thompson SW, Malcangio M (2004) Brain-derived neurotrophic factor induces NMDA receptor subunit one phosphorylation via ERK and PKC in the rat spinal cord. Eur J Neurosci 20: 1769–1778.

Snider WD, McMahon SB (1998) Tackling pain at the source: new ideas about nociceptors. Neuron 20: 629–632.

Story GM, Peier AM, Reeve AJ, Eid SR, Mosbacher J, Hricik TR, Earley TJ, Hergarden AC, Andersson DA, Hwang SW, McIntyre P, Jegla T, Bevan S, Patapoutian A (2003) ANKTM1, a TRP-like channel expressed in nociceptive neurons, is activated by cold temperatures. Cell 112: 819–829.

Stucky CL, Lewin GR (1999) Isolectin B4 -positive and -negative nociceptors are functionally distinct. J Neurosci 19: 6497–6505.

Subang MC, Richardson PM (2001) Influence of injury and cytokines on synthesis of monocyte chemoattractant protein-1 mRNA in peripheral nervous tissue. Eur J Neurosci 13: 521–528.

Summer GJ, Romero-Sandoval EA, Bogen O, Dina OA, Khasar SG, Levine JD (2008) Proinflammatory cytokines mediating burn-injury pain. Pain 135: 98–107.

Sun Y, Zigmond RE (1996) Leukemia inhibitory factor-induced in the sciatic-nerve after axotomy is involved in the induction of galanin in sensory neurons. Eur J Neurosci 8: 2213–2220.

Szallasi A, Cortright DN, Blum CA, Eid SR (2007) The vanilloid receptor TRPV1: 10 years from channel cloning to antagonist proof-of-concept. Nat Rev Drug Discov 6: 357–372.

Tamura S, Morikawa Y, Miyajima A, Senba E (2003) Expression of oncostatin M receptor beta in a specific subset of nociceptive sensory neurons. Eur J Neurosci 17: 2287–2298.

Tamura S, Morikawa Y, Senba E (2005) TRPV2, a capsaicin receptor homologue, is expressed predominantly in the neurotrophin-3-dependent subpopulation of primary sensory neurons. Neurosci 130: 223–228.

Thacker MA, Clark AK, Bishop T, Grist J, Yip PK, Moon LD, Thompson SW, Marchand F, McMahon SB (2009) CCL2 is a key mediator of microglia activation in neuropathic pain states. Eur J Pain 13: 263–272.

Thompson SWN, Priestley JV, Southall A (1998) Gp130 cytokines, leukemia inhibitory factor and interleukin-6, induce neuropeptide expression in intact adult rat sensory neurons in vivo: time course, specificity and comparison with sciatic nerve axotomy. Neurosci 84: 1247–1255.

Thompson SWN, Vernallis AB, Heath JK, Priestley JV (1997) Leukaemia inhibitory factor is retrogradely transported by a distinct population of adult rat sensory neurons: co-localization with trka and other neurochemical markers. Eur J Neurosci 9: 1244–1251.

Trevisani M, Siemens J, Materazzi S, Bautista DM, Nassini R, Campi B, Imamachi N, Andre E, Patacchini R, Cottrell GS, Gatti R, Basbaum AI, Bunnett NW, Julius D, Geppetti P (2007) 4-Hydroxynonenal, an endogenous aldehyde, causes pain and neurogenic inflammation through activation of the irritant receptor TRPA1. Proc Natl Acad Sci USA 104: 13519–13524.

Vellani V, Zachrisson O, McNaughton PA (2004) Functional bradykinin B1 receptors are expressed in nociceptive neurones and are upregulated by the neurotrophin GDNF. J Physiol 560: 391–401.

Verge VM, Richardson PM, Wiesenfeld-Hallin Z, Hokfelt T (1995) Differential influence of nerve growth factor on neuropeptide expression in vivo: a novel role in peptide suppression in adult sensory neurons. J Neurosci 15: 2081–2096.

Villar MJ, Cortes R, Theodorsson E, Wiesenfeld-Hallin Z, Schalling M, Fahrenkrug J, Emson PC, Hokfelt T (1989) Neuropeptide expression in rat dorsal root ganglion cells and spinal cord after peripheral nerve injury with special reference to galanin. Neurosci 33: 587–604.

Vulchanova L, Olson TH, Stone LS, Riedl MS, Elde R, Honda CN (2001) Cytotoxic targeting of isolectin IB4-binding sensory neurons. Neurosci 108: 143–155.

Vydyanathan A, Wu ZZ, Chen SR, Pan HL (2005) A-type voltage-gated K+ currents influence firing properties of isolectin B4-positive but not isolectin B4-negative primary sensory neurons. J Neurophysiol 93: 3401–3409.

Wang R, Guo W, Ossipov MH, Vanderah TW, Porreca F, Lai J (2003) Glial cell line-derived neurotrophic factor normalizes neurochemical changes in injured dorsal root ganglion neurons and prevents the expression of experimental neuropathic pain. Neurosci 121: 815–824.

Woodbury CJ, Ritter AM, Koerber HR (2000) On the problem of lamination in the superficial dorsal horn of mammals: a reappraisal of the substantia gelatinosa in postnatal life. J Comp Neurol 417: 88–102.

Woodbury CJ, Zwick M, Wang S, Lawson JJ, Caterina MJ, Koltzenburg M, Albers KM, Koerber HR, Davis BM (2004) Nociceptors lacking TRPV1 and TRPV2 have normal heat responses. J Neurosci 24: 6410–6415.

Woolf CJ, Costigan M (1999) Transcriptional and posttranslational plasticity and the generation of inflammatory pain. Proc Natl Acad Sci USA 96: 7723–7730.

Woolf CJ, Safieh-Garabedian B, Ma QP, Crilly P, Winter J (1994) Nerve growth factor contributes to the generation of inflammatory sensory hypersensitivity. Neurosci 62: 327–331.

Wotherspoon G, Winter J (2000) Bradykinin B1 receptor is constitutively expressed in the rat sensory nervous system. Neurosci Lett 294: 175–178.

Wu D, Zhang Y, Bo X, Huang W, Xiao F, Zhang X, Miao T, Magoulas C, Subang MC, Richardson PM (2007) Actions of neuropoietic cytokines and cyclic AMP in regenerative conditioning of rat primary sensory neurons. Exp Neurol 204: 66–76.

Wynick D, Thompson SW, McMahon SB (2001) The role of galanin as a multi-functional neuropeptide in the nervous system. Curr Opin Pharmacol 1: 73–77.

Zhong J, Dietzel ID, Wahle P, Kopf M, Heumann R (1999) Sensory impairments and delayed regeneration of sensory axons in interleukin-6-deficient mice. J Neurosci 19: 4305–4313.

Zhu M, Oishi K, Lee SC, Patterson PH (2001) Studies using leukemia inhibitory factor (LIF) knockout mice and a LIF adenoviral vector demonstrate a key anti-inflammatory role for this cytokine in cutaneous inflammation. J Immunol 166: 2049–2054.

Zwick M, Davis BM, Woodbury CJ, Burkett JN, Koerber HR, Simpson JF, Albers KM (2002) Glial cell line-derived neurotrophic factor is a survival factor for isolectin B4-positive, but not vanilloid receptor 1-positive, neurons in the mouse. J Neurosci 22: 4057–4065.

Zylka MJ, Rice FL, Anderson DJ (2005) Topographically distinct epidermal nociceptive circuits revealed by axonal tracers targeted to Mrgprd. Neuron 45: 17–25.

Part II
Fast Synaptic Transmission in the Dorsal Horn

Chapter 3
Fast Inhibitory Transmission of Pain in the Spinal Cord

Hanns Ulrich Zeilhofer, Robert Witschi, and Torbjörn Johansson

Abstract The gate-control-theory of pain attributes a pivotal role in nociceptive processing to inhibitory interneurons in the superficial dorsal horn of the spinal cord. Loss of synaptic inhibition is a contributing factor to the generation and maintenance of chronic pain. Different signaling pathways involved in inflammatory and neuropathic pain converge onto diminished synaptic inhibition in the spinal dorsal horn. Accordingly, restoring inhibition through drugs that facilitate GABAergic or glycinergic neurotransmission should reverse exaggerated pain sensitivity. Indeed, subtype-selective $GABA_A$ receptor ligands and inhibitors of glycine transporters might constitute new treatments for chronic pain.

Abbreviations

BDNF	brain-derived neurotrophic factor
COX	cyclooxygenase
GABA	γ-aminobutyric acid
GAD	glutamic acid decarboxylase
GlyR	glycine receptor
GlyT	glycine transporter
EGFP	enhanced green fluorescent protein
mIPSP	miniature inhibitory postsynaptic potentials
PGE2	prostaglandin E2
PKA	protein kinase A
PKC	protein kinase C

H.U. Zeilhofer (✉)
Institute of Pharmacology and Toxicology, University of Zurich, Winterthurerstrasse 190, CH-8057 Zurich, Switzerland
e-mail: zeilhofer@pharma.uzh.ch

M. Malcangio (ed.), *Synaptic Plasticity in Pain*,
DOI 10.1007/978-1-4419-0226-9_3, © Springer Science+Business Media, LLC 2009

3.1 Introduction

More than 40 years ago, the gate-control-theory of pain attributed to
inhibitory interneurons located in the superficial dorsal horn a pivotal role
in nociceptive processing. In the original model proposed by Melzack and
Wall (1965), the activity of these inhibitory neurons was driven by excitatory input from thick, low threshold mechanosensitive fibers and reduced
by input from small unmyelinated nociceptors (Fig. 3.1). The inhibitory
interneurons themselves reduced synaptic transmission between both types
of primary afferent nerve fibers and dorsal horn neurons. Details of the
gate-control-theory have been the subject of intense scientific discussions,
however the basic principle of a profound inhibitory pain control mediated
by superficial dorsal horn interneurons has impressively been confirmed by
experimental research. Loss of synaptic inhibition is meanwhile widely
accepted as an important factor contributing to the generation and maintenance of chronic pain and its neurobiological basis is subject to extensive
investigations.

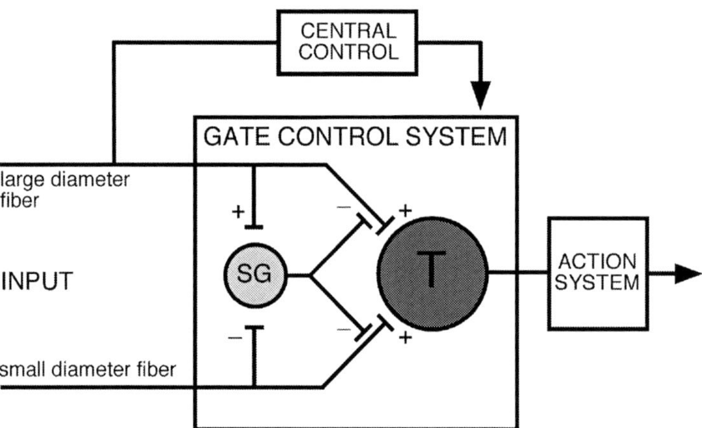

Fig. 3.1 Gate-control-theory. Schematic diagram illustrating the circuitry proposed by
Melzack and Wall (1965). Small nociceptive and large low-threshold-mechanosensitive
fibers tune the activity of substantia gelatinosa (SG) neurons in opposite directions. Inhibitory SG neurons reduce synaptic transmission between both types of primary afferent
nerve fibers. The efficacy of the spinal gate-control-system is in addition tuned by descending input from supraspinal sites (modified from Melzack and Wall, 1965, with permission
from AAAS)

Inhibitory interneurons in the spinal dorsal horn use the two amino acids,
γ-aminobutyric acid (GABA) and glycine, for fast inhibitory neurotransmission.
After being released into the synaptic cleft, GABA and glycine bind to their
postsynaptic receptors to open chloride permeable ion channels. They thereby

inhibit neuronal excitation through hyperpolarization of the postsynaptic neuron and activation of a shunting conductance, which impairs the propagation of excitatory signals along dendrites. Experiments performed already in the 1980s demonstrated that blockade of spinal $GABA_A$ or glycine receptors with bicuculline or strychnine dramatically lowers the thresholds of nocifensive reflexes in rats, while spinal injection of GABA or glycine was antinociceptive under most circumstances (Beyer et al., 1985; Roberts et al., 1986; Yaksh, 1989).

Since these early discoveries, our knowledge on inhibitory pain control in the spinal dorsal horn and its contribution to the generation of pathological pain states has dramatically increased. In this chapter, we will discuss how inhibitory interneurons are integrated in dorsal horn pain controlling circuits and how plasticity of fast inhibitory neurotransmission contributes to the generation and maintenance of chronic pain states.

3.2 Physiology of Inhibitory Neurotransmission in the Spinal Dorsal Horn

3.2.1 Distribution of GABAergic and Glycinergic Neurons in the Spinal Dorsal Horn

Antisera against GABA and glycine, which label neuronal somata of GABA and glycine releasing (GABAergic and glycinergic) neurons, demonstrate a wide-spread distribution of inhibitory neurons throughout the grey matter of the spinal cord. About 30% of lamina I and II neurons and 46% of lamina III neurons exhibited GABA-like immunoreactivity in a seminal study by Todd and Sullivan (1990). Many of these GABAergic neurons (33, 43, and 64%, in lamina I, II and III, respectively) were also immunoreactive for glycine. Purely glycinergic neurons were rarely found in the superficial layers (I–III), but are frequently seen in the deeper dorsal horn (laminae IV and V). More recently, different genetic engineering techniques have been employed to generate mice, which express enhanced green fluorescent protein (EGFP) under the transcriptional control of marker genes such as the glutamic acid decarboxylase (GAD)67 (*gad1*) gene for GABAergic neurons (Oliva et al., 2000; Tamamaki et al., 2003) or the GlyT2 (*scl6a5*) gene for glycinergic neurons (Zeilhofer et al., 2005). Gross sections of spinal cords from GAD67-EGFP mice show numerous EGFP labeled neurons in laminae I–III and around the central canal but only few neurons in the deeper dorsal horn (Fig. 3.2). GlyT2-EGFP-positive neurons show a somewhat different distribution (Zeilhofer et al., 2005) with a high abundance in the deeper layers (laminae III–V) and relatively few neurons in laminae I and II, a distribution, which is also shown in in situ hybridization studies (Hossaini et al., 2007).

GAD67-EGFP   GlyT2-EGFP

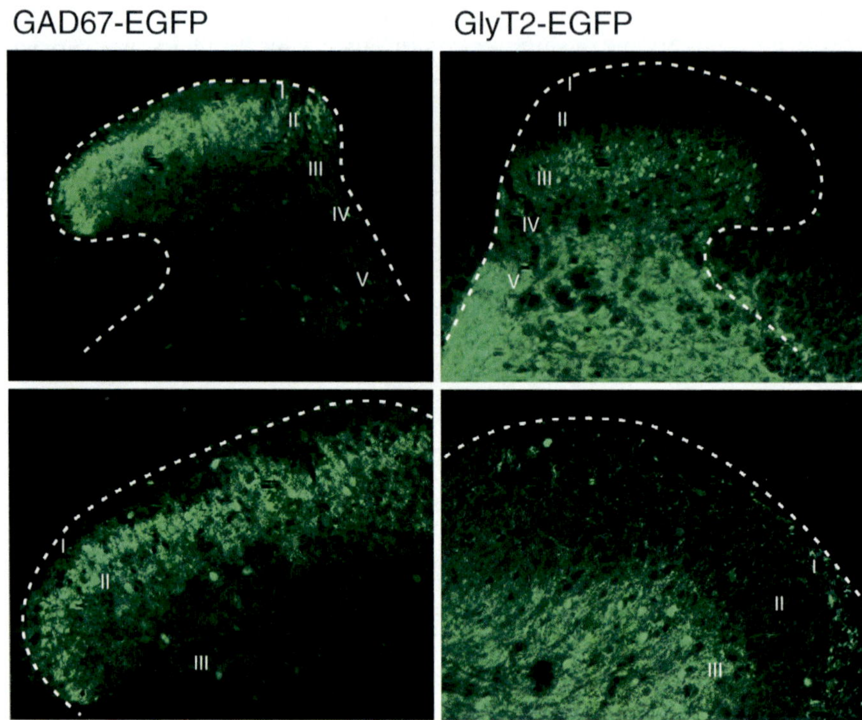

Fig. 3.2 Distribution of GABAergic and glycinergic neurons. Transverse sections of the lumbar spinal cord of GAD67-EGFP (Tamamaki et al., 2003) and GlyT2-EGFP (Zeilhofer et al., 2005) transgenic mice at different magnification. GAD67-EGFP tagged somata are most prominent in lamina II, whereas GlyT2-EGFP positive somata are most abundant in the deeper dorsal horn (lamina III and V)

3.2.2 Integration of Inhibitory Dorsal Horn Neurons Dorsal Horn Circuits

Figure 3.3 shows a simplified scheme illustrating the integration of inhibitory interneurons into dorsal horn neuronal circuits. As proposed by the gate-control-theory, GABAergic input to substantia gelatinosa neurons can be activated by stimulation of primary afferent nerve fibers (Yoshimura and Nishi, 1995). Activation of GABAergic neurons in the superficial dorsal horn has been demonstrated directly in spinal cord slices, where GABAergic neurons could be identified either by their green fluorescence in GAD67-EGFP trans-genic mice (Heinke et al., 2004) or by the GABAergic IPSCs that their stimulation evoked in synaptically connected cells (Lu and Perl, 2003). Many but not all GABAergic neurons in lamina II show morphological characteristics of islet cells (Grudt and Perl, 2002) and exhibit a tonic firing pattern (Lu and Perl, 2003;

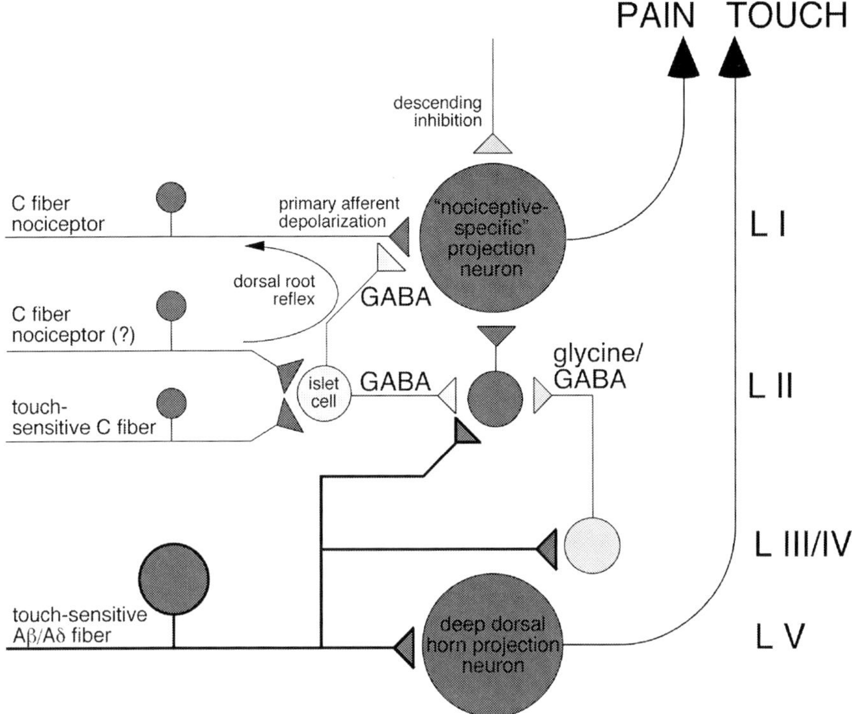

Fig. 3.3 Possible integration of inhibitory dorsal horn neurons in spinal pain processing circuits. GABAergic islet cells in the substantia gelatinosa (L II) receive monosynaptic input from C fibers, which are believed to be mainly non-nociceptive touch-sensitive (Bennett et al., 1980). These lamina II GABAergic cells form different synaptic connections, with presynaptic terminals of primary afferent fibers, where they induce primary afferent depolarization, and with intrinsic superficial dorsal horn neurons, where they cause classical postsynaptic inhibition. They probably also synapse to lamina I (L I) projection neurons. Mixed GABA/ glycinergic cells and pure glycinergic cells are located mainly in the deeper laminae (L III and V). They are probably excited primarily by mechanosensitive fibers and cause postsynaptic inhibition of excitatory interneurons and possibly also of projection neurons. Removal of this inhibition leads to polysynaptic excitation of normally nociceptive specific neurons in lamina I and induces touch-evoked pain (allodynia)

Hantman and Perl, 2005). The majority of GABAergic neurons, in particular those with an islet cell- or central cell-like morphology (Grudt and Perl, 2002), were excited by monosynaptic input only from C fibers (Hantman and Perl, 2005; Yasaka et al., 2007), while other GABAergic neurons also received polysynaptic input from Aδ fibers (Yasaka et al., 2007). However, this does not necessarily mean that GABAergic neurons are activated by nociceptive stimuli. There is indeed evidence that the C fibers which project to these GABAergic neurons have a relatively large diameter (Lu and Perl, 2003) and may convey tactile rather than nociceptive information (Light et al., 1979; Bennett et al., 1980). In vivo recordings have indeed shown that inhibitory synaptic events in the superficial dorsal horn are primarily triggered by light mechanical (brush) stimulation of the skin (Narikawa et al., 2000). On the other hand, in slices local application of capsaicin to islet cells triggered action potentials in these cells (Maxwell et al., 2007). At present, glycinergic neurons in the deeper dorsal horn have not yet been investigated in similar studies. According to well-established patterns of primary afferent innervation of the dorsal horn, these neurons should primarily receive input from low threshold mechanosensitive fibers.

Part of the inhibitory input to the spinal dorsal horn comes from descending fibers, which originate from the rostral ventromedial medulla (Antal et al., 1996) (see Chapter 19) and form monosynaptic connections with dorsal horn neurons (Kato et al., 2006). In addition, slow transmitters such as norepinephrine and serotonin depolarize identified GABAergic neurons in the dorsal horn through activation of G-protein coupled receptors (Hantman and Perl, 2005), an action, which may contribute to the antinociceptive action of descending serotoninergic and noradrenergic fiber tracts.

3.2.3 Inhibitory Input to Dorsal Horn Neurons

Several studies have characterized inhibitory postsynaptic currents in superficial dorsal horn neurons evoked by spinal cord field stimulation. Such studies have shown that the majority of spinal dorsal horn neurons receive GABAergic and glycinergic synaptic input, but the relative contribution of both transmitters varies between the different laminae and changes during postnatal maturation. Shortly after birth, mixed GABAergic and glycinergic miniature IPSCs, which indicate co-release of both transmitters from a single vesicle, occur frequently, but become less frequent during development and disappear completely by P30 (at least in laminae I and II) (Keller et al., 2001). In adult rats (at postnatal day P30), mIPSCs were purely glycinergic in lamina I, while in lamina II and in the deeper dorsal horn (laminae III–IV) glycinergic and GABAergic mIPSCs occurred with comparable frequencies in the vast majority of the neurons recorded (Keller et al., 2001, see also Baccei and Fitzgerald, 2004) (see Chapter 4). The shift towards pure glycinergic events in lamina I appears to be mainly due to a postsynaptic specialization, since a GABAergic component could still be unmasked by application of benzodiazepines, which increase the responsiveness of $GABA_A$ receptors for GABA (Chery and de Koninck, 1999; Keller et al., 2001). Furthermore, the GABAergic component becomes progressively bigger with increasing presynaptic stimulation strength (Chery and De Koninck, 1999), suggesting that inhibitory boutons contacting lamina I neurons still release both GABA and glycine also in the adult, but only glycine receptors are located synaptically whereas $GABA_A$ receptors are mainly extrasynaptic. Indeed, it has been proposed that the first target of synaptically released GABA in lamina I neurons are $GABA_B$ and not $GABA_A$ receptors (Chery and De Koninck, 2000). The combined storage (and release) of GABA and glycine is due to the non-selective transport of both amino acids into presynaptic vesicles by the same vesicular inhibitory aminoacid transporter (VIAAT, also called vesicular GABA transporter, VGAT) (Chaudhry et al., 1998; Dumoulin et al., 1999; Wojcik et al., 2006).

3.2.4 Presynaptic Inhibition

GABAergic, but not glycinergic, neurons also form axo-axonic synapses with spinal primary afferent nerve terminals to cause primary afferent depolarization (Rudomin and Schmidt, 1999). Unlike most CNS neurons, which are hyperpolarized by GABA, primary sensory neurons are usually depolarized by $GABA_A$ receptor activation due to an unusually high intracellular chloride concentration. This originates from a particular expression pattern of chloride transporters with high NKCC1 expression (transports chloride, sodium and potassium into the cell) (Sung et al., 2000; Price et al., 2006) and low expression of KCC2 (Rivera et al., 1999), which transports chloride and potassium out of the cell. Sub-threshold primary afferent depolarization causes presynaptic inhibition, which reduces transmitter release from primary afferent nerve terminals through voltage-dependent inactivation of Ca^{2+} channels. By contrast, supra-threshold depolarizations can elicit action potentials in spinal primary afferent terminals and possibly trigger so called dorsal root reflexes, which may contribute to neurogenic inflammation (Willis, 1999). To date, no recordings from individual primary afferent terminals have been reported, yet such experiments were necessary to identify the exact circuit underlying GABAergic primary afferent depolarization. Furthermore, it is still not possible to specifically block presynaptic $GABA_A$ receptors in behavioral experiments and hence difficult to determine the role of presynaptic inhibition under different conditions.

3.3 Functional Consequences of Reduced Inhibitory Neurotransmission in the Spinal Dorsal Horn

Cellular correlates of diminished synaptic inhibition in the dorsal horn have been studied in vitro in dorsal horn slices of rodents. Baba et al. (2003) showed a dramatic increase in excitatory input to substantia gelatinosa neurons after blockade of $GABA_A$ receptors with bicuculline. This increased synaptic input was mainly polysynaptic in nature and depended on the activation of NMDA receptors. Interestingly, slices obtained from neuropathic rats showed an increase in excitatory inputs similar to those induced by bicuculline in naïve slices. Furthermore the effect of bicuculline was dramatically reduced in slices obtained from neuropathic rats suggesting that neuropathy may produce itself a reduction in GABAergic inhibition (see also below).

Other in vitro and in vivo studies have looked specifically at the excitation of lamina I projection neurons, which are normally activated by monosynaptic input from high threshold C and $A\delta$ fibers only (Torsney and MacDermott, 2006; Keller et al., 2007). In these studies, blockade of GABAergic or glycinergic neurotransmission induced significant polysynaptic input from $A\beta$ fibers in these neurons. In other words, "nociceptive specific" neurons became excitable by input from low threshold mechanosensitive fibers. These experiments and modeling studies by Prescott et al. (2006) have shown that diminished synaptic inhibition of lamina I projection neurons is sufficient to induce hyperalgesia and allodynia. However, additional signaling cascades downstream of glycine receptor inhibition may also be involved. A recent study by Miraucourt et al. (2007) suggests that reduced glycinergic inhibition triggers the downstream activation of PKC_γ, an enzyme, which is critically involved in pathological pain states (Malmberg et al., 1997a). Although Miraucourt et al. (2007) relate their findings primarily to neuropathic pain, spinal PKC_γ expression is also induced after peripheral inflammation (Martin et al., 1999) and its induction correlates well with pain sensitization.

3.3.1 Does a Loss of Synaptic Inhibition Occur Naturally In Vivo?

Hyperalgesia and allodynia are two key symptoms of inflammatory and neuropathic pain syndromes, which can reliably be induced by blockade of spinal $GABA_A$ and glycine receptors. Several lines of evidence indicate that a similar loss of synaptic inhibition also occurs naturally and significantly contributes to the generation of pathological pain. Distinct mechanisms underlying this loss of inhibition have been identified for inflammatory and neuropathic diseases, which will be discussed in the following three sections.

3.3.1.1 Inflammatory Pain

Peripheral inflammation induces a pronounced increase in the spinal production of prostaglandin E_2 (PGE_2), a key mediator of central inflammatory hyperalgesia (Fig. 3.4A). The two enzymes required for inflammation-induced PGE_2 production, cyclooxygenase-2 (COX-2) and inducible microsomal prostaglandin E synthase (mPGES-1), are up-regulated in the dorsal horn within hours after induction of a peripheral inflammation (Beiche et al., 1996; Murakami et al., 2000; Samad et al., 2001; Claveau et al., 2003). A major down-stream effect of spinally produced PGE_2 is the reduction of glycinergic transmission in the superficial dorsal horn (Ahmadi et al., 2002). This inhibition occurs through a postsynaptic mechanism involving the activation of PGE_2 receptors of the EP2

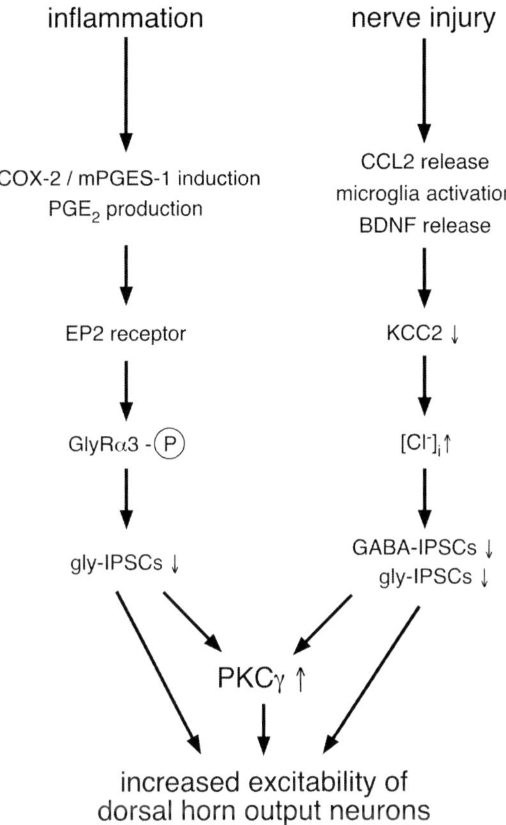

Fig. 3.4 Molecular pathways leading to synaptic disinhibition in inflammatory and neuropathic pain syndromes. The central component of inflammatory hyperalgesia depends largely on the spinal induction of COX-2 and inducible (microsomal) PGE_2 synthase-1 (mPGES1) and a subsequent block of glycine receptors in the superficial dorsal horn. In neuropathic pain following peripheral nerve injury, disinhibition is initiated by microglia activation through primary afferent nerve fiber-derived cytokines such as CCL2 acting on CCR2 receptors (Zhang et al., 2007; Thacker et al., 2008) and the purinergic agonist ATP acting on P2X4 (Tsuda et al., 2003) or P2X7 (Chessell et al., 2005) receptors, which finally leads to a disruption of the chloride homeostasis in dorsal horns neurons. Both pathways converge onto diminished synaptic inhibition and increased excitability of output neurons in the spinal dorsal horn. Activation of PKCγ is likely a consequence of reduced synaptic inhibition and contributes to increased excitability of dorsal horn output neurons

subtype, subsequent cAMP production and activation of protein kinase A (PKA). Activated PKA phosphorylates and inhibits a specific isoform of glycine receptors containing the $\alpha 3$ subunit, which in the spinal cord is distinctly expressed in the superficial layers (Harvey et al., 2004). Interestingly, PGE_2 mediated inhibition of glycine receptors occurs in the majority of excitatory superficial dorsal horn neurons (Ahmadi et al., 2002; Reinold et al., 2005), in a pattern which is reminiscent of the PKCγ activation seen in vivo after blockade of glycine receptors with strychnine (Miraucourt et al., 2007). Work in EP2 receptor-deficient mice and in mice lacking the glycine receptor α3 (GlyRα3) subunit has shown that the pro-nociceptive actions of PGE_2 are virtually absent in these mice and that inflammatory pain is strongly reduced (Harvey et al., 2004; Reinold et al., 2005; for a direct comparison of both mutant mice see Zeilhofer, 2005). Interestingly, the development of neuropathic pain in chronic constriction injury model is not altered in EP2 or GlyRα3 deficient mice as compared to wild type mice (Hosl et al., 2006). A similar pattern has been reported earlier in mice carrying a null mutation in the regulatory subunit of neuronal protein kinase A (Malmberg et al., 1997b). These mice also showed reduced nociceptive responses to intrathecal PGE_2, but exhibited normal pain responses in the chronic constriction injury model.

It should be noted that inflammation can also upregulate GABAergic inhibition in the spinal dorsal horn through production of endogenous 3α5α-neurosteroids (such as allopregnanolone and tetrahydro-deoxycorticosterone). These neurosteroids are positive allosteric modulators of $GABA_A$ receptors (Hosie et al., 2006), which, at least in the spinal dorsal horn, target primarily extrasynaptic $GABA_A$ receptors (Mitchell et al., 2007). They are regularly produced in the spinal cord during development, but are normally absent in the adult (Keller et al., 2004). In response to a peripheral inflammatory stimulus, they reappear in the spinal cord and limit inflammatory thermal hyperalgesia (Poisbeau et al., 2005; Vergnano et al., 2007).

3.3.1.2 Neuropathic Pain

At least two mechanisms have been proposed that would cause synaptic disinhibition after peripheral nerve injury. The group of Clifford Woolf firstly demonstrated that GABAergic synaptic transmission was diminished in spinal cord slices of neuropathic rats (Moore et al., 2002). This loss was accompanied by a reduction in GAD65 immunoreactivity on the neuropathic side. Whether or not this loss is due to an apoptotic death of GABAergic interneurons, as suggest by the authors, is still controversial (Polgar et al., 2005; Scholz et al., 2005). Since GAD antisera mainly stain synaptic boutons (Mackie et al., 2003), it is conceivable that only GABAergic boutons degenerate, while their neuronal somata remain intact.

Another pathway leading to diminished synaptic inhibition depends on the activation of spinal microglia, which has been recognized as a key event in the generation of neuropathic pain following peripheral nerve damage (Tsuda et al., 2003; Scholz and Woolf, 2007). A possible link between microglia activation and altered neuronal processing of sensory information is the release of brain-derived neurotrophic factor (BDNF) from microglia cells and the subsequent trkB-mediated down-regulation of the potassium chloride cotransporter KCC2 in dorsal horn neurons (Coull et al., 2003, 2005) (Fig. 3.4B), which is required for the maintenance of a low intracellular chloride concentration (Rivera et al., 1999). Reduced expression of KCC2 hence increases intracellular chloride concentrations and renders GABAergic and glycinergic input less inhibitory, or can even turn inhibition into excitation. Details of this signal transduction are discussed in Chapter 22.

3.3.1.3 Activity-Dependent Sensitization

A loss of dorsal horn synaptic inhibition can also occur in the absence of inflammation and nerve injury. Selective activation of C fiber nociceptors with capsaicin induces mechanical and thermal hyperalgesia at the site of injection (primary hyperalgesia) and in addition an exclusively mechanical sensitization (pin-prick hyperalgesia and touch-evoked pain) in a surrounding healthy skin area (Treede and Magerl, 2000). Plenty of evidence indicates that this secondary hyperalgesia is of central (spinal) origin (Woolf, 1983), involves diminished synaptic inhibition (Sivilotti and Woolf, 1994) and thus can be considered as a form of heterosynaptic plasticity (Zieglgansberger and Herz, 1971). It is at present not known what links intense C fiber input to reduced synaptic inhibition. Available evidence suggests that COX-derived prostaglandins are not involved (Schattschneider et al., 2007), microglia activation and KCC2 down-regulation are probably too slow to fit with the short delay between C fiber excitation and the development of secondary hyperalgesia.

3.4 Restoring Synaptic Inhibition in Pathological Pain States

Different signaling pathways involved in inflammatory and neuropathic pain apparently converge onto diminished synaptic inhibition in the spinal dorsal horn. Accordingly, restoring inhibition through drugs that facilitate GABAergic or glycinergic neurotransmission should reverse exaggerated pain sensitivity. There is clear evidence that increasing GABAergic neurotransmission in the spinal dorsal horn exerts a profound analgesic or antihyperalgesic effect. A major challenge for the translation of these findings into clinical therapy, however comes from the ubiquitous presence of GABA receptors in the central nervous system, which makes it difficult to separate desired spinal analgesia from unwanted side effects, such as sedation, motor and memory impairment, liability to tolerance development and addiction.

3.4.1 Subtype-Selective GABA$_A$ Receptor Ligands

Benzodiazepines, which facilitate the activation of most GABA$_A$ receptors are generally not considered as analgesics. However, when applied to the spinal cord, they exert clear analgesic or antihyperalgesic actions both in animal models of hyperalgesia (Malan et al., 2002; Scholz et al., 2005) and in patients

(Tucker et al., 2004), for a review see (Jasmin et al., 2004). Although anecdotal reports also suggest an anti-hyperalgesic effect of systemic benzodiazepines in chronic pain syndromes with an inflammatory or neuropathic component (Jasmin et al., 2004), unwanted effects originating mainly from supraspinal sites preclude their routine use as analgesics. Recent results have now raised the hope that a separation of desired and undesired effects might become possible through the development of benzodiazepine site ligands selective for specific $GABA_A$ receptor subtypes. Classical benzodiazepines exert their effects through at least four different subtypes of $GABA_A$ receptors, which contain one or two of the following α subunits ($\alpha1$, $\alpha2$, $\alpha3$, $\alpha5$) in addition to one $\gamma2$ subunit (Wieland et al., 1992). With the use of $GABA_A$ receptor point mutated ("knock-in") mice (Rudolph and Möhler, 2004), it became possible to attribute the spinal analgesic (anti-hyperalgesic) effects of benzodiazepines to $GABA_A$ receptors containing the $\alpha2$ and / or $\alpha3$ subunit (Knabl et al., 2008) (Fig. 3.5). Both subunits are concentrated in the superficial dorsal horn, while $\alpha1$ and $\alpha5$ are almost absent there (Bohlhalter et al., 1996). Interestingly, a similar subunit specificity has previously been found for the anxiolytic properties of benzodiazepines (Löw et al., 2000). Several drug companies meanwhile run programs aiming at the development of subtype-selective benzodiazepines for the use as non-sedative anxiolytics. An experimental drug (L-838,417; McKernan et al., 2000), which is a partial agonist at the benzodiazepine binding site of $\alpha2$, $\alpha3$ and $\alpha5$, but an antagonist at the $\alpha1$ subunit, causes profound anti-hyperalgesia in rat inflammatory and neuropathic pain models (Knabl et al., 2008), but, unfortunately, possesses unfavorable pharmacokinetic properties in humans

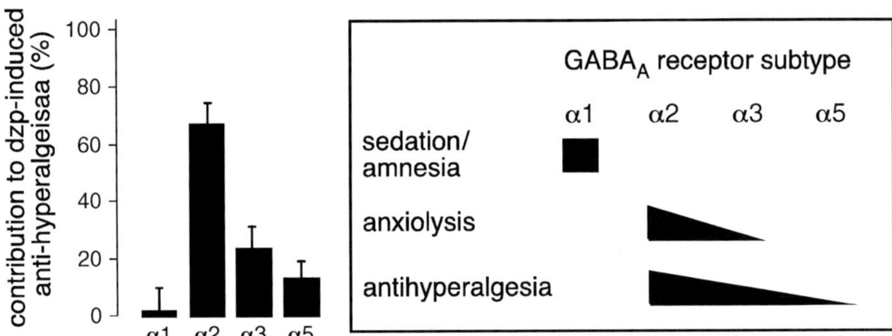

Fig. 3.5 GABA$_A$receptor subtypes and spinal anti-hyperalgesic effects of diazepam. Left, Contribution of different benzodiazepine-sensitive $GABA_A$ receptor α subunits to the anti-hyperalgesic effect (reversal to thermal hyperalgesia) of intrathecal diazepam (dzp, 0.09 mg/kg) in the mouse chronic constriction model of neuropathic pain. Data were obtained with the use of $GABA_A$ receptor point mutated mice in which specific $GABA_A$ receptor α subunits had been rendered diazepam insensitive (Rudolph and Möhler, 2004). Modified from Knabl et al. (2008) with permission. **Right,** Comparison of the different subunit contributions of the different $GABA_A$ receptor α subunits to systemic actions of diazepam. Data on sedation, amnesia and anxiety adapted from Möhler et al. (2001)

(Rogawski, 2006). Available evidence suggests that such subtype-selective benzodiazepine site ligands lack most of the unwanted effects of classical benzodiazepines (Ator, 2005).

Interestingly, both intrathecal diazepam and systemic L-838,417 normalize a pathologically increased pain sensitivity without being active against acute nociceptive pain (Knabl et al., 2008). The reason for this different effect is not exactly known. It has been suggested that the GABAergic tone in the nociceptive system is low under physiological conditions (Dirig and Yaksh, 1995). This, however, is unlikely as spinal application of bicuculline is clearly pro-allodynic (Roberts et al., 1986; Sivilotti and Woolf, 1994). Alternatively, different GABAergic pathways might exist in the spinal dorsal horn, with one of them being activated by input from mechanosensitive fibers and saturated already under normal condition, and a second one being activated only during tonic nociceptive stimulation or in sensitized disease states. Blockade of the first one would cause allodynia, while increasing the efficacy of the second one, e.g. with benzodiazepines, would be antinociceptive (see also Fig. 3.3).

3.4.2 Glycine Transporter Inhibitors

Selective facilitation of glycinergic synaptic transmission may be another strategy, which would avoid unwanted effects typical for non-selective GABAergic drugs. At present, there are no specific glycine receptors modulators available which would facilitate glycinergic inhibition in way analogous to what benzodiazepines do to $GABA_A$ receptors (Laube et al., 2002). However, inhibition of glycine transporters might be an alternative way to selectively facilitate glycinergic inhibition. In the CNS, glycine is taken up by two sodium chloride-dependent transporters: GlyT1 is mainly, but perhaps not exclusively, expressed by glia cells and GlyT2 is specifically expressed in glycinergic neurons where it mediates the recycling of glycine into the presynaptic terminals (Eulenburg et al., 2005). Two recent reports suggest analgesic effects of GlyT1 inhibitors (ORG 25935 and sarcosine) (Morita et al., 2008; Tanabe et al., 2008) and GlyT2 inhibitors (ALX 1393 and ORG 25543) (Morita et al., 2008) in different mouse models of neuropathic pain.

3.5 Concluding Remarks

In the last 10 years, it has become increasingly clear that a loss of synaptic inhibition in the dorsal horn is a major contributing factor to pathological pain of different origin including inflammatory pain and at least some forms of neuropathic pain. The pathways responsible for this loss of inhibition may vary significantly between pain syndromes, but they appear to converge onto diminished GABAergic or glycinergic inhibition. Since chronic pain in patients often involves both inflammatory and neuropathic components, therapeutic interventions, which specifically target only one of these aspects, will often yield only limited pain relief. Recent advances in our understanding of the molecular and cellular basis of spinal inhibitory pain control have provided a rational basis for the development of new therapeutic strategies against chronic pain.

References

Ahmadi S, Lippross S, Neuhuber WL, Zeilhofer HU (2002) PGE$_2$ selectively blocks inhibitory glycinergic neurotransmission onto rat superficial dorsal horn neurons. Nat Neurosci 5:34–40.

Antal M, Petko M, Polgar E, Heizmann CW, Storm-Mathisen J (1996) Direct evidence of an extensive GABAergic innervation of the spinal dorsal horn by fibres descending from the rostral ventromedial medulla. Neuroscience 73:509–518.

Ator NA (2005) Contributions of GABA$_A$ receptor subtype selectivity to abuse liability and dependence potential of pharmacological treatments for anxiety and sleep disorders. CNS Spectr 10:31–39.

Baba H, Ji RR, Kohno T, Moore KA, Ataka T, Wakai A, Okamoto M, Woolf CJ (2003) Removal of GABAergic inhibition facilitates polysynaptic A fiber-mediated excitatory transmission to the superficial spinal dorsal horn. Mol Cell Neurosci 24:818–830.

Baccei ML, Fitzgerald M (2004) Development of GABAergic and glycinergic transmission in the neonatal rat dorsal horn. J Neurosci 24:4749–4757.

Beiche F, Scheuerer S, Brune K, Geisslinger G, Goppelt-Struebe M (1996) Up-regulation of cyclooxygenase-2 mRNA in the rat spinal cord following peripheral inflammation. FEBS Lett 390:165–169.

Bennett GJ, Abdelmoumene M, Hayashi H, Dubner R (1980) Physiology and morphology of substantia gelatinosa neurons intracellularly stained with horseradish peroxidase. J Comp Neurol 194:809–827.

Beyer C, Roberts LA, Komisaruk BR (1985) Hyperalgesia induced by altered glycinergic activity at the spinal cord. Life Sci 37:875–882.

Bohlhalter S, Weinmann O, Mohler H, Fritschy JM (1996) Laminar compartmentalization of GABAA-receptor subtypes in the spinal cord: an immunohistochemical study. J Neurosci 16:283–297.

Chaudhry FA, Reimer RJ, Bellocchio EE, Danbolt NC, Osen KK, Edwards RH, Storm-Mathisen J (1998) The vesicular GABA transporter, VGAT, localizes to synaptic vesicles in sets of glycinergic as well as GABAergic neurons. J Neurosci 18:9733–9750.

Chery N, De Koninck Y (1999) Junctional versus extrajunctional glycine and GABA$_A$ receptor-mediated IPSCs in identified lamina I neurons of the adult rat spinal cord. J Neurosci 19:7342–7355.

Chery N, De Koninck Y (2000) GABA$_B$ receptors are the first target of released GABA at lamina I inhibitory synapses in the adult rat spinal cord. J Neurophysiol 84:1006–1011.

Chessell IP, Hatcher JP, Bountra C, Michel AD, Hughes JP, Green P, Egerton J, Murfin M, Richardson J, Peck WL, Grahames CB, Casula MA, Yiangou Y, Birch R, Anand P, Buell GN (2005) Disruption of the P2X7 purinoceptor gene abolishes chronic inflammatory and neuropathic pain. Pain 114:386–396.

Claveau D, Sirinyan M, Guay J, Gordon R, Chan CC, Bureau Y, Riendeau D, Mancini JA (2003) Microsomal prostaglandin E synthase-1 is a major terminal synthase that is selectively up-regulated during cyclooxygenase-2-dependent prostaglandin E2 production in the rat adjuvant-induced arthritis model. J Immunol 170:4738–4744.

Coull JA, Boudreau D, Bachand K, Prescott SA, Nault F, Sik A, De Koninck P, De Koninck Y (2003) Trans-synaptic shift in anion gradient in spinal lamina I neurons as a mechanism of neuropathic pain. Nature 424:938–942.

Coull JA, Beggs S, Boudreau D, Boivin D, Tsuda M, Inoue K, Gravel C, Salter MW, De Koninck Y (2005) BDNF from microglia causes the shift in neuronal anion gradient underlying neuropathic pain. Nature 438:1017–1021.

Dirig DM, Yaksh TL (1995) Intrathecal baclofen and muscimol, but not midazolam, are antinociceptive using the rat-formalin model. J Pharmacol Exp Ther 275:219–227.

Dumoulin A, Rostaing P, Bedet C, Levi S, Isambert MF, Henry JP, Triller A, Gasnier B (1999) Presence of the vesicular inhibitory amino acid transporter in GABAergic and glycinergic synaptic terminal boutons. J Cell Sci 112 (Pt 6):811–823.

Eulenburg V, Armsen W, Betz H, Gomeza J (2005) Glycine transporters: essential regulators of neurotransmission. Trends Biochem Sci 30:325–333.

Grudt TJ, Perl ER (2002) Correlations between neuronal morphology and electrophysiological features in the rodent superficial dorsal horn. J Physiol 540:189–207.

Hantman AW, Perl ER (2005) Molecular and genetic features of a labeled class of spinal substantia gelatinosa neurons in a transgenic mouse. J Comp Neurol 492:90–100.

Harvey RJ, Depner UB, Wässle H, Ahmadi S, Heindl C, Reinold H, Smart TG, Harvey K, Schütz B, Abo-Salem OM, Zimmer A, Poisbeau P, Welzl H, Wolfer DP, Betz H, Zeilhofer HU, Müller U (2004) GlyRα3: an essential target for spinal PGE2-mediated inflammatory pain sensitization. Science 304:884–887.

Heinke B, Ruscheweyh R, Forsthuber L, Wunderbaldinger G, Sandkühler J (2004) Physiological, neurochemical and morphological properties of a subgroup of GABAergic spinal lamina II neurones identified by expression of green fluorescent protein in mice. J Physiol 560:249–266.

Hosie AM, Wilkins ME, da Silva HM, Smart TG (2006) Endogenous neurosteroids regulate $GABA_A$ receptors through two discrete transmembrane sites. Nature 444:486–489.

Hösl K, Reinold H, Harvey RJ, Müller U, Narumiya S, Zeilhofer HU (2006) Spinal prostaglandin E receptors of the EP2 subtype and the glycine receptor α3 subunit, which mediate central inflammatory hyperalgesia, do not contribute to pain after peripheral nerve injury or formalin injection. Pain 126:46–53.

Hossaini M, French PJ, Holstege JC (2007) Distribution of glycinergic neuronal somata in the rat spinal cord. Brain Res 1142:61–69.

Jasmin L, Wu MV, Ohara PT (2004) GABA puts a stop to pain. Curr Drug Targets CNS Neurol Disord 3:487–505.

Kato G, Yasaka T, Katafuchi T, Furue H, Mizuno M, Iwamoto Y, Yoshimura M (2006) Direct GABAergic and glycinergic inhibition of the substantia gelatinosa from the rostral ventromedial medulla revealed by in vivo patch-clamp analysis in rats. J Neurosci 26:1787–1794.

Keller AF, Breton JD, Schlichter R, Poisbeau P (2004) Production of 5alpha-reduced neurosteroids is developmentally regulated and shapes $GABA_A$ miniature IPSCs in lamina II of the spinal cord. J Neurosci 24:907–915.

Keller AF, Beggs S, Salter MW, De Koninck Y (2007) Transformation of the output of spinal lamina I neurons after nerve injury and microglia stimulation underlying neuropathic pain. Mol Pain 3:27.

Keller AF, Coull JA, Chery N, Poisbeau P, De Koninck Y (2001) Region-specific developmental specialization of GABA-glycine cosynapses in laminas I-II of the rat spinal dorsal horn. J Neurosci 21:7871–7880.

Knabl J, Witschi R, Hösl K, Reinold H, Zeilhofer UB, Ahmadi S, Brockhaus J, Sergejeva M, Hess A, Brune K, Fritschy J-M, Rudolph U, Möhler H, Zeilhofer HU (2008) Reversal of pathological pain through specific spinal $GABA_A$ receptor subtypes. Nature 451:330–334.

Laube B, Maksay G, Schemm R, Betz H (2002) Modulation of glycine receptor function: a novel approach for therapeutic intervention at inhibitory synapses? Trends Pharmacol Sci 23:519–527.

Light AR, Trevino DL, Perl ER (1979) Morphological features of functionally defined neurons in the marginal zone and substantia gelatinosa of the spinal dorsal horn. J Comp Neurol 186:151–171.

Löw K, Crestani F, Keist R, Benke D, Brünig I, Benson JA, Fritschy JM, Rülicke T, Bluethmann H, Möhler H, Rudolph U (2000) Molecular and neuronal substrate for the selective attenuation of anxiety. Science 290:131–134.

Lu Y, Perl ER (2003) A specific inhibitory pathway between substantia gelatinosa neurons receiving direct C-fiber input. J Neurosci 23:8752–8758.

Mackie M, Hughes DI, Maxwell DJ, Tillakaratne NJ, Todd AJ (2003) Distribution and colocalisation of glutamate decarboxylase isoforms in the rat spinal cord. Neuroscience 119:461–472.

Malan TP, Mata HP, Porreca F (2002) Spinal GABA$_A$ and GABA$_B$ receptor pharmacology in a rat model of neuropathic pain. Anesthesiology 96:1161–1167.

Malmberg AB, Chen C, Tonegawa S, Basbaum AI (1997a) Preserved acute pain and reduced neuropathic pain in mice lacking PKCgamma. Science 278:279–283.

Malmberg AB, Brandon EP, Idzerda RL, Liu H, McKnight GS, Basbaum AI (1997b) Diminished inflammation and nociceptive pain with preservation of neuropathic pain in mice with a targeted mutation of the type I regulatory subunit of cAMP-dependent protein kinase. J Neurosci 17:7462–7470.

Martin WJ, Liu H, Wang H, Malmberg AB, Basbaum AI (1999) Inflammation-induced up-regulation of protein kinase Cgamma immunoreactivity in rat spinal cord correlates with enhanced nociceptive processing. Neuroscience 88:1267–1274.

Maxwell DJ, Belle MD, Cheunsuang O, Stewart A, Morris R (2007) Morphology of inhibitory and excitatory interneurons in superficial laminae of the rat dorsal horn. J Physiol 584:521–533.

McKernan RM, Rosahl TW, Reynolds DS, Sur C, Wafford KA, Atack JR, Farrar S, Myers J, Cook G, Ferris P, Garrett L, Bristow L, Marshall G, Macaulay A, Brown N, Howell O, Moore KW, Carling RW, Street LJ, Castro JL, Ragan CI, Dawson GR, Whiting PJ (2000) Sedative but not anxiolytic properties of benzodiazepines are mediated by the GABA$_A$ receptor α1 subtype. Nat Neurosci 3:587–592.

Melzack R, Wall PD (1965) Pain mechanisms: a new theory. Science 150:971–979.

Miraucourt LS, Dallel R, Voisin DL (2007) Glycine inhibitory dysfunction turns touch into pain through PKCγ interneurons. PLoS ONE 2:e1116.

Mitchell EA, Gentet LJ, Dempster J, Belelli D (2007) GABA$_A$ and glycine receptor-mediated transmission in rat lamina II neurones: relevance to the analgesic actions of neuroactive steroids. J Physiol 583:1021–1040.

Möhler H, Crestani F, Rudolph U (2001) GABA$_A$-receptor subtypes: a new pharmacology. Curr Opin Pharmacol 1:22–25.

Moore KA, Kohno T, Karchewski LA, Scholz J, Baba H, Woolf CJ (2002) Partial peripheral nerve injury promotes a selective loss of GABAergic inhibition in the superficial dorsal horn of the spinal cord. J Neurosci 22:6724–6731.

Morita K, Motoyama N, Kitayama T, Morioka N, Kifune K, Dohi T (2008) Spinal anti-allodynia action of glycine transporter inhibitors in neuropathic pain models in mice. J Pharmacol Exp Ther 326:633–645.

Murakami M, Naraba H, Tanioka T, Semmyo N, Nakatani Y, Kojima F, Ikeda T, Fueki M, Ueno A, Oh S, Kudo I (2000) Regulation of prostaglandin E2 biosynthesis by inducible membrane-associated prostaglandin E2 synthase that acts in concert with cyclooxygenase-2. J Biol Chem 275:32783–32792.

Narikawa K, Furue H, Kumamoto E, Yoshimura M (2000) In vivo patch-clamp analysis of IPSCs evoked in rat substantia gelatinosa neurons by cutaneous mechanical stimulation. J Neurophysiol 84:2171–2174.

Oliva AA, Jr., Jiang M, Lam T, Smith KL, Swann JW (2000) Novel hippocampal interneuronal subtypes identified using transgenic mice that express green fluorescent protein in GABAergic interneurons. J Neurosci 20:3354–3368.

Poisbeau P, Patte-Mensah C, Keller AF, Barrot M, Breton JD, Luis-Delgado OE, Freund-Mercier MJ, Mensah-Nyagan AG, Schlichter R (2005) Inflammatory pain upregulates spinal inhibition via endogenous neurosteroid production. J Neurosci 25:11768–11776.

Polgar E, Hughes DI, Arham AZ, Todd AJ (2005) Loss of neurons from laminas I–III of the spinal dorsal horn is not required for development of tactile allodynia in the spared nerve injury model of neuropathic pain. J Neurosci 25:6658–6666.

Prescott SA, Sejnowski TJ, de Koninck Y (2006) Reduction of anion reversal potential subverts the inhibitory control of firing rate in spinal lamina I neurons: towards a biophysical basis for neuropathic pain. Mol Pain 2:32.

Price TJ, Hargreaves KM, Cervero F (2006) Protein expression and mRNA cellular distribution of the NKCC1 cotransporter in the dorsal root and trigeminal ganglia of the rat. Brain Res 1112:146–158.

Reinold H, Ahmadi S, Depner UB, Layh B, Heindl C, Hamza M, Pahl A, Brune K, Narumiya S, Muller U, Zeilhofer HU (2005) Spinal inflammatory hyperalgesia is mediated by prostaglandin E receptors of the EP2 subtype. J Clin Invest 115:673–679.

Rivera C, Voipio J, Payne JA, Ruusuvuori E, Lahtinen H, Lamsa K, Pirvola U, Saarma M, Kaila K (1999) The K + /Cl- co-transporter KCC2 renders GABA hyperpolarizing during neuronal maturation. Nature 397:251–255.

Roberts LA, Beyer C, Komisaruk BR (1986) Nociceptive responses to altered GABAergic activity at the spinal cord. Life Sci 39:1667–1674.

Rogawski MA (2006) Diverse mechanisms of antiepileptic drugs in the development pipeline. Epilepsy Res 69:273–294.

Rudolph U, Möhler H (2004) Analysis of $GABA_A$ receptor function and dissection of the pharmacology of benzodiazepines and general anesthetics through mouse genetics. Annu Rev Pharmacol Toxicol 44:475–498.

Rudomin P, Schmidt RF (1999) Presynaptic inhibition in the vertebrate spinal cord revisited. Exp Brain Res 129:1–37.

Samad TA, Moore KA, Sapirstein A, Billet S, Allchorne A, Poole S, Bonventre JV, Woolf CJ (2001) Interleukin-1beta-mediated induction of Cox-2 in the CNS contributes to inflammatory pain hypersensitivity. Nature 410:471–475.

Schattschneider J, zum Buttel I, Binder A, Wasner G, Hedderich J, Baron R (2007) Mechanisms of adrenosensitivity in capsaicin induced hyperalgesia. Eur J Pain 11:756–763.

Scholz J, Woolf CJ (2007) The neuropathic pain triad: neurons, immune cells and glia. Nat Neurosci 10:1361–1368.

Scholz J, Broom DC, Youn DH, Mills CD, Kohno T, Suter MR, Moore KA, Decosterd I, Coggeshall RE, Woolf CJ (2005) Blocking caspase activity prevents transsynaptic neuronal apoptosis and the loss of inhibition in lamina II of the dorsal horn after peripheral nerve injury. J Neurosci 25:7317–7323.

Sivilotti L, Woolf CJ (1994) The contribution of $GABA_A$ and glycine receptors to central sensitization: disinhibition and touch-evoked allodynia in the spinal cord. J Neurophysiol 72:169–179.

Sung KW, Kirby M, McDonald MP, Lovinger DM, Delpire E (2000) Abnormal $GABA_A$ receptor-mediated currents in dorsal root ganglion neurons isolated from Na-K-2Cl cotransporter null mice. J Neurosci 20:7531–7538.

Tamamaki N, Yanagawa Y, Tomioka R, Miyazaki J, Obata K, Kaneko T (2003) Green fluorescent protein expression and colocalization with calretinin, parvalbumin, and somatostatin in the GAD67-GFP knock-in mouse. J Comp Neurol 467:60–79.

Tanabe M, Takasu K, Yamaguchi S, Kodama D, Ono H (2008) Glycine transporter inhibitors as a potential therapeutic strategy for chronic pain with memory impairment. Anesthesiology 108:929–937.

Thacker, MA Clark AK, Bishop T, Grist J, Yip PK, Moon LD, Thompson SW, Marchand F, McMahon SB (2009) CCL2 is a key mediator of microglia activation in neuropathic pain states. Eur J Pain 13:263–272.

Todd AJ, Sullivan AC (1990) Light microscope study of the coexistence of GABA-like and glycine-like immunoreactivities in the spinal cord of the rat. J Comp Neurol 296:496–505.

Torsney C, MacDermott AB (2006) Disinhibition opens the gate to pathological pain signaling in superficial neurokinin 1 receptor-expressing neurons in rat spinal cord. J Neurosci 26:1833–1843.

Treede RD, Magerl W (2000) Multiple mechanisms of secondary hyperalgesia. Prog Brain Res 129:331–341.

Tsuda M, Shigemoto-Mogami Y, Koizumi S, Mizokoshi A, Kohsaka S, Salter MW, Inoue K (2003) P2X4 receptors induced in spinal microglia gate tactile allodynia after nerve injury. Nature 424:778–783.

Tucker AP, Mezzatesta J, Nadeson R, Goodchild CS (2004) Intrathecal midazolam II: combination with intrathecal fentanyl for labor pain. Anesth Analg 98:1521–1527.

Vergnano AM, Schlichter R, Poisbeau P (2007) PKC activation sets an upper limit to the functional plasticity of GABAergic transmission induced by endogenous neurosteroids. Eur J Neurosci 26:1173–1182.

Wieland HA, Luddens H, Seeburg PH (1992) A single histidine in $GABA_A$ receptors is essential for benzodiazepine agonist binding. J Biol Chem 267:1426–1429.

Willis WD, Jr. (1999) Dorsal root potentials and dorsal root reflexes: a double-edged sword. Exp Brain Res 124:395–421.

Wojcik SM, Katsurabayashi S, Guillemin I, Friauf E, Rosenmund C, Brose N, Rhee JS (2006) A shared vesicular carrier allows synaptic corelease of GABA and glycine. Neuron 50:575–587.

Woolf CJ (1983) Evidence for a central component of post-injury pain hypersensitivity. Nature 306:686–688.

Yaksh TL (1989) Behavioral and autonomic correlates of the tactile evoked allodynia produced by spinal glycine inhibition: effects of modulatory receptor systems and excitatory amino acid antagonists. Pain 37:111–123.

Yasaka T, Kato G, Furue H, Rashid MH, Sonohata M, Tamae A, Murata Y, Masuko S, Yoshimura M (2007) Cell-type-specific excitatory and inhibitory circuits involving primary afferents in the substantia gelatinosa of the rat spinal dorsal horn in vitro. J Physiol 581:603–618.

Yoshimura M, Nishi S (1995) Primary afferent-evoked glycine- and GABA-mediated IPSPs in substantia gelatinosa neurones in the rat spinal cord in vitro. J Physiol 482 (Pt 1):29–38.

Zeilhofer HU (2005) The glycinergic control of spinal pain processing. Cell Mol Life Sci 62:2027–2035.

Zeilhofer HU, Studler B, Arabadzisz D, Schweizer C, Ahmadi S, Layh B, Bösl MR, Fritschy JM (2005) Glycinergic neurons expressing enhanced green fluorescent protein in bacterial artificial chromosome transgenic mice. J Comp Neurol 482:123–141.

Zieglgänsberger W, Herz A (1971) Changes of cutaneous receptive fields of spino-cervical-tract neurones and other dorsal horn neurones by microelectrophoretically administered amino acids. Exp Brain Res 13:111–126.

Zhang J, Shi XQ, Echeverry S, Mogil JS, De Koninck Y, Rivest S (2007) Expression of CCR2 in both resident and bone marrow-derived microglia plays a critical role in neuropathic pain. J Neurosci 27:12396–12406.

Chapter 4
Synaptic Transmission of Pain in the Developing Spinal Cord

Fast Synaptic Transmission in the Dorsal Horn

Rachel A. Ingram, Mark L. Baccei, and Maria Fitzgerald

Abstract The altered nociceptive behaviour of neonatal animals implies that there are underlying differences in pain transmission between young and mature individuals. One important location where these differences have been shown to occur is at the level of the spinal cord dorsal horn where sensory information from the periphery is first integrated at the synaptic level. The maturation of synaptic transmission in this region will therefore have a profound effect on pain behaviour. In addition, the development of spinal pain processing is modulated by incoming sensory activity and immature dorsal horn synapses play a key role in postnatal activity-dependent shaping of pain circuitry. This chapter will explore the maturation of dorsal horn synaptic transmission at anatomical, molecular and functional levels. The competing forces of excitatory and inhibitory neurotransmission will be focussed upon separately and then finally the integration of these synaptic drives will be discussed in relation to the emergence of mature spinal pain processing.

4.1 Introduction – Development of Pain Transmission in Neonates

Neonatal animals are in many respects more sensitive to noxious stimulation than adults. Lower intensities of stimulation are required to induce a cutaneous reflex withdrawal and these reflexes are often exaggerated and widespread, involving the movement of several muscle groups. This is in contrast to the more specific, individuated response seen in the adult which is usually restricted to a single limb or paw (Ekholm 1967; Fitzgerald et al. 1988). Dorsal horn neurones in the spinal cords of neonates are more excitable than their mature counterparts with lower mechanical and thermal thresholds and a greater propensity for afterdischarge upon repetitive stimulation (Falcon et al. 1996; Jennings and Fitzgerald 1998). Immature neurones also have larger excitatory

R.A. Ingram (✉)
Department of Anatomy and Developmental Biology, University College London, London, UK
e-mail: rachel.ingram@ucl.ac.uk

M. Malcangio (ed.), *Synaptic Plasticity in Pain*,
DOI 10.1007/978-1-4419-0226-9_4, © Springer Science+Business Media, LLC 2009

cutaneous receptive fields and diffuse inhibitory fields (Bremner and Fitzgerald 2008; Torsney and Fitzgerald 2002). These correlations between behavioural changes and spinal cord response properties indicate that the dorsal horn is a key site for the postnatal development of pain signalling.

The anatomical organisation of the dorsal horn undergoes many developmental changes over the first weeks of life. The terminals of low threshold, tactile A-fibre primary afferents are present in the dorsal horn from well before birth (E13), with some A-fibre projections extending up into laminae I and II, a region associated with only nociceptive transmission in the adult (Pignatelli et al. 1989; Woodbury and Koerber 2003). These terminals form functional connections with immature superficial dorsal horn neurones as monosynaptic responses to low threshold stimulation can be recorded in young animals as well as the expression of c-fos (Jennings and Fitzgerald 1996; Park et al. 1999).

C-fibre terminals enter the dorsal horn a few days before birth at E18–19 but only peptidergic afferents are seen before P5, at which time IB4 positive cells begin to appear (Jackman and Fitzgerald 2000). The somatotopic pattern of C-fibre terminations in laminae I and II is precise from the outset and the expression of most neuropeptides can also be detected at this stage (Fitzgerald and Swett 1983; Marti et al. 1987). However, stimulation of C-fibres fails to evoke action potential firing in dorsal horn neurones before P10 (Jennings and Fitzgerald 1998), suggesting a delayed maturation of C-fiber synapses occurs during the early postnatal period. This may explain the observation that mustard oil, an irritant which specifically activates C-fibres, evokes a weak reflex and low c-fos expression in the first week of life despite still being able to excite peripheral nociceptors (Fitzgerald and Gibson 1984; Soyguder et al. 1994). In addition, capsaicin increases the frequency of glutamatergic mEPSCs in neurones from spinal cord slices taken at P0, but not to the same extent as in more mature cells (Baccei et al. 2003). Therefore it would appear that at least some functional synapses between nociceptive primary afferents and dorsal horn neurones are present before birth, but the efficacy of these synapses may be insufficient to faithfully transmit evoked responses.

The postnatal reorganisation of primary afferent terminals can be modulated by treatments such as systemic neonatal capsaicin (to destroy C-fibres) and chronic intrathecal application of NMDA antagonists (Beggs et al. 2002; Torsney et al. 2000). This indicates that the process is activity-dependent, a theme common to other aspects of sensory development such as the postnatal tuning of the tail flick response (Waldenstrom et al. 2003). Neonates often flick their tail in the wrong direction when thermal stimulation is applied but the frequency of such errors decreases as the animals mature. This process was shown to depend upon the presence of low threshold tactile input in early life and may be the result of the presence of A-fibres in superficial laminae of the immature dorsal horn.

The specification of dorsal horn neurones themselves mainly occurs prenatally. Neurones are generated in a ventro-dorsal pattern with motor neurones being born first and the substansia gelatinosa maturing last, just prior to birth

(Kitao et al. 1996). The late generation of lamina II neurones means that their axodendritic growth is still occurring postnatally whereas lamina I projection neurones complete their supraspinal projections earlier (Bice and Beal 1997a, b).Despite this immature connectivity the intrinsic membrane excitability of superficial dorsal horn neurones does not change significantly over postnatal development (Baccei and Fitzgerald 2005). The observed changes in threshold and firing properties of these neurones are likely to be due to alterations in excitatory or inhibitory synaptic activity or in the integration of these two inputs.

4.2 Excitatory Synaptic Transmission in the Developing Spinal Cord

The majority of fast excitatory transmission that occurs in the dorsal horn is mediated by the amino acid neurotransmitter glutamate acting on postsynaptic ionotropic receptors. These receptors fall into three categories according to which exogenous agonists they are activated by.

4.2.1 AMPA (α-amino-3-hydroxy-5-methyl-4-isoxazolepropionic acid) Receptors

AMPARs mediate the majority of fast excitatory transmission in the spinal cord. They are widely and highly expressed in the dorsal horn at birth before their numbers are reduced to around a fifth of neonatal levels in the adult (Brown et al. 1992; Jakowec et al. 1995b, a). Although increased expression of AMPARs in the immature spinal cord may suggest an increase in excitatory drive, no change is seen in the amplitude of glutamatergic mEPSCs recorded in lamina II neurones over the first 10 postnatal days, despite a five fold increase in the frequency of these currents over the same period. However, AMPARs are heteromers of four subunits (GluR1–4), the precise combination of which will have a powerful effect on the desensitisation, ion permeability and current-voltage relationship exhibited by the receptor. At birth GluR1, 2 and 4 are the most highly expressed subunits. While GluR 1 and 4 halve in numbers over postnatal development, GluR2 is only reduced to 70% of neonatal levels in the adult spinal cord (Jakowec et al. 1995b, a). This results in a relatively higher percentage of calcium impermeable GluR2-containing AMPARs in the mature dorsal horn (Burnashev et al. 1992). AMPAR-mediated calcium entry may be utilised for synaptic strengthening and growth in neonatal neurones (Gu and Spitzer 1993). The presence of calcium-permeable AMPARs has been shown on both GABAergic inhibitory neurones and NK1-expressing excitatory neurones in cultures of embryonic dorsal horn neurones (Albuquerque et al. 1999) and in vitro studies in slices of spinal cord from postnatal rats showed that this type of AMPAR is expressed in the superficial laminae of the dorsal horn (Engelman et al. 1999).

4.2.2 KA (Kainate) Receptors

KARs are also responsible for excitatory synaptic transmission in the spinal cord of neonates where they have been shown to mediate high threshold, C-fibre transmission onto dorsal horn neurones (Li et al. 1999). The subunits which make up KARs (GluR5-7 and KAR1&2) are highly expressed in the dorsal horn in the newborn pup but virtually absent after 21 days, indicating that these receptors are important for the development of nociceptive transmission (Stegenga and Kalb 2001). Kainate receptors can also be found presynaptically in the neonatal spinal cord and are able to negatively modulate the release of glutamate (possibly through primary afferent depolarisation) in the immature spinal cord (Agrawal and Evans 1986; Kerchner et al. 2001). In the first week of postnatal development these presynaptic KARs can also be calcium-permeable, especially those found on the terminals of non-peptidergic C-fibres (Lee et al. 2001). Calcium entry into C fibre terminals at this stage of development may be involved in directing their growth into the dorsal horn to form connections with second order neurones.

4.2.3 NMDA (N-methyl-D-aspartate) Receptors

NMDARs are not involved in rapid glutamatergic transmission under normal conditions due to the presence of magnesium ions blocking the channel pore at resting membrane potential. Repetitive stimulation allows sufficient charge through AMPAR and KARs to depolarise the post-synaptic cell and relieve the magnesium block, permitting the flow of cations through the NMDAR channel (Nowak et al. 1984). Because NMDARs have far greater calcium permeability than most AMPAR and KARs, their opening is more likely to lead to the activation of calcium-dependent intracellular cascades that affect synaptic strength (Brown et al. 1988). Overall levels of NMDAR expression in the dorsal horn are greater at birth than in the adult and appear in a uniform manner across all laminae (Gonzalez et al. 1993). The calcium response to NMDAR activation gradually declines during the second postnatal week and also becomes more focussed on the substantia gelatinosa, as is the case in the mature spinal cord. This developmental maturation of NMDAR expression can be delayed by neonatal capsaicin treatment to destroy C fibres, indicating that activity in these primary afferents neurones is important for the tuning of this system (Hori and Kanda 1994).

NMDARs, like other ionotropic glutamate receptors, are heteromers of their constituent subunits (NR1 to 3), four of which make up each channel. The precise combination of subunits will affect channel properties such as receptor kinetics and magnesium sensitivity (Paoletti and Neyton 2007). One particular subunit, NR2B, is widely expressed in the rat dorsal horn at birth and is known to cause slower EPSC decay kinetics than the NR2A subunit, which gradually becomes more prevalent over the first three weeks of life (Watanabe et al. 1994). This developmental shift in subunit stoichiometry has previously been demonstrated in the visual cortex and shown to be an activity-dependent process (Philpot et al. 2001). However, a straightforward switch in subunit expression is less likely to be the case in the superficial dorsal horn as these neurones have been shown to possess NMDAR EPSCs with rapid decay kinetics even in the first two weeks after birth (Bardoni et al. 1998). This, along with the fact that neonatal NMDARs show a very high sensitivity to block by magnesium ions, implies that a novel subunit composition may be present at these ages (Green and Gibb 2001). Another aspect of NMDAR function under tight developmental control is the existence of "silent synapses" containing NMDA but not AMPA and KARs, which can be converted to a fully-functioning state via the rapid insertion of AMPARs into the synaptic membrane. First described in the hippocampus, these synapses have now been found in the dorsal horn of neonatal rats but are down-regulated by the end of the second postnatal week (Baba et al. 2000; Bardoni et al. 1998; Isaac et al. 1995; Li and Zhuo 1998). Recent evidence demonstrates that activation of spinal NMDARs during the early postnatal period is critical for the tuning of dorsal horn neurone peripheral receptive fields and the retraction of A fibre connections from the superficial laminae (Beggs et al. 2002). When animals were chronically exposed to a low dose intrathecal dose of an NMDAR antagonist from birth, receptive fields remained in a large, unrefined postnatal state and A fibre evoked responses could still be measured in superficial dorsal horn neurones at an age when C fibre responses should dominate.

In summary, excitatory neurotransmission undergoes considerable postnatal maturation with many changes seen in glutamatergic systems with respect to receptor expression patterns, subunit composition and channel properties. Overall these changes appear to allow increased calcium entry into the postsynaptic cell even without the intense stimulation required for this to occur in the adult. In combination with the presence of A-fibre terminals in the superficial dorsal horn, this may act to increase synaptic strengthening and neuronal outgrowth in the maturing pain pathway while the neonate is only exposed to innocuous tactile stimulation.

4.3 Inhibitory Synaptic Transmission in the Developing Spinal Cord

While the increased sensitivity of immature dorsal horn neurones may in part be due to postnatal changes in the organisation of excitatory transmission, it is now becoming clear that inhibitory circuits play an equally vital role in shaping the response to nociceptive stimuli. The two main neurotransmitters that mediate fast inhibitory synaptic transmission in the spinal cord are γ–aminobutyric acid (GABA), acting mainly through $GABA_A Rs$, and glycine. Both the synthesis of the transmitters themselves and the expression of their receptors undergo developmental changes over the early postnatal period which will affect the way they gate excitatory input.

4.3.1 Synthesis and Localisation of Inhibitory Neurotransmitters

While glycine is a simple amino acid, GABA must be manufactured via the decarboxylation of glutamate. The enzyme responsible for this process, GAD (glutamic acid decarboxylase) has two subtypes, GAD65 and GAD67, which show around 70% sequence homology. Despite this similarity the two isoforms have different cellular locations in the mature spinal cord; GAD67 is soluble and remains cytoplasmic whereas GAD65 is membrane-bound, usually seen in vesicular membranes and plays a greater role in synthesising GABA for rapid synaptic transmission (Erlander et al. 1991). The mRNA for both subtypes is highly expressed throughout the embryonic spinal cord and then undergoes a three-fold decrease over the first two postnatal weeks whilst also becoming fairly restricted to the dorsal horn in the case of GAD67 (Somogyi et al. 1995). The intracellular distribution of the GAD isoforms was also found to be developmentally regulated in studies of cultured spinal cord slices (Tran et al. 2003). Both subtypes were seen in the somata during neuronal proliferation at embryonic stages before GAD65 became restricted to terminal-like varicosities in older slices. GABA itself cannot be detected in the substantia gelatinosa of the rat spinal cord until the late stages of embryonic life (E17–18) (Ma et al. 1992). After this time there is a large increase in the proportion of spinal cord neurones expressing GABA, reaching a peak of 50% of all neurones in the first two postnatal weeks before dropping to 20% by P21 (Schaffner et al. 1993).

The expression pattern of glycine is mature well before birth and coincides precisely with immunoreactivity for the glycine transporter GlyT2 (Berki et al. 1995; Poyatos et al. 1997). In the adult rat GlyT2 is found at greater densities in deep dorsal horn, and c-fos studies following the application of the GlyR antagonist strychnine show that glycinergic neurons exert the majority of their tonic inhibitory effects in this region of the spinal cord (Cronin et al. 2004; Hossaini et al. 2007). Around half of GABAergic neurones co-express glycine and the proportion of double-labelled GABA/glycine neurones appears to stay constant throughout early development (Berki et al. 1995; Todd and Sullivan 1990).

4.3.2 Developmental Regulation of GABA$_A$R and GlyR Stoichiometry

GABA$_A$Rs are heteromeric complexes of five subunits arranged around a central pore. Subunits classified in the human are α (six subtypes), β (three subtypes), γ (three subtypes) and one of each of δ, ε, π, & θ (Barnard et al. 1998). The most common assembly of these subunits is $\alpha_2\beta_2\gamma$ but the wide variety of subtypes and additional subunits allows a vast range of functional properties to be expressed at different synapses around the CNS and at different stages of development. The subunits with the highest expression in the neonatal dorsal horn are $\alpha2$, $\alpha3$, $\beta3$ and $\gamma2$ and their numbers peak in the first week after birth before decreasing slightly to adult levels (Ma et al. 1993). Two of these subunits, $\alpha2$ and $\alpha3$, are of particular interest in pain signalling pathways as their specific activation has recently been shown to significantly reduce both inflammatory and neuropathic pain in the absence of $\alpha1$-mediated sedation (Knabl et al. 2008).

Glycine receptor complexes are also pentamers but have a smaller repertoire of subunits, consisting of four α subtypes and a single accessory β subunit, which modulates channel function, assembly and tethers the channel in the cell substructure by binding to the cytoplasmic protein gephyrin (Legendre 2001). The expression of these different subunits and the composition of the resulting GlyRs is regulated over pre- and postnatal development (Aguayo et al. 2004; Malosio et al. 1991; Watanabe and Akagi 1995). A very low density of $\alpha1$ subunits is seen in the embryo but it reaches high levels of expression by P15. $\alpha2$ shows the opposite profile and is abundant embryonically before undergoing a large decrease on or after birth leaving few of these subunits in the adult spinal cord. The $\alpha3$ subunit cannot be detected in the cord until after P21, although in the adult it is thought to play an important role in gating nociceptive signalling due to its location in the superficial dorsal horn and modulation by inflamma-tory prostaglandins (Harvey et al. 2004). However, more recent reports have questioned the importance of this link as the $\alpha3$ knock-out mouse still develops full hyperalgesia in response to formalin injection or peripheral nerve injury (Hosl et al. 2006; Racz et al. 2005). The $\alpha4$ subunit is thus far only predicted from genomic studies but is thought have similar properties to $\alpha1$ (Matzenbach et al. 1994). The expression of the β subunit mirrors that of $\alpha1$, increasing postnatally and peaking after the second postnatal week (Aguayo et al. 2004). In the mature spinal cord the dominant GlyR stoichiometry is $\alpha1\beta$, consisting of three $\alpha1$ and two β subunits. In the embryo and early neonates $\alpha2$ homomers or heteromers which also include β subunits dominate instead (Aguayo et al. 2004). Because the opening time of the $\alpha2$ subunit is almost two orders of magnitude greater than that of the $\alpha1$, glycinergic IPSCs in the youngest animals have significantly longer decay time constants than their mature coun-terparts (Takahashi et al. 1992).

4.3.3 $GABA_A$ R and GlyR Synaptic Function in the Developing Dorsal Horn

In the days immediately after birth, superficial dorsal horn neurones show low frequency GABAergic miniature IPSCs but a complete absence of spontaneous glycinergic synaptic activity. However, the application of exogenous glycine evokes a strychnine-sensitive current in all SG neurones at P0, implying that functional GlyR are indeed present in the membrane at this early age (Baccei and Fitzgerald 2004). A possible explanation for this disparity is that immature GlyRs may not be tethered at synaptic sites due to either an absence of the gephyrin-binding β subunit or a lack of the structural binding protein itself (Kirsch et al. 1993). Studies of the developmental expression of gephyrin support the latter theory as levels of the protein are very low in the dorsal horn in the first week of life (Fig. 4.1). At P10 diffuse staining can be seen in the deeper laminae and by P21 this has formed a concentrated band in lamina III with some punctuate expression visible in lamina I. The mature pattern of expression in the adult consists of a dense band of staining extending from lamina III outer to lamina II and into lamina I (Fig. 4.1).

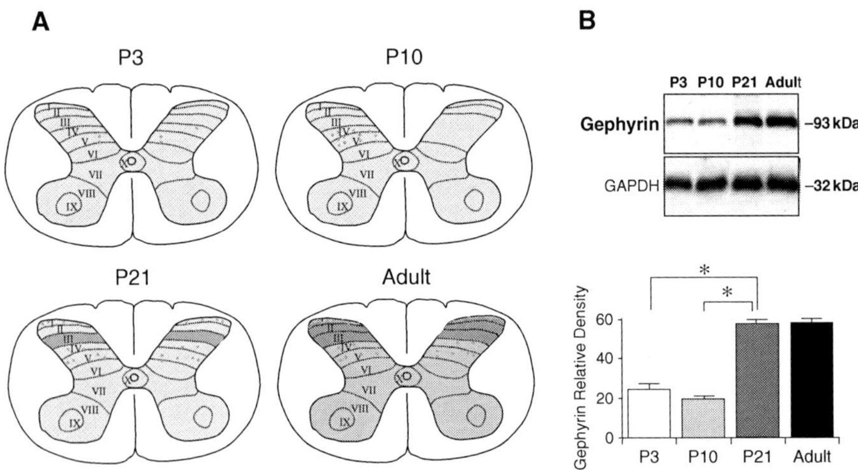

Fig. 4.1 Gephyrin expression in the postnatal spinal cord. **A** – Diagrams showing the expression of gephyrin (as determined by Immunohistochemistry) at 3, 10 and 21 days after birth and in the adult spinal cord. **B** – Quantitative analysis of gephyrin levels in the dorsal horn of the spinal cord over postnatal development as determined by western blot. A significant upregulation of gephyrin expression is seen between P10 and P21 ($*P < 0.05$) (Harrop E & Fitzgerald M, unpublished data)

As a large proportion of superficial dorsal horn neurones express both GABA and glycine, they are often packaged into the same synaptic vesicles by their common transporter, VIATT (Gasnier 2000). This can lead to co-release of the two inhibitory neurotransmitters which in young animals results in mixed mIPSCS which have both GABAergic and glycinergic components (Jonas et al. 1998; Keller et al. 2001). The appearance of these mixed currents is developmentally down-regulated so that by P30 all mIPSCs are mediated by either GABA or glycine and never both (Keller et al. 2001). Interestingly, this developmental change occurs in a region specific manner with only glycinergic mIPSCs seen in lamina I neurones in the adult whereas mature lamina II neurones showed either GlyR (48% of all events) or GABA$_A$R (52%) mediated mIPSCs (Chery and De Koninck 1999; Keller et al. 2001). In laminae III–IV both GABA$_A$R and GlyR mIPSCs can be recorded at 10–15 days after birth but none of these events are mediated by a mixture of the two transmitters (Inquimbert et al. 2007). Positively modulating the sensitivity of GABA$_A$Rs pharmacologically with benzodiazepines reveals mixed mIPSCs in adult lamina I neurones (Chery and De Koninck 1999), suggesting that a relocalisation of GABA$_A$Rs to the extrasynaptic space may occur during postnatal development.

The kinetics of GABA$_A$R and GlyR currents will also have a strong influence on the efficacy of inhibitory synaptic transmission in the dorsal horn. As discussed above, the postnatal GlyR subunit switch from α2 to α1 results in a major decrease in the decay time constant of these currents (Takahashi et al. 1992). A shorter duration of channel opening means that less negative charge will be able to enter the post-synaptic cell and the hyperpolarising effect will be reduced. GABA$_A$R mediated currents in lamina II neurones also show a significant acceleration in their decay kinetics between P8 and P23 (Keller et al. 2001). This was later shown to be caused by the tonic production of 5α-neurosteroids in the immature dorsal horn (Keller et al. 2004). Blockade of neurosteroidgenesis, either via direct inhibition of the 5α-reductase enzyme or by specific antagonism of the Peripheral benzodiazepine receptor (PBR), results in a shorter decay rate of GABA$_A$R mIPSCs in the immature spinal cord but does not affect GABA$_A$R signalling in adult neurones or GlyR currents. Increasing 5α-neurosteroid production by stimulation of PBR slows GABA$_A$R kinetics at all ages and also increases the relative number of mixed GABA$_A$/Glycine mIPSCs in lamina II (Keller et al. 2004). Neurosteroid production can also be increased by incubating slices in progesterone before recording and this procedure leads to longer GABA$_A$R opening times in lamina III–IV neurones from the neonatal dorsal horn but does not cause the appearance of mixed mIPSCs (Inquimbert et al. 2007). Neurosteroids therefore appear to shape GABA$_A$R kinetics over postnatal development through a developmental decrease in their local production within the dorsal horn. They may also play a role in the modulation of pathological pain states in the adult (Schlichter et al. 2006).

4.3.4 Is GABA an Excitatory Neurotransmitter in the Immature Dorsal Horn?

In the adult CNS the opening of GABA$_A$R and GlyRs allows chloride ions (Cl$^-$) to flow into the neurones, down the electrochemical gradient, causing hyperpolarisation of the cell membrane. This normal train of events can be disrupted in immature neurones where the intracellular concentration of Cl⁻ ions ([Cl⁻]$_i$) is high. This can lead to the reversal potential (E$_{Cl}$) becoming more positive than the resting membrane potential resulting in chloride efflux, depolarisation and sometimes even action potential firing upon GABA$_A$R (or GlyR) activation (Akerman and Cline 2006; Ben Ari et al. 1989; Rivera et al. 1999). GABA$_A$R mediated excitation is vital for activity-dependent tuning in areas of the CNS where glutamatergic signalling does not appear until later in development (Ben Ari 2002). However, in the dorsal horn elevated [Cl⁻]$_i$ could impair the ability of GABA and glycine to gate noxious input from the periphery and thus result in greater pain sensation.

Increased [Cl⁻]$_i$ in developing neurones is the result of the lower expression of the K⁺-Cl⁻ cotransporter KCC2 which normally pumps chloride ions out of the cell in exchange for potassium ions entering (Ehrlich et al. 1999; Rivera et al. 1999). The expression of KCC2 protein in the dorsal horn, while virtually nil at P3, increases significantly over postnatal development and reaches adult levels by P21 (Fig. 4.2). Immunohistochemical studies over the same period show that in the youngest neonates KCC2 is expressed ventrally in motor neurones but it does not appear in the superficial dorsal horn until P10, when a thin band can be seen. By P21 the dorsal band of KCC2 expands to encompass Laminae I, II and III outer (Fig. 4.2). In cultured embryonic dorsal horn neurones, application of GABA or glycine causes membrane depolarisation and increased intracellular

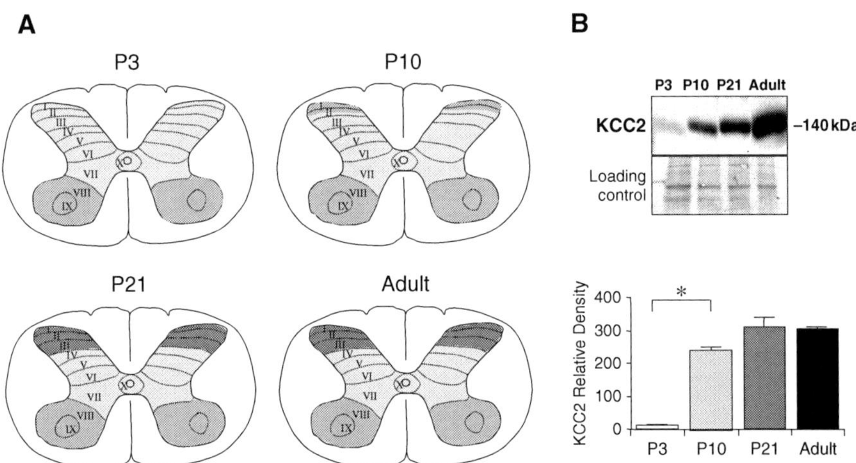

Fig. 4.2 KCC2 expression in the postnatal spinal cord. A – Diagrams showing the expression of KCC2 (as determined by Immunohistochemistry) at 3, 10 and 21 days after birth and in the adult spinal cord. **B** – Quantitative analysis of KCC2 levels in the dorsal horn of the spinal cord over postnatal development as determined by western blot. KCC2 expression is low at P3 and then increases significantly by P10 (*P < 0.05) (Harrop E & Fitzgerald M, unpublished data)

calcium (Reichling et al. 1994; Wang et al. 1994). Similar results were found in gramicidin perforated patch-clamp experiments performed in spinal cord slices from newborn rats where $GABA_AR$ activation caused depolarisation in 40% of superficial dorsal horn neurones (Baccei and Fitzgerald 2004). However, these depolarisations were never sufficient to produce action potential firing and by the end of the first postnatal week all $GABA_AR$-mediated responses were hyperpolarising. Even high concentrations of exogenous GABA could not evoke spiking in neurones from very young rats (P0–P1) (Baccei and Fitzgerald 2004). $GABA_AR/GlyR$ mediated depolarisations may still have facilitatory effects in the neonatal spinal cord but they will be limited by the temporal and spatial organisation of other inputs in the network (Gulledge and Stuart 2003; Jean-Xavier et al. 2007). The physical opening of these channels alone will reduce the excitatory effect of coincident glutamatergic signals by reducing the resistance of the cell membrane and therefore lowering the voltage change produced by the incoming current, a phenomenon known as shunting (Jean-Xavier et al. 2007).

Studies of spinal GABAergic transmission in vivo have confirmed that this system remains inhibitory in the neonate (Bremner et al. 2006). Extracellular electrophysiological recordings were made from dorsal horn neurones before and after application of the specific $GABA_AR$ antagonist gabazine. In both P3 and P21 rats AP discharge was facilitated and cutaneous receptive fields expanded to a similar degree. Electromyograph (EMG) studies of the effects of gabazine on hindlimb reflexes in the whole animal also showed that $GABA_ARs$ are consistently inhibitory at all postnatal ages when the spinal cord was isolated via spinalisation (Hathway et al. 2006). Conversely, the intrathecal application of gabazine in intact animals had opposite effects in neonates and adults, indicating that a postnatal switch occurs in the GABAergic control of cutaneous reflexes which is mediated by supraspinal circuits rather than those in the spinal cord itself. Consistent results were obtained in a study of the systemic actions of the benzodiazepine midazolam, which acted as an analgesic and a sedative in adults but potentiated nociceptive behaviour and sensitised cutaneous reflexes in the neonate without producing any sedation (Koch et al. 2008).

4.3.5 Short Term Plasticity of GABAergic Transmission

A simple developmental switch in spinal GABAergic function now seems an unlikely explanation for the baseline cutaneous excitability in the immature dorsal horn. Inhibitory transmission may, however, still be less efficacious in young animals when examined under conditions of repetitive high frequency stimuli which are common during nociceptive signaling. The reduced chloride extrusion capacity of immature neurones due to low KCC2 expression can lead to postsynaptic Cl^- accumulation during repetitive stimulation, which may cause a transient reversal in the direction of the chloride current (Cordero-Erausquin et al. 2005) and thus compromise the strength of GABAergic inhibition in neonates when faced with prolonged or intense stimulation.

Other recent experiments have examined the presynaptic properties of GABAergic signaling to establish whether there are any developmental changes in the short term plasticity of inhibitory transmission in the dorsal horn (Ingram et al. 2008). Both the amplitude and fidelity of GABAergic IPSCs were shown to be significantly lower in younger animals, reducing the reliability of synaptic inhibition. Although the extent of short term depression following trains of stimuli did not change over postnatal development, the rate of recovery accelerated over the first three postnatal weeks (Ingram et al. 2008). The combination of these properties will affect the integration of synaptic inputs within immature dorsal horn neurones and may explain some of the increased sensitivity to sensory stimuli seen in neonatal animals.

To summarise, although dorsal horn inhibitory transmission has been shown to be functional in even the youngest neonates it may still have flaws, including very low levels of tonic glycinergic transmission, increased variability and a poor recovery from repetitive stimulation. While these deficiencies may contribute to increased pain sensation in young animals, they may be developmentally necessary to allow sufficient excitatory synaptic activity to stimulate the maturation of these pathways. Lower thresholds and increased responsiveness to noxious stimuli in early life could be key in shaping the adult nervous system's response to pain without the need for neonates to repeatedly experience potentially tissue damaging stimuli. The integration of excitatory and inhibitory synaptic inputs during the early postnatal period will therefore be crucial for normal sensory development.

4.4 Integration of Synaptic Inputs

While a full understanding of the details of both excitatory and inhibitory synaptic transmission is important, to gain a complete view of spinal cord development we must also study the integration of these inputs at a cellular and network level. It can be difficult to dissect out these separate forces but one approach that has been successful is to observe the contralateral inhibitory receptive fields (RFs) of dorsal horn neurones over postnatal development (Bremner and Fitzgerald 2008). By removing descending controls (via spinalisation) the activity of spinal networks can be observed in detail. In neonates contralateral inhibitory RFs, like their ipsilateral excitatory counterparts, were relatively larger and more diffuse than in adulthood. Inhibition could often be evoked by stimulation of the whole ipsilateral hind paw rather than a more focused "mirror" of the excitatory RF as seen in older animals. Interestingly, in the neonate both innocuous touch as well as noxious pinch on the opposite foot could provide inhibition whereas in the adult only high-intensity stimuli could decrease the firing of dorsal horn neurones. These results demonstrate that inhibition and excitation are not as precisely balanced in young animals and this may impact upon the integration of these systems.

Postnatal changes in the integration of synaptic inputs may result from an age-dependent reorganisation in the pattern of connections between neurones in the spinal cord. The application of new technologies now allows the study of local anatomical wiring of inhibitory and excitatory synaptic inputs onto single dorsal horn neurones at an extremely high resolution. Laser scanning photo-stimulation has been used to elucidate the precise locations of inhibitory and excitatory interneurones and their synapses onto islet cells in the SG of the adult rat (Kato et al. 2007). The results images show a precise arrangement of cells in the dorsal horn with interneurones providing inhibitory input onto islet cells located more proximally to the islet cell soma. Synapses from the two types of interneurone were also segregated in this manner with transmitter release sites for excitatory connections spread around the dendritic tree while those that were inhibitory remained concentrated in the peri-somatic region. If this meti-culous arrangement of synaptic connections is not yet fully in place in neonates then it could easily affect the integration of excitatory and inhibitory signalling in the dorsal horn. Evidence for spatial disorganisation in the immature dorsal horn can be found in primary afferent back-labelling experiments which track the patterns of central terminations over postnatal development (Granmo et al. 2008). Here it was confirmed, as in earlier experiments (Jackman and Fitzgerald 2000), that terminals were spread over a wider area in younger animals but a new finding was that these fields also showed greater inter-individual variation in location in neonates. The postnatal refinement of primary afferent termina-tion patterns was shown to be activity-dependent as it was blocked by intrathe-cal NMDA antagonism and could also be modulated by an overexposure to tactile stimuli by housing litters of rat pups in a vibrating cage (Granmo et al. 2008).

4.5 Conclusion

Fast synaptic transmission in the dorsal horn of neonatal animals is not merely a less efficient version of that present in the adult. To gain a full appreciation of how pain is sensed in young animals we must consider differences in both excitatory and inhibitory transmission and the integration of these synaptic inputs (see Fig. 4.3). This will allow us to form better hypotheses about how the system functions in normal and pathological states. Many of the properties of immature synapses in the neonatal dorsal horn may be designed to amplify peripheral sensory inputs, in order to facilitate the synaptic reorganization which underlies the maturation of pain circuitry. As a result, the increased

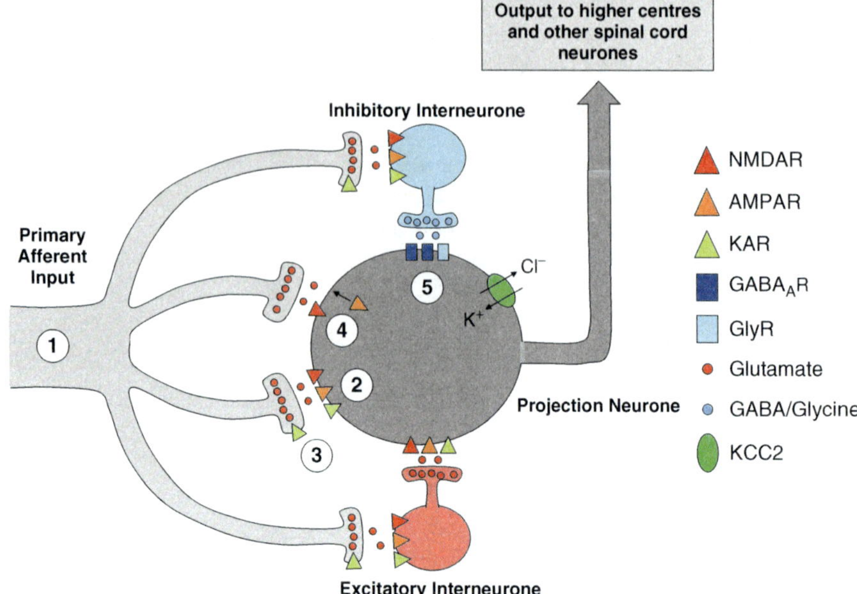

Fig. 4.3 Postnatal changes in synaptic transmission in the dorsal horn. Diagram illustrating a number of developmental changes in synaptic properties that could contribute to altered sensory perception in neonates: 1 C-fibre inputs are weak and sparse in the neonate whereas A-fibres make strong connections and are relatively abundant. 2 AMPA and KARs may be permeable to calcium in early life due to changes in subunit composition. 3 KARs are found presynaptically in the neonate but not the adult and modulate neurotransmitter release. 4 Silent synapses which contain only NMDARs are found in early postnatal life. AMPAR insertion is evoked by intense stimulation. 5 Immature neurones have high [Cl⁻]i due to low levels of KCC2. GABAA/GlyR activation can therefore cause depolarisation

excitability of the immature dorsal horn network may in fact reflect a precise weighting of excitatory and inhibitory synaptic signalling which is optimal for shaping the development of pain pathways in the CNS.

References

Agrawal SG and Evans RH (1986) The primary afferent depolarizing action of kainate in the rat. Br J Pharmacol. 87:345–355

Aguayo LG, van Zundert B, Tapia JC et al (2004) Changes on the properties of glycine receptors during neuronal development. Brain Research Reviews. 47:33–45

Akerman CJ and Cline HT (2006) Depolarizing GABAergic conductances regulate the balance of excitation to inhibition in the developing retinotectal circuit in vivo. J Neurosci. 26:5117–5130

Albuquerque C, Lee CJ, Jackson AC et al (1999) Subpopulations of GABAergic and non-GABAergic rat dorsal horn neurons express Ca2+-permeable AMPA receptors. Eur J Neurosci. 11:2758–2766

Baba H, Doubell TP, Moore KA et al (2000) Silent NMDA receptor-mediated synapses are developmentally regulated in the dorsal horn of the rat spinal cord. J Neurophysiol. 83:955–962

Baccei ML, Bardoni R, and Fitzgerald M (2003) Development of nociceptive synaptic inputs to the neonatal rat dorsal horn: glutamate release by capsaicin and menthol. J Physiol. 549:231–242

Baccei ML and Fitzgerald M (2004) Development of GABAergic and glycinergic transmission in the neonatal rat dorsal horn. J Neurosci. 24:4749–4757

Baccei ML and Fitzgerald M (2005) Intrinsic firing properties of developing rat superficial dorsal horn neurons. Neuroreport. 16:1325–1328

Bardoni R, Magherini PC, and MacDermott AB (1998) NMDA EPSCs at glutamatergic synapses in the spinal cord dorsal horn of the postnatal rat. J Neurosci. 18:6558–6567

Barnard EA, Skolnick P, Olsen RW et al (1998) International Union of Pharmacology. XV. Subtypes of gamma – Aminobutyric AcidA receptors: classification on the basis of subunit structure and receptor function. Pharmacol Rev. 50:291–314

Beggs S, Torsney C, Drew LJ et al (2002) The postnatal reorganization of primary afferent input and dorsal horn cell receptive fields in the rat spinal cord is an activity-dependent process. Eur J Neurosci. 16:1249–1258

Ben Ari Y (2002) Excitatory actions of gaba during development: the nature of the nurture. Nat Rev Neurosci. 3:728–739

Ben Ari Y, Cherubini E, Corradetti R et al (1989) Giant synaptic potentials in immature rat CA3 hippocampal neurones. J Physiol. 416:303–325

Berki AC, O'Donovan MJ, and Antal M (1995) Developmental expression of glycine immunoreactivity and its colocalization with GABA in the embryonic chick lumbosacral spinal cord. J Comp Neurol. 362:583–596

Bice TN and Beal JA (1997a) Quantitative and neurogenic analysis of neurons with supraspinal projections in the superficial dorsal horn of the rat lumbar spinal cord. J Comp Neurol. 388:565–574

Bice TN and Beal JA (1997b) Quantitative and neurogenic analysis of the total population and subpopulations of neurons defined by axon projection in the superficial dorsal horn of the rat lumbar spinal cord. J Comp Neurol. 388:550–564

Bremner L, Fitzgerald M, and Baccei M (2006) Functional GABAA-receptor-mediated inhibition in the neonatal dorsal horn. J Neurophysiol. 95:3893–3897

Bremner LR and Fitzgerald M (2008) Postnatal tuning of cutaneous inhibitory receptive fields in the rat. J Physiol. 586:1529–1537

Brown KM, Wrathall JR, Yasuda RP et al (2002) Quantitative measurement of glutamate receptor subunit protein expression in the postnatal rat spinal cord. Dev Brain Res. 137:127–133

Brown TH, Chapman PF, Kairiss EW et al (1988) Long-term synaptic potentiation. Science. 242:724–728

Burnashev N, Monyer H, Seeburg PH et al (1992) Divalent ion permeability of AMPA receptor channels is dominated by the edited form of a single subunit. Neuron. 8:189–198

Chery N and De Koninck Y (1999) Junctional versus extrajunctional glycine and GABA(A) receptor-mediated IPSCs in identified lamina I neurons of the adult rat spinal cord. J Neurosci. 19:7342–7355

Cordero-Erausquin M, Coull JAM, Boudreau D et al (2005) Differential maturation of GABA action and anion reversal potential in spinal lamina I neurons: impact of chloride extrusion capacity. J Neurosci. 25:9613–9623

Cronin JN, Bradbury EJ, and Lidierth M (2004) Laminar distribution of GABAA- and glycine-receptor mediated tonic inhibition in the dorsal horn of the rat lumbar spinal cord: effects of picrotoxin and strychnine on expression of Fos-like immunoreactivity. Pain. 112:156–163

Ehrlich I, Lohrke S, and Friauf E (1999) Shift from depolarizing to hyperpolarizing glycine action in rat auditory neurones is due to age-dependent Cl- regulation. J Physiol. 520 Pt 1:121–137.

Ekholm J (1967) Postnatal changes in cutaneous reflexes and in the discharge pattern of cutaneous and articular sense organs. A morphological and physiological study in the cat. Acta Physiol Scand Suppl. 297:1–130.:1–130

Engelman HS, Allen TB, and MacDermott AB (1999) The Distribution of Neurons Expressing Calcium-Permeable AMPA Receptors in the Superficial Laminae of the Spinal Cord Dorsal Horn. J Neurosci. 19:2081–2089

Erlander MG, Tillakaratne NJK, Feldblum S et al (1991) Two genes encode distinct glutamate decarboxylases. Neuron. 7:91–100

Falcon M, Guendellman D, Stolberg A et al (1996) Development of thermal nociception in rats. Pain. 67:203–208

Fitzgerald M and Gibson S (1984) The postnatal physiological and neurochemical development of peripheral sensory C fibres. Neuroscience. 13:933–944

Fitzgerald M, Shaw A, and MacIntosh N (1988) Postnatal development of the cutaneous flexor reflex: comparative study of preterm infants and newborn rat pups. Dev Med Child Neurol. 30:520–526

Fitzgerald M and Swett J (1983) The termination pattern of sciatic nerve afferents in the substantia gelatinosa of neonatal rats. Neurosci Lett. 43:149–154

Gasnier B (2000) The loading of neurotransmitters into synaptic vesicles. Biochimie. 82:327–337

Gonzalez DL, Fuchs JL, and Droge MH (1993) Distribution of NMDA receptor binding in developing mouse spinal cord. Neurosci Lett. 151:134–137

Granmo M, Petersson P, and Schouenborg J (2008) Action-based body maps in the spinal cord emerge from a transitory floating organization. J Neurosci. 28:5494–5503

Green GM and Gibb AJ (2001) Characterization of the single-channel properties of NMDA receptors in laminae I and II of the dorsal horn of neonatal rat spinal cord. Eur J Neurosci. 14:1590–1602

Gu X and Spitzer NC (1993) Low-threshold Ca2 + current and its role in spontaneous elevations of intracellular Ca2 + in developing Xenopus neurons. J Neurosci. 13:4936–4948

Gulledge AT and Stuart GJ (2003) Excitatory actions of GABA in the cortex. Neuron. 37:299–309

Harvey RJ, Depner UB, Wassle H et al (2004) GlyR {alpha}3: An essential target for spinal PGE2-mediated inflammatory pain sensitization. Science. 304:884–887

Hathway G, Harrop E, Baccei M et al (2006) A postnatal switch in GABAergic control of spinal cutaneous reflexes. Eur J Neurosci. 23:112–118

Hori Y and Kanda K (1994) Developmental alterations in NMDA receptor-mediated [Ca2+]i elevation in substantia gelatinosa neurons of neonatal rat spinal cord. Brain Res Dev Brain Res. 80:141–148

Hosl K, Reinold H, Harvey RJ et al (2006) Spinal prostaglandin E receptors of the EP2 subtype and the glycine receptor [alpha]3 subunit, which mediate central inflammatory hyperalgesia, do not contribute to pain after peripheral nerve injury or formalin injection. Pain. 126:46–53

Hossaini M, French PJ, and Holstege JC (2007) Distribution of glycinergic neuronal somata in the rat spinal cord. Brain Res. 1142:61–69

Ingram RA, Fitzgerald M, and Baccei ML (2008) Developmental changes in the fidelity and short term plasticity of GABAergic synapses in the neonatal rat dorsal horn. J Neurophysiol. 99:3144–3150

Inquimbert P, Rodeau JL, and Schlichter R (2007) Differential contribution of GABAergic and glycinergic components to inhibitory synaptic transmission in lamina II and laminae III–IV of the young rat spinal cord. Eur J Neurosci. 26:2940–2949

Isaac JT, Nicoll RA, and Malenka RC (1995) Evidence for silent synapses: implications for the expression of LTP. Neuron. 15:427–434

Jackman A and Fitzgerald M (2000) Development of peripheral hindlimb and central spinal cord innervation by subpopulations of dorsal root ganglion cells in the embryonic rat. J Comp Neurol. 418:281–298

Jakowec MW, Fox AJ, Martin LJ et al (1995a) Quantitative and qualitative changes in AMPA receptor expression during spinal cord development. Neuroscience. 67:893–907

Jakowec MW, Yen L, and Kalb RG (1995b) In situ hybridization analysis of AMPA receptor subunit gene expression in the developing rat spinal cord. Neuroscience. 67: 909–920

Jean-Xavier C, Mentis GZ, O'Donovan MJ et al (2007) Dual personality of GABA/glycine-mediated depolarizations in immature spinal cord. PNAS. 104:11477–11482

Jennings E and Fitzgerald M (1996) C-fos can be induced in the neonatal rat spinal cord by both noxious and innocuous peripheral stimulation. Pain. 68:301–306

Jennings E and Fitzgerald M (1998) Postnatal changes in responses of rat dorsal horn cells to afferent stimulation: a fibre-induced sensitization. J Physiol. 509:859–868

Jonas P, Bischofberger J, and Sandkuhler J (1998) Corelease of two fast neurotransmitters at a central synapse. Science. 281:419–424

Kato G, Kawasaki Y, Ji RR et al (2007) Differential wiring of local excitatory and inhibitory synaptic inputs to islet cells in rat spinal lamina II demonstrated by laser scanning photostimulation. J Physiol (Lond). 580:815–833

Keller AF, Breton JD, Schlichter R et al (2004) Production of 5alpha-reduced neurosteroids is developmentally regulated and shapes GABA(A) miniature IPSCs in lamina II of the spinal cord. J Neurosci. 24:907–915

Keller AF, Coull JA, Chery N et al (2001) Region-specific developmental specialization of GABA-glycine cosynapses in laminas I–II of the rat spinal dorsal horn. J Neurosci. 21: 7871–7880

Kerchner GA, Wilding TJ, Li P et al (2001) Presynaptic kainate receptors regulate spinal sensory transmission. J Neurosci. 21:59–66

Kirsch J, Wolters I, Triller A et al (1993) Gephyrin antisense oligonucleotides prevent glycine receptor clustering in spinal neurons. Nature. 366:745–748

Kitao Y, Robertson B, Kudo M et al (1996) Neurogenesis of subpopulations of rat lumbar dorsal root ganglion neurons including neurons projecting to the dorsal column nuclei. J Comp Neurol. 371:249–257

Knabl J, Witschi R, Hosl K et al (2008) Reversal of pathological pain through specific spinal GABAA receptor subtypes. Nature. 451:330–334

Koch SC, Fitzgerald M, and Hathway GJ (2008) Midazolam potentiates nociceptive behavior, sensitizes cutaneous reflexes, and is devoid of sedative action in neonatal rats. Anesthesiology. 108:122–129

Lee CJ, Kong H, Manzini MC et al (2001) Kainate receptors expressed by a subpopulation of developing nociceptors rapidly switch from high to low Ca2+ permeability. J Neurosci. 21:4572–4581

Legendre P (2001) The glycinergic inhibitory synapse. Cell Mol Life Sci. 58:760–793

Li P and Zhuo M (1998) Silent glutamatergic synapses and nociception in mammalian spinal cord. Nature. 393:695–698

Li P, Wilding TJ, Kim SJ et al (1999) Kainate-receptor-mediated sensory synaptic transmission in mammalian spinal cord. Nature. 397:161–164

Ma W, Behar T, and Barker JL (1992) Transient expression of GABA immunoreactivity in the developing rat spinal cord. J Comp Neurol. 325:271–290

Ma W, Saunders PA, Somogyi R et al (1993) Ontogeny of GABAA receptor subunit mRNAs in rat spinal cord and dorsal root ganglia. J Comp Neurol. 338:337–359

Malosio ML, Marqueze-Pouey B, Kuhse J et al (1991) Widespread expression of glycine receptor subunit mRNAs in the adult and developing rat brain. EMBO J. 10:2401–2409

Marti E, Gibson SJ, Polak JM et al (1987) Ontogeny of peptide- and amine-containing neurones in motor, sensory, and autonomic regions of rat and human spinal cord, dorsal root ganglia, and rat skin. J Comp Neurol. 266:332–359

Matzenbach B, Maulet Y, Sefton L et al (1994) Structural analysis of mouse glycine receptor alpha subunit genes. Identification and chromosomal localization of a novel variant. J Biol Chem. 269:2607–2612

Nowak L, Bregestovski P, Ascher P et al (1984) Magnesium gates glutamate-activated channels in mouse central neurones. Nature. 307:462–465

Paoletti P and Neyton J (2007) NMDA receptor subunits: function and pharmacology. Curr Opinion Pharmacol. 7:39–47

Park JS, Nakatsuka T, Nagata K et al (1999) Reorganization of the primary afferent termination in the rat spinal dorsal horn during post-natal development. Brain Res Dev Brain Res. 113:29–36

Philpot BD, Sekhar AK, Shouval HZ et al (2001) Visual experience and deprivation bidirectionally modify the composition and function of NMDA receptors in visual cortex. Neuron. 29:157–169

Pignatelli D, Ribeiro-da-Silva A, and Coimbra A (1989) Postnatal maturation of primary afferent terminations in the substantia gelatinosa of the rat spinal cord. An electron microscopic study. Brain Res. 491:33–44

Poyatos I, Ponce J, Aragon C et al (1997) The glycine transporter GLYT2 is a reliable marker for glycine-immunoreactive neurons. Brain Res Mol Brain Res. 49:63–70

Racz I, Schutz B, Abo-Salem OM et al (2005) Visceral, inflammatory and neuropathic pain in glycine receptor alpha 3-deficient mice. Neuroreport. 16:2025–2028

Reichling DB, Kyrozis A, Wang J et al (1994) Mechanisms of GABA and glycine depolarization-induced calcium transients in rat dorsal horn neurons. J Physiol. 476:411–421

Rivera C, Voipio J, Payne JA et al (1999) The K + /Cl- co-transporter KCC2 renders GABA hyperpolarizing during neuronal maturation. Nature. 397:251–255

Schaffner AE, Behar T, Nadi S et al (1993) Quantitative analysis of transient GABA expression in embryonic and early postnatal rat spinal cord neurons. Brain Res Dev Brain Res. 72:265–276

Schlichter R, Keller AF, De Roo M et al (2006) Fast nongenomic effects of steroids on synaptic transmission and role of endogenous neurosteroids in spinal pain pathways. J Mol Neurosci. 28:33–51

Somogyi R, Wen X, Ma W et al (1995) Developmental kinetics of GAD family mRNAs parallel neurogenesis in the rat spinal cord. J Neurosci. 15:2575–2591

Soyguder Z, Schmidt HH, and Morris R (1994) Postnatal development of nitric oxide synthase type 1 expression in the lumbar spinal cord of the rat: a comparison with the induction of c-fos in response to peripheral application of mustard oil. Neurosci Lett. 180:188–192

Stegenga SL and Kalb RG (2001) Developmental regulation of N-methyl – aspartate- and kainate-type glutamate receptor expression in the rat spinal cord. Neuroscience. 105:499–507

Takahashi T, Momiyama A, Hirai K et al (1992) Functional correlation of fetal and adult forms of glycine receptors with developmental changes in inhibitory synaptic receptor channels. Neuron. 9:1155–1161

Todd AJ and Sullivan AC (1990) Light microscope study of the coexistence of GABA-like and glycine-like immunoreactivities in the spinal cord of the rat. J Comp Neurol. 296:496–505

Torseney C and Fitzgerald M (2002) Age-dependent effects of peripheral inflammation on the electrophysiological properties of neonatal rat dorsal horn neurons. J Neurophysiol. 87:1311–1317

Torsney C, Meredith-Middleton J, and Fitzgerald M (2000) Neonatal capsaicin treatment prevents the normal postnatal withdrawal of A fibres from lamina II without affecting fos responses to innocuous peripheral stimulation. Brain Res Dev Brain Res. 121:55–65

Tran TS, Alijani A, and Phelps PE (2003) Unique developmental patterns of GABAergic neurons in rat spinal cord. J Comp Neurol. 456:112–126

Waldenstrom A, Thelin J, Thimansson E et al (2003) Developmental learning in a pain-related system: evidence for a cross-modality mechanism. J Neurosci. 23:7719–7725

Wang J, Reichling DB, Kyrozis A et al (1994) Developmental loss of GABA- and glycine-induced depolarization and Ca2+ transients in embryonic rat dorsal horn neurons in culture. Eur J Neurosci. 6:1275–1280
Watanabe E and Akagi H (1995) Distribution patterns of mRNAs encoding glycine receptor channels in the developing rat spinal cord. Neurosci Res. 23:377–382
Watanabe M, Mishina M, and Inoue Y (1994) Distinct spatiotemporal distributions of the N-methyl-D-aspartate receptor channel subunit mRNAs in the mouse cervical cord. J Comp Neurol. 345:314–319
Woodbury CJ and Koerber HR (2003) Widespread projections from myelinated nociceptors throughout the substantia gelatinosa provide novel insights into neonatal hypersensitivity. J Neurosci. 23:601–610

Part III
Slow Synaptic Transmission in the Dorsal Horn

Chapter 5
BDNF and TrkB Mediated Mechanisms in the Spinal Cord

Rita Bardoni and Adalberto Merighi

Abstract The neurotrophin brain-derived neurotrophic factor (BDNF) plays an essential role during development, promoting the survival of specific populations of central and peripheral neurons. During adulthood, BDNF also acts as a synaptic modulator in several areas of the central nervous system (CNS), including the spinal cord, and is involved in short and long term changes of synaptic efficacy. In spinal cord dorsal horn BDNF is expressed in the peptidergic terminals originating from primary afferent fibres, while its high affinity receptor trkB has been detected on both primary afferent terminals and dorsal horn neurons. In superficial dorsal horn, exogenous BDNF modulates fast excitatory (glutamatergic) and inhibitory (GABAergic/glycinergic) signals, as well as slow peptidergic neurotransmission. Conditions of inflammatory and neuropathic pain alter the expression of BDNF and trkB receptors in dorsal horn. In experimental pain models, modulation of synaptic transmission by BDNF plays an important role in the induction and maintenance of central sensitization.

Abbreviations

AMPA	α-amino-methyl-isoxazole-propionic acid
BDNF	brain derived neurotrophic factor
CCI	chronic constriction injury
CFA	complete Freund's adjuvant
CGRP	calcitonin gene-related peptide
CNS	central nervous system
CREB	cyclic AMP response element binding protein

R. Bardoni (✉)
Department of Biomedical Sciences, University of Modena and Reggio Emilia,
Modena, Italy
e-mail: bardoni@unimo.it

M. Malcangio (ed.), *Synaptic Plasticity in Pain*,
DOI 10.1007/978-1-4419-0226-9_5, © Springer Science+Business Media, LLC 2009

DCV   dense core vesicle
DRG   dorsal root ganglion
EPSCs   excitatory post-synaptic currents
ERK   extracellular signal-regulated kinase
fl-trkB   full-length trkB receptor
GABA   γ-amino butyric acid
KCC2   potassium-chloride co-transporter 2
MAPK   mitogen-activated protein kinase
MEK   MAPK/ERK kinase
NGF   nerve growth factor
NMDA   N-methyl-D-aspartate
NT   neurotrophin
p75NTR   common neurotrophin low affinity receptor
PAF   primary afferent fibre
PKC   protein kinase C
PLC   phospholipase C
PNS   peripheral nervous system
PSD   post-synaptic density
SST   spinothalamic tract
TNF-α   tumour necrosis factor α
trkA   tropomyosine receptor kinase A
trkB   tropomyosine receptor kinase B
trkC   tropomyosine receptor kinase C
tr-trkB   truncated trkB receptor

5.1 Introduction

Brain-derived neurotrophic factor (BDNF) is a 12.4 kDa basic protein belonging to the family of the neurotrophins (NTs). Like the other growth factors of this family, it plays a fundamental role during development, promoting the survival of several populations of neurons of the central and peripheral divisions of the nervous system.

In adulthood, BDNF becomes an important modulator of synapses in several areas of the brain and spinal cord, being clearly involved in short and long term changes of synaptic strength (reviewed, e.g. in Schuman, 1999, Malenka and Nicoll, 1999, Schinder and Poo, 2000, Poo, 2001, Lessmann et al., 2003; Malcangio and Lessmann, 2003; Carvalho et al., 2008).

This chapter summarizes current data on the cellular and molecular mechanisms as well as the circuitries by which BDNF modulates sensory neurotransmission in nociceptive pathways at the spinal cord level and mediates the central sensitization that underlies many forms of hyperalgesia and allodynia.

5.2 Synthesis, Storage and Pattern of Expression of BDNF

NTs, including BDNF, are synthesized in the cell body as pre-pro-NT which undergo post-translational modifications prior to the generation of the mature homodimeric protein. In neurons, the expression of the bdnf gene is regulated by activity and depends on the intracellular calcium concentration (reviewed in Mellstrom et al., 2004). When acting according to the prototype for survival factors, BDNF is taken up by peripheral targets and retrogradely transported to the cell body where it influences gene expression and several major cellular functions. However, when synthesized by neurons, BDNF (and the immature form pro-BDNF) is packed in secretory granules, and delivered to pre-synaptic axon terminals by anterograde transport (Altar et al., 1997; von Bartheld, 2004). Synthesis and anterograde transport of BDNF are not limited to CNS. In the peripheral nervous system (PNS), the peptidergic small- to medium-sized dark neurons in dorsal root ganglia (DRGs) represent an additional class of neurons which constitutively synthesize and anterogradely transport BDNF to their central terminals in the dorsal horn of the spinal cord (Kerr et al., 1999; Michael et al., 1997).

BDNF has also been detected in neurons lacking mRNA transcripts, indicating that, in neurons unable to synthesize BDNF, the extracellular NT could be endocytosed at neuronal somatodendritic domains (transcytosis) and then targeted to terminals by anterograde axonal transport (von Bartheld, 2004). However, the uptake of BDNF by transcytosis has been convincingly demonstrated only in vitro. The terminals of the central and peripheral neurons that synthesize BDNF contain the NT packaged in dense core vesicles (DCVs), which, in spinal cord, display a concentration of the BDNF about 30 times higher than that observed in agranular vesicles (Salio et al., 2007). BDNF-containing DCVs also contain the neuropeptides substance P and calcitonin gene-related peptide (CGRP). In central axon terminals of DRG neurons, BDNF is co-stored in DCVs together with substance P and CGRP, following remarkably constant stoichiometric ratios (Salio et al., 2007).

In the spinal cord dorsal horn BDNF is mostly localized in primary afferent fibre (PAF) terminals in laminae I-II (Zhou et al., 1993, 2004), whereas there are no second order neurons expressing the NT in this location (Salio et al., 2005, 2007).

5.3 Release of BDNF

The neuronal release of high molecular weight substances stored in DCVs, such as BDNF and neuropeptides, is profoundly different from the release of classical low molecular weight transmitters such as glutamate contained in agranular vesicles. The release of DCVs content is characterized by slow emptying, lack of physical docking at synaptic sites and dependence on prolonged intracellular Ca^{2+} elevation in the release compartment (Balkowiec and Katz, 2000; Lessmann et al., 2003).

Although BDNF and neuropeptides are co-stored within the same DCVs, they are not necessarily co-released. Whilst electrically-evoked neuropeptide release is dependent on stimulation frequency (Mansvelder and Kits, 2000), the release of BDNF also depends on stimulation patterns, where high frequency burst stimulations seem to be particularly efficient (Balkowiec and Katz, 2000; Lever et al., 2001). Recent studies suggest that DCVs in primary afferent terminals contain a cocktail of high molecular weight transmitters (Salio et al., 2007), and thus differential release of BDNF and co-stored peptides in vivo could rely on differences in the relative rate of their dissolution from the DCV core, since this is the critical determinant of the speed of peptide/NT secretion (Brigadski et al., 2005).

In the superficial dorsal horns basal release of BDNF has been measured using antibody-coated microprobes under anaesthesia (Walker et al., 2001). Electrical activity can trigger secretion of BDNF from DRG axon terminals into dorsal horn only upon burst stimulation of C fibres, while release of substance P (co-stored in the same DCVs) is observed also during constant low frequency or tetanic high frequency stimulation (Lever et al., 2001). Release of BDNF is dependent on extracellular calcium and mediated by pre-synaptic NMDA receptors; capsaicin (the pungent ingredient of chilly pepper) can also induce a dose-dependent release of BDNF (Lever et al., 2001).

In pathological pain states, cellular expression, content and release of BDNF in dorsal horn have been found to be altered. In models of inflammatory pain, mimicked by systemic or intrathecal treatment with nerve growth factor (NGF), BDNF content in PAF terminals and BDNF release can increase significantly. In the peripheral nerve injury model of sciatic nerve transection, the basal release of the NT is increased and detected throughout the whole dorsal horn, instead of being limited to superficial laminae (Walker et al., 2001).

In models of neuropathic pain following peripheral nerve injury, activated microglia, rather than PAF terminals, become the source for BDNF in the spinal cord (Coull et al., 2005). Specifically, Coull and colleagues have shown that intrathecal administration of microglia activated by ATP as well as intrathecal injection of BDNF reproduce the mechanical allodynia observed after peripheral nerve injury.

5.4 Expression of BDNF Receptors (trkB) in Spinal Cord

Most of BDNF cellular actions are mediated by its high-affinity receptor, the tropomyosine receptor kinase B (trkB). This receptor is abundant during development, but also widely distributed in the CNS of adult animals, suggesting a continuing role for BDNF on the adult nervous system. BDNF (and the other members of the NT family) also binds, albeit with much lower affinity, to a different receptor generally referred to as the common NT low affinity receptor (p75NTR). The biological functions mediated by p75NTR remain, for the most, elusive and appear to be mainly related to cell survival (Barrett, 2000) rather than (putative) neurotransmitter function, and therefore are not taken into consideration here.

In developing and adult CNS, alternative splicing of the trkB mRNA generates three different isoforms (Middlemas et al., 1991): the full-length trkB (fl-trkB) receptor and two truncated receptor forms (tr-trkB). All trkB isoforms share a common extracellular domain, whereas the truncated isoforms lack the signal transducing intracellular tyrosine kinase domain, and thus do not appear to be able to trigger the intracellular signal transduction pathways that are commonly utilized by BDNF to exert its biological functions. Tr-trkB is prevalently expressed in choroid plexus, ependymal cells and astrocytes, i.e. the non neuronal cells within the nervous tissue, whereas both fl-trkB and tr-trkB are expressed in neurons (Altar et al., 1994; Armanini et al., 1995; Beck et al., 1993; Ernfors et al., 1992; Frisén et al., 1993; Klein et al., 1990a, 1990; Yan et al., 1997).

Initial light microscopy studies on the distribution of trkB receptors in the spinal cord have led to the localization of mRNA (Ernfors et al., 1993; Mannion et al., 1999) and protein (Zhou et al., 1993; Garraway et al., 2003) in neurons (and glia) throughout the dorsal horn, including in neurons of the spinothalamic tract (SST - Slack et al., 2005). It was originally assumed that trkB was unlikely to be present on PAF terminals since immunocytochemical staining was limited to the dorsal root entry zone and only a small proportion of medium to large sized DRG neurons expressed the receptor under normal conditions (McMahon et al., 1994; Wright and Snider, 1995; Koltzenburg et al., 1999). Although in situ hybridization experiments provided evidence that neurons expressing the trkB mRNA were very numerous in the dorsal horn, immunocytochemistry revealed that the trkB protein was considerably less abundant or even absent (Mannion et al., 1999; Michael et al., 1999). In keeping with these observations, high levels of fl-trkB expression were only observed following electrical stimulation of C fibres or in pain models (see below). Salio et al. (2005) have provided the first ultrastructural description of fl-trkB localization at synapses between first and second order sensory neurons in rat and mouse spinal lamina II, showing that fl-trkB is present not only in somatodendritic membranes of lamina II neurons, but also in axon terminals (Figs. 5.1 and 5.2). About 90% of these terminals belong to PAFs. Interestingly, a subpopulation of peptidergic small- to medium-sized DRG neurons (about 10% of the total DRG neurons) co-express BDNF + trkB, suggesting that trkB could act as an autoreceptor at their central terminals in dorsal horn. These neurons represent 1/3 of the DRG cells containing BDNF, whereas the remaining 2/3 express the NGF receptor tropomyosine receptor kinase A (trkA) and are thus NGF-sensitive (see below). The central terminals of the DRG neurons expressing BDNF + trkB belong, for the most part, to the complex synaptic structures located in lamina II, commonly referred to as glomeruli. Quantitative analysis

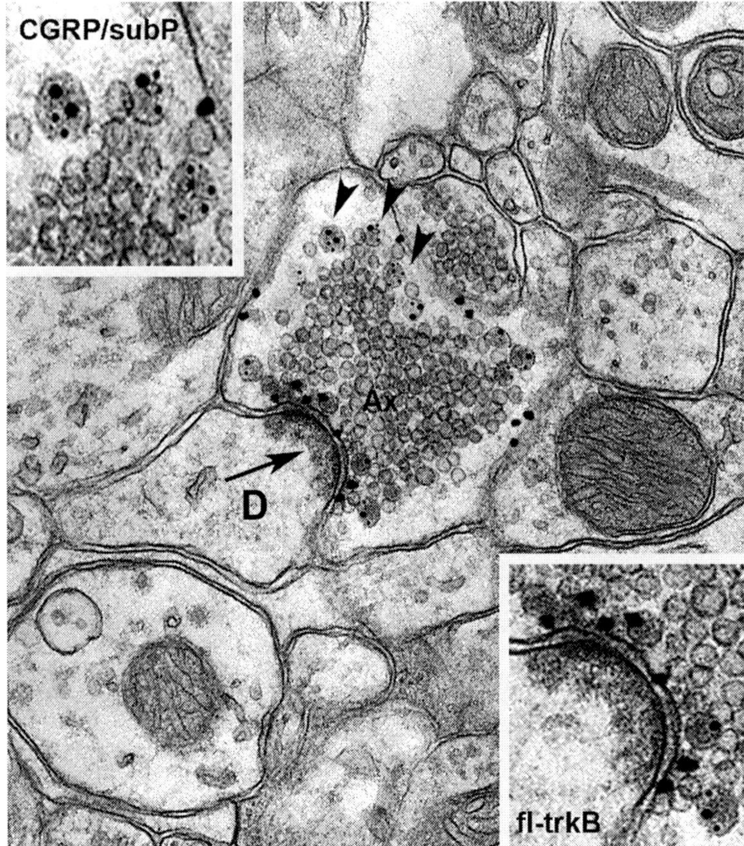

Fig. 5.1 Triple immunogold labeling of an axo-dendritic synapse in the spinal cord lamina II showing the localization of fl-trkB. The axon terminal (Ax) belongs to a PAF since it contains a number of DCVs immunolabeled for CGRP and substance P. Three of them are marked by arrowheads and shown at higher magnification in the insert at top left. The dendrite (D) post-synaptic density (arrow) is marked by gold enhanced-gold particles indicative of fl-trkB immunoreactivity (see insert at bottom right)

has revealed that BDNF + trkB immunoreactive terminals form the central boutons of type Ib glomeruli, i.e. the glomeruli originating from nociceptive peptidergic PAFs (see Merighi et al., 2008b).

Peptidergic Ib glomeruli are apparently the key structures where exogenous BDNF exerts a pre-synaptic modulatory role of the excitatory synapses between I to II order neurons. Immunocytochemistry, electrophysiology, real time calcium imaging at the confocal microscope and release/depletion measurements on acute slices demonstrated that BDNF is capable to evoke a sustained release of peptides and glutamate from PAFs by acting on pre-synaptic fl-trkB (Merighi et al., 2008a). Glomeruli represent multi-synaptic

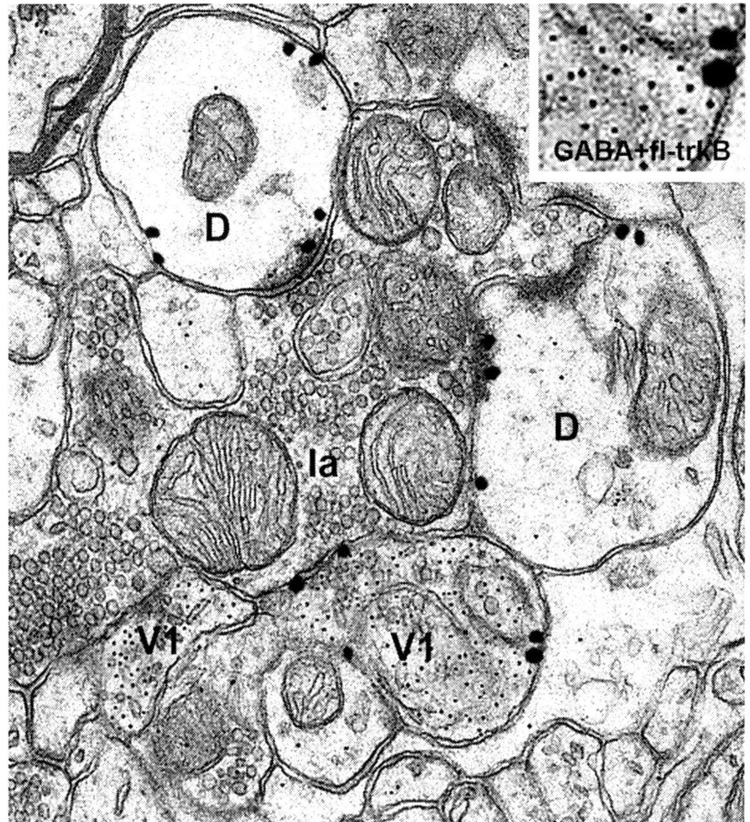

Fig. 5.2 Double immunogold labeling of a type Ia synaptic glomerulus in the spinal cord lamina II showing the localization of fl-trkB and GABA. Two plain dendrites (D) in the upper part of the micrograph display gold enhanced-gold particles indicative of fl-trkB immunor-eactivity. The other two dendrites (V1) in the lower part of the micrograph are filled with synaptic vesicles immunoreactive for GABA, but only the one at right is positive for fl-trkB (see the two large gold enhanced-gold particles in the insert at top right)

sites which are ideally structured to amplify the response of II order neurons to activation of individual PAF endings, and it is thus conceivable that a pre-synaptic modulation of these synapses also occurs following release of the endogenous NT under normal and/or pathological conditions.

As regarding inhibitory synapses, the expression of post-synaptic fl-trkB on GABAergic interneurons has been detected in all types of non-peptidergic glo-meruli (Bardoni et al., 2007), in other words in all types except Ib (Fig. 5.2). Volume diffusion of BDNF from nearby type Ib glomeruli could be important for the activation of the receptors expressed in GABAergic interneurons.

The mechanisms related to modulation of glomerular synapses by BDNF are further discussed below in paragraph 5.6.1.

5.5 Regulation of BDNF and Trk B Receptor Expression After Inflammation and Nerve Injury

Peripheral inflammation and nerve injury alter BDNF gene transcription and protein synthesis in DRG neurons (for reviews see Pezet et al., 2002b; Merighi et al., 2004; Pezet and McMahon, 2006; Obata and Noguchi, 2006).

In particular, peripheral inflammation strongly up-regulates BDNF mRNA in small- and medium-sized DRG neurons, in adult and neonatal animals (see, for example, Cho et al., 1997a, b; Lee et al., 1999; Mannion et al.1999; Chien et al., 2007) and causes de novo synthesis of BDNF in DRG neurons with myelinated axons (Mannion et al., 1999). BDNF synthesis in DRGs is strongly stimulated by NGF, whose concentration is enhanced at the site of inflammation (McMahon and Priestley 1995). The retrograde transport of NGF to the DRGs from target tissues enhances the expression of the BDNF gene in trkA-expressing sensory neurons (Apfel et al., 1996; Michael et al., 1997; Kerr et al., 1999). Models of inflammatory pain (such as the intraplantar injection of complete Freund's adjuvant - CFA) significantly up-regulate BDNF mRNA expression in DRGs in a NGF-dependent way, followed by an increase of BDNF synthesis in DRG neurons and of anterograde transport of BDNF to central terminals (Cho et al., 1997a, b).

The regulation of trkB expression in response to inflammation has also been investigated, and it has been shown that intraplantar CFA as well as C-fiber electrical activity increase fl-trkB receptor levels in dorsal horn (Mannion et al., 1999).

In models of neuropathic pain following peripheral nerve injury, different effects on BDNF expression have been reported, depending on methodological issues, such as the type of injury (transection versus ligation), the extent of the lesion (single versus multiple levels), the time interval and the distance of the axonal injury from the DRG.

Axotomy or dorsal rhizotomy cause an increase of both BDNF expression and anterograde transport to central terminals. These effects have been mainly observed in the medium and large-size DRG neurons, while the NT content decreases in small damaged neurons (reviewed in Obata and Noguchi, 2006; Pezet and McMahon 2006). Therefore, following peripheral nerve injury, a phenotypic switch in the subpopulations of DRG neurons expressing BDNF does occur, whereby BDNF expression is reduced in the trkA population and elevated in trkB- and in tropomyosine receptor kinase C (trkC)-positive neurons.

In keeping with these observations, BDNF-containing nerve terminals in the dorsal horn ipsilateral to nerve injury are reduced in the central region of lamina II after axotomy, but increased in more medial regions or deeper laminae III/IV (Zhou et al., 1999).

Furthermore, several studies report that axotomy or ligation of the L5 spinal nerve induces an increase of BDNF content in the intact ipsilateral L4 DRG neurons (Obata et al., 2006; Fukuoka et al., 2001). Products of Wallerian degeneration of injured fibres, such as growth factors, and in particular NGF, chemokines and cytokines, could be responsible for these alterations in neighbouring intact L4 fibres.

Injury models involving an inflammatory component (chronic constriction injury or experimental disk herniation) produce a significant increase in the percentage of small, medium and large BDNF-immunoreactive neurons (Ha et al., 2001; Onda et al., 2003, 2004). The pro-inflammatory cytokine tumour necrosis factor α (TNF-α) could be involved in the up-regulation of BDNF in these pain models. Notably, beside regulating BDNF expression in dorsal horn, nerve injury induces also an increase in trkB expression in dorsal horn: partial sciatic nerve ligation causes the up-regulation of the protein level of trkB receptor, that is completely reversed by concomitant intrathecal injection of BDNF antibody (Yajima et al., 2002).

5.6 Modulation of Synaptic Transmission by BDNF in Normal Animals and in Animal Models of Pain

5.6.1 Normal Animals

The role of endogenous BDNF in mediating basal, physiological pain is still contentious. As mentioned, BDNF is expressed in spinal cord somato-sensory pathways of normal animals (see Merighi et al., 2008b) and a significant, basal release of the NT has been detected in superficial dorsal horn (Walker et al., 2001). Noxious mechanical, chemical and thermal stimuli induce phosphorylation of trkB receptors, implying a role of endogenous BDNF in normal pain (Pezet et al., 2002a). Accordingly, the nociceptive reflex (Figs. 5.3 and 5.4) is impaired in an in vitro spinal cord preparation from neonate BDNF-deficient mice (Heppenstall and Lewin 2001). On the other hand, a study by Kerr et al. (1999) reports that the BDNF sequestering antibody trkB-IgG does not alter the reflex evoked by A- and C-fibre stimulation in naïve rats, whereas a significant inhibition of the nociceptive reflex is observed only in animals pretreated with NGF. Consistently with this observation, Mannion et al. (1999) have shown that sequestering endogenous BDNF with a TrkB-Fc fusion protein does not alter basal mechanical and thermal pain, nor affects the mechanical hypersensitivity induced by capsaicin administration.

BDNF release is triggered in dorsal horn by electrical stimulation of C fibres with high frequency bursts (Lever et al., 2001, see also above). Since these fibres display a burst firing pattern when the intensity of nociceptive stimuli increases over a certain threshold (Adelson et al., 1997), it is possible that the role of BDNF in modulating pain transmission in normal animals becomes significant only during intense and repetitive noxious stimulation.

Application of exogenous BDNF in untreated animals causes significant changes in pain behaviour, and most studies report that intrathecal administration

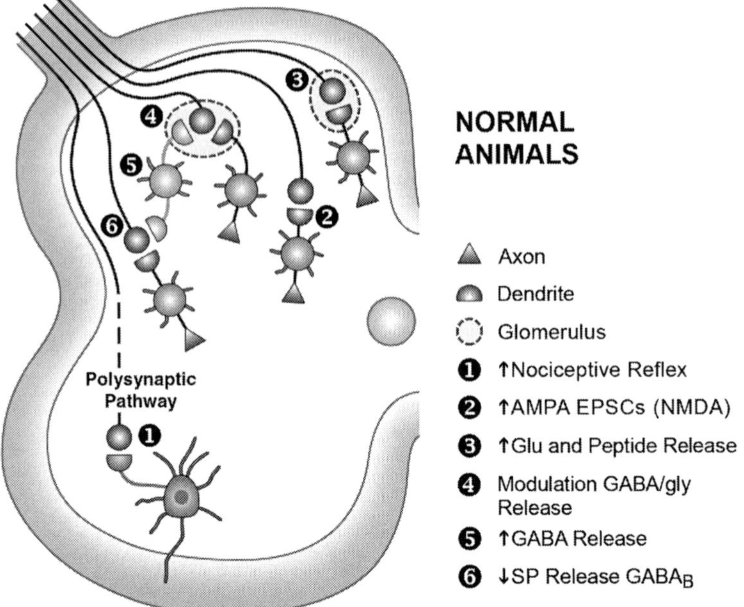

Fig. 5.3 Schematic representation of the modulation of BDNF at synapses in the spinal cord under normal conditions. 1: modulation of the nociceptive reflex through a yet unidentified polysynaptic pathway that links the NT released from PAF terminals to ventral horn motor-neurons; 2: potentiation of AMPA-mediated EPSCs due to the activation post-synaptic trkB receptors at synapses in lamina II. The glomerular and/or not-glomerular nature of these synapses has not been determined yet; 3: pre-synaptic modulation of glutamate and peptide release at type IIb glomeruli; 4: increase of inhibitory neurotransmission at non-peptidergic glomeruli; 5: increase of GABA release from lamina II interneurons; 6: reduction of substance P release from PAFs by an indirect mechanism involving the activation of GABAB receptors

of the NT causes thermal hyperalgesia and mechanical allodynia in rat and mice (see for reviews Pezet and McMahon 2006; Merighi et al., 2008a).

In keeping with this observation, several studies have described a variety of modulatory effects, exerted by exogenous BDNF, on excitatory and inhibitory synaptic transmission in the spinal cord (Fig. 5.3). In particular, glutamatergic synaptic transmission is rapidly facilitated by BDNF, interacting with post-synaptic trkB receptors. Application of the NT increases the nociceptive reflex in the isolated spinal cord (Kerr et al., 1999) and potentiates AMPA-mediated post-synaptic currents (EPSCs), evoked by C fibre stimulation, in lamina II neurons in spinal cord slices (Garraway et al., 2003, 2005).

N-methyl-D-aspartate (NMDA) receptors represent a major target of BDNF-mediated synaptic modulation: intracellular block of these receptors prevents the increase of AMPA EPSCs in lamina II and the ventral root response evoked by NMDA is facilitated by the NT (Garraway et al., 2003; Kerr et al., 1999).

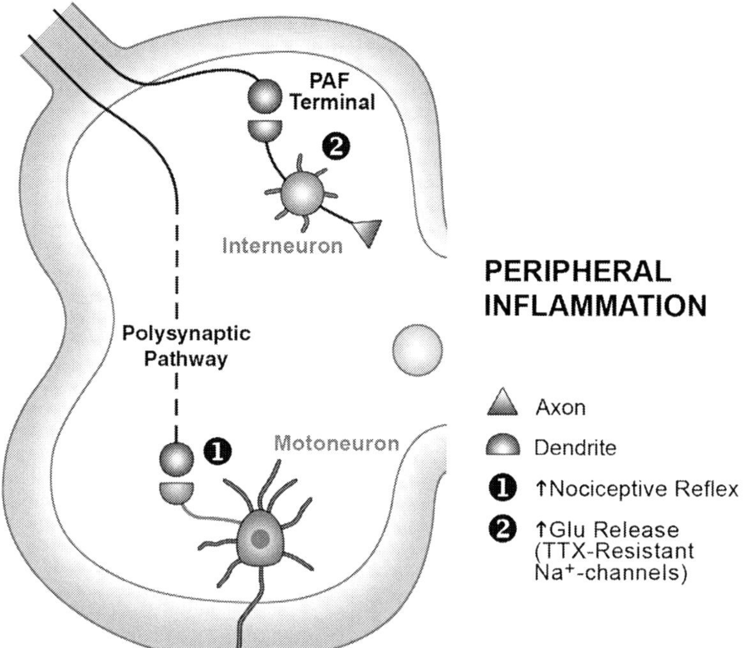

Fig. 5.4 Schematic representation of the modulation of BDNF at synapses in the spinal cord under peripheral inflammation. 1: modulation of the nociceptive reflex through a yet unidentified polysynaptic pathway that links the NT released from PAF terminals to ventral horn motorneurons. The reflex is exacerbated as a consequence of an increased BDNF release; 2: activation of post-synaptic trkB and lidocaine-sensitive TTX-resistant, Na+ channels expressed on nociceptors. These channels undergo an up-regulation in inflammation

The intracellular mechanisms by which BDNF potentiates NMDA receptors have been thoroughly investigated. As shown in several CNS areas, including the spinal cord, binding of the NT with trkB causes the autophosphorylation of the receptors and activates several intracellular pathways (reviewed by Carvalho et al., 2008). One of the major consequences of trkB activation in spinal cord dorsal horn is the phosphorylation of NMDA receptors and their subsequent modulation. Superfusion of isolated hemisected spinal cord in BDNF causes a significant increase of phosphorylation in the NMDA receptor subunit NR1, observed in lamina I, II and V of the dorsal horn (Slack and Thompson 2002). Slack et al. (2004) report that electrical stimulation of roots attached to isolated dorsal horn induces a concomitant release of BDNF and an enhanced NR1 phosphorylation. They also demonstrate that signalling mechanisms involved in BDNF-induced phosphorylation are the two intracellular pathways mitogen-activated protein kinase (MAPK)/extracellular signal-regulated kinase (ERK) kinase (MEK) and phospholipase C (PLC)/protein kinase C (PKC). Consistently with these results, Pezet et al. (2002a, c) report that incubation of spinal cord slices with BDNF or intrathecal administration of the NT induce the increase of ERK phosphorylation in superficial dorsal horn. Activation of ERK by BDNF has been particularly investigated in STT neurons (Slack et al., 2005). The PLC/PKC pathway is involved in the facilitation of excitatory transmission shown by Garraway et al. (2003), since PLC and PKC inhibitors abolish the potentiation of AMPA EPSCs induced by BDNF in lamina II neurons. Furthermore, PKC activation contributes to long-lasting thermal hyperalgesia and tactile allodynia induced in mice by intrathecal administration of BDNF (Yajima et al., 2005).

Apart from NR1 subunit, other phosphorylation sites exist on NMDA receptors. Incubation with BDNF of post-synaptic densities (PSDs) purified from rat spinal cord induces the phosphorylation of NR2A and NR2B subunits and this effect is dependent on the presence of functional trkB on PSDs (Di Luca et al., 2001).

As described in several CNS areas, and particularly in hippocampus, BDNF regulates the activity of NMDA receptors in numerous ways (for a review see Carvalho et al., 2008). The NT may, for example, promote the delivery of NMDA receptors to the plasma membrane or modulate, through receptor phosphorylation, the biophysical properties of the channel, by increasing the open probability. In addition to these "short term effects" BDNF upregulates the expression of NMDA receptor subunits by transcription-dependent mechanisms.

Long-term effects of BDNF on gene transcription have been also observed in spinal cord: the intrathecal administration of the NT increases the expression of the early gene products c-fos, c-jun and Krox-24 in adult rat, particularly in superficial dorsal horn (Kerr et al., 1999; Jongen et al., 2005). BDNF-induced signalling activates the transcription factor CREB (cyclic AMP response element binding protein). Local application of BDNF onto dorsal horn in vivo induces the phosphorylation of CREB and administration of k-252a (a general antagonist of kinase activity) reduces CREB phosphorylation observed after loose ligation of the sciatic nerve (Miletic et al., 2004).

Besides the post-synaptic facilitation of glutamatergic transmission by BDNF, a pre-synaptic modulation of glutamate release has also been described. Application of exogenous BDNF causes a rapid increase of frequency of miniature EPSCs in lamina II neurons from neonatal rats (Merighi et al., 2008b). In the same study, incubation of spinal cord slices with BDNF for two hours increases the release of glutamate and peptides from PAFs, evoking long-lasting calcium oscillations in lamina II neurons. Glutamate depletion of nociceptive fibers by capsaicin abolishes BDNF-induced oscillations. These data suggest that BDNF increases the release of neurotransmitters by acting on pre-synaptic trkB receptors, expressed on sensory fibre terminals, in keeping with the ultrastructural studies showing the presence of the receptors on peptidergic PAF terminals (Salio et al., 2005).

Modulatory effects of BDNF on inhibitory synaptic transmission, mediated by GABA and glycine, have been studied in different spinal cord preparations.

Studies performed on the isolated dorsal horn have shown that the NT can regulate GABA release and, indirectly, modulate primary sensory neuron synaptic efficacy (Meyer-Tuve et al., 2001; Pezet et al., 2002c). In particular, application of exogenous BDNF to the isolated dorsal horn rapidly and reversibly facilitates the potassium-stimulated release of GABA. This, in turn, inhibits substance P evoked release from primary sensory neurons, by activating pre-synaptic $GABA_B$ receptors.

Evidence suggests that BDNF acts pre-synaptically to modulate GABA and glycine release from superficial dorsal horn interneurons. Application of BDNF to spinal cord slices from neonatal rats induces an increase of spontaneous GABA and glycine release, while the evoked release of the two neurotransmitters is depressed (Bardoni et al., 2007). The mechanisms responsible for these effects, still not investigated, are probably quite complex and may be related to different patterns of BDNF release and activity in dorsal horn. However, the presence of fl-trkB receptors at the dendritic membranes of the inhibitory interneurons containing GABA immunoreactive vesicles, described in the same study, gives further support to the role of BDNF as pre-synaptic modulator of GABA and glycine release. These GABAergic interneurons expressing trkB are islet cells, one of the two main neuronal types in lamina II. They are engaged in different type of glomeruli with the exception of type IIb (Merighi et al., 2008a). It seems therefore reasonable that the modulation of inhibitory neurotransmission is not directly related to a pre-synaptic release of endogenous BDNF from PAF terminals, since, as mentioned, fl-trkB+BDNF immunoreactive terminals form the core of type IIb glomeruli (Merighi et al., 2008a).

5.6.2 Inflammatory Pain

Peripheral inflammation causes up-regulation of BDNF and trkB receptor expression in superficial dorsal horn. Behavioural experiments have demonstrated that BDNF plays a significant role in mediating central sensitization to inflammatory pain. Inactivation of BDNF or trkB receptors by using different techniques decrease thermal and tactile hyperalgesia produced by the subcutaneous injection of different inflammatory agents (reviewed in Merighi et al., 2008a). Deletion of BDNF gene in Nav1.8-positive, mainly nociceptive, neurons, determines the inhibition of carrageenan- and NGF-induced hyperalgesia and attenuation of formalin-induced pain behaviour in the second phase (Zhao et al., 2006).

In conditions of peripheral inflammation, BDNF contributes to the increase of dorsal horn excitability by facilitating excitatory glutamatergic transmission (Fig. 5.4). After treatment with NGF, a procedure that mimics peripheral inflammatory states and raises BDNF levels in dorsal horn, the nociceptive reflex is potentiated. Sequestration of BDNF significantly reduces the increased central excitability induced by the NT (Kerr et al., 1999). Matayoshi et al.

(2005) have investigated BDNF pre-synaptic action on an inflammatory model. They have found that BDNF rapidly increases the release of glutamate in lamina II through activation of receptor tyrosine kinase and lidocaine-sensitive, but TTX-resistant, Na^+ channels. These channels are expressed on nociceptors and are up-regulated in pathological states. They also report that the NT induces an increase of the monosynaptic Aβ inputs to lamina II, thereby contributing to the development of hyperalgesia and/or allodynia during inflammation.

An additional mechanism involved in the increase of neuronal excitability induced by peripheral inflammation has been recently proposed by Zhang et al. (2008). Subcutaneous injection of CFA causes the down-regulation of the potassium-chloride co-transporter 2 (KCC2); the trk receptor antagonist k-252a inhibits this effect, suggesting that BDNF is involved. Decrease of KCC2 synthesis produces a shift of chloride reversal potential toward positive potentials and, as a consequence, GABA effects become depolarizing, rather than hyperpolarizing. This mechanism has been originally described for neuropathic pain (Coull et al., 2005, see below).

5.6.3 Neuropathic Pain

The role of BDNF in mediating neuropathic pain is still under debate. As mentioned, BDNF expression in the spinal cord is highly regulated in nerve injury models, nevertheless behavioural experiments have provided controversial results. In the spinal nerve ligation model, protracted BDNF sequestering by a TrkB/Fc chimera protein completely suppresses the induction of allodynia and thermal hyperalgesia (Yajima et al., 2005). However, Lever and colleagues (2003) describe a short-lasting anti-nociceptive effect of intrathecal BDNF in the same pain model, related to the facilitatory action of the NT on GABA release. In this study it appears that the NT restores the normal levels of GABA release, significantly decreased after the injury.

The already mentioned study by Zhao et al. (2006) reports that conditional deletion of BDNF in nociceptors (expressing Nav 1.8 channels) does not alter the expression of pain behaviour induced by spinal nerve ligation, suggesting the involvement of BDNF released from other DRG neuron subpopulations and/or from other cell types (such as support and immune cells). Consistently with this observation, Coull et al. (2005) have demonstrated that intrathecal administration of microglia activated by ATP induces mechanical allodynia by releasing BDNF. The same pain behaviour can be observed after peripheral injury or upon intrathecal administration of the NT. BDNF exogenously applied or endogenously released by microglia, produces an alteration of the anion concentrations (principally chloride) in lamina I neurons, likely through the down-regulation of KCC2. The anion reversal potential is shifted toward values more positive than the neuron resting potential, and the effect of GABA

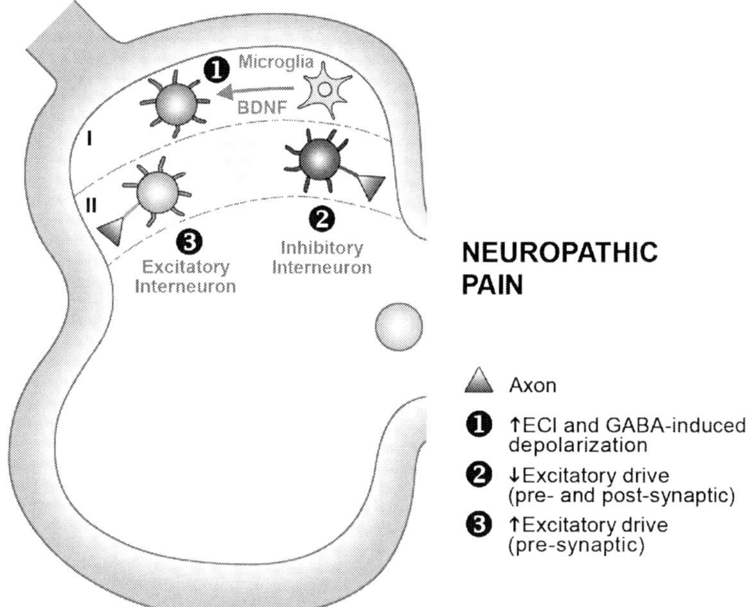

Fig. 5.5 Schematic representation of the modulation of BDNF at synapses in the spinal cord under neuropathic pain. 1: dishinibition of lamina I neurons by BDNF released from microglia; 2 and 3: dishinibition of the dorsal horn network and central sensitization

receptor activation becomes depolarizing, causing the disinhibition of lamina I neurons (Fig. 5.5). Blockade of this microglial-neuronal pathway alleviates chronic neuropathic pain in the rat model.

Decrease of KCC2 protein levels has also been described in rat dorsal horn after ligation of the sciatic nerve, where the effect coincides with a significant ipsilateral increase in BDNF protein (Miletic and Miletic 2008).

Increased pain sensitivity induced by another experimental model, the chronic constriction injury (CCI), is reduced by viral vector-driven expression of BDNF and by grafts of BDNF-expressing cells to the spinal cord, consistently with an antinociceptive role of the NT (Cejas et al., 2000; Eaton et al., 2002). In contrast with these data, a recent study by Lu et al. (2007) shows that long-term exposures to BDNF of dorsal horn neurons in organotypic cultures determines a global increase in dorsal horn network excitability, by producing a pattern of neuron-specific synaptic changes that resembles that observed with CCI in vivo (Balasubramanyan et al., 2006). In particular, BDNF increases the excitatory drive to putative excitatory neurons, by exerting a pre-synaptic modulation, while the excitatory input to putative inhibitory neurons is decreased by the NT, through pre- and post-synaptic actions (Fig. 5.5). These effects would generate a disinhibition of the dorsal horn network, possibly involved in CCI-induced central sensitization.

5.7 Concluding Remarks

The currently available data on the modulatory role of BDNF in the superficial dorsal horns describe several cellular mechanisms by which the NT exerts its function at synapses. However, the synaptic circuitries involved in these effects have not been clearly understood. Furthermore, the interpretation of the role played by BDNF in the different pain models is still very complex. A multidisciplinary approach involving both functional and morphological techniques will be therefore required, in order to fully elucidate the contribution of the NT to the transmission of pain at the spinal cord level and to develop future therapeutic tools.

Acknowledgments The work described in this paper has been funded by the Italian Ministry of Education and Scientific Research (Fondi PRIN), Compagnia di San Paolo, Torino, and Fondazione CRT, Torino. We are greatly indebted to Dr. Chiara Salio for her kind gift of unpublished electron micrographs, and to Mr. Gianfranco Zanutto for his excellent graphic artwork.

References

Adelson DW, Wei JY, Kruger L (1997) Warm-sensitive afferent splanchnic C-fiber units in vitro. J Neurophysiol 77: 2989–3002

Altar CA, Siuciak JA, Wright P et al (1994) In situ hybridization of trkB and trkC receptor mRNA in rat forebrain and association with high-affinity binding of [125I]BDNF, [125I]NT-4/5 and [125I]NT-3. Eur J Neurosci 6: 1389–1405

Altar CA, Cai N, Bliven T et al (1997) Anterograde transport of brain-derived neurotrophic factor and its role in the brain. Nature 389: 856–860

Apfel SC, Wright DE, Wiideman AM et al (1996) Nerve growth factor regulates the expression of brain-derived neurotrophic factor mRNA in the peripheral nervous system. Mol Cell Neurosci 7: 134–142

Armanini MP, McMahon SB, Sutherland J et al (1995) Truncated and catalytic isoforms of trkB are co-expressed in neurons of rat and mouse CNS. Eur J Neurosci 7:1403–1409

Balasubramanyan S, Stemkowski PL, Stebbing MJ et al (2006) Sciatic chronic constriction injury produces cell-type-specific changes in the electrophysiological properties of rat substantia gelatinosa neurons. J Neurophysiol 96: 579–590

Balkowiec A, Katz DM (2000) Activity-dependent release of endogenous brain-derived neurotrophic factor from primary sensory neurons detected by ELISA in situ. J Neurosci 20: 7417–7423

Bardoni R, Ghirri A, Salio C et al (2007) BDNF-mediated modulation of GABA and glycine release in dorsal horn lamina II from postnatal rats. Dev Neurobiol 67: 960–975

Barrett GL (2000) The p75 neurotrophin receptor and neuronal apoptosis. Prog Neurobiol 61: 205–229

Beck KD, Lamballe F, Klein R, et al (1993) Induction of noncatalytic TrkB neurotrophin receptors during axonal sprouting in the adult hippocampus. J Neurosci 13:4001–4014

Brigadski T, Hartmann M, Lessmann V (2005) Differential vesicular targeting and time course of synaptic secretion of the mammalian neurotrophins. J Neurosci 25: 7601–7614

Carvalho AL, Caldeira MV, Santos SD et al (2008) Role of the brain-derived neurotrophic factor at glutamatergic synapses. Br J Pharmacol 153 Suppl 1: S310–S324

Cejas PJ, Martinez M, Karmally S et al (2000) Lumbar transplant of neurons genetically modified to secrete brain-derived neurotrophic factor attenuates allodynia and hyperalgesia after sciatic nerve constriction. Pain 86: 195–210

Chien CC, Fu WM, Huang HI et al (2007) Expression of neurotrophic factors in neonatal rats after peripheral inflammation. J Pain 8: 161–167

Cho HJ, Kim JK, Zhou XF et al (1997a) Increased brain-derived neurotrophic factor immunoreactivity in rat dorsal root ganglia and spinal cord following peripheral inflammation. Brain Res 764: 269–272

Cho HJ, Kim SY, Park MJ et al (1997b) Expression of mRNA for brain-derived neurotrophic factor in the dorsal root ganglion following peripheral inflammation. Brain Res 749: 358–362

Coull JA, Beggs S, Boudreau D et al (2005) BDNF from microglia causes the shift in neuronal anion gradient underlying neuropathic pain. Nature 438: 1017–1021

Di Luca M, Gardoni F, Finardi A et al (2001) NMDA receptor subunits are phosphorylated by activation of neurotrophin receptors in PSD of rat spinal cord. Neuroreport 12: 1301–1305

Eaton MJ, Blits B, Ruitenberg MJ et al (2002) Amelioration of chronic neuropathic pain after partial nerve injury by adeno-associated viral (AAV) vector-mediated over-expression of BDNF in the rat spinal cord. Gene Ther 9: 1387–1395

Ernfors P, Merlio JP, Persson H (1992) Cells expressing mRNA for Neurotrophins and their receptors during embryonic rat development. Eur J Neurosci 4: 1140–1158

Ernfors P, Rosario CM, Merlio JP et al (1993) Expression of mRNAs for neurotrophin receptors in the dorsal root ganglion and spinal cord during development and following peripheral or central axotomy. Brain Res Mol Brain Res 17: 217–226

Frisén J, Verge VM, Fried K et al (1993) Characterization of glial trkB receptors: differential response to injury in the central and peripheral nervous systems. Proc Natl Acad Sci USA 90: 4971–4975

Fukuoka T, Kondo E, Dai Y et al (2001) Brain-derived neurotrophic factor increases in the uninjured dorsal root ganglion neurons in selective spinal nerve ligation model. J Neurosci 21: 4891–4900

Garraway SM, Petruska JC, Mendell LM (2003) BDNF sensitizes the response of lamina II neurons to high threshold primary afferent inputs. Eur J Neurosci 18: 2467–2476

Garraway SM, Anderson AJ, Mendell LM (2005) BDNF-induced facilitation of afferent-evoked responses in lamina II neurons is reduced after neonatal spinal cord contusion injury. J Neurophysiol 94: 1798–1804

Ha SO, Kim JK, Hong HS et al (2001) Expression of brain-derived neurotrophic factor in rat dorsal root ganglia, spinal cord and gracile nuclei in experimental models of neuropathic pain. Neuroscience 107: 301–309

Heppenstall PA, Lewin GR (2001) BDNF but not NT-4 is required for normal flexion reflex plasticity and function. Proc Natl Acad Sci USA 98: 8107–8112

Jongen JL, Haasdijk ED, Sabel-Goedknegt H et al (2005) Intrathecal injection of GDNF and BDNF induces immediate early gene expression in rat spinal dorsal horn. Exp Neurol 194: 255–266

Kerr BJ, Bradbury EJ, Bennett DL et al (1999) Brain-derived neurotrophic factor modulates nociceptive sensory inputs and NMDA-evoked responses in the rat spinal cord. J Neurosci 19: 5138–5148

Klein R, Martin-Zanca D, Barbacid M et al (1990a) Expression of the tyrosine kinase receptor gene trkB is confined to the murine embryonic and adult nervous system. Development 109: 845–850

Klein R, Conway D, Parada LF et al (1990b) The trkB tyrosine protein kinase gene codes for a second neurogenic receptor that lacks the catalytic kinase domain. Cell 61: 647–656.

Koltzenburg M, Bennett DL, Shelton DL et al (1999) Neutralization of endogenous NGF prevents the sensitization of nociceptors supplying inflamed skin. Eur J Neurosci 11: 1698–1704

Lee SL, Kim JK, Kim DS et al (1999) Expression of mRNAs encoding full-length and truncated TrkB receptors in rat dorsal root ganglia and spinal cord following peripheral inflammation. Neuroreport 10: 2847–2851

Lessmann V, Gottmann K, Malcangio M (2003) Neurotrophin secretion: current facts and future prospects. Prog Neurobiol 69: 341–374

Lever IJ, Bradbury EJ, Cunningham JR et al (2001) Brain-derived neurotrophic factor is released in the dorsal horn by distinctive patterns of afferent fiber stimulation. J Neurosci 21: 4469–4477

Lever I, Cunningham J, Grist J et al (2003) Release of BDNF and GABA in the dorsal horn of neuropathic rats. Eur J Neurosci 18: 1169–1174

Lu VB, Ballanyi K, Colmers WF et al (2007) Neuron type-specific effects of brain-derived neurotrophic factor in rat superficial dorsal horn and their relevance to 'central sensitization'. J Physiol 584: 543–563

Malcangio M, Lessmann V (2003) A common thread for pain and memory synapses? Brain-derived neurotrophic factor and trkB receptors. Trends Pharmacol Sci 24: 116–121

Malenka RC, Nicoll RA (1999) Long-term potentiation - a decade of progress? Science 285: 1870–1874

Mannion RJ, Costigan M, Decosterd I et al (1999) Neurotrophins: peripherally and centrally acting modulators of tactile stimulus-induced inflammatory pain hypersensitivity. Proc Natl Acad Sci USA 96: 9385–9390

Mansvelder HD, Kits KS (2000) Calcium channels and the release of large dense core vesicles from neuroendocrine cells: spatial organization and functional coupling. Prog Neurobiol 62: 427–441

Matayoshi S, Jiang N, Katafuchi T et al (2005) Actions of brain-derived neurotrophic factor on spinal nociceptive transmission during inflammation in the rat. J Physiol 569: 685–695

McMahon SB, Armanini MP, Ling LH et al (1994) Expression and coexpression of Trk receptors in subpopulations of adult primary sensory neurons projecting to identified peripheral targets. Neuron 12: 1161–1171

McMahon SB, Priestley JV (1995) Peripheral neuropathies and neurotrophic factors: animal models and clinical perspectives. Curr Opin Neurobiol 5: 616–624

Mellstrom B, Torres B, Link WA et al (2004) The BDNF gene: exemplifying complexity in Ca2+ -dependent gene expression. Crit Rev Neurobiol 16: 43–49

Merighi A, Carmignoto G, Gobbo S et al (2004) Neurotrophins in spinal cord nociceptive pathways. Prog Brain Res 146: 291–321

Merighi A, Salio C, Ghirri A et al (2008a) BDNF as a pain modulator. Prog Neurobiol 85: 297–317

Merighi A, Bardoni R, Salio C et al (2008b) Presynaptic functional trkB receptors mediate the release of excitatory neurotransmitters from primary afferent terminals in lamina II (substantia gelatinosa) of postnatal rat spinal cord. Dev Neurobiol 68: 457–475

Meyer-Tuve A, Malcangio M, Ebersberger A et al (2001) Effect of brain-derived neurotrophic factor on the release of substance P from rat spinal cord. Neuroreport 12: 21–24

Michael GJ, Averill S, Nitkunan A et al (1997) Nerve growth factor treatment increases brain-derived neurotrophic factor selectively in TrkA-expressing dorsal root ganglion cells and in their central terminations within the spinal cord. J Neurosci 17: 8476–8490

Michael GJ, Averill S, Shortland PJ et al (1999) Axotomy results in major changes in BDNF expression by dorsal root ganglion cells: BDNF expression in large trkB and trkC cells, in pericellular baskets, and in projections to deep dorsal horn and dorsal column nuclei. Eur J Neurosci 11: 3539–3551

Middlemas DS, Lindberg RA, Hunter T (1991) trkB, a neural receptor protein-tyrosine kinase: evidence for a full-length and two truncated receptors. Mol Cell Biol 11: 143–153

Miletic G, Hanson EN, Miletic V (2004) Brain-derived neurotrophic factor-elicited or sciatic ligation-associated phosphorylation of cyclic AMP response element binding protein in the rat spinal dorsal horn is reduced by block of tyrosine kinase receptors. Neurosci Lett 361: 269–271

Miletic G, Miletic V (2008) Loose ligation of the sciatic nerve is associated with TrkB receptor-dependent decreases in KCC2 protein levels in the ipsilateral spinal dorsal horn. Pain 137: 532–539.

Obata K, Noguchi K (2006) BDNF in sensory neurons and chronic pain. Neurosci Res 55: 1–10

Obata K, Yamanaka H, Kobayashi K, Day Y, Mizushima T, Katsura H, Fukuoka T, Tokunaga A, Noguchi K (2006) The effect of site and type of nerve injury on the expression of brain-derived neurotrophic factor in the dorsal root ganglion and on neuropathic behavior. Neurosci 37: 961–970.

Onda A, Murata Y, Rydevik B et al (2003) Immunoreactivity of brain-derived neurotrophic factor in rat dorsal root ganglion and spinal cord dorsal horn following exposure to herniated nucleus pulposus. Neurosci Lett 352: 49–52

Onda A, Murata Y, Rydevik B et al (2004) Infliximab attenuates immunoreactivity of brain-derived neurotrophic factor in a rat model of herniated nucleus pulposus. Spine 29: 1857–1861

Pezet S, Malcangio M, Lever IJ et al (2002a) Noxious stimulation induces Trk receptor and downstream ERK phosphorylation in spinal dorsal horn. Mol Cell Neurosci 21: 684–695

Pezet S, Malcangio M, McMahon SB (2002b) BDNF: a neuromodulator in nociceptive pathways? Brain Res Brain Res Rev 40: 240–249

Pezet S, Cunningham J, Patel J et al (2002c) BDNF modulates sensory neuron synaptic activity by a facilitation of GABA transmission in the dorsal horn. Mol Cell Neurosci 21: 51–62

Pezet S, McMahon SB (2006) Neurotrophins: mediators and modulators of pain. Annu Rev Neurosci 29: 507–538

Poo MM (2001) Neurotrophins as synaptic modulators. Nat Rev Neurosci 2: 24–32

Salio C, Lossi L, Ferrini F et al (2005) Ultrastructural evidence for a pre- and postsynaptic localization of full-length trkB receptors in substantia gelatinosa (lamina II) of rat and mouse spinal cord. Eur J Neurosci 22: 1951–1966

Salio C, Averill S, Priestley JV et al (2007) Costorage of BDNF and neuropeptides within individual dense-core vesicles in central and peripheral neurons. Dev Neurobiol 67: 326–338

Schuman EM (1999) Neurotrophin regulation of synaptic transmission. Curr Opin Neurobiol 9: 105–109

Schinder AF, Poo MM (2000) The neurotrophin hypothesis for synaptic plasticity. Trends Neurosci 23: 639–645

Slack SE, Thompson SW (2002) Brain-derived neurotrophic factor induces NMDA receptor 1 phosphorylation in rat spinal cord. Neuroreport 13: 1967–1970

Slack SE, Pezet S, McMahon SB et al (2004) Brain-derived neurotrophic factor induces NMDA receptor subunit one phosphorylation via ERK and PKC in the rat spinal cord. Eur J Neurosci 20: 1769–1778

Slack SE, Grist J, Mac Q et al (2005) TrkB expression and phospho-ERK activation by brain-derived neurotrophic factor in rat spinothalamic tract neurons. J Comp Neurol 489: 59-68

Von Bartheld CS (2004) Axonal transport and neuronal transcytosis of trophic factors, tracers, and pathogens. J Neurobiol 58: 295–314

Walker SM, Mitchell VA, White DM et al (2001) Release of immunoreactive brain-derived neurotrophic factor in the spinal cord of the rat following sciatic nerve transection. Brain Res 899: 240–247

Wright DE, Snider WD (1995) Neurotrophin receptor mRNA expression defines distinct populations of neurons in rat dorsal root ganglia. J Comp Neurol 351: 329–338

Yajima Y, Narita M, Narita M et al (2002) Involvement of a spinal brain-derived neurotrophic factor/full-length TrkB pathway in the development of nerve injury-induced thermal hyperalgesia in mice. Brain Res 958: 338–346

Yajima Y, Narita M, Usui A et al (2005) Direct evidence for the involvement of brain-derived neurotrophic factor in the development of a neuropathic pain-like state in mice. J Neurochem 93: 584–594

Yan Q, Radeke MJ, Matheson CR et al (1997) Immunocytochemical localization of TrkB in the central nervous system of the adult rat. J Comp Neurol 378:135–157.

Zhang W, Liu LY, Xu TL (2008) Reduced potassium-chloride co-transporter expression in spinal cord dorsal horn neurons contributes to inflammatory pain hypersensitivity in rats. Neuroscience 152: 502–510

Zhao J, Seereeram A, Nassar MA et al (2006) Nociceptor-derived brain-derived neurotrophic factor regulates acute and inflammatory but not neuropathic pain. Mol Cell Neurosci 31: 539–548

Zhou XF, Parada LF, Soppet D et al (1993) Distribution of trkB tyrosine kinase immunoreactivity in the rat central nervous system. Brain Res 622: 63–70

Zhou XF, Chie ET, Deng YS et al (1999) Injured primary sensory neurons switch phenotype for brain-derived neurotrophic factor in the rat. Neuroscience 92: 841–853

Zhou XF, Song XY, Zhong JH et al (2004) Distribution and localization of pro-brain-derived neurotrophic factor-like immunoreactivity in the peripheral and central nervous system of the adult rat. J Neurochem 91: 704–715

Chapter 6
Dorsal Horn Substance P and NK1 Receptors: Study of a Model System in Spinal Nociceptive Processing

Xiao-Ying Hua and Tony L. Yaksh

Abstract Small unmyelinated fibers are the primary afferents by which input initiated by high intensity thermal and mechanical stimuli or by chemical products generated secondary to tissue injury, is communicated to the spinal cord. Events which increase small afferent terminal excitability and neurotransmitter release enhance the nociceptive message, while events which diminish small afferent terminal excitability/release diminish the magnitude of the post-synaptic depolarization and attenuate the pain message. The regulation of the activity of small afferent terminals which contain and release SP at the spinal level can be reasonably interpreted as representing effects (direct or indirect) on this family of peptidergic C-fiber terminals. Thus, SP release provides a well-defined model system for characterizing modulatory components involved in the *in vivo* regulation of pain behavior at this critical first order link.

Abbreviations

AMPA	α-amino-3-hydroxy-5-methyl-4-isoxazole propionic acid
BDNF	brain-derived neurotrophic factor
CGRP	calcitonin gene-related peptide
COX	cyclooxygenase
GPCR	G protein-coupled receptors
LDCV	large dense core vesicles
MAPK	mitogen-activated protein kinase
NGF	nerve growth factor
NK1	neurokinin 1
NKA	neurokinin A
NKB	neurokinin B
NMDA	N-methyl-D-aspartate

X.-Y. Hua (✉)
Department of Anesthesiology, University of California San Diego, La Jolla,
CA 92093-0818, USA
e-mail: xyhua@ucsd.edu

M. Malcangio (ed.), *Synaptic Plasticity in Pain*,
DOI 10.1007/978-1-4419-0226-9_6, © Springer Science+Business Media, LLC 2009

NPY   neuropeptide Y
PGE$_2$ prostaglandin E$_2$
PLA$_2$ phosphlipase A$_2$
PLC   phospholipase C
PKA   protein kinase A
PKC   protein kinase C
PPT-A   preprotachykinin A
PPT-B   preprotachykinin B
ROS   reactive oxygen species
SP   substance P
TRP   transient receptor potential receptor
TRPV1   transient receptor potential vanilloid 1 receptors

6.1 Introduction

Small unmyelinated fibers are the primary afferents by which input initiated by
high intensity thermal and mechanical stimuli or by chemical products generated secondary to tissue injury, is communicated to the spinal cord. This
functional association with high intensity input, together with the attenuation
of pain sensation produced by differential block of small afferent transmission,
provides the historical focus for considering the role of these sensory afferents in
nociception. An important element in this linkage is the release of excitatory
neurotransmitters from the small afferent terminals. Detailed analysis of the
small afferent system resulted in the identification of several candidate neurotransmitters including excitatory amino acids, such as glutamate and aspartate
(Battaglia and Rustioni, 1988; De Biasi and Rustioni, 1988; Merighi et al.,
1991), growth factors, such as brain-derived neurotrophic factor (BDNF)
(Zhou and Rush, 1996; Michael et al., 1997), adenosine triphosphate (Fyffe
and Perl, 1984; Salter et al., 1993; Nakatsuka and Gu, 2001) and a variety of
peptides, such as substance P (SP), calcitonin gene-related peptide (CGRP),
vasoactive intestinal polypeptide and cholecystokinin (Hokfelt et al., 1975;
Pernow, 1983; Otsuka and Yoshioka, 1993). These molecules are present in
synaptic vesicles and released upon afferent terminal depolarization. After
release, transmitters can interact with eponymous receptors that typically lead
to depolarization of second-order neurons. Pharmacological interventions in
conjunction with recording from second-order neurons, activated by monosynaptic input from primary afferents, have revealed the importance of glutamate
acting though α-amino-3-hydroxy-5-methyl-4-isoxazole propionic acid
(AMPA) channels as being the critical element for acute excitation initiated
by all populations of primary afferents (large and small). In the face of repetitive
afferent input, however, there is evidence that peptides such as SP may produce
a persistent depolarization of the membrane that enables activation of the

N-methyl-D-aspartate (NMDA) receptor channel, leading to an augmentation of the discharge (Urban et al., 1994).

These studies have generated considerable interest in the factors governing primary afferent terminal excitability. This focus is predicated on the fact that events which increase small afferent terminal excitability and neurotransmitter release will enhance the nociceptive message, while events which diminish small afferent terminal excitability/release will diminish the magnitude of the post-synaptic depolarization and attenuate the pain message. In this chapter, we wish to focus on the regulation of small afferent terminals which contain and release SP at the spinal level. This focus is appropriate for three reasons. First, it is a model system that can be identified and studied in the intact organism. As will be reviewed, extracellular levels of dorsal horn SP derive principally from primary afferent terminals, the majority of which are associated with small high threshold polymodal nociceptors. Accordingly manipulations that alter the dorsal horn SP release can be reasonably interpreted as representing effects (direct or indirect) on this family of peptidergic C-fiber terminals. In contrast, changes in the levels of dorsal horn glutamate represent release from virtually any primary afferent, from interneurons, or even alteration of non-neuronal cell uptake processes. Second, a role of SP-containing fibers in nociceptive transmission has been suggested by the fact that such afferents bear transient receptor potential (TRP) vanilloid 1 receptors (TRPV1, i.e. capsaicin receptor, see below) and that degeneration of these axons has robust effects on pain behavior (Holzer, 1991). Third, SP acts post-synaptically upon neurokinin 1 (NK1) receptors and the behavioral effects resulting from the destruction of such NK1-bearing cells with intrathecal neurotoxins has linked the functional contribution of NK1 positive neurons to the events associated with the activation of these TRPV1 afferents. Thus, SP release provides a well-defined model system for characterizing modulatory components involved in the *in vivo* regulation of pain behavior at this critical first order link.

6.2 Substance P and Its Receptor in the Dorsal Horn

6.2.1 *Substance P Synthesis*

Substance P was first described by von Euler and Gaddum (1931) (von Euler and Gaddum, 1931) and identified by Susan Leeman's group in the early 1970s as an undecapeptide (Chang et al., 1971). SP belongs to the tachykinin family, which represents a group of biologically active peptides with a similar sequence of amino acids in the C-terminal regions (see (Severini et al., 2002; Pennefather et al., 2004) for review). SP, together with two well-known tachykinins, neuro-kinin A (NKA) and neurokinin B (NKB) (Kangawa et al., 1983), are encoded by two genes, preprotachykinin A (PPT-A) and preprotachykinin B (PPT-B) (Nawa et al., 1983; Kotani et al., 1986; Carter and Krause, 1990). Through alternative splicing, four individual forms of mRNA (α, β, γ and δ) are expressed by PPT-A. All of them encode SP, whilst the β and γ forms encode both SP and NKA (Nawa et al., 1983; Carter and Krause, 1990). The PPT-B gene expresses only NKB (Kotani et al., 1986). Translation of the mature PPT-A/B mRNA generates pre-protachykinins, which pass through the endoplasmic reticulum where they are processed into protachykinin. Final active peptide is formed and packed in the Golgi apparatus (Krause et al., 1987; Holmgren and Jensen, 2001). SP (and NKA) is synthesized in small and medium sized neurons of dorsal root ganglia (DRG), stored in dense core vesicles and distributed to both spinal and peripheral nerve terminals by fast axonal transport (Brimijoin et al., 1980; Yamasaki et al., 1984). In contrast, NKB is not expressed by sensory neurons but by a population of excitatory neurons in superficial dorsal horn (Too and Maggio, 1991; McLeod et al., 2000; Polgar et al., 2006).

6.2.2 Origin of SP in Dorsal Horn

Pioneering studies carried independently by Bengt Pernow and Fred Lembeck in 1950s revealed that SP in spinal cord is concentrated in dorsal, but not ventral horn (Lembeck, 1953; Pernow, 1953). This, together with expression of SP in dorsal root ganglia, led to the early hypothesis that SP may be one of neurotransmitters in primary sensory neurons for central transmission of afferent information (Lembeck, 1953; Pernow, 1953). The anatomical distribution of SP in spinal cord has been extensively studied in various species including rat, cat and monkey. In the trigeminal and spinal dorsal horn, SP is largely present in nerve terminals of unmyelinated C-fibers and to a much lesser degree in finely myelinated Aδ fibers (Hokfelt et al., 1975; Tuchscherer and Seybold, 1985; McCarthy and Lawson, 1989). The highest density of SP fibers is found in superficial layers of dorsal horn lamina I and outer lamina II, and innervation decreases substantially in the deeper laminae (III–V), the lateral spinal nucleus and around the central canal (Ribeiro-da-Silva and Hokfelt, 2000; Todd, 2002). SP positive terminals in laminae I/II form axo-dendritic synapses with dorsal horn neurons, and specifically with a considerable number of NK1 receptor bearing neurons which project to various parts of the brain such as the caudal ventrolateral medulla, parabrachial area, periaqueductal grey and thalamus (Todd, 2002). SP co-exists with glutamate (Battaglia and Rustioni, 1988; De Biasi and Rustioni, 1988; Merighi et al., 1991) in addition to peptides like NKA and CGRP (Gibson et al., 1984; Hua et al., 1985; Skofitsch and Jacobowitz, 1985). As mentioned above, this group of C-fibers also expresses TRPV1 receptor and TrkA receptor for nerve growth factor (NGF) (Hunt and Mantyh, 2001). Note there is another group of TRPV1 bearing C-fibers, which is non-peptidergic and express surface proteins such as IB-4 lectin-binding protein and receptors for glial-cell-derived neurotrophic factor (Hunt and Mantyh, 2001).

In addition to sensory afferent axons, a limited pool of SP is present in terminals of some bulbospinal projection neurons containing 5-HT, that arise from the caudal raphe and travel in the dorsolateral funiculus. Although many of these SP-containing descending fibers terminate in ventral horn, some may innervate the superficial laminae of the dorsal horn (Hokfelt et al., 2000). Finally, there is a small population of SP interneurons; their cell bodies are located mainly in superficial dorsal horn and many co-express enkephalin (Ribeiro-da-Silva and Hokfelt, 2000).

6.2.3 Tachykinin Receptors

Tachykinins (e.g. SP, NKA and NKB) target specific membrane receptors belonging to a family of G protein-coupled receptors (GPCR). Three distinct tachykinin receptors, NK1, NK2 and NK3, have been cloned in various species including human (Ohkubo and Nakanishi, 1991). Early binding and pharmacological studies indicated that SP, NKA and NKB are endogenous ligands for NK1, NK2 and NK3 receptors, respectively. The rank order of potency of the three tachykinins for these receptors is: $SP \geq NKA > NKB$ on NK1 receptor, $NKA > NKB > SP$ on NK2 receptor, and $NKB > NKA > SP$ on NK3 receptor (Ohkubo and Nakanishi, 1991; Regoli et al., 1994; Maggi, 1995).

After ligand binding many GPCR undergo rapid phosphorylation, endosomal internalization, dissociation from ligand in endosome, dephosphorylation, and finally receptor recycling back to the plasma membrane (Caron and Lefkowitz, 1993). In 1995 Mantyh and colleagues (Mantyh et al., 1995b) demonstrated by immunohistochemistry that NK1 receptors normally confined to the cell membrane of dorsal horn neurons underwent dramatic internalization in response to receptor occupancy by an NK1 agonist. Thus, intrathecally delivered tachykinin or high intensity afferent stimulation results in a marked increase in the percentage of cells showing an elevated level of NK1 immunoreactivity in intracellular endosome (Trafton et al., 2001). Afferent-evoked NK1 internalization is rapid and reversible (Mantyh et al., 1995b). The majority of NK1 positive neurons in superficial laminae display endocytosis within 5 minutes after stimulation, and then the number of NK1 positive endosomes return to baseline within 60 min. Such internalization is blocked by NK1 antagonists (Mantyh et al., 1995a). Elegant *ex vivo* spinal slice work has indeed demonstrated that the degree of internalization correlates to the increase in extracellular SP concentrations in superficial laminae following afferent stimulation (Marvizon et al., 2003). Accordingly, spinal NK1 internalization has been often used as a marker of the presence of NK1 ligands in the extracellular space (see below).

Cellular responses to tachykinins largely depend on a receptor-mediated second messenger system. Despite some differences on the interaction with β-arrestins (Schmidlin et al., 2003), all three tachykinins trigger phosphoinositide hydrolysis via a G protein and phospholipase C (PLC) pathway, which mobilizes intracellular calcium (Akasu et al., 1996; Schmidlin et al., 2003), and initiates a number of signal cascades including activation of kinases, like protein kinase A (PKA), protein kinase C (PKC), mitogen-activated protein kinase (MAPK) (Barber and Vasko, 1996; Smith et al., 2000; Fehrenbacher et al., 2003; Svensson et al., 2003), and transcription factors such as, Fos (Trafton et al., 1999). Activation of the NK1 receptor is also able to depolarize the membrane through activation of non-selective cation channels (Ito et al., 2002).

Spinal NK1 receptors are expressed most densely in lamina I and less so in laminae II (Liu et al., 1994a). Dorsally-directed dendrites of neurons in laminae III–IV also bear NK1 receptors, some of which form synaptic contact with SP afferents (Naim et al., 1997; Todd, 2002). NK2 and NK3 receptors have also been reported in spinal dorsal horn (McCarson and Krause, 1994; Seybold et al., 1997; Zerari et al., 1998). NK2 receptors are present in a population of astrocytes in superficial dorsal horn (Zerari et al., 1998), while NK3 receptors are present in interneurons in laminae I–III (McCarson and Krause, 1994; Seybold et al., 1997). Some evidence suggests that NK1 and NK3 may also be expressed on presynaptic terminals of SP containing C-fibers, thereby regulating SP release (NK1: inhibition; NK3 facilitation) (Malcangio and Bowery, 1999; Zaratin et al., 2000).

6.3 Primary Afferent SP Release

The central terminals of nociceptors, via the release of transmitters such as glutamate, neuropeptides, and proteins like BDNF, drive synaptic input to second-order neurons and transmit afferent-evoked excitation (Hunt and Mantyh, 2001). Assessing SP release from these terminals has been recognized as an important read-out for assessing C-fiber terminal excitability. Several approaches have been used to measure the release of SP in spinal cord: (1) measuring SP content in synaptic overflow into spinal superfusates *in vivo* or slices/cells *ex vivo*; (2) measuring immunoreactive SP with an antibody coated microprobe inserted into the dorsal horn to monitor the pattern of extracellular SP in parenchyma; and (3) assessing internalized NK1 receptors in dorsal horn neurons by immunocytochemistry (see reviews, (Pernow, 1983; Otsuka and Yoshioka, 1993; Hokfelt et al., 2001). All these approaches have generated particular insights on the systems regulating spinal afferent SP release.

It has been long established that SP (together with other peptides, e.g. NKA and CGRP) is released in the spinal cord upon stimulation of sensory afferents (Pernow, 1983; Otsuka and Yoshioka, 1993). *Ex vivo* studies on isolated spinal cord, dorsal horn slices, spinal synaptosomes and cultured DRG neurons have demonstrated that SP release can be evoked (1) by depolarization of membrane (increasing potassium concentration) (Otsuka and Konishi, 1976; Gamse et al., 1979; Pang and Vasko, 1986), (2) by high intensity stimulation of small afferent fibers in dorsal roots (Malcangio and Bowery, 1993; Marvizon et al., 1997), and (3) by capsaicin which activates TRPV1 receptors on C-fiber terminals (Jessell and Iversen, 1977; Mauborgne et al., 1987; Huang et al., 2003; Marvizon et al., 2003).

An important property of transmitter release is its calcium dependence. Dememes et al. (2000) reported that there is a close anatomical association between voltage-gated calcium channels, the synaptic vesicles and synaptic membrane-associated proteins on afferent nerve calyces and probably in afferent boutons (Dememes et al., 2000). Consistent with depolarization evoked calcium-mediated vesicular exocytosis, depolarization initiated increases in extracellular concentrations of SP are blocked by omission of calcium from the perfusion media (Vasko et al., 1994; Marvizon et al., 1997). Activation of voltage-gated Ca^{2+} channels is critical for afferent transmitter release. The role of several Ca^{2+} channel subtypes in regulation of spinal SP release is discussed below (see Section 6.4).

In 1980 Yaksh and colleagues demonstrated *in vivo* an increase in SP release from superfused cat spinal cord following stimulation of small (Aδ and C), but not large (Aβ) fibers in sciatic nerve. The effects were observed in acutely spinal-transected animals, excluding a contribution by bulbospinal pathways (Yaksh et al., 1980). This was the first evidence revealing that SP is released from spinal cord when nociceptive afferents are activated. Since then, numerous studies have shown that activation of somatic afferents with noxious heat, mechanical and chemical stimuli, as well as local tissue injury and inflammation are also able to produce reliable increases in the extracellular concentrations of SP in spinal dorsal horn (Go and Yaksh, 1987; Duggan et al., 1988; Linderoth and Brodin, 1988; Aimone and Yaksh, 1989; Kuraishi et al., 1989; Duggan et al., 1990; Neugebauer et al., 1994; Mantyh et al., 1995a, Abbadie et al., 1997; Allen et al., 1997; Honor et al., 1999). Similarly, stimulation of cardiac nociceptive sensory neurons or injection of capsaicin into pancreatic duct (e.g. visceral afferents) also trigger SP (and CGRP) release in spinal cord (Hua et al., 2004; Wick et al., 2006).

6.4 Modulation of SP Release

Release of neurotransmitters from any nerve terminal is typically subject to modulation by a variety of ionotropic and metabotropic (G protein-coupled) receptors. Spinal C-fiber terminals are no exception. Support linking any given receptor protein and small afferent fibers is typically based on several observations: (1) The target protein or respective message can be identified in small type B dorsal root ganglion cells and the protein can be detected in the superficial layers of the dorsal horn. (2) Following root / nerve section or treatment with high doses of capsaicin there is a loss of terminals distal to the axon section or more selectively those bearing TRPV1 receptors, respectively. Thus, this injury results in a loss of protein otherwise contained in the terminal (Gamse et al., 1979; Jhamandas et al., 1984; Holzer, 1991). Depending upon the presence of such receptor sites, the effects of activating these receptors may be associated with an increase in terminal excitability, resulting enhanced release; or a decrease in terminal excitability that will translate into reduced or diminished release. Importantly, as reviewed above, because of the principal localization of SP to the primary afferent terminals, SP release in dorsal horn is considered to be indicative the excitation of small afferent terminals. In the following sections, we discuss the role of several such spinal regulatory receptors in modulating spinal SP release (see Fig. 6.1).

6.4.1 Receptors Increasing SP Release

6.4.1.1 EP Receptors

Antihyperalgesic actions produced by inhibition of spinal cyclooxygenase (COX) indicate that COX products, such as prostaglandin E_2 (PGE_2) play a significant role in spinal nociceptive transmission (Malmberg and Yaksh, 1992; Samad et al., 2001; Svensson and Yaksh, 2002). Several subtypes of EP receptors (the receptors for PGE_2) are expressed on presynaptic terminals of primary sensory afferents in addition to their location on dorsal horn neurons (Southall and Vasko, 2001). In parallel with the observation that PGE_2 augments the firing of sensory neurons in response to noxious stimuli (Martin et al., 1987; Kumazawa et al., 1996), exposing cultured DRG neurons or spinal cord slices to PGE_2 facilitate stimulation-evoked SP release (Hingtgen and Vasko, 1994; Vasko et al., 1994). Apparently EP_3C and EP_4 receptors, which are expressed on sensory neurons, mediate this facilitated release (Southall and Vasko, 2001). This increase in release is believed to result from a receptor-mediated facilitation of the opening of voltage sensitive calcium channels (Nicol et al., 1992; Evans et al., 1996). Importantly, intrathecal treatment with COX_2 inhibitors attenuates injury- and inflammation-induced release of SP as measured by radioimmunoassay in spinal superfusates (Southall et al., 1998) or by assessment of dorsal horn NK1 internalization (Ghilardi et al., 2004).

6.4.1.2 P2X and P2Y

ATP can be released in the dorsal horn from a variety of sources including interneurons, primary afferents and astrocytes (Sawynok et al., 1993; Jo and Schlichter, 1999; Sawynok and Liu, 2003; Werry et al., 2006). In primary DRG cell cultures, this purine potentiates capsaicin-evoked SP release, and the effect is likely via activation of $P2Y_2$ receptors (Huang et al., 2003). It is well documented that ATP, via P2Y receptors, lead to activation of PLC and other signal transduction pathways such as PKC, and phosphlipase A_2 (PLA_2) as well as

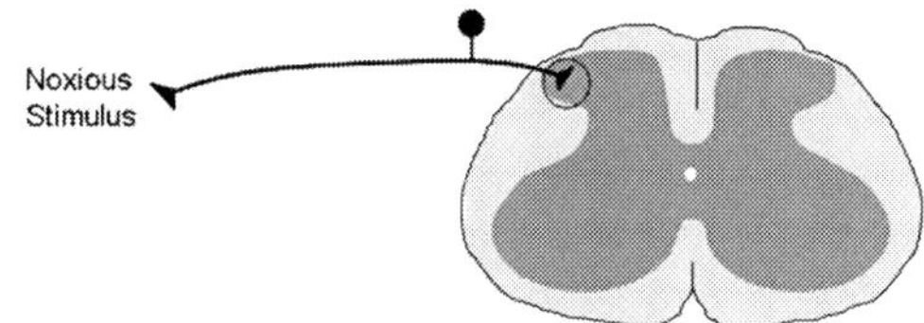

A. Excitatory modulation

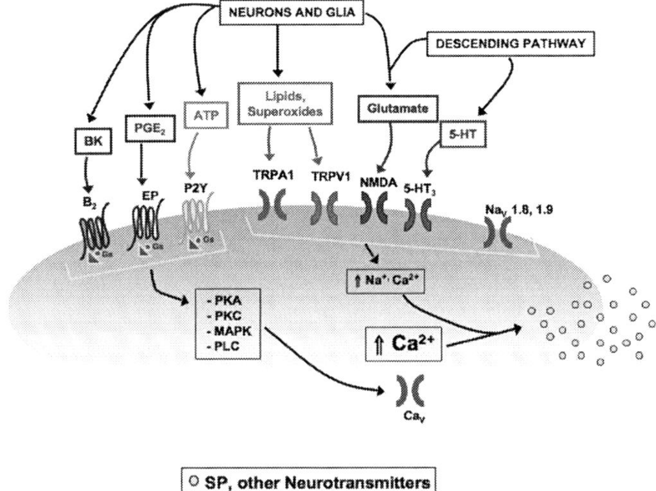

B. Inhibitory modulation

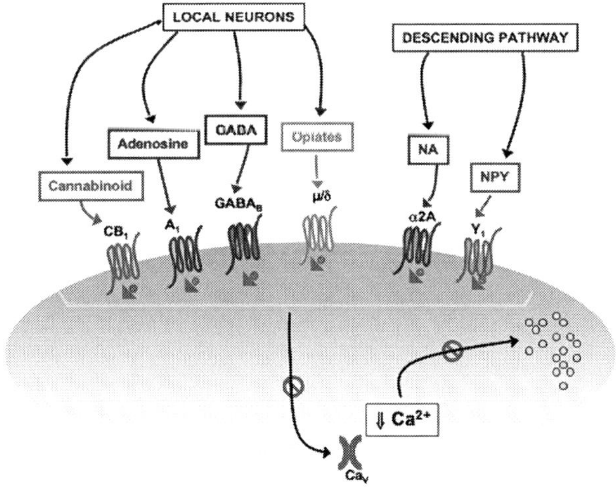

Fig. 6.1 Schematic illustration of the molecules, receptors/channels and related signal events involved in excitatory and inhibitory modulation of afferent terminal release in dorsal horn (details see text)

calcium-sensitive ion channels (Ralevic and Burnstock, 1998). Indeed, $P2Y_2$-mediated facilitation of afferent SP release is blocked by inhibition of PKC (Huang et al., 2003). The role of P2X channels on afferent SP release is controversial (Nakatsuka and Gu, 2001; Huang et al., 2003). While it is known that P2X channels, such as $P2X_{2/3}$, are not expressed on SP containing C-fibers, some data suggest that $P2X_1$ may play a role in this release (Petruska et al., 2000; Nakatsuka et al., 2001).

6.4.1.3 TRPV1/TRPA1

Primary sensory neurons express several members of the TRP family of ion channels including TRPV1 and TRPA1. TRPV1 is a well-characterized non-selective cation channel, activated by heat, lipid mediators and protons. In the DRG, this receptor is expressed uniquely in type B ganglion cells (not all of which contain SP) and on their respective central terminals in the superficial dorsal horn. Activation of TRPV1 on the central terminals by intrathecal capsaicin and its homologues *in vivo* produces a profound dose-dependent increase in extracellular SP concentrations in *in vivo* spinal perfusates (Jhamandas et al., 1984), in *ex vivo* spinal slice perfusates (Marvizon et al., 2003) and in DRG cell cultures (Huang et al., 2003). In the *in vivo* work, the potency of producing release paralleled their *in vivo* irritant potency (Jhamandas et al., 1984). This terminal activation reflects the increase in terminal cation permeability leading to depolarization. Although it has been assumed that blocking TRPV1 on peripheral terminals of afferent C-fibers would be a major target for antinociception, a recent study by Cui et al. (2006) showed that TRPV1 antagonists with better CNS penetrating profiles display higher antihyperalgesic potency than those restricted to the periphery (Cui et al., 2006). This argues strongly for a facilitatory role of endogenous ligands in spinal cord, mediated via presynaptic TRPV1 receptor which augments C-fiber input. This raises the question as to what are endogenous ligands. Some work suggests that fatty acid amides (e.g. anandamide) or lipoxygenase products (e.g. 12- and 15-HPETE) could be candidates (Zygmunt et al., 1999; Smart and Jerman, 2000), but they are only weak agonists to TRPV1. Nevertheless, anandamide increases SP release in DRG cells, possibly via TRPV1 (Tognetto et al., 2001). Although these lipid messengers may act synergistically with other pro-nociceptive agents, such as bradykinin, NGF or protons, to facilitate TRPV1 gating, it is likely that endo-genous ligands with greater potency at TRPV1 may still be found.

The TRPA1 channel is frequently co-expressed with TRPV1 in sensory neurons, and this channel can be activated by a broad array of chemical irritants including pungent extracts from garlic, mustard oil, acrolein as well as by lipid products from reactive oxygen species (ROS) released in response to tissue injury and stress (Barber and Vasko, 1996; Trevisani et al., 2007). Like TRPV1, TRPA1 also acts as a receptor-operated channel for ligands such as bradykinin, activating GPCR coupled to PLC and leading to a form of gating of TRPA1. TRPA1-mediated SP and CGRP release from spinal cord has been demonstrated recently (Trevisani et al., 2007).

6.4.1.4 Bradykinin Receptors (B2)

Resembling its sensitizing effect on peripheral afferent terminals, Bradykinin is also involved in pain transmission at the spinal level (Wang et al., 2005). Bradykinin is released in the spinal cord upon afferent stimulation, and brady-kinin receptor, B2, is expressed by both DRG and dorsal horn neurons (Wang et al., 2005). This peptide facilitates SP release in cultured DRG cells, and the effect is mediated though PKC and PKA (Barber and Vasko, 1996). It should be mentioned that bradykinin also increases calcium influx via the TRPA1 channel (Bautista et al., 2006), and that function serves to regulate transmitter release (Trevisani et al., 2007).

6.4.1.5 NMDA

NMDA receptors are expressed on SP containing primary afferent terminals (Shigemoto et al., 1992; Liu et al., 1994b). As a result, presynaptic NMDA may intrinsically regulate primary afferent transmitter release by permitting direct entry of calcium into presynaptic terminals. However, there are several conflicting reports and this issue is not yet resolved. Direct activation of spinal NMDA receptors has been shown either to increase SP release *in vivo* and *in vitro* (Liu et al., 1997; Marvizon et al., 1997), or to have no effect (Afrah et al., 2001; Nazarian et al., 2008b) as measured by NK1 internalization, or to display a dose-dependent biphasic effects as measured by extracellular levels in spinal slice superfusates (Malcangio et al., 1998). Studies with NMDA antagonists are also controversial (Malcangio et al., 1998; Afrah et al., 2001). Nazarian et al. (2008b) compared AP-5 and morphine at analgesic doses on nociception-induced spinal NK1 internalization *in vivo*, and found that morphine blocks the release effec-tively, while MK801 or AP-5 had at most a modest or no effect upon release. These data suggest that in physiological setting presynaptic NMDA receptors in the spinal cord play only a limited role on afferent release of SP.

6.4.1.6 Voltage-Gated Calcium Channels

Ca^{2+} channels are divided into two groups: high-voltage-activated and low voltage-activated (T-type) channels. High-voltage-activated Ca^{2+} channels include L-, N-, R and P/Q-types. Pharmacological evidence suggests that L-, N- and P/Q- type Ca^{2+} channels are involved in spinal pain signal processing and afferent-evoked transmitter release including SP (Del Bianco et al., 1991; Vedder and Otten, 1991; Evans et al., 1996; Smith et al., 2002). The major Ca^{2+} channel in afferent nociceptors appears to be N-type, i.e. Ca_v 2.2 channel (Bell et al., 2004), which is a key mediator of nociceptive signal transmission (Altier and Zamponi, 2004; Heinke et al., 2004). Ca_v 2.2 RNA undergoes alternative splicing, producing multiple Cav2.2 channels isoforms, and one such splicing event occurs through the alternate use of exons e37a and e37b to produce proximal C termini that differ by 14 amino acids (Bell et al., 2004). Though both e37a and e37b isoforms are present in sensory neurons, e37a-containing $Ca_v2.2$ channels are preferentially expressed by TRPV1-positive DRG neurons (Bell et al., 2004). More intriguingly, e37a and e37b isoforms have different sensitivity to G-protein dependent inhibition (Castiglioni et al., 2006; Raingo et al., 2007). The voltage-dependent GPCR inhibitory pathway requires $G\beta\gamma$, coupled to both e37a and e37b channels, while voltage-independent inhibition by tyrosine phosphorylation is only coupled to the e37a $Ca_v2.2$ channel (Castiglioni et al., 2006; Raingo et al., 2007). In agreement with the preferred location of e37a on TRPV1-positive DRG neurons, Altier et al. (2007) have recently demonstrated that the $Ca_v2.2$. e37a channel isoform contributes more extensively to SP release from capsaicin-sensitive afferents, whereas SP release from capsaicin-insensitive neurons is mediated predominantly by e37b channels (Altier et al., 2007).

The $\alpha2\delta$ subunit of N-type calcium channel has been recently identified as a binding site for gabapentin and pregabalin, drugs that are efficient in treating neuropathic pain (Field et al., 2006). A major effect of gabapentin apparently involves some degree of inhibition of N-type Ca^{2+} currents in DRG neurons (Sutton et al., 2002). In parallel, it has been shown that gabapentin attenuates spinal SP release, though this effect is limited during a chronic state of inflammation (Fehrenbacher et al., 2003).

T- type channels are also important for pain transmission (see review (Altier and Zamponi, 2004)). They are expressed in small and medium sized DRG neurons and in the neurons of superficial dorsal horn (Ryu and Randic, 1990; Talley et al., 1999; Ikeda et al., 2003). T-type channels on lamina I projection neurons selectively regulate NK1 receptor-mediated signal transduction (Ikeda et al., 2003). Although some evidence suggests that T-type channels may participate in fast transmitter release (Carbone et al., 2006), the role of this channel in modulation of afferent SP release is not clear.

6.4.1.7 5-HT$_3$ Receptor

Dorsal horn serotonin derives from terminals of bulbospinal projections. A variety of 5-HT receptors have been identified within the spinal dorsal horn. The 5-HT$_3$ receptors, ligand-gated cation channels, have been identified on small peptidergic primary afferents and are considered to be excitatory. A facilitatory role of 5-HT$_3$ on afferent SP release has been suggested (Saria et al., 1990; Inoue et al., 1997), and the effect is apparently mediated via nitric oxide by an increase in cyclic GMP production (Inoue et al., 1997). However, immunohistochemical findings showed that only small fraction of 5-HT$_3$ receptors are expressed on SP containing C-fibers, while the majority seem to be located on myelinated $A\beta$ fibers (Zeitz et al., 2002). Current work also points to the activation of the bulbospinal pathways by outflow from lamina I projection neurons that contribute to facilitated discharges of lamina V neurons though a population of 5-HT$_3$ receptors (Suzuki et al., 2002; Suzuki et al., 2005). Other serotonin receptors may be involved, as activation of 5-HT$_{2A/2C}$ receptors also enhances TRPV1-mediated SP release in spinal cord (Kjorsvik Bertelsen et al.,2003).

6.4.1.8 TrkA

Growth factors such as NGF are released from a variety of inflammatory cells and schwann cells. TrkA receptor protein has been identified as being co-localized in SP/TRPV1 positive DRG cells (Hunt and Mantyh, 2001). However, acute application of NGF to isolated spinal cord slices did not change either basal or evoked SP release (Malcangio et al., 1997), while chronic treatment with NGF, either IT or systemically, increased both endpoints (Malcangio et al., 1997, 2000). It is known that chronic inflammation increases synthesis of SP in DRG neurons by NGF via action on the trkA receptor (Honore et al., 2000; Pezet et al., 2001; Yang et al., 2007). This suggests that the functional role of trkA receptors on sensory neurons is related more to regulation of gene expression, but not directly to synaptic excitability and neurotransmitter release.

6.4.2 Receptors Reducing SP Release

6.4.2.1 μ/δ Opiate Receptors

These receptors are Gi/o coupled receptors that are know to be negatively coupled to the opening of voltage sensitive calcium channels (Schroeder et al., 1991; Soldo and Moises, 1998) (see also Marvizon, Chapter 7). Immunocytochemistry and receptor autoradiography show that opioid receptors are located on small dorsal root ganglion cells and in laminae I–II of dorsal horn (Fields et al., 1980; Cheng et al., 1996; Abbadie et al., 2002). In addition to their postsynaptic location on dorsal horn neurons, both μ and δ receptors are present presynaptically on C-fiber terminals, and the latter is confirmed by fact of reduction in receptor numbers after rhizotomy or neonatal capsaicin (Gamse et al., 1979; Abbadie et al., 2002). Postsynaptically, opioid receptors decrease neuronal excitability by opening potassium channels (Yoshimura and North, 1983; Grudt and Williams, 1994). Presynaptically, opioids inhibit neurotransmitter release by inactivating voltage-gated calcium channels (Schroeder et al., 1991; Soldo and Moises, 1998). Indeed, spinal application of μ and δ receptor agonists reduces the increase in extracellular concentration of SP and CGRP otherwise evoked by local depolarization *in vitro* (Jessell and Iversen, 1977; Chang et al., 1989; Pohl et al., 1989) or by C-fiber stimulation *in vivo* (Yaksh et al., 1980; Go and Yaksh, 1987; Aimone and Yaksh, 1989). Studies using NK1 internalization as a marker have shown that several intrathecal opiates at behaviorally defined analgesic doses inhibit noxious stimuli-induced SP release from primary afferents in a naloxone reversible fashion (Kondo et al., 2005). Similar results were observed using slices with attached dorsal roots (Kondo et al., 2005). Chronic spinal infusion of opiates leads to tolerance, e.g. continued infusion of morphine blocks evoked SP release on day 1, but has no effect after 5 days of infusion. Interestingly, in the spinal morphine tolerant animals, naloxone produces a pronounced increase in spinal SP release (Gu et al., 2005). These results suggest that the opiate receptors on the afferent terminal display the profile of release associated tolerance (loss of effect) and withdrawal (e.g. increased release with acute receptor blockade).

The δ *receptor* is expressed by primary sensory neurons, however, the receptor is concentrated in large dense core vesicles (LDCV) and is notably absent from normal plasma membrane (Zhang et al., 1998). Upon afferent activation, these LDCV-associated δ receptors are released by exocytosis via interaction with pro-tachykinin SP domain and then inserted into the membrane (Guan et al., 2005). Inhibition of SP release from spinal cord after exposure to δ agonists has been demonstrated *in vivo* and *in vitro* (Yaksh et al., 1980; Go and Yaksh, 1987; Aimone and Yaksh, 1989; Chang et al., 1989; Pohl et al., 1989; Kondo et al., 2005).

6.4.2.2 Adrenergic α_2 Receptors

The family of α_2 receptors is also Gi/o coupled, and can attenuate the opening of voltage sensitive calcium channels. Immunohistochemistry has indicated that in spinal dorsal horn the predominant α_2 subtypes are α_{2A} and α_{2C} (Zeng and Lynch, 1991; Nicholas et al., 1993). Histological evidence indicates that α_{2A} adrenoceptors are mainly located on SP containing C-fiber terminals; although a smaller population may also exists on other sites within laminae I–II (Stone et al., 1998). Previous findings have shown that dexmedetomidine, clonidine and ST-91 reduce the evoked release of SP from *ex vivo* spinal cord preparations (Pang and Vasko, 1986; Ono et al., 1991; Bourgoin et al., 1993; Takano et al., 1993). More recently, in contrast to the previous release studies, IT administration of dexmedetomidine failed to reduce intraplantar formalin-induced NK1 internalization when examined over a wide range of doses (Nazarian et al., 2008a). Importantly, in the same experiments, spinal morphine blocked evoked NK1 internalization. Moreover, dexmedetomidine at the doses employed indeed prevent the stimulus evoked increase in c-Fos expression in dorsal horn. The origin of this discordance between the well-controlled release and internalization studies is not clear.

6.4.2.3 GABA$_{A/B}$

About one third of neurons in laminae I–II are GABAergic (Todd and McKenzie, 1989, 1990), and some of these neurons form synapses with C-fiber terminals (Todd and Lochhead, 1990; Todd, 1996). Both GABA$_A$ and GABA$_B$ receptors are present in primary afferents containing SP (Price et al., 1984; Todd and Lochhead, 1990; Malcangio and Bowery, 1996). Activation of pre-synaptic GABA$_B$receptors produces a G protein-mediated inactivation of N-type Ca^{2+} channels, resulting in reduction of neurotransmitter release (Dolphin and Scott, 1987; Menon-Johansson et al., 1993). Several studies have demonstrated that GABA$_B$ receptors inhibit SP (and NKA) release from C-fiber terminals in spinal cord (Malcangio and Bowery, 1993, 1996; Marvizon et al., 1999; Riley et al., 2001). $GABA_A$ receptors are ionophores and their activation increases permeability of Cl^- to allow Cl^- efflux. Increase of Cl^- conductance on sensory afferents may result in a modest depolarization that is paradoxically inhibitory, as it inactivates voltage sensitive calcium channels. Such primary afferent depolarization is considered to be a classic form of presynaptic inhibition. Not surprisingly, under different conditions, $GABA_A$ activation can facilitate release (Lao and Marvizon, 2005).

6.4.2.4 Adenosine A_1

Adenosine A_1 and A_{2A} are expressed on both SP-positive and SP-negative C fiber terminals (Sawynok and Liu, 2003). The adenosine A_1 receptor is a G_i protein coupled receptor that has a negative effect on pain transmission in dorsal horn (Sawynok and Liu, 2003). It is unclear whether this behavioral effect involves inhibition of afferent SP release.

6.4.2.5 CB_1

Cannabinoid CB_1 receptors are G_i protein coupled receptors and they are present in central terminals of C-fibers containing SP (Khasabova et al., 2004). Activation of these receptors produces antinociception (Agarwal et al., 2007) and inhibits basal and capsaicin-evoked CGRP release in DRG within a proscribed dose range (Ahluwalia et al., 2003).

6.4.2.6 NPY (Y_1)

Intrathecal administration of neuropeptide Y (NPY) and analogues produces antinociception (Hua et al., 1991). Indeed, injection of NPY into superficial dorsal horn reduces stimulation-evoked SP release (Duggan et al., 1991). Knockout of NPY receptor Y_1 shows reduction of NPY-mediated analgesia (Naveilhan et al., 2001), suggesting that Y_1 receptor may regulate SP release from C-fiber terminals.

6.5 Role of SP/NK1 in Spinal Nociceptive Processing

The preceding discussion has focused on the factors that regulate the release of SP from small peptidergic primary afferents. This evident association of SP in small sensory afferents leads to the important question of what role spinal SP release and its associated postsynaptic NK1 receptors play in pain transmission (see also Todd, Chapter 1). In the following we summarize some major findings obtained with various pharmacological and generic manipulations on SP/NK1 system to address this issue (also see Table 6.1).

Table 6.1 Summary of experimental findings supporting a role of SP-NK1 system in nociception

IT treatment/Gene disruption	Dorsal horn activation/wind-up	Thermal threshold	Mechanical threshold	Inflammation (Carrageenan / CFA) hyperalgesia	Formalin-flinches	Neuropathic pain
SP (PPT-A) Knockout		⇑/⇔*	⇑/⇔ *	⇔	⇓/⇔ **	
NK1 Knockout	⇓			Tactile ⇔	⇓	
IT NK1 antagonists	⇓	⇔	⇔	Thermal ⇓	⇓	Mechanical ⇓
IT SP-saporin	⇓	⇔	⇔	Tactile ⇓	⇓	Tactile ⇓
IT capsaicin[@] / IT TRPV1 antagonists	⇓	⇑	⇔	Thermal ⇓ Tactile ⇓		Thermal ⇔
TRPV1 Knockout	⇓	⇑	⇔	Thermal ⇓ Tactile ⇔		Thermal ⇔ Tactile ⇔

⇔ no change in response or thresholds, ⇓ decrease in response, ⇑ increase in thresholds.

* The responses depend on certain intensities of stimulation (Cao et al., 1998).

** There were conflicting data obtained with PPT-A mutant mice: Cao et al. (1998) reported no change in formalin-induced 2nd phase flinching, while Zimmer et al. (1998) observed the decreased flinches.

[@] IT capsaicin with high dose (e. g. 30–75 µg), which is known to deplete peptides including SP in TRPV1 afferent terminals (Holzer, 1991).

References: PPT-A knockout (Cao et al., 1998; Zimmer et al., 1998); NK1 knockout (De Felipe et al., 1998); IT NK1 antagonists (Yamamoto and Yaksh, 1991, 1992; Chapman and Dickenson, 1993, Iyengar et al., 1997; Campbell et al., 1998; Okano et al., 1998; Cahill and Coderre, 2002); IT SP-saporin (Mantyh et al., 1997, Nichols et al., 1999, Khasabov et al., 2002); IT capsaicin/TRPV1 antagonists (Holzer, 1991; Yamamoto and Yaksh, 1992; Cui et al., 2006); TRPV1 knockout (Caterina et al., 2000).

6.5.1 Activation of Spinal NK1 Receptor

Intrathecal delivery of SP produces agitation (Wilcox, 1988) which, though evident, is mild compared to the aggressive vocalization produced by, for example, activation of glutamatergic AMPA or NMDA receptors (Malmberg and Yaksh, 1992). A robust effect of NK1 receptor activation is, however, the development of a transient hyperalgesia which is blocked by a spinally administrated NK1 antagonist, and corresponds with the activation of dorsal horn MAPK (e.g. p38) and the release of spinal prostaglandins (Hua et al., 1999; Svensson et al., 2003, 2005).

6.5.2 Inhibition of Spinal NK1 Receptor

The implication of SP in nociception led to concerted efforts to develop highly specific NK1 receptor blockers starting in the 1980s. Numerous studies with intrathecal NK1 antagonists have demonstrated that these agents display minimal effects upon acute nociceptive thresholds, but reduce the hyperalgesic state initiated by tissue/nerve injury and inflammation in a variety of models including intraplantar carrageenan and formalin (Ohkubo and Nakanishi, 1991; Yamamoto and Yaksh, 1991, 1992; Chapman and Dickenson, 1993; Iyengar et al., 1997; Campbell et al., 1998; Cahill and Coderre, 2002). Comparable results on the formalin model of phase 2 flinching were observed with specific knockdown of spinal NK1 receptors in rat models using intrathecal antisense which produced a robust knock down of spinal NK1 receptors (Hua et al., 1998). These results are mirrored at the level of spinal dorsal horn neurons where local spinal NK1 blockade typically had little or no effect upon acute neuron activation, but appeared to be necessary for the induction C-fiber evoked spinal facilitation (Chapman and Dickenson, 1993; Liu and Sandkuhler, 1997). Development of appropriate NK1 receptor targeted drugs led to their administration to humans. With the exception of one clinical trial in dental pain (Dionne et al., 1998), which was positive, all other trials indicated no clinical benefit in terms of analgesia (Hill, 2000). This difference between preclinical and the several clinical results remains puzzling. It is interesting to note that the development of NK1 antagonists is largely predicated on their presumed spinal mechanisms of action, while the clinical trials were all performed with systemic drug. It is hypothesized that the lack of effect in humans in pain may be related to the higher concentrations that can be achieved by the intrathecal route in the preclinical studies (Campbell et al., 1998; Urban and Fox, 2000; Cahill and Coderre, 2002).

6.5.3 Studies on Knockout Animals

Several groups have reported results on pain phonotype obtained from mice in which either PPT-A gene (encoding SP and NKA) or NK1 gene was disrupted

(Cao et al., 1998; De Felipe et al., 1998; Zimmer et al., 1998). Although there are some controversies between these studies, all mutant mice showed reduced responses to painful stimuli. The evidence that characteristic amplification ('wind up') (see Thompson, Chapter 11) (De Felipe et al., 1998) and formalin phase 2 flinches (De Felipe et al., 1998; Zimmer et al., 1998) were absent in PPT-A or NK1 knockout animals suggests a role for SP/NK1 system in central sensitization. The observation that tactile hyperalgesia secondary to paw inflammation (CFA) is still fully expressed in mice either lacking SP or NK1 receptors, however, is apparently contradictory to this hypothesis (Cao et al., 1998; De Felipe et al., 1998). Since it is known that the CFA inflammatory model involves extensive systemic infection, and that the pro-nociceptive cytokines (e.g. IL-1β, TNF) play predominant roles in spinal nociceptive transmission (McMahon et al., 2005), it is possible that the SP/NK1 system may be overridden in such a scenario. Indeed the study on PPT-A knockout mice indicates that importance of SP seems to apply only to a certain 'window' of pain intensities, such as thermal pain evoked by 55.5°C, but not by 58.5°C (Cao et al., 1998).

6.5.4 Ablation of NK1 Bearing Cells

The modest effects upon central facilitation seen with SP/NK1 knockout mice bring up an important issue that selective depletion of either SP or NK1 receptor does not eliminate the excitatory drive associated with that synapse. This raises the question of what role in pain is played by the cells which bear NK1 receptors. A direct strategy for assessing that question is the use of a targeted neurotoxin, such as substance P-saporin (Mantyh et al., 1997). Exposure of the membrane to saporin has no effect. Linking saporin to the carboxyl terminal of SP results in this peptide being able to bind to the NK1 site, undergo internalization, as described previously, leading to the delivery of this anti-ribosylating agent into the cell and ultimately causing its demise. Such intrathecal delivery has been shown to produce a robust abolition of NK1 bearing cells in superficial dorsal horn including laminae I/II in several species (Allen et al., 2006). These animals display only a modest decrease in acute nociceptive thresholds, but reduced agitation induced by intraplantar capsaicin (Mantyh et al., 1997) and reversal of hyperalgesia in several models of inflammation and nerve injury (Nichols et al., 1999; Suzuki et al., 2002; Vierck et al., 2003; Suzuki et al., 2005). It has also been demonstrated electrophysiologically that elimination of NK1-bearing cells prevents development of central sensitization (Khasabov et al., 2002). As mentioned above, many of these lamina I neurons give rise to spinofugal axons which project to the medulla, though the brainstem and to diencephalic regions (Suzuki et al., 2002, 2005). There is no doubt that the findings with the SP-SAP technique support the role of the NK1 bearing cells in spinal nociceptive processing. However, it is important to stress that ablation of NK1 bearing neurons does not uniquely reflect the effect of blocking NK1 receptor, a similar case can be made that the elimination of TRPV1 positive primary afferents does not selectively indicate the effect of SP (Holzer, 1991; Caterina et al., 2000). Significant anti-hyperalgesic effects seen with both pharmacological approaches emphasize that pain information conveyed to higher centers is due to the excitatory drive linked to the SP-NK1 pathway which involves multiple neurotransmitters released from primary afferent terminals and multiple receptors expressed on NK1 bearing neurons.

6.6 Concluding Remarks

This chapter has discussed the properties of the circuit presented by the SP containing primary afferent terminals and the NK1 bearing second-order neurons. The ability to assess release at this synapse by virtue of the specific association of SP with those afferents provides an important tool whereby the pharmacology of a specific population of C-fiber afferent terminals can be studied. This *in vivo* capability is of particular importance when there is a desire to investigate the linkage between a particular pharmacology with its effect upon terminal activity and a behavioral phenotype. Similarly, the second-order NK1 bearing cell represents an important spinal projection. It is possible to demonstrate the behavioral relevance of that cell by employing targeted neurotoxins. It should be stressed that this model system does not imply that block of receptor will necessarily lead to changes in pain behavior. The modest effects observed in humans with NK1 antagonists merely emphasize that a linkage at any synapse is not necessarily mediated by a single transmitter when others (such as glutamate) are released from that terminal. Indeed as noted, NK1 antagonists rarely have a pronounced effect upon the excitation of the second-order neurons as compared to AMPA antagonists. Nevertheless, it seems evident that the preclinical data support a role for NK1 receptor in initiating a facilitated state. This leaves open the possibility that under certain specific pain conditions, should the degree of NK1 blockade be sufficiently high, a condition that might only be achieved by an intrathecal dosing, the role of the NK1 receptor in pain processing will be revealed.

Acknowledgments The authors wish to thank Bethany Fitzsimmons for the artwork and Dr. Linda Sorkin for reading the manuscript. This work was supported by NIH Grant NS 16541 and DA02110.

References

Abbadie, C., Lombard, M. C., Besson, J. M., Trafton, J. A. and Basbaum, A. I., 2002. Mu and delta opioid receptor-like immunoreactivity in the cervical spinal cord of the rat after dorsal rhizotomy or neonatal capsaicin: an analysis of pre- and postsynaptic receptor distributions. Brain Res. 930, 150–162.

Abbadie, C., Trafton, J., Liu, H., Mantyh, P. W. and Basbaum, A. I., 1997. Inflammation increases the distribution of dorsal horn neurons that internalize the neurokinin-1 receptor in response to noxious and non-noxious stimulation. J Neurosci. 17, 8049–8060.

Afrah, A. W., Stiller, C. O., Olgart, L., Brodin, E. and Gustafsson, H., 2001. Involvement of spinal N-methyl-D-aspartate receptors in capsaicin-induced in vivo release of substance P in the rat dorsal horn. Neurosci Lett. 316, 83–86.

Agarwal, N., Pacher, P., Tegeder, I., Amaya, F., Constantin, C. E., Brenner, G. J., Rubino, T., Michalski, C. W., Marsicano, G., Monory, K., Mackie, K., Marian, C., Batkai, S., Parolaro, D., Fischer, M. J., Reeh, P., Kunos, G., Kress, M., Lutz, B., Woolf, C. J. and Kuner, R., 2007. Cannabinoids mediate analgesia largely via peripheral type 1 cannabinoid receptors in nociceptors. Nat Neurosci. 10, 870–879.

Ahluwalia, J., Urban, L., Bevan, S. and Nagy, I., 2003. Anandamide regulates neuropeptide release from capsaicin-sensitive primary sensory neurons by activating both the cannabinoid 1 receptor and the vanilloid receptor 1 in vitro. Eur J Neurosci. 17, 2611–2618.

Aimone, L. D. and Yaksh, T. L., 1989. Opioid modulation of capsaicin-evoked release of substance P from rat spinal cord in vivo. Peptides. 10, 1127–1131.

Akasu, T., Ishimatsu, M. and Yamada, K., 1996. Tachykinins cause inward current through NK1 receptors in bullfrog sensory neurons. Brain Res. 713, 160–167.

Allen, B. J., Rogers, S. D., Ghilardi, J. R., Menning, P. M., Kuskowski, M. A., Basbaum, A. I., Simone, D. A. and Mantyh, P. W., 1997. Noxious cutaneous thermal stimuli induce a graded release of endogenous substance P in the spinal cord: imaging peptide action in vivo. J Neurosci. 17, 5921–5927.

Allen, J. W., Mantyh, P. W., Horais, K., Tozier, N., Rogers, S. D., Ghilardi, J. R., Cizkova, D., Grafe, M. R., Richter, P., Lappi, D. A. and Yaksh, T. L., 2006. Safety evaluation of intrathecal substance P-saporin, a targeted neurotoxin, in dogs. Toxicol Sci. 91, 286–298.

Altier, C., Dale, C. S., Kisilevsky, A. E., Chapman, K., Castiglioni, A. J., Matthews, E. A., Evans, R. M., Dickenson, A. H., Lipscombe, D., Vergnolle, N. and Zamponi, G. W., 2007. Differential role of N-type calcium channel splice isoforms in pain. J Neurosci. 27, 6363–6373.

Altier, C. and Zamponi, G. W., 2004. Targeting Ca2 + channels to treat pain: T-type versus N-type. Trends Pharmacol Sci. 25, 465–470.

Barber, L. A. and Vasko, M. R., 1996. Activation of protein kinase C augments peptide release from rat sensory neurons. J Neurochem. 67, 72–80.

Battaglia, G. and Rustioni, A., 1988. Coexistence of glutamate and substance P in dorsal root ganglion neurons of the rat and monkey. J Comp Neurol. 277, 302–312.

Bautista, D. M., Jordt, S. E., Nikai, T., Tsuruda, P. R., Read, A. J., Poblete, J., Yamoah, E. N., Basbaum, A. I. and Julius, D., 2006. TRPA1 mediates the inflammatory actions of environmental irritants and proalgesic agents. Cell. 124, 1269–1282.

Bell, T. J., Thaler, C., Castiglioni, A. J., Helton, T. D. and Lipscombe, D., 2004. Cell-specific alternative splicing increases calcium channel current density in the pain pathway. Neuron. 41, 127–138.

Bourgoin, S., Pohl, M., Mauborgne, A., Benoliel, J. J., Collin, E., Hamon, M. and Cesselin, F., 1993. Monoaminergic control of the release of calcitonin gene-related peptide- and substance P-like materials from rat spinal cord slices. Neuropharmacology. 32, 633–640.

Brimijoin, S., Lundberg, J. M., Brodin, E., Hokfelt, T. and Nilsson, G., 1980. Axonal transport of substance P in the vagus and sciatic nerves of the guinea pig. Brain Res. 191, 443–457.

Cahill, C. M. and Coderre, T. J., 2002. Attenuation of hyperalgesia in a rat model of neuropathic pain after intrathecal pre- or post-treatment with a neurokinin-1 antagonist. Pain. 95, 277–285.

Campbell, E. A., Gentry, C. T., Patel, S., Panesar, M. S., Walpole, C. S. and Urban, L., 1998. Selective neurokinin-1 receptor antagonists are anti-hyperalgesic in a model of neuropathic pain in the guinea-pig. Neuroscience. 87, 527–532.

Cao, Y. Q., Mantyh, P. W., Carlson, E. J., Gillespie, A. M., Epstein, C. J. and Basbaum, A. I., 1998. Primary afferent tachykinins are required to experience moderate to intense pain. Nature. 392, 390–394.

Carbone, E., Giancippoli, A., Marcantoni, A., Guido, D. and Carabelli, V., 2006. A new role for T-type channels in fast "low-threshold" exocytosis. Cell Calcium. 40, 147–154.

Caron, M. G. and Lefkowitz, R. J., 1993. Catecholamine receptors: structure, function, and regulation. Recent Prog Horm Res. 48, 277–290.

Carter, M. S. and Krause, J. E., 1990. Structure, expression, and some regulatory mechanisms of the rat preprotachykinin gene encoding substance P, neurokinin A, neuropeptide K, and neuropeptide gamma. J Neurosci. 10, 2203–2214.

Castiglioni, A. J., Raingo, J. and Lipscombe, D., 2006. Alternative splicing in the C-terminus of CaV2.2 controls expression and gating of N-type calcium channels. J Physiol. 576, 119–134.

Caterina, M. J., Leffler, A., Malmberg, A. B., Martin, W. J., Trafton, J., Petersen-Zeitz, K. R., Koltzenburg, M., Basbaum, A. I. and Julius, D., 2000. Impaired nociception and pain sensation in mice lacking the capsaicin receptor. Science. 288, 306–313.

Chang, H. M., Berde, C. B., Holz, G. G. t., Steward, G. F. and Kream, R. M., 1989. Sufentanil, morphine, met-enkephalin, and kappa-agonist (U-50,488H) inhibit substance P release from primary sensory neurons: a model for presynaptic spinal opioid actions. Anesthesiology. 70, 672–677.

Chang, M. M., Leeman, S. E. and Niall, H. D., 1971. Amino-acid sequence of substance P. Nat New Biol. 232, 86–87.

Chapman, V. and Dickenson, A. H., 1993. The effect of intrathecal administration of RP67580, a potent neurokinin 1 antagonist on nociceptive transmission in the rat spinal cord. Neurosci Lett. 157, 149–152.

Cheng, P. Y., Moriwaki, A., Wang, J. B., Uhl, G. R. and Pickel, V. M., 1996. Ultrastructural localization of mu-opioid receptors in the superficial layers of the rat cervical spinal cord: extrasynaptic localization and proximity to Leu5-enkephalin. Brain Res. 731, 141–154.

Cui, M., Honore, P., Zhong, C., Gauvin, D., Mikusa, J., Hernandez, G., Chandran, P., Gomtsyan, A., Brown, B., Bayburt, E. K., Marsh, K., Bianchi, B., McDonald, H., Niforatos, W., Neelands, T. R., Moreland, R. B., Decker, M. W., Lee, C. H., Sullivan, J. P. and Faltynek, C. R., 2006. TRPV1 receptors in the CNS play a key role in broad-spectrum analgesia of TRPV1 antagonists. J Neurosci. 26, 9385–9393.

De Biasi, S. and Rustioni, A., 1988. Glutamate and substance P coexist in primary afferent terminals in the superficial laminae of spinal cord. Proceedings of the National Academy of Science USA. 85, 7820–7824.

De Felipe, C., Herrero, J. F., O'Brien, J. A., Palmer, J. A., Doyle, C. A., Smith, A. J., Laird, J. M., Belmonte, C., Cervero, F. and Hunt, S. P., 1998. Altered nociception, analgesia and aggression in mice lacking the receptor for substance P. Nature. 392, 394–397.

Del Bianco, E., Santicioli, P., Tramontana, M., Maggi, C. A., Cecconi, R. and Geppetti, P., 1991. Different pathways by which extracellular Ca2+ promotes calcitonin gene-related peptide release from central terminals of capsaicin-sensitive afferents of guinea pigs: effect of capsaicin, high K+ and low pH media. Brain Res. 566, 46–53.

Dememes, D., Seoane, A., Venteo, S. and Desmadryl, G., 2000. Efferent function of vestibular afferent endings? Similar localization of N-type calcium channels, synaptic vesicle and synaptic membrane-associated proteins. Neuroscience. 98, 377–384.

Dionne, R. A., Max, M. B., Gordon, S. M., Parada, S., Sang, C., Gracely, R. H., Sethna, N. F. and MacLean, D. B., 1998. The substance P receptor antagonist CP-99,994 reduces acute postoperative pain. Clin Pharmacol Ther. 64, 562–568.

Dolphin, A. C. and Scott, R. H., 1987. Calcium channel currents and their inhibition by (–)-baclofen in rat sensory neurones: modulation by guanine nucleotides. J Physiol. 386, 1–17.

Duggan, A. W., Hendry, I. A., Morton, C. R., Hutchison, W. D. and Zhao, Z. Q., 1988. Cutaneous stimuli releasing immunoreactive substance P in the dorsal horn of the cat. Brain Res. 451, 261–273.

Duggan, A. W., Hope, P. J., Jarrott, B., Schaible, H. G. and Fleetwood-Walker, S. M., 1990. Release, spread and persistence of immunoreactive neurokinin A in the dorsal horn of the cat following noxious cutaneous stimulation. Studies with antibody microprobes. Neuroscience. 35, 195–202.

Duggan, A. W., Hope, P. J. and Lang, C. W., 1991. Microinjection of neuropeptide Y into the superficial dorsal horn reduces stimulus-evoked release of immunoreactive substance P in the anaesthetized cat. Neuroscience. 44, 733–740.

Evans, A. R., Nicol, G. D. and Vasko, M. R., 1996. Differential regulation of evoked peptide release by voltage-sensitive calcium channels in rat sensory neurons. Brain Res. 712, 265–273.

Fehrenbacher, J. C., Taylor, C. P. and Vasko, M. R., 2003. Pregabalin and gabapentin reduce release of substance P and CGRP from rat spinal tissues only after inflammation or activation of protein kinase C. Pain. 105, 133–141.

Field, M. J., Cox, P. J., Stott, E., Melrose, H., Offord, J., Su, T. Z., Bramwell, S., Corradini, L., England, S., Winks, J., Kinloch, R. A., Hendrich, J., Dolphin, A. C., Webb, T. and Williams, D., 2006. Identification of the alpha2-delta-1 subunit of voltage-dependent calcium channels as a molecular target for pain mediating the analgesic actions of pregabalin. Proc Natl Acad Sci USA. 103, 17537–17542.

Fields, H. L., Emson, P. C., Leigh, B. K., Gilbert, R. F. and Iversen, L. L., 1980. Multiple opiate receptor sites on primary afferent fibres. Nature. 284, 351–353.

Fyffe, R. E. and Perl, E. R., 1984. Is ATP a central synaptic mediator for certain primary afferent fibers from mammalian skin? Proc Natl Acad Sci USA. 81, 6890–6893.

Gamse, R., Molnar, A. and Lembeck, F., 1979. Substance P release from spinal cord slices by capsaicin. Life Sci. 25, 629–636.

Ghilardi, J. R., Svensson, C. I., Rogers, S. D., Yaksh, T. L. and Mantyh, P. W., 2004. Constitutive spinal cyclooxygenase-2 participates in the initiation of tissue injury-induced hyperalgesia. J Neurosci. 24, 2727–2732.

Gibson, S. J., Polak, J. M., Bloom, S. R., Sabate, I. M., Mulderry, P. M., Ghatei, M. A., McGregor, G. P., Morrison, J. F., Kelly, J. S., Evans, R. M. and et al., 1984. Calcitonin gene-related peptide immunoreactivity in the spinal cord of man and of eight other species. J Neurosci. 4, 3101–3111.

Go, V. L. and Yaksh, T. L., 1987. Release of substance P from the cat spinal cord. J Physiol. 391, 141–167.

Grudt, T. J. and Williams, J. T., 1994. mu-Opioid agonists inhibit spinal trigeminal substantia gelatinosa neurons in guinea pig and rat. J Neurosci. 14, 1646–1654.

Gu, G., Kondo, I., Hua, X. Y. and Yaksh, T. L., 2005. Resting and evoked spinal substance P release during chronic intrathecal morphine infusion: parallels with tolerance and dependence. J Pharmacol Exp Ther. 314, 1362–1369.

Guan, J. S., Xu, Z. Z., Gao, H., He, S. Q., Ma, G. Q., Sun, T., Wang, L. H., Zhang, Z. N., Lena, I., Kitchen, I., Elde, R., Zimmer, A., He, C., Pei, G., Bao, L. and Zhang, X., 2005. Interaction with vesicle luminal protachykinin regulates surface expression of delta-opioid receptors and opioid analgesia. Cell. 122, 619–631.

Heinke, B., Balzer, E. and Sandkuhler, J., 2004. Pre- and postsynaptic contributions of voltage-dependent Ca2+ channels to nociceptive transmission in rat spinal lamina I neurons. Eur J Neurosci. 19, 103–111.

Hill, R., 2000. NK1 (substance P) receptor antagonists – why are they not analgesic in humans? Trends Pharmacol Sci. 21, 244–246.

Hingtgen, C. M. and Vasko, M. R., 1994. Prostacyclin enhances the evoked-release of substance P and calcitonin gene-related peptide from rat sensory neurons. Brain Res. 655, 51–60.

Hokfelt, T., Arvidsson, U., Cullheim, S., Millhorn, D., Nicholas, A. P., Pieribone, V., Seroogy, K. and Ulfhake, B., 2000. Multiple messengers in descending serotonin neurons: localization and functional implications. J Chem Neuroanat. 18, 75–86.

Hokfelt, T., Kellerth, J. O., Nilsson, G. and Pernow, B., 1975. Substance P localization in the central nervous system and in some primary sensory neurons. Science. 190, 889–890.

Hokfelt, T., Pernow, B. and Wahren, J., 2001. Substance P: a pioneer amongst neuropeptides. J Intern Med. 249, 27–40.

Holmgren, S. and Jensen, J., 2001. Evolution of vertebrate neuropeptides. Brain Res Bull. 55, 723–735.

Holzer, P., 1991. Capsaicin: cellular targets, mechanisms of action, and selectivity for thin sensory neurons. Pharmacol Rev. 43, 143–201.

Honor, P., Menning, P. M., Rogers, S. D., Nichols, M. L., Basbaum, A. I., Besson, J. M. and Mantyh, P. W., 1999. Spinal substance P receptor expression and internalization in acute, short-term, and long-term inflammatory pain states. J Neurosci. 19, 7670–7678.

Honore, P., Menning, P. M., Rogers, S. D., Nichols, M. L. and Mantyh, P. W., 2000. Neurochemical plasticity in persistent inflammatory pain. Prog Brain Res. 129, 357–363.

Hua, F., Ricketts, B. A., Reifsteck, A., Ardell, J. L. and Williams, C. A., 2004. Myocardial ischemia induces the release of substance P from cardiac afferent neurons in rat thoracic spinal cord. Am J Physiol Heart Circ Physiol. 286, H1654–1664.

Hua, X. Y., Boublik, J. H., Spicer, M. A., Rivier, J. E., Brown, M. R. and Yaksh, T. L., 1991. The antinociceptive effects of spinally administered neuropeptide Y in the rat: systematic studies on structure-activity relationship. J Pharmacol Exp Ther. 258, 243–248.

Hua, X. Y., Chen, P., Marsala, M. and Yaksh, T. L., 1999. Intrathecal substance P-induced thermal hyperalgesia and spinal release of prostaglandin E2 and amino acids. Neuroscience. 89, 525–534.

Hua, X. Y., Chen, P., Polgar, E., Nagy, I., Marsala, M., Phillips, E., Wollaston, L., Urban, L., Yaksh, T. L. and Webb, M., 1998. Spinal neurokinin NK1 receptor down-regulation and antinociception: effects of spinal NK1 receptor antisense oligonucleotides and NK1 receptor occupancy. J Neurochem. 70, 688–698.

Hua, X. Y., Theodorsson-Norheim, E., Brodin, E., Lundberg, J. M. and Hokfelt, T., 1985. Multiple tachykinins (Neurokinin A, Neuropeptide K and Substance P) in capsaicin-sensitive sensory neurons in the guinea-pig. Regul Pept. 13, 1–19.

Huang, H., Wu, X., Nicol, G. D., Meller, S. and Vasko, M. R., 2003. ATP augments peptide release from rat sensory neurons in culture through activation of P2Y receptors. J Pharmacol Exp Ther. 306, 1137–1144.

Hunt, S. P. and Mantyh, P. W., 2001. The molecular dynamics of pain control. Nat Rev Neurosci. 2, 83–91.

Ikeda, H., Heinke, B., Ruscheweyh, R. and Sandkuhler, J., 2003. Synaptic plasticity in spinal lamina I projection neurons that mediate hyperalgesia. Science. 299, 1237–1240.

Inoue, A., Hashimoto, T., Hide, I., Nishio, H. and Nakata, Y., 1997. 5-Hydroxytryptamine-facilitated release of substance P from rat spinal cord slices is mediated by nitric oxide and cyclic GMP. J Neurochem. 68, 128–133.

Ito, K., Rome, C., Bouleau, Y. and Dulon, D., 2002. Substance P mobilizes intracellular calcium and activates a nonselective cation conductance in rat spiral ganglion neurons. Eur J Neurosci. 16, 2095–2102.

Iyengar, S., Hipskind, P. A., Gehlert, D. R., Schober, D., Lobb, K. L., Nixon, J. A., Helton, D. R., Kallman, M. J., Boucher, S., Couture, R., Li, D. L. and Simmons, R. M., 1997. LY303870, a centrally active neurokinin-1 antagonist with a long duration of action. J Pharmacol Exp Ther. 280, 774–785.

Jessell, T. M. and Iversen, L. L., 1977. Opiate analgesics inhibit substance P release from rat trigeminal nucleus. Nature. 268, 549–551.

Jhamandas, K., Yaksh, T. L., Harty, G., Szolcsanyi, J. and Go, V. L., 1984. Action of intrathecal capsaicin and its structural analogues on the content and release of spinal substance P: selectivity of action and relationship to analgesia. Brain Res. 306, 215–225.

Jo, Y. H. and Schlichter, R., 1999. Synaptic corelease of ATP and GABA in cultured spinal neurons. Nat Neurosci. 2, 241–245.

Kangawa, K., Minamino, N., Fukuda, A. and Matsuo, H., 1983. Neuromedin K: a novel mammalian tachykinin identified in porcine spinal cord. Biochem Biophys Res Commun. 114, 533–540.

Khasabov, S. G., Rogers, S. D., Ghilardi, J. R., Peters, C. M., Mantyh, P. W. and Simone, D. A., 2002. Spinal neurons that possess the substance P receptor are required for the development of central sensitization. J Neurosci. 22, 9086–9098.

Khasabova, I. A., Harding-Rose, C., Simone, D. A. and Seybold, V. S., 2004. Differential effects of CB1 and opioid agonists on two populations of adult rat dorsal root ganglion neurons. J Neurosci. 24, 1744–1753.

Kjorsvik Bertelsen, A., Warsame Afrah, A., Gustafsson, H., Tjolsen, A., Hole, K. and Stiller, C. O., 2003. Stimulation of spinal 5-HT(2A/2C) receptors potentiates the capsaicin-induced in vivo release of substance P-like immunoreactivity in the rat dorsal horn. Brain Res. 987, 10–16.

Kondo, I., Marvizon, J. C., Song, B., Salgado, F., Codeluppi, S., Hua, X. Y. and Yaksh, T. L., 2005. Inhibition by spinal mu- and delta-opioid agonists of afferent-evoked substance P release. J Neurosci. 25, 3651–3660.

Kotani, H., Hoshimaru, M., Nawa, H. and Nakanishi, S., 1986. Structure and gene organization of bovine neuromedin K precursor. Proc Natl Acad Sci USA. 83, 7074–7078.

Krause, J. E., Chirgwin, J. M., Carter, M. S., Xu, Z. S. and Hershey, A. D., 1987. Three rat preprotachykinin mRNAs encode the neuropeptides substance P and neurokinin A. Proc Natl Acad Sci USA. 84, 881–885.

Kumazawa, T., Mizumura, K., Koda, H. and Fukusako, H., 1996. EP receptor subtypes implicated in the PGE2-induced sensitization of polymodal receptors in response to bradykinin and heat. J Neurophysiol. 75, 2361–2368.

Kuraishi, Y., Hirota, N., Sato, Y., Hanashima, N., Takagi, H. and Satoh, M., 1989. Stimulus specificity of peripherally evoked substance P release from the rabbit dorsal horn in situ. Neuroscience. 30, 241–250.

Lao, L. and Marvizon, J. C., 2005. GABA(A) receptor facilitation of neurokinin release from primary afferent terminals in the rat spinal cord. Neuroscience. 130, 1013–1027.

Lembeck, F., 1953. Zur Frage der zentralen Ubertragung afferenter Impulse III. Mitteilung. Das Vorommen und die Bedeutung der Substanz P in den dorsalen Wurzeln des Ruckenmarks. Naunyn Schmiedebergs Arch Pharmacol. 219, 197–213.

Linderoth, B. and Brodin, E., 1988. Tachykinin release from rat spinal cord in vitro and in vivo in response to various stimuli. Regul Pept. 21, 129–140.

Liu, H., Brown, J. L., Jasmin, L., Maggio, J. E., Vigna, S. R., Mantyh, P. W. and Basbaum, A. I., 1994a. Synaptic relationship between substance P and the substance P receptor: light and electron microscopic characterization of the mismatch between neuropeptides and their receptors. Proc Natl Acad Sci USA. 91, 1009–1013.

Liu, H., Mantyh, P. W. and Basbaum, A. I., 1997. NMDA-receptor regulation of substance P release from primary afferent nociceptors. Nature. 386, 721–724.

Liu, H., Wang, H., Sheng, M., Jan, L. Y., Jan, Y. N. and Basbaum, A. I., 1994b. Evidence for presynaptic N-methyl-D-aspartate autoreceptors in the spinal cord dorsal horn. Proc Natl Acad Sci USA. 91, 8383–8387.

Liu, X. and Sandkuhler, J., 1997. Characterization of long-term potentiation of C-fiber-evoked potentials in spinal dorsal horn of adult rat: essential role of NK1 and NK2 receptors. J Neurophysiol. 78, 1973–1982.

Maggi, C. A., 1995. The mammalian tachykinin receptors. Gen Pharmacol. 26, 911–944.

Malcangio, M. and Bowery, N. G., 1993. Gamma-aminobutyric acidB, but not gamma-aminobutyric acidA receptor activation, inhibits electrically evoked substance P-like immunoreactivity release from the rat spinal cord in vitro. J Pharmacol Exp Ther. 266, 1490–1496.

Malcangio, M. and Bowery, N. G., 1996. GABA and its receptors in the spinal cord. Trends Pharmacol Sci. 17, 457–462.

Malcangio, M. and Bowery, N. G., 1999. Peptide autoreceptors: does an autoreceptor for substance P exist? Trends Pharmacol Sci. 20, 405–407.

Malcangio, M., Fernandes, K. and Tomlinson, D. R., 1998. NMDA receptor activation modulates evoked release of substance P from rat spinal cord. Br J Pharmacol. 125, 1625–1626.

Malcangio, M., Garrett, N. E., Cruwys, S. and Tomlinson, D. R., 1997. Nerve growth factor- and neurotrophin-3-induced changes in nociceptive threshold and the release of substance P from the rat isolated spinal cord. J Neurosci. 17, 8459–8467.

Malcangio, M., Ramer, M. S., Boucher, T. J. and McMahon, S. B., 2000. Intrathecally injected neurotrophins and the release of substance P from the rat isolated spinal cord. Eur J Neurosci. 12, 139–144.

Malmberg, A. B. and Yaksh, T. L., 1992. Hyperalgesia mediated by spinal glutamate or substance P receptor blocked by spinal cyclooxygenase inhibition. Science. 257, 1276–1279.

Mantyh, P. W., Allen, C. J., Ghilardi, J. R., Rogers, S. D., Mantyh, C. R., Liu, H., Basbaum, A. I., Vigna, S. R. and Maggio, J. E., 1995a. Rapid endocytosis of a G protein-coupled receptor: substance P evoked internalization of its receptor in the rat striatum in vivo. Proc Natl Acad Sci USA. 92, 2622–2626.

Mantyh, P. W., DeMaster, E., Malhotra, A., Ghilardi, J. R., Rogers, S. D., Mantyh, C. R., Liu, H., Basbaum, A. I., Vigna, S. R., Maggio, J. E. and et al., 1995b. Receptor endocytosis and dendrite reshaping in spinal neurons after somatosensory stimulation. Science. 268, 1629–1632.

Mantyh, P. W., Rogers, S. D., Honore, P., Allen, B. J., Ghilardi, J. R., Li, J., Daughters, R. S., Lappi, D. A., Wiley, R. G. and Simone, D. A., 1997. Inhibition of hyperalgesia by ablation of lamina I spinal neurons expressing the substance P receptor. Science. 278, 275–279.

Martin, H. A., Basbaum, A. I., Kwiat, G. C., Goetzl, E. J. and Levine, J. D., 1987. Leukotriene and prostaglandin sensitization of cutaneous high-threshold C- and A-delta mechanonociceptors in the hairy skin of rat hindlimbs. Neuroscience. 22, 651–659.

Marvizon, J. C., Grady, E. F., Stefani, E., Bunnett, N. W. and Mayer, E. A., 1999. Substance P release in the dorsal horn assessed by receptor internalization: NMDA receptors counteract a tonic inhibition by GABA(B) receptors. Eur J Neurosci. 11, 417–426.

Marvizon, J. C., Martinez, V., Grady, E. F., Bunnett, N. W. and Mayer, E. A., 1997. Neurokinin 1 receptor internalization in spinal cord slices induced by dorsal root stimulation is mediated by NMDA receptors. J Neurosci. 17, 8129–8136.

Marvizon, J. C., Wang, X., Matsuka, Y., Neubert, J. K. and Spigelman, I., 2003. Relationship between capsaicin-evoked substance P release and neurokinin 1 receptor internalization in the rat spinal cord. Neuroscience. 118, 535–545.

Mauborgne, A., Bourgoin, S., Benoliel, J. J., Hirsch, M., Berthier, J. L., Hamon, M. and Cesselin, F., 1987. Enkephalinase is involved in the degradation of endogenous substance P released from slices of rat substantia nigra. J Pharmacol Exp Ther. 243, 674–680.

McCarson, K. E. and Krause, J. E., 1994. NK-1 and NK-3 type tachykinin receptor mRNA expression in the rat spinal cord dorsal horn is increased during adjuvant or formalin-induced nociception. J Neurosci. 14, 712–720.

McCarthy, P. W. and Lawson, S. N., 1989. Cell type and conduction velocity of rat primary sensory neurons with substance P-like immunoreactivity. Neuroscience. 28, 745–753.

McLeod, A. L., Krause, J. E. and Ribeiro-Da-Silva, A., 2000. Immunocytochemical localization of neurokinin B in the rat spinal dorsal horn and its association with substance P and GABA: an electron microscopic study. J Comp Neurol. 420, 349–362.

McMahon, S. B., Cafferty, W. B. and Marchand, F., 2005. Immune and glial cell factors as pain mediators and modulators. Exp Neurol. 192, 444–462.

Menon-Johansson, A. S., Berrow, N. and Dolphin, A. C., 1993. G(o) transduces GABAB-receptor modulation of N-type calcium channels in cultured dorsal root ganglion neurons. Pflugers Arch. 425, 335–343.

Merighi, A., Polak, J. M. and Theodosis, D. T., 1991. Ultrastructural visualization of glutamate and aspartate immunoreactivities in the rat dorsal horn, with special reference to the co-localization of glutamate, substance P and calcitonin-gene related peptide. Neuroscience. 40, 67–80.

Michael, G. J., Averill, S., Nitkunan, A., Rattray, M., Bennett, D. L., Yan, Q. and Priestley, J. V., 1997. Nerve growth factor treatment increases brain-derived neurotrophic factor selectively in TrkA-expressing dorsal root ganglion cells and in their central terminations within the spinal cord. J Neurosci. 17, 8476–8490.

Naim, M., Spike, R. C., Watt, C., Shehab, S. A. and Todd, A. J., 1997. Cells in laminae III and IV of the rat spinal cord that possess the neurokinin-1 receptor and have dorsally directed dendrites receive a major synaptic input from tachykinin-containing primary afferents. J Neurosci. 17, 5536–5548.

Nakatsuka, T. and Gu, J. G., 2001. ATP P2X receptor-mediated enhancement of glutamate release and evoked EPSCs in dorsal horn neurons of the rat spinal cord. J Neurosci. 21, 6522–6531.

Nakatsuka, T., Mena, N., Ling, J. and Gu, J. G., 2001. Depletion of substance P from rat primary sensory neurons by ATP, an implication of P2X receptor-mediated release of substance P. Neuroscience. 107, 293–300.

Naveilhan, P., Hassani, H., Lucas, G., Blakeman, K. H., Hao, J. X., Xu, X. J., Wiesenfeld-Hallin, Z., Thoren, P. and Ernfors, P., 2001. Reduced antinociception and plasma extra-vasation in mice lacking a neuropeptide Y receptor. Nature. 409, 513–517.

Nawa, H., Hirose, T., Takashima, H., Inayama, S. and Nakanishi, S., 1983. Nucleotide sequences of cloned cDNAs for two types of bovine brain substance P precursor. Nature. 306, 32–36.

Nazarian A., Christianson, C. A., Hua, X.-Y. and Yaksh, T. L., 2008a. Dexmedetomidine and ST91 anlgesia in the formalin model is mediated by alpha 2A adrenoceptors: a mechanism of action distinct from morphine. Neuroscience. 152, 119–127.

Nazarian, A., Gu, G., Gracias, N. G., Wilkinson, K., Hua, X. Y., Vasko, M. R. and Yaksh, T. L., 2008b. Spinal N-methyl-D-aspartate receptors and nociception-evoked release of primary afferent substance P. Neuroscience. 152, 119–127.

Neugebauer, V., Schaible, H. G., Weiretter, F. and Freudenberger, U., 1994. The involvement of substance P and neurokinin-1 receptors in the responses of rat dorsal horn neurons to noxious but not to innocuous mechanical stimuli applied to the knee joint. Brain Res. 666, 207–215.

Nicholas, A. P., Pieribone, V. A. and Hokfelt, T., 1993. Cellular localization of messenger RNA for beta-1 and beta-2 adrenergic receptors in rat brain: an in situ hybridization study. Neuroscience. 56, 1023–1039.

Nichols, M. L., Allen, B. J., Rogers, S. D., Ghilardi, J. R., Honore, P., Luger, N. M., Finke, M. P., Li, J., Lappi, D. A., Simone, D. A. and Mantyh, P. W., 1999. Transmission of chronic nociception by spinal neurons expressing the substance P receptor. Science. 286, 1558–1561.

Nicol, G. D., Klingberg, D. K. and Vasko, M. R., 1992. Prostaglandin E2 increases calcium conductance and stimulates release of substance P in avian sensory neurons. J Neurosci. 12, 1917–1927.

Ohkubo, H. and Nakanishi, S., 1991. Molecular characterization of the three tachykinin receptors. Ann NY Acad Sci. 632, 53–62.

Ono, H., Mishima, A., Ono, S., Fukuda, H. and Vasko, M. R., 1991. Inhibitory effects of clonidine and tizanidine on release of substance P from slices of rat spinal cord and antagonism by alpha-adrenergic receptor antagonists. Neuropharmacology. 30, 585–589.

Okano,K., Kuraishi,Y. and Satoh, M., 1998. Involvement of spinal substance P and excitatory amino acids in inflammatory hyperalgesia in rats. Jpn J Pharmacol.76, 15-22.

Otsuka, M. and Konishi, S., 1976. release of substance P-like immunoreactivity from isolated spinal cord of newborn rat. Nature. 264, 83–84.

Otsuka, M. and Yoshioka, K., 1993. Neurotransmitter functions of mammalian tachykinins. Physiol Rev. 73, 229–308.

Pang, I. H. and Vasko, M. R., 1986. Morphine and norepinephrine but not 5-hydroxytryptamine and gamma-aminobutyric acid inhibit the potassium-stimulated release of substance P from rat spinal cord slices. Brain Res. 376, 268–279.

Pennefather, J. N., Lecci, A., Candenas, M. L., Patak, E., Pinto, F. M. and Maggi, C. A., 2004. Tachykinins and tachykinin receptors: a growing family. Life Sci. 74, 1445–1463.

Pernow, B., 1953. Studies on substance P purification, occurrence, and biological actions. Acta Physiol Scand. 29, 1–90.

Pernow, B., 1983. Substance P. Pharmacol Rev. 35, 85–141.

Petruska, J. C., Cooper, B. Y., Gu, J. G., Rau, K. K. and Johnson, R. D., 2000. Distribution of P2X1, P2X2, and P2X3 receptor subunits in rat primary afferents: relation to population markers and specific cell types. J Chem Neuroanat. 20, 141–162.

Pezet, S., Onteniente, B., Jullien, J., Junier, M. P., Grannec, G., Rudkin, B. B. and Calvino, B., 2001. Differential regulation of NGF receptors in primary sensory neurons by adjuvant-induced arthritis in the rat. Pain. 90, 113–125.

Pohl, M., Mauborgne, A., Bourgoin, S., Benoliel, J. J., Hamon, M. and Cesselin, F., 1989. Neonatal capsaicin treatment abolishes the modulations by opioids of substance P release from rat spinal cord slices. Neurosci Lett. 96, 102–107.

Polgar, E., Furuta, T., Kaneko, T. and Todd, A., 2006. Characterization of neurons that express preprotachykinin B in the dorsal horn of the rat spinal cord. Neuroscience. 139, 687–697.

Price, G. W., Wilkin, G. P., Turnbull, M. J. and Bowery, N. G., 1984. Are baclofen-sensitive GABAB receptors present on primary afferent terminals of the spinal cord? Nature. 307, 71–74.

Raingo, J., Castiglioni, A. J. and Lipscombe, D., 2007. Alternative splicing controls G protein-dependent inhibition of N-type calcium channels in nociceptors. Nat Neurosci. 10, 285–292.

Ralevic, V. and Burnstock, G., 1998. Receptors for purines and pyrimidines. Pharmacol Rev. 50, 413–492.

Regoli, D., Nguyen, Q. T. and Jukic, D., 1994. Neurokinin receptor subtypes characterized by biological assays. Life Sci. 54, 2035–2047.

Ribeiro-da-Silva, A. and Hokfelt, T., 2000. Neuroanatomical localisation of Substance P in the CNS and sensory neurons. Neuropeptides. 34, 256–271.

Riley, R. C., Trafton, J. A., Chi, S. I. and Basbaum, A. I., 2001. Presynaptic regulation of spinal cord tachykinin signaling via GABA(B) but not GABA(A) receptor activation. Neuroscience. 103, 725–737.

Ryu, P. D. and Randic, M., 1990. Low- and high-voltage-activated calcium currents in rat spinal dorsal horn neurons. J Neurophysiol. 63, 273–285.

Salter, M. W., De Koninck, Y. and Henry, J. L., 1993. Physiological roles for adenosine and ATP in synaptic transmission in the spinal dorsal horn. Prog Neurobiol. 41, 125–156.

Samad, T. A., Moore, K. A., Sapirstein, A., Billet, S., Allchorne, A., Poole, S., Bonventre, J. V. and Woolf, C. J., 2001. Interleukin-1beta-mediated induction of Cox-2 in the CNS contributes to inflammatory pain hypersensitivity. Nature. 410, 471–475.

Saria, A., Javorsky, F., Humpel, C. and Gamse, R., 1990. 5-HT3 receptor antagonists inhibit sensory neuropeptide release from the rat spinal cord. Neuroreport. 1, 104–106.

Sawynok, J., Downie, J. W., Reid, A. R., Cahill, C. M. and White, T. D., 1993. ATP release from dorsal spinal cord synaptosomes: characterization and neuronal origin. Brain Res. 610, 32–38.

Sawynok, J. and Liu, X. J., 2003. Adenosine in the spinal cord and periphery: release and regulation of pain. Prog Neurobiol. 69, 313–340.

Schmidlin, F., Roosterman, D. and Bunnett, N. W., 2003. The third intracellular loop and carboxyl tail of neurokinin 1 and 3 receptors determine interactions with beta-arrestins. Am J Physiol Cell Physiol. 285, C945–958.

Schroeder, J. E., Fischbach, P. S., Zheng, D. and McCleskey, E. W., 1991. Activation of mu opioid receptors inhibits transient high- and low-threshold Ca2+ currents, but spares a sustained current. Neuron. 6, 13–20.

Severini, C., Improta, G., Falconieri-Erspamer, G., Salvadori, S. and Erspamer, V., 2002. The tachykinin peptide family. Pharmacol Rev. 54, 285–322.

Seybold, V. S., Grkovic, I., Portbury, A. L., Ding, Y. Q., Shigemoto, R., Mizuno, N., Furness, J. B. and Southwell, B. R., 1997. Relationship of NK3 receptor-immunoreactivity to subpopulations of neurons in rat spinal cord. J Comp Neurol. 381, 439–448.

Shigemoto, R., Ohishi, H., Nakanishi, S. and Mizuno, N., 1992. Expression of the mRNA for the rat NMDA receptor (NMDAR1) in the sensory and autonomic ganglion neurons. Neurosci Lett. 144, 229–232.

Skofitsch, G. and Jacobowitz, D. M., 1985. Calcitonin gene-related peptide coexists with substance P in capsaicin sensitive neurons and sensory ganglia of the rat. Peptides. 6, 747–754.

Smart, D. and Jerman, J. C., 2000. Anandamide: an endogenous activator of the vanilloid receptor. Trends Pharmacol Sci. 21, 134.

Smith, J. A., Davis, C. L. and Burgess, G. M., 2000. Prostaglandin E2-induced sensitization of bradykinin-evoked responses in rat dorsal root ganglion neurons is mediated by cAMP-dependent protein kinase A. Eur J Neurosci. 12, 3250–3258.

Smith, M. T., Cabot, P. J., Ross, F. B., Robertson, A. D. and Lewis, R. J., 2002. The novel N-type calcium channel blocker, AM336, produces potent dose-dependent antinociception after intrathecal dosing in rats and inhibits substance P release in rat spinal cord slices. Pain. 96, 119–127.

Soldo, B. L. and Moises, H. C., 1998. mu-opioid receptor activation inhibits N- and P-type Ca2+ channel currents in magnocellular neurones of the rat supraoptic nucleus. J Physiol. 513 (Pt 3), 787–804.

Southall, M. D., Michael, R. L. and Vasko, M. R., 1998. Intrathecal NSAIDS attenuate inflammation-induced neuropeptide release from rat spinal cord slices. Pain. 78, 39–48.

Southall, M. D. and Vasko, M. R., 2001. Prostaglandin receptor subtypes, EP3C and EP4, mediate the prostaglandin E2-induced cAMP production and sensitization of sensory neurons. J Biol Chem. 276, 16083–16091.

Stone, L. S., Broberger, C., Vulchanova, L., Wilcox, G. L., Hokfelt, T., Riedl, M. S. and Elde, R., 1998. Differential distribution of alpha2A and alpha2C adrenergic receptor immunoreactivity in the rat spinal cord. J Neurosci. 18, 5928–5937.

Sutton, K. G., Martin, D. J., Pinnock, R. D., Lee, K. and Scott, R. H., 2002. Gabapentin inhibits high-threshold calcium channel currents in cultured rat dorsal root ganglion neurones. Br J Pharmacol. 135, 257–265.

Suzuki, R., Morcuende, S., Webber, M., Hunt, S. P. and Dickenson, A. H., 2002. Superficial NK1-expressing neurons control spinal excitability through activation of descending pathways. Nat Neurosci. 5, 1319–1326.

Suzuki, R., Rahman, W., Rygh, L. J., Webber, M., Hunt, S. P. and Dickenson, A. H., 2005. Spinal-supraspinal serotonergic circuits regulating neuropathic pain and its treatment with gabapentin. Pain. 117, 292–303.

Svensson, C. I., Fitzsimmons, B., Azizi, S., Powell, H. C., Hua, X. Y. and Yaksh, T. L., 2005. Spinal p38beta isoform mediates tissue injury-induced hyperalgesia and spinal sensitization. J Neurochem. 92, 1508–1520.

Svensson, C. I., Marsala, M., Westerlund, A., Calcutt, N. A., Campana, W. M., Freshwater, J. D., Catalano, R., Feng, Y., Protter, A. A., Scott, B. and Yaksh, T. L., 2003. Activation of p38 mitogen-activated protein kinase in spinal microglia is a critical link in inflammation-induced spinal pain processing. J Neurochem. 86, 1534–1544.

Svensson, C. I. and Yaksh, T. L., 2002. The spinal phospholipase-cyclooxygenase-prostanoid cascade in nociceptive processing. Annu Rev Pharmacol Toxicol. 42, 553–583.

Takano, M., Takano, Y. and Yaksh, T. L., 1993. Release of calcitonin gene-related peptide (CGRP), substance P (SP), and vasoactive intestinal polypeptide (VIP) from rat spinal cord: modulation by alpha 2 agonists. Peptides. 14, 371–378.

Talley, E. M., Cribbs, L. L., Lee, J. H., Daud, A., Perez-Reyes, E. and Bayliss, D. A., 1999. Differential distribution of three members of a gene family encoding low voltage-activated (T-type) calcium channels. J Neurosci. 19, 1895–1911.

Todd, A. J., 1996. GABA and glycine in synaptic glomeruli of the rat spinal dorsal horn. Eur J Neurosci. 8, 2492–2498.

Todd, A. J., 2002. Anatomy of primary afferents and projection neurones in the rat spinal dorsal horn with particular emphasis on substance P and the neurokinin 1 receptor. Exp Physiol. 87, 245–249.

Todd, A. J. and Lochhead, V., 1990. GABA-like immunoreactivity in type I glomeruli of rat substantia gelatinosa. Brain Res. 514, 171–174.

Todd, A. J. and McKenzie, J., 1989. GABA-immunoreactive neurons in the dorsal horn of the rat spinal cord. Neuroscience. 31, 799–806.

Tognetto, M., Amadesi, S., Harrison, S., Creminon, C., Trevisani, M., Carreras, M., Matera, M., Geppetti, P. and Bianchi, A., 2001. Anandamide excites central terminals of dorsal root ganglion neurons via vanilloid receptor-1 activation. J Neurosci. 21, 1104–1109.

Too, H. P. and Maggio, J. E., 1991. Immunocytochemical localization of neuromedin K (neurokinin B) in rat spinal ganglia and cord. Peptides. 12, 431–443.

Trafton, J. A., Abbadie, C. and Basbaum, A. I., 2001. Differential contribution of substance P and neurokinin A to spinal cord neurokinin-1 receptor signaling in the rat. J Neurosci. 21, 3656–3664.

Trafton, J. A., Abbadie, C., Marchand, S., Mantyh, P. W. and Basbaum, A. I., 1999. Spinal opioid analgesia: how critical is the regulation of substance P signaling? J Neurosci. 19, 9642–9653.

Trevisani, M., Siemens, J., Materazzi, S., Bautista, D. M., Nassini, R., Campi, B., Imamachi, N., Andre, E., Patacchini, R., Cottrell, G. S., Gatti, R., Basbaum, A. I., Bunnett, N. W., Julius, D. and Geppetti, P., 2007. 4-Hydroxynonenal, an endogenous aldehyde, causes pain and neurogenic inflammation through activation of the irritant receptor TRPA1. Proc Natl Acad Sci USA. 104, 13519–13524.

Tuchscherer, M. M. and Seybold, V. S., 1985. Immunohistochemical studies of substance P, cholecystokinin-octapeptide and somatostatin in dorsal root ganglia of the rat Neuroscience. 14, 593–605.

Urban, L., Thompson, S. W. and Dray, A., 1994. Modulation of spinal excitability: co-operation between neurokinin and excitatory amino acid neurotransmitters. Trends Neurosci. 17, 432–438.

Urban, L. A. and Fox, A. J., 2000. NK1 receptor antagonists—are they really without effect in the pain clinic? Trends Pharmacol Sci. 21, 462–464; author reply 465.

Vasko, M. R., Campbell, W. B. and Waite, K. J., 1994. Prostaglandin E2 enhances bradykinin-stimulated release of neuropeptides from rat sensory neurons in culture. J Neurosci. 14, 4987–4997.

Vedder, H. and Otten, U., 1991. Biosynthesis and release of tachykinins from rat sensory neurons in culture. J Neurosci Res. 30, 288–299.

Vierck, C. J., Jr., Kline, R. H. and Wiley, R. G., 2003. Intrathecal substance p-saporin attenuates operant escape from nociceptive thermal stimuli. Neuroscience. 119, 223–232.

von Euler, U. S. and Gaddum, J. H., 1931. An unidentified depressor substance in certain tissue extracts. J Physiol. 72, 74–87.

Wang, H., Kohno, T., Amaya, F., Brenner, G. J., Ito, N., Allchorne, A., Ji, R. R. and Woolf, C. J., 2005. Bradykinin produces pain hypersensitivity by potentiating spinal cord glutamatergic synaptic transmission. J Neurosci. 25, 7986–7992.

Werry, E. L., Liu, G. J. and Bennett, M. R., 2006. Glutamate-stimulated ATP release from spinal cord astrocytes is potentiated by substance P. J Neurochem. 99, 924–936.

Wick, E. C., Pikios, S., Grady, E. F. and Kirkwood, K. S., 2006. Calcitonin gene-related peptide partially mediates nociception in acute experimental pancreatitis. Surgery. 139, 197–201.

Wilcox, G. L., 1988. Pharmacological studies of grooming and scratching behavior elicited by spinal substance P and excitatory amino acids. Ann NY Acad Sci. 525, 228–236.

Yaksh, T. L., Jessell, T. M., Gamse, R., Mudge, A. W. and Leeman, S. E., 1980. Intrathecal morphine inhibits substance P release from mammalian spinal cord in vivo. Nature. 286, 155–157.

Yamamoto, T. and Yaksh, T. L., 1991. Stereospecific effects of a nonpeptidic NK1 selective antagonist, CP-96,345: antinociception in the absence of motor dysfunction. Life Sci. 49, 1955–1963.

Yamamoto, T. and Yaksh, T. L., 1992. Effects of intrathecal capsaicin and an NK-1 antagonist, CP,96–345, on the thermal hyperalgesia observed following unilateral constriction of the sciatic nerve in the rat. Pain. 51, 329–334.

Yamasaki, H., Kubota, Y., Takagi, H. and Tohyama, M., 1984. Immunoelectron-microscopic study on the fine structure of substance-P-containing fibers in the taste buds of the rat. J Comp Neurol. 227, 380–392.

Yang, X. D., Liu, Z., Liu, H. X., Wang, L. H., Ma, C. H. and Li, Z. Z., 2007. Regulatory effect of nerve growth factor on release of substance P in cultured dorsal root ganglion neurons of rat. Neurosci Bull. 23, 215–220.

Yoshimura, M. and North, R. A., 1983. Substantia gelatinosa neurones hyperpolarized in vitro by enkephalin. Nature. 305, 529–530.

Zaratin, P., Angelici, O., Clarke, G. D., Schmid, G., Raiteri, M., Carita, F. and Bonanno, G., 2000. NK3 receptor blockade prevents hyperalgesia and the associated spinal cord substance P release in monoarthritic rats. Neuropharmacology. 39, 141–149.

Zeitz, K. P., Guy, N., Malmberg, A. B., Dirajlal, S., Martin, W. J., Sun, L., Bonhaus, D. W., Stucky, C. L., Julius, D. and Basbaum, A. I., 2002. The 5-HT3 subtype of serotonin receptor contributes to nociceptive processing via a novel subset of myelinated and unmyelinated nociceptors. J Neurosci. 22, 1010–1019.

Zeng, D. W. and Lynch, K. R., 1991. Distribution of alpha 2-adrenergic receptor mRNAs in the rat CNS. Brain Res Mol Brain Res. 10, 219–225.

Zerari, F., Karpitskiy, V., Krause, J., Descarries, L. and Couture, R., 1998. Astroglial distribution of neurokinin-2 receptor immunoreactivity in the rat spinal cord. Neuroscience. 84, 1233–1246.

Zhang, X., Bao, L., Arvidsson, U., Elde, R. and Hokfelt, T., 1998. Localization and regulation of the delta-opioid receptor in dorsal root ganglia and spinal cord of the rat and monkey: evidence for association with the membrane of large dense-core vesicles. Neuroscience. 82, 1225–1242.

Zhou, X. F. and Rush, R. A., 1996. Endogenous brain-derived neurotrophic factor is anterogradely transported in primary sensory neurons. Neuroscience. 74, 945–953.

Zimmer, A., Zimmer, A. M., Baffi, J., Usdin, T., Reynolds, K., Konig, M., Palkovits, M. and Mezey, E., 1998. Hypoalgesia in mice with a targeted deletion of the tachykinin 1 gene. Proc Natl Acad Sci USA. 95, 2630–2635.

Zygmunt, P. M., Petersson, J., Andersson, D. A., Chuang, H., Sorgard, M., Di Marzo, V., Julius, D. and Hogestatt, E. D., 1999. Vanilloid receptors on sensory nerves mediate the vasodilator action of anandamide. Nature. 400, 452–457.

Chapter 7
Opioidergic Transmission in the Dorsal Horn

Juan Carlos Marvizon

Abstract The potent analgesia produced by opiate drugs is induced, at least in part, in the spinal cord. The three "classical" opioid receptors, μ, δ and κ, are found in dorsal horn neurons and primary afferent terminals. Dorsal horn neurons expressing opioid receptors are mostly excitatory, and their inhibition by opioids decreases pain intensity. In primary afferents, opioid receptors inhibit the release of the pro-nociceptive neuropeptides substance P and CGRP. The spinal cord also contains "atypical" opioid receptors: the nociceptin receptor, the opioid growth factor receptor and toll-like receptors, which modulate pain in ways still not well understood. Enkephalins and dynorphins are the main opioid peptides in the dorsal horn, and are expressed by different neuronal populations. Endorphins are not found in the dorsal horn, and recent studies question whether endomorphins are indeed endogenous. Enkephalins and dynorphins are highly susceptible to peptidase degradation, which has prompted the use of peptidase inhibitors as analgesics. Endogenous peptidase inhibitors with analgesic properties have also been found. Opioid release in the spinal cord is inhibited by several neurotransmitter receptors, including adrenergic α_{2C} receptors, serotonin 5-HT$_{1A}$ receptors and NMDA receptors. Spinal opioid release appears to be driven by signals originating in both in primary afferents and supraspinally. Pain modality appears to determine whether pain induces spinal opioid release through local or supraspinal circuits. Some forms of stress-induced analgesia are also mediated by spinal opioid release. This involves a circuit originating in the dorsal raphe nucleus involved in the fear/anxiety response. Spinal opioids also mediate the analgesia induced by acupuncture.

J.C. Marvizon (✉)
Center for Neurobiology of Stress, Division of Digestive Diseases, Department of Medicine, David Geffen School of Medicine at UCLA, Los Angeles, CA 90095, USA;
Veteran Affairs Greater Los Angeles Healthcare System, Los Angeles, CA 90073, USA
e-mail: marvizon@ucla.edu

M. Malcangio (ed.), *Synaptic Plasticity in Pain*,
DOI 10.1007/978-1-4419-0226-9_7, © Springer Science+Business Media, LLC 2009

Abbreviations

5-HT	5-hydroxytriptamine
ACTH	adrenocorticotropin hormone
BK channels	Ca^{2+}-sensitive large conductance potassium channels
$Ca_V2.2$	N-type voltage-gated Ca^{2+} channels
CCK	cholecystokinin
CGRP	calcitonin gene-related peptide
CNS	central nervous system
CTAP	D-Phe-Cys-Tyr-D-Trp-Arg-Thr-Pen-Thr-NH_2
CRF	corticotrophin-releasing factor
DAMGO	[D-Ala^2, N-methyl-Phe^4, Gly-ol^5]enkephalin
DOR	δ-opioid receptor
DLF	dorsolateral funiculus
DPDPE	[D-penicillamine^2, D- penicillamine^5]enkephalin; DRN, dorsal raphe nucleus
DRG	dorsal root ganglia
GPCR	G protein-coupled receptor
KOR	κ-opioid receptor
HPLC	high pressure liquid chromatography
MOR	μ-opioid receptor
NRM	nucleus raphe magnus
NK1R	neurokinin 1 receptor
NOP_1	nociceptin receptor
OFQ/N	orphanin FQ/nociceptin
OGF	opioid growth factor
PAG	periaqueductal gray
POMC	proopiomelanocortin
RVM	rostral-ventral medulla
SIA	stress-induced analgesia
VGLUT2	vesicular glutamate transporter 2

7.1 Introduction

Opioid peptides, commonly known as "endorphins", have captured the popular imagination since their discovery. For example, the "endorphin rush" and "runner high" myths are commonplace in popular culture. One recent study (Boecker et al., 2008) does indeed support the idea that the runner's high is mediated by opioid release in the prefrontal cortex and limbic areas. While these myths push some people to extremes of their physical endurance with promises of feelings of euphoria and transcendence, others turn to opiate drugs in search of similar experiences. Opiate abuse can also result from the misuse of drugs to treat pain. Regardless of its cause, opiate abuse has catastrophic results and is still a growing and unresolved health problem worldwide. Perhaps the endogenous opioids can help solve the dual, interrelated problems of pain treatment and opiate abuse.

This chapter addresses the role of opioid peptides and their receptors in pain modulation in the spinal dorsal horn. There is evidence that part of the potent analgesia produced by opiate drugs is induced in the spinal cord, although supraspinal brain regions do contribute to it (Zorman et al., 1982; Jensen and Yaksh, 1984; Morgan et al., 1991; Budai and Fields, 1998). In fact, it is becoming increasingly clear that supraspinal and spinal opioids work together to control pain, since it appears that spinal opioid release is triggered by supraspinal mechanisms. However, the specific mechanisms involved are far form clear. This chapter reviews what is already known about the actions of spinal opioids and their receptors, with special emphasis in controversial issues and unresolved questions.

This book is about synapses and synaptic plasticity. Opioids are indeed released from synapses, and the modulation of their release by presynaptic receptors is addressed in details in Section 7.6. However, it is unlikely that opioid act in a classic synaptic manner, that is, by binding to receptors across the synaptic cleft. Instead, the presence of opioid receptors in the soma and the whole dendritic tree of dorsal horn neurons strongly suggests that opioids act by volume transmission (Fuxe and Agnati, 1991; Zoli and Agnati, 1996). Further evidence for this is provided by the fact that μ-opioid receptors (MORs) in a large number of dorsal horn neurons undergo internalization when opioid peptides are released in slices (Song and Marvizon, 2003b, a) or in vivo (Lao et al., 2008).

7.2 "Classical" Opioid Receptors in the Dorsal Horn

There are three "classical" opioid receptors: μ (MOR), δ (DOR) and κ (KOR). All are G protein-coupled receptors (GPCRs); they belong to the sub-family of rhodopsin receptors and have seven transmembrane domains. They also signal in similar ways, coupling to $\alpha_{i/o}$ G proteins to inhibit adenylyl cyclase, open inner-rectifying K^+ channels, and inactivate L-type and N-type Ca^{2+} channels. All three of these receptors mediate analgesia in the spinal cord (Zorman et al., 1982; Jensen and Yaksh, 1984; Russell et al., 1987; Morgan et al., 1991; Watkins et al., 1992; Takemori and Portoghese, 1993; Budai and Fields, 1998; Chen and Pan, 2006; Chen et al., 2007), although the participation of DORs in analgesia has been questioned recently (Scherrer et al., 2004).

7.2.1 Localization: Dorsal Horn Neurons and Primary Afferent Terminals

In the dorsal horn, opioid receptors are present at two locations: dorsal horn neurons and primary afferents terminals (Arvidsson et al., 1995; Mansour et al., 1995; Kemp et al., 1996).

Dorsal horn neurons with MORs are of small size and are located primarily in lamina II but also found in lamina I. Morphologically, these neurons are identified as islet or stalked cells (Eckert et al., 2003). They have dendrites oriented in the rostro-caudal direction (Mansour et al., 1995; Marvizon et al., 1999; Song and Marvizon, 2003b) and axons projecting ventrally to lamina III–V or dorsally to lamina I. MOR neurons are mostly excitatory (Kemp et al., 1996) and may form part of a polysynaptic excitatory pathway from primary afferents to neurons in the deep dorsal horn or the superficial dorsal horn. Hence, when MORs decrease the firing of these neurons by hyperpolarizing them, this results in analgesia.

DORs and KORs appear to be less abundant than MORs in the dorsal horn. KORs are mostly found in lamina II (Morris and Herz, 1987; Mansour et al., 1988), whereas DORs are spread over laminae I–VI (Minami and Satoh, 1995). KORs are present in dendrites, axon and axon terminals and in a few neuronal somata and glia. There seem to be sex differences in the expression of KORs in the dorsal horn: they are more abundant in estrous and proestrous female rats than in male rats (Harris et al., 2004).

Primary afferents contain all three receptors: MORs, DORs and KORs, which colocalize in them in different combinations (Fields et al., 1980). Of the primary afferent somata in dorsal root ganglia (DRG), 55% contain MOR mRNA, 20% contain DOR mRNA and 18% contain KOR mRNA (Minami and Satoh, 1995). Most of the primary afferents that express opioid receptors are nociceptive C-fibers (Fields et al., 1980; Besse et al., 1990; Grudt and Williams, 1994). Primary afferent fibers in the dorsal horn account for approximately one half of the MOR immunoreactivity and two thirds of the DOR immunoreactivity (Abbadie et al., 2002). KORs are also present in primary afferent fibers, since their density decreased after rhizotomy (Besse et al., 1990). Interestingly, DORs and KORs in C-fibers are associated with dense core vesicles (Besse et al., 1990; Zhang et al., 1998).

There is some controversy on whether the analgesic effect of opioids in the spinal cord (Yaksh, 1981) is mediated by opioid receptors located in C-fiber terminals or in dorsal horn neurons. Despite previous studies showing that MORs inhibit substance P release (Jessell and Iversen, 1977; Yaksh et al., 1980; Aimone and Yaksh, 1989), Trafton et al. (1999) reported that morphine and the selective MOR agonist DAMGO showed only a limited ability to inhibit substance P release induced by noxious stimuli. They used the internalization of neurokinin 1 receptors (NK1Rs) as a measure of both substance P release and NK1R activation. Based on these results, they concluded that MORs in C-fiber terminals reduced substance P release, but only to concentrations that are still supra-threshold to activate NK1Rs. If this is true, then MOR inhibition of substance P release would not play a role in inducing analgesia. Nevertheless, a follow-up study by Kondo et al. (2005) found that analgesic doses of morphine and DAMGO readily inhibited NK1R internalization evoked by noxious stimuli. The DOR selective agonist DPDPE also inhibited the evoked NK1R internalization, albeit at higher doses, indicating that DOR in C-fiber terminals contribute to the inhibition of substance P release. The reason for the discrepancy between these two studies is not totally clear, but Kondo et al. found that the percutaneous method for intrathecal injection used by Trafton et al. was less effective than intrathecal injections performed through chronic catheters, which could be the reason why MOR agonists were found to be less effective. The issue of whether analgesia is produced by MORs in C-fiber terminals or in dorsal horn neurons was addressed by Kline and Wiley (2008) using a different approach. They used dermorphin-saporin to eliminate MOR-expressing neurons in the dorsal horn without affecting MOR-expressing primary afferents. This treatment reduced the antinociceptive action of morphine, administered either systemically or intrathecally, indicating that MOR-expressing dorsal horn neurons do contribute to opioid analgesia. However, the size of this effect relative to the contribution of the MORs in C-fibers remains unclear.

7.2.2 *Opioid Receptor Signaling*

In dorsal horn neurons, the effect of opioid receptors involves mainly a hyperpolarization produced by the opening of inwardly-rectifying K^+ channels (Yoshimura and North, 1983; Grudt and Williams, 1993, 1994).

In primary afferent terminals, MORs inhibit the release of substance P and CGRP (Jessell and Iversen, 1977; Yaksh et al., 1980; Go and Yaksh, 1987; Pohl et al., 1989; Kondo et al., 2005) (see Hua and Yaksh, Chapter 6). DORs, but not KORs, contribute to the inhibition of substance P release (Kondo et al., 2005). However, KORs decreased Ca^{2+} currents in DRG (Macdonald and Werz, 1986; Gross and Macdonald, 1987), suggesting that they may inhibit neurotransmitter release from other primary afferents. Inhibition of substance P release suppresses the activation of NK1Rs (Kondo et al., 2005), which are located in a population of lamina 1 projection neurons (Todd et al., 2000) and play a key role in the induction of chronic pain states (Traub, 1996, 1997; Ikeda et al., 2003). MORs inhibit neurotransmitter release by inactivating voltage-gated Ca^{2+} channels (Schroeder et al., 1991). Receptors that couple to $\alpha_{i/o}$ G proteins are known to inhibit voltage gated Ca^{2+} channels by a voltage-sensitive mechanism involving binding of the $\beta\gamma$ subunits of the G proteins to the channels (Adamson et al., 1989; Li and Bayliss, 1998; Dolphin, 2003). However, a recent study (Raingo et al., 2007) showed that the N-type Ca^{2+} channels ($Ca_V2.2$) expressed in primary afferents have a splice isoforms of the $\alpha1$ subunit containing exon 37a. These $Ca_V2.2$ channels can be inhibited by MORs and $GABA_B$ receptors in a voltage-insensitive fashion by tyrosine kinase phosphorylation (Diverse-Pierluissi et al., 1997; Strock and Diverse-Pierluissi, 2004; Raingo et al., 2007). This mechanism seems to be unique to primary afferents and may have important implications for pain control.

7.2.3 Opioid Receptor Internalization, Trafficking and Synergism

MORs are usually present at the cell surface in neurons in the dorsal horn and other CNS regions (Eckersell et al., 1998; Sinchak and Micevych, 2001; Mills et al., 2004). Upon agonist binding, MORs are internalized into endosomes and in about one hour are recycled back to the cell surface (Marvizon et al., 1999; Trafton et al., 2000; Song and Marvizon, 2003b). However, not all agonists of MORs cause their internalization: morphine and other opiate alkaloids (heroin, codeine and buprenorphine) do not produce MOR internalization (Keith et al., 1996, 1998). It has been suggested that the inability of some opiates to induce MOR internalization is related to their propensity to induce tolerance (Whistler and von Zastrow, 1998; Whistler et al., 1999), based on the observation that enkephalins produce little tolerance and even block the tolerance induced by morphine (Graf et al., 1979). However, this hypothesis is not consistent with the fact that endomorphins and other opioids produce both tolerance and MOR internalization (Soignier et al., 2004). Another study (He et al., 2002) showed that morphine was able to induce MOR internalization in the presence of low doses of the MOR agonist DAMGO, suggesting that the simultaneous binding of these two agonists to MOR dimers induces internalization.

MOR internalization provides an ideal way to measure opioid release, analogous to the way NK1R internalization is used to measure substance P release (Mantyh et al., 1995; Abbadie et al., 1997; Allen et al., 1997; Marvizon et al., 1997; Honore et al., 1999; Marvizon et al., 2003a; Kondo et al., 2005). This technique takes advantage of the prominent change in the intracellular localization of the MOR after it is activated by agonists. Thus, the MOR itself takes over the role of the antibody in an immunoassay as detector of the opioids, but there is no need to extract the peptides because the receptor is present in the tissue. Advantages of this method over conventional methods like immunoassay include: (1) it detects the release of all the opioids able to activate MOR (Song and Marvizon, 2003b), whereas immunoassays usually detect only one (i.e. Met-enkephalin) among the many opioids released in the spinal cord (Nyberg et al., 1983; Yaksh et al., 1983); (2) MOR internalization and pain responses can be measured in the same animals (Chen et al., 2007); (3) MOR internalization is an in situ measure of opioid release, providing spatial information about the sites of release (i.e., spinal segment and side) (Lao et al., 2008); (4) it also provides information about the activation of MORs by the released opioids. This method has been used to measure opioid release from several tissues, including the brain (Eckersell et al., 1998; Sinchak and Micevych, 2001; Mills et al., 2004), the spinal cord (Trafton et al., 2000; Song and Marvizon, 2003b, a, 2005) and the intestine (Patierno et al., 2005).

MOR internalization is not only a measure of opioid release, but also of the activation of MORs by the released opioids. Evidence for this includes: (1) MOR internalization correlated with the analgesia induced by the MOR-selective agonists DAMGO (Trafton et al., 2000) and endomorphin-2 (Chen et al., 2007); (2) the dose-responses of DAMGO to elicit MOR internalization (Marvizon et al., 1999), adenylyl cyclase inhibition (Keith et al., 1996, 1998) and $[\gamma^{-35}S]$GTP binding (Yabaluri and Medzihradsky, 1997) are virtually identical; (3) MOR internalization increased with the intensity of the stimulus delivered to spinal cord slices to evoke opioid release (Song and Marvizon, 2003a).

7.2.4 Synergism Between MORs and DORs

In contrast to MORs, DORs appear to be constitutively internalized, both in spinal cord neurons and in primary afferent terminals. In dorsal horn neurons, presence of DORs at the cell surface increased after prolonged (48 h) treatment with morphine (Cahill et al., 2001), an effect that was reversible and dependent on the presence of MORs (Morinville et al., 2003). Unilateral rhizotomy also increased DOR expression at the cell surface in both the ipsilateral and the contralateral dorsal horns (Morinville et al., 2004). The mechanisms involved in these effects are still unknown. In primary afferent terminals, DORs in are associated with dense core vesicles and are inserted in the membrane when substance P and other neuropeptides are released from these vesicles (Zhang et al., 1998; Bao et al., 2003). Moreover, unlike MORs, DORs in DRG neurons fail to inhibit voltage-gated Ca^{2+} channels (Schroeder et al., 1991; Moises et al., 1994; Liu et al., 1995; Walwyn et al., 2005); quite likely because they were not present at the cell surface and thus not accessible to the agonist. However, several stimuli increase trafficking of DORs to the plasma membrane and the analgesic efficacy of DOR agonists. These include DOR agonists, chronic inflammation, forced swimming and prolonged exposure to morphine (Cahill et al., 2001; Bao et al., 2003; Cahill et al., 2003; Commons, 2003). Other treatments that induced DOR surface expression were exposure to a DOR antagonist, and prolonged treatment with a MOR antagonist (CTAP) followed by a brief application of a DOR agonist (DPDPE) (Walwyn et al., 2005). It should be kept in mind, however, that many of these experiments were done using DRG neuron cultures and may not reflect the physiological state of the presynaptic terminal.

The fact that MORs control the trafficking of DORs may form the basis for a synergistic interaction between these two receptors. Thus, enkephalins induce analgesia through a synergism between MORs and DORs, as indicated by pharmacological analysis of the interaction between MOR and DOR agonists (Malmberg and Yaksh, 1992) and by the fact that enkephalin-induced analgesia is abolished by both MOR and DOR antagonists (Chen et al., 2007). Moreover, analgesia induced by intrathecal Leu-enkephalin lasts more than 25 min (Chen et al., 2007), while MORs internalize within 10 min (Marvizon et al., 1999). In contrast with enkephalin, the MOR-selective agonist endomorphin-2 produced analgesia with lower potency and for a shorter time. This suggests that opioid receptors induce analgesia sequentially: MOR are activated first and internalize in about 10 min (Marvizon et al., 1999) and then DORs are trafficked to the cell surface (Cahill et al., 2001; Morinville et al., 2003) in time to replace them.

7.2.5 Opioid Receptor Heterodimers

Another feature that can mediate the synergy between different opioid receptors subtypes is the formation of heterodimers. MORs (Hazum et al., 1982) and DORs (Cvejic and Devi, 1997) can exist as dimers. The formation of MOR-DOR heterodimers was first shown in cells co-expressing these two receptors (Gomes et al., 2000). These heterodimers display peculiar pharmacological properties: a DOR antagonist increased the binding of a MOR agonist, a MOR antagonist increased the binding of a DOR agonist, and a DOR agonist increased the binding of a MOR agonist. Since MORs and DORs colocalize in DRG neurons and the central terminals of primary afferent (Fields et al., 1980; Arvidsson et al., 1995), it is possible that these heterodimers occur physiologically. Indeed, a DOR antagonist potentiated morphine analgesia in vivo, indicating that these heterodimers do have a functional role (Gomes et al., 2004).

Similarly, DOR and KOR can form heterodimers that have novel pharmacological and signaling properties (Jordan and Devi, 1999). However, increase of agonist binding by antagonists was not observed in this case. Recently, a selective agonist of DOR-KOR heterodimers has been discovered and shown to produce analgesia in the spinal cord (Waldhoer et al., 2005).

Opioid receptors can also form heterodimers with non-opioid receptors: for example, DORs appear to be able to form heterodimers with α_{2A} adrenergic receptors (Rios et al., 2004).

7.3 Atypical Opioid Receptors

7.3.1 Nociceptin Receptor

The nociceptin receptor NOP_1 (previously termed ORL1) was discovered in 1994 by several groups as a clone with about 50% homology to the classical opioid receptors (Mogil and Pasternak, 2001). Like the classical opioid receptors, NOP_1 receptors couple to $\alpha_{i/o}$ G proteins and inhibit adenylyl cyclase, inactivate voltage-gated Ca^{2+} channels and open K^+ channels. In electrophysiological studies, NOP1 produced mostly inhibitory effects. This receptor does not bind the known opioid peptides, but soon after the discovery of NOP_1 receptors, an endogenous ligand was isolated by two independent groups, who named it nociceptin (Meunier et al., 1995) and orphanin FQ (Reinscheid et al., 1995). These two names are still used interchangeably and often combined as orphanin FQ/nociceptin (OFQ/N). OFQ/N is a 17 amino acid peptide with some homology to dynorphin A. Its N-terminus has the motif FGGF instead of the YGGF motif of the other endogenous opioids. Both the NOP_1 receptor and OFQ/N are present in the spinal cord (Neal et al., 1999b, a). The effect of the $OFQ/N - NOP_1$ system on pain is complex and still controversial. Initial experiments using intracerebroventricular injections of OFQ/N found increased responses to thermal stimuli (tail-flick and hot plate tests) (Meunier et al., 1995; Reinscheid et al., 1995) and were interpreted as a hyperalgesic effect. Later studies uncovered both hyperalgesic and analgesic effects of supraspinal OFQ/N. However, it has become increasingly clear that OFQ/N blocks the analgesia produced by the other opioid receptors (Mogil et al., 1996; Mogil and Pasternak, 2001), by α_2 adrenergic receptors (King et al., 1998) and by GABA$_B$ receptors (Citterio et al., 2000). At the spinal cord level, the effects of OFQ/N are equally complex. Low doses of OFQ/N produced spontaneous pain that was suppressed by NK1R antagonists (Sakurada et al., 1999). Higher doses produced anti-hyperalgesic effects, but not in all the studies.

7.3.2 Opioid Growth Factor (OGF) Receptor

The OGF receptor mediates the potent effects on cellular growth and development of Met-enkephalin (known in this context as "opioid growth factor"). Met-enkephalin acts as a negative growth regulator not only in neural and non-neural tissues of many animals, but also in prokaryotes (Zagon et al., 2002). Its functions include control of development, cellular renewal, anti-tumor, wound healing and angiogenesis. Like the MOR, the OGF receptor is blocked by naloxone in a stereospecific fashion, but it differs from MORs in most of its other features. The OGF receptor binds Met-enkephalin at nanomolar concentrations, but not dynorphins, alkaloid opiates and most other ligands of MORs and DORs. It has no sequence homology to the classical opioid receptors, and is located in the nuclear membrane and in the cytoplasm next to the nucleus. In the CNS, OGF receptors are present in both glia and neurons. Its function is to modulate DNA synthesis when it is transported into the nucleus after binding Met-enkephalin. Although the OGF receptor is thought to be important mostly during development, it is possible that it participates in the genotype switching that takes place in neuropathic pain.

7.3.3 Toll-Like Receptors as Receptors for Opiate Drugs

There is currently a lot of excitement about the role of glial cells in the spinal cord in the maintenance of neuropathic pain (see Abbadie and Sullivan, Chapter 15; Tawfik and DeLeo, Chapter 17; Milligan, Soderquist, Mahoney, Chapter 17; Ji, Chapter 20; Beggs, Chapter 22). Glia are also involved in some of the undesirable effects of opiate drugs, like tolerance and morphine-induced anti-analgesia (reviewed in Watkins et al., 2005; Hutchinson et al., 2007). These phenomena are not mediated by the classic opioid receptors, as was first suggested by the ability of the inactive stereoisomer of naloxone, (+)-naloxone, to reverse allodynia in a model of neuropathic pain. Morphine activates glia, increasing their production of nitric oxide and pro-inflammatory cytokines. The resulting hyperalgesia counters the analgesic effects of morphine mediated by neuronal MORs and DORs. Glial activation by morphine and de-activation by naloxone presents an anomalous pharmacological profile when compared to that of the classical opioid receptors: there is no stereoselectivity for morphine, methadone or naloxone, and the potent MOR agonist etorphine has no effect. Indeed, it now seems that these effects are mediated by the binding of the opiate drugs to toll-like receptor 4. It is not known whether endogenous opioid peptides activate this receptor.

7.4 Opioid Peptides in the Dorsal Horn

Opioid peptides are encoded by three genes, which are translated into protein precursors where the active peptide is flanked on both sides by pairs of basic amino acids (Costa et al., 1987). These basic amino acids serve as signals for the cleavage of the precursor into the opioid peptides. The proopiomelanocortin (POMC) gene encodes the endorphins, as well as adrenocorticotropin (ACTH) and melanocyte-stimulating hormone. The proenkephalin (or proenkephalin A) gene contains six copies of Met-enkephalin and one copy of Leu-enkephalin. The prodynorphin (or proenkephalin B) gene encodes for dynorphins of different length and α-neoendorphin, all of which contain the Leu-enkephalin sequence. These three genes are generally expressed in different types of neurons.

7.4.1 Endorphins

Endorphins are not present in the dorsal horn (Tsou et al., 1986), and therefore are not the opioid peptides that activate opioid receptors in this region. Nevertheless, immunoreactivity to endorphins and other peptides encoded by the POMC gene was found in the area around the central canal (lamina X) and in the ventral horn. Since lamina X appears to be involved in some forms of analgesia, it is possible that endorphins contribute to pain control there. It was initially thought that all the endorphin-containing terminals in the spinal cord were of supraspinal origin, because no POMC gene products were found in cell bodies in the spinal cord. Furthermore, transection of the spinal cord resulted in loss of immunoreactivity to the POMC peptides below the level of the transection. However, a later study by the same group (Gutstein et al., 1992), using radioimmunoassay to measure β-endorphin content, revealed that about one third of the β-endorphin persisted below the level of a spinal transection. This suggests that some spinal cord neurons express the POMC gene. In any case, the amount of β-endorphin released from the cat or the rat spinal cord was very small compared with the amounts of enkephalins and dynorphins measured in the same experiments (Yaksh et al., 1983). Therefore, endorphins likely play only a minor role, if any, in pain modulation in the spinal cord.

7.4.2 Enkephalins

Enkephalins are the most abundant opioid peptides in the dorsal horn. The proenkephalin and the prodynorphin genes are expressed by different dorsal horn neurons (Cruz and Basbaum, 1985; Standaert et al., 1986). Laminae I–II contain numerous enkephalin-positive presynaptic terminals and axons (Todd and Spike, 1993) and also neuronal somata expressing pre-proenkephalin (Harlan et al., 1987). Enkephalin-immunoreactive fibers are also present in the dorsolateral funiculus (Song and Marvizon, 2003a), and are probably axons from supraspinal regions or from other spinal segments. Opioids released from the spinal cord include the Met-enkephalin and Leu-enkephalin penta-peptides, as well as longer enkephalin peptides with 6–8 amino acids (Yaksh et al., 1983).

Enkephalin coexists with GABA in some dorsal horn neurons (Todd et al., 1992), indicating that enkephalins may reinforce the inhibitory action of GABA. Some enkephalinergic neurons also express somatostatin (Todd and Spike, 1992), another important inhibitory neuropeptide. However, many enkephali-nergic neurons in lamina II are not GABAergic and may be in fact excitatory. In many presynaptic terminals, enkephalins colocalize with vesicular glutamate transporter 2 (VGLUT2, a marker of glutamatergic terminals) and adrenergic α_{2C} receptors (Stone et al., 1998; Olave and Maxwell, 2002, 2004; Marvizon et al., 2007). These excitatory interneurons make synapses with nociceptive projections neurons (Olave and Maxwell, 2003a, b), suggesting that they belong to a pro-algesic pathway despite the fact that they release enkephalins. Importantly, some dorsal horn enkephalinergic neurons project supraspinally, in particular to the parabrachial nucleus (Standaert et al., 1986). The exact role of enkephalinergic neurons in modulating pain in the dorsal horn is therefore still unclear.

Primary afferent terminals do not contain enkephalins. Most dorsal root ganglion (DRG) neurons do not express enkephalins (Pohl et al., 1994), and even the few that do (about 3.5%) appear to transport enkephalins to their peripheral terminals and not their central terminals (Bras et al., 2001). In spite of this, some studies reported colocalization of enkephalins and substance P (Senba et al., 1988; Ribeiro-da-Silva et al., 1991). These are probably not primary afferent terminals, but dorsal horn neurons containing substance P (Hunt et al., 1981). In fact, stimuli that produced abundant substance P release in the dorsal horn (detected with NK1R internalization) produced no MOR internalization (Trafton et al., 2000; Song and Marvizon, 2003a), indicating that enkephalins and substance P are not co-released in the dorsal horn.

7.4.3 Dynorphins

Dynorphins include several peptides encoded by the prodynorphin gene, including dynorphin A, dynorphin B and α-neoendorphin. The prodynorphin gene is expressed by a different set of dorsal horn neurons than the proenke-phalin gene (Cruz and Basbaum, 1985; Standaert et al., 1986; Todd and Spike, 1993). These neurons and prodynorphin-containing terminals are located more superficially than the enkephalinergic neurons, primarily in lamina I and outer lamina II. Dynorphin-containing neurons are also more sparsely located in the deeper laminae of the dorsal horn. In addition, dynorphins are present in some primary afferents (Botticelli et al., 1981; Basbaum et al., 1986). Although dynorphin colocalizes with substance P in presynaptic terminals in the dorsal horn, these seem to originate from spinal cord neurons and not from primary afferents (Tuchscherer and Seybold, 1989). As in the case of enkephalins, some of the dynorphin neurons in the dorsal horn project to the parabrachial nucleus (Standaert et al., 1986).

Importantly, preprodynorphin mRNA and dynorphin peptides increase substantially in a variety of chronic pain models, including inflammation, polyarthritis and peripheral nerve injury (Cho and Basbaum, 1988; Ruda et al., 1988; Todd and Spike, 1993). These increased dynorphin levels may contribute for the hyperalgesia encountered in chronic pain disorders and after sustained spinal administration of opiates (Vanderah et al., 2000). This action of dynorphin does not involve its interaction with opioid receptors but with NMDA receptors (Tang et al., 1999), although non-NMDA mechanisms may also be involved (Tang et al., 2000) (see Lai and Porreca, Chapter 21).

7.4.4 Endomorphins: Are They Really Endogenous?

Endomorphin-1 and endomorphin-2 are two tetrapeptides (YPWF-NH$_2$ and YPFF-NH$_2$, respectively) that are potent and selective agonists of MORs. They were synthesized in 1997 by Zadina et al., who later isolated them from bovine and human brain (Hackler et al., 1997; Zadina et al., 1997). Since then, a great deal of studies have been performed on the assumption that endomorphins are endogenous (reviewed in Fichna et al., 2007). However, a dozen years after their discovery, a gene encoding endomorphins has not been found. This has prevented the use of molecular biology techniques (PCR, in situ hybridiza-tion, etc) to detect them. Instead, practically all the studies in which endo-morphins are detected have relied on antibodies raised against them. Even the original reports on their isolation (Hackler et al., 1997; Zadina et al., 1997) relied partially on radioimmunoassay with an endomorphin-1 antibody to identify the peptides. It is important to keep this in mind, because the use of polyclonal antibodies can lead to artifacts when they recognize other peptides. Thus, the amidated phenylalanine C-terminus of the endomorphins is a significant source of antigenicity, and many other neuropeptides possess it. This include CGRP, neuropeptide-FF, gonadotropin inhibitory hormone and prolactin-releasing peptide. Indeed, anti-endomorphin antibodies have shown some cross-reactivity with CGRP (Pierce et al., 1998). One study

(Lisi and Sluka, 2006) attempted to detect endogenous endomorphins without using antibodies, by using HPLC and electrochemical detection instead. It detected no release of endomorphin-1 from the spinal cord and only a small peak corresponding to the endomorphin-2 elution time, which could have been due to contamination by their standards.

In the case of the spinal cord, there are important inconsistencies between studies using anti-endomorphin antibodies and other approaches. Immunohistochemistry studies by several groups (Martin-Schild et al., 1998; Pierce et al., 1998; Spike et al., 2002; Nydahl et al., 2004) concur in that endomorphins are present in substance P-containing primary afferent terminals. However, endomorphin immunoreactivity was not found in DRG (Schreff et al., 1998). More significantly, several functional studies cannot be reconciled with the idea that endomorphins are co-released with substance P from primary afferent terminals. First, dorsal root stimulation of rat spinal cord slices, which produced extensive NK1R internalization by releasing substance P (Song and Marvizon, 2003a), did not produce any MOR internalization. Capsaicin and NMDA also induce substance P release and NK1R internalization (Marvizon et al., 1997; Malcangio et al., 1998; Lever et al., 2001; Lever and Malcangio, 2002; Lao et al., 2003; Marvizon et al., 2003a), but did not induce any MOR internalization (Song and Marvizon, 2003a). Since exogenous endomorphins potently induced MOR internalization when applied to the slices (Song and Marvizon, 2003b), these results demonstrate that endomorphins are not co-released with substance P. Second, if endomorphins are co-released with substance P they should bind to the MORs that are present in those same terminals and inhibit substance P release. However, these MORs seem to be vacant, because exogenous opiates are able to inhibit substance P release (Jessell and Iversen, 1977; Yaksh et al., 1980; Go and Yaksh, 1987; Pohl et al., 1989; Lever et al., 2001; Kondo et al., 2005). Third, while noxious stimulation produces abundant substance P release (Abbadie et al., 1997; Allen et al., 1997), it did not induce MOR internalization (Trafton et al., 2000), as would be expected if endomorphins are co-released with substance P. Although we were able to observe MOR internalization induced by noxious stimuli (Lao et al., 2008), this occurred only after intrathecal injection of peptidase inhibitors, showing that the opioids released were susceptible to peptidase degradation, while endomorphins are not (Song and Marvizon, 2003b).

In view of all these problems, it remains dubious that endomorphins play a role in pain modulation in the spinal cord. In fact, it is still uncertain that endomorphins are indeed endogenous. It is imperative at this point to use powerful analytical techniques, like mass spectrometry, to resolve this issue while avoiding the uncertainties associated with the use of anti-endomorphin antibodies.

7.4.5 Receptor Specificity: Is It Important?

It has been assumed that ligands for the opioid receptors have to be receptor-specific. However, most endogenous opioid peptides are able to activate different opioid receptors with only minor differences in their potencies. Thus, enkephalins and endorphins are able to activate both MORs and DORs with relatively small differences in potencies, while dynorphins, considered to be KOR-specific, also activate MORs (Raynor et al., 1993). Moreover, the different susceptibility of the opioids to peptidase degradation makes it hard to predict their concentration in the receptor microenvironment, and may cancel differences in potencies. For example, in the absence of peptidase inhibitors, the apparent potency of dynorphin A to induce MOR internalization is greater than that of Leu-enkephalin, due to the more effective degradation of enkephalins by peptidases (Song and Marvizon, 2005). Therefore, it is likely that both enkephalins and dynorphins activate MORs in physiological conditions. Indeed, co-activation of the different opioid receptors by opioids may be functionally important. Thus, in the presence of peptidase inhibitors, Leu-enkephalin was very potent and effective to produce analgesia by co-activating MORs and DORs, whereas the selective MOR agonist endomorphin-2 was two orders of magnitude less potent (Chen et al., 2007). Hence, synergism between MORs and DORs produces more potent analgesia than activation of MORs alone.

7.5 Opioid Degradation by Peptidases

7.5.1 Peptidases that Degrade Opioids

The high susceptibility of the opioid peptides to peptidase degradation has been known for some time (Guyon et al., 1979). Three peptidases are responsible or opioid degradation (Oka et al., 1986; Numata et al., 1988): aminopeptidases (E.C. 3.4.11.-), dipeptidyl carboxypeptidase (E.C. 3.4.15.1), and neutral endopeptidase (E.C. 3.4.24.11). Opioid peptides (except the endomorphins) have a common four amino acid sequence at the N-terminus (YGGF), which is required for binding to opioid receptors. Aminopeptidases cleave one amino acid at the time starting at the amino terminus; therefore, just one cleavage results in the inactivation of the opioid peptide. Neutral endopeptidase cleaves the GF bond, which also results in the immediate inactivation of the peptide. In contrast, dipeptidyl carboxypeptidase cleaves two amino acids at the time starting at the carboxyl terminus. Hence, whilst this peptidase would immediately inactivate short peptides like Met- and Leu-enkephalin, longer peptides would have to be cleaved repeatedly for them to be inactivated. Indeed, dynorphin A is less susceptible than Leu-enkephalin to degradation by dipeptidyl carboxypeptidase (Song and Marvizon, 2003b).

Selective inhibitors of these three peptidases have been used to study their ability to degrade opioids in different tissues. Amastatin or actinonin are used to inhibit aminopeptidase, captopril to inhibit dipeptidyl carboxypeptidase, and phosphoramidon or thiorphan to inhibit neutral endopeptidase. This strategy revealed that peptidases are very effective to degrade both enkephalins and dynorphins in guinea pig intestine and brain (Aoki et al., 1984; Hiranuma et al., 1997; 1998). In the rat spinal cord, inhibitors of these three peptidases increased 100-fold the potency of Leu-enkephalin to elicit MOR internalization, and 10-fold that of dynorphin A (Song and Marvizon, 2003b). Peptidase inhibitors also enabled Met-enkephalin and α-neoendorphin to produce MOR internalization.

The aminopeptidase that degrades opioids is probably aminopeptidase N, which is present in the dorsal horn (Noble et al., 2001), judging by its pharmacological profile (Tieku and Hooper, 1992; Suzuki et al., 1997). Thus, aminopeptidase N is inhibited by amastatin, actinonin and high concentrations of bestatin, but not by puromycin (Song and Marvizon, 2003b), whereas other enkephalin-degrading aminopeptidases are puromycin-sensitive (Shimamura et al., 1983; Dyer et al., 1990; Hui et al., 1998).

Endomorphins do not seem to be appreciably degraded by peptidases in the spinal cord, because they produced MOR internalization with the same high potencies in the presence and absence of peptidase inhibitors (Song and Marvizon, 2003b), including an inhibitor of dipeptidyl peptidase IV, considered an endomorphin-degrading enzyme (Shane et al., 1999). Hence, since the opioids released in the spinal cord are degraded by peptidases (Song and Marvizon, 2003b, a, 2005; Lao et al., 2008), they are not endomorphins. Like endomorphins, β-endorphin produced MOR internalization in the absence of peptidase inhibitors. It is likely that the length of β-endorphin puts its amino terminus out of reach of the carboxypeptidases, while folds in its tertiary structure protect it against aminopeptidases (Bewley and Li, 1985). Neutral endopeptidase preferentially cleaves β-endorphin at the Leu[17]-Phe[18] bond (Burbach and De Kloet, 1982; Graf et al., 1985), producing γ-endorphin.

7.5.2 *Peptidase Inhibitors Used as Analgesics*

Since enkephalins and dynorphins are the main opioids in the dorsal horn and they are rapidly degraded by peptidases, peptidase inhibitors delivered to the spinal cord should produce analgesia by increasing their availability. There is substantial evidence that this is the case. Thus, the analgesia induced by intrathecal Met-enkephalin or dynorphin, or by electroacupuncture, was substantially increased by co-injection of peptidase inhibitors (Kishioka et al., 1994). We recently compared analgesia and MOR internalization in dorsal horn neurons produced by several intrathecal doses Leu-enkephalin, with and without peptidase inhibitors (Chen et al., 2006). Without peptidase inhibitors, Leu-enkephalin produced absolutely no analgesia or MOR internalization at doses up to 100 nmol. In contrast, with peptidase inhibitors Leu-enkephalin produced maximal analgesia and MOR internalization at 10 nmol; and substantial analgesia was still found with 0.3 nmol Leu-enkephalin.

Dipeptidyl carboxypeptidase and neutral endopeptidase also degrade substance P (Duggan et al., 1992; Marvizon et al., 2003b). Since substance P participates in the induction of chronic pain (Traub, 1996, 1997; Ikeda et al., 2003), inhibiting these peptidases could also have pro-algesic effects. However, peptidases appear to have a more pronounced effect on opioids than on substance P. Thus, captopril and phosphoramidon increased only three-fold the potency of substance P to induce NK1R internalization in spinal cord slices, and amastatin had no effect (Marvizon et al., 2003b). In contrast, amastatin, captopril and phosphoramidon increased 100-fold the potency of Leu-enkephalin to induce MOR internalization (Song and Marvizon, 2003b). Therefore, the effect of peptidase inhibitors on the availability of opioids is likely to overcome their effect on the availability of substance P.

The analgesic effect of peptidase inhibitors prompted the development of RB 101, a pro-drug that generates both an aminopeptidase inhibitor and a neutral endopeptidase inhibitor (Noble et al., 1992b, 1997). RB 101 produced analgesia in the mouse and the rat (hot plate, writhing and tail-flick tests) after systemic administration. Its effects were suppressed by naloxone, but not by the DOR antagonist naltrindole, indicating that they were mediated by MORs. Remarkably, RB 101 did not induce tolerance (Noble et al., 1992a) or physical dependence (Noble et al., 1992a, 1993).

7.5.3 The Opioid-Peptidase Paradox

The cleavage of neuropeptides by peptidases is generally regarded as a mechanism to terminate their action. However, in the case of the opioids the action of the peptidases seems to be so effective that it prevents them from activating opioid receptors. This presents the paradox that opioid release would be futile, an unlikely event in a biological system.

This paradox was first encountered when using MOR internalization to measure opioid release in the spinal cord. In a first attempt to produce MOR internalization in the spinal cord by releasing endogenous opioids, Trafton et al. (2000) administered a variety of noxious stimuli to rats. None of them were able to evoke MOR internalization. In contrast, intrathecal injections of MOR agonists did induce MOR internalization. The fact that large doses of Met-enkephalin (>10 μM) were necessary to produce this effect suggested that peptidases may have hindered the effect of the opioids released by the noxious stimuli. Indeed, opioids released from spinal cord slices (Song and Marvizon, 2003b, 2003a; Chen et al., 2008b) were able to produce MOR internalization in the presence of peptidase inhibitors, but not in their absence. In a more recent study (Lao et al., 2008), we showed that a noxious stimulus accompanied by intrathecal peptidase inhibitors does induce MOR internalization. The requirement for peptidase inhibitors is not an artifact of MOR internalization, but represents a genuine inability of opioids to activate MORs because of cleavage by peptidases. Thus, intrathecal Leu-enkephalin only produced analgesia in the presence of peptidases inhibitors (Chen et al., 2007).

In contrast to what happens in the spinal cord, opioids released in the hypothalamus and amygdala produce MOR internalization without requiring peptidase inhibitors (Eckersell et al., 1998; Sinchak and Micevych, 2001; Mills et al., 2004). This is probably because, unlike the spinal cord, these areas are rich in β-endorphin, which is resistant to cleavage by peptidases (Bewley and Li, 1985; Song and Marvizon, 2003a). Alternatively, these areas may contain less peptidases than the spinal cord.

Opioids released from the spinal cord can be readily detected by immunoassay of spinal superfusates (Yaksh and Elde, 1981; Cesselin et al., 1985; Le Bars et al., 1987a, b; Cesselin et al., 1989; Bourgoin et al., 1990). However, peptidase inhibitors did increase 10-fold the amount of Met-enkephalin released from the spinal cord (Yaksh and Chipkin, 1989). It is likely that when opioids diffuse out of the spinal cord they are degraded less readily than when they approach MORs, because peptidases are bound to the extracellular surface of neurons (Roques, 2000) and some of them may even associate with MORs (Hui et al., 1985).

There is evidence that MOR receptors are indeed activated by released opioids in the absence of peptidase inhibitors. Stimulation of the rostral-ventral medulla (RVM) (Zorman et al., 1982) or the periaqueductal gray (PAG) (Jensen and Yaksh, 1984; Morgan et al., 1991) produced analgesia that was blocked by spinal MOR antagonists (Budai and Fields, 1998) (see Bee and Dickenson, Chapter 19).

Possible explanations for the opioid-peptidase paradox include:

1. Enkephalins produce analgesia at very low doses by co-activating MORs and DORs. Therefore, although peptidases keep the released opioids under their level of detection by MOR internalization, their concentrations are probably high enough to produce analgesia.
2. Some stimuli may release opioids in quantities large enough to overcome peptidase degradation. It is possible that noxious stimuli are not the primary inducers of spinal opioid release; for example, larger amounts may be released in physiological situations like stress or partum.
3. Peptidase activity may be decreased in some conditions. Peptidase activity is known to change by switching between cytosolic and membrane-bound forms (Dyer et al., 1990). However, to avoid enkephalin degradation three peptidases would have to be switched off, which seems unlikely.
4. There may be endogenous peptidase inhibitors that protect opioids from degradation.

7.5.4 Endogenous Peptidase Inhibitors

Three endogenous inhibitors of two of the enzymes that degrade enkephalin, aminopeptidases and neutral endopeptidase, have been identified so far.

Spinorphin (LVVYPWT) was isolated from bovine spinal cord (Nishimura and Hazato, 1993); sialorphin (QHNPR) was found in the rat submandibular gland and prostate (Rougeot et al., 2003), and opiorphin (QRFSR) was isolated from human saliva (Wisner et al., 2006). Spinorphin increases the antinociceptive effects of Leu-enkephalin (Honda et al., 2001), and sialorphin and opiorphin have analgesic properties. Sialorphin secretion is stimulated by stress (Rougeot et al., 2003), which produces opioid-mediated analgesia (Lewis et al., 1981). Therefore, endogenous peptidase inhibitors may act as enablers of the action of endogenous opioids, so that the induction of opioid analgesia would require the simultaneous release of opioids and these peptidase-inhibiting peptides. If proven, the existence of such a coincidence mechanism for the activation of opioid receptors would be of great physiological significance.

7.6 Neurotransmitter Receptors that Control Spinal Opioid Release

7.6.1 Adrenergic Receptors

Adrenergic α_{2C} receptors colocalize extensively with enkephalins in the presynaptic terminals of excitatory dorsal horn interneurons (Stone et al., 1998; Olave and Maxwell, 2002, 2003b, a, 2004; Marvizon et al., 2007). We recently found that these receptors do inhibit opioid release evoked by veratridine and measured through MOR receptor internalization (Chen et al., 2008a). The α_2 agonists clonidine and guanfacine abolished the evoked MOR internalization, while medetomidine behaved as a partial agonist, albeit with high potency. However, inhibition of opioid release by α_{2C} receptors does not appear to induce hyperalgesia, since these receptors contribute to the analgesic effect of norepinephrine in the spinal cord (Fairbanks et al., 2002). It is possible that the inhibition by α_{2C} receptors serves to shut down the spinal opioid system whenever the spinal adrenergic system is active.

7.6.2 Serotonin Receptors

5-HT_{1A} serotonin receptors also inhibit spinal opioid release, since the selective 5-HT_{1A} receptor agonist 8-OH-DPAT inhibited MOR internalization evoked by electrical stimulation of rat spinal cord slices (Song et al., 2007). Inhibition by 8-OH-DPAT was reversed by the 5-HT_{1A} antagonist WAY100135. However, the inhibition of MOR internalization by 8-OH-DPAT was not complete, suggesting that 5-HT_{1A} receptors act only on some of the opioid terminals in the dorsal horn.

7.6.3 NMDA Receptors

NMDA receptors are also able to completely inhibit opioid release from rat spinal cord slices, assessed by MOR internalization (Song and Marvizon, 2005). This inhibition was independent of the method used to induce opioid release: high K^+, veratridine or electrical stimulation. Since NMDA receptors generally produce excitatory effects by depolarizing the neurons and increasing intracellular Ca^{2+}, this effect is somewhat paradoxical. It was explained by a functional coupling between NMDA receptors and Ca^{2+}-sensitive large conductance (BK or maxi-K) potassium channels similar to the one found in the olfactory bulb (Isaacson and Murphy, 2001): Ca^{2+} influx through the NMDA receptors causes the opening of the BK channels, which in turn leads to inhibition of neuronal firing. Thus, inhibition by NMDA was reversed not only by NMDA receptor antagonists, but also by BK channel blockers like paxilline and iberiotoxin. Moreover, the BK channel opener NS-1619 also inhibited opioid release, and its effect was reversed by iberiotoxin. These NMDA receptors are probably located extra-synaptically.

7.6.4 Receptors with No Effect or Unclear Effects on Opioid Release

Cesselin et al. (1984) found that Met-enkephalin release from rat spinal cord slices was facilitated by cholecystokinin (CCK) and partially inhibited by $GABA_B$ receptors. Other studies found that DORs inhibit enkephalin release in the spinal cord (Collin et al., 1994) and that neuropeptide FF suppresses this action (Ballet et al., 1999; Mauborgne et al., 2001). However, measuring spinal opioid release with MOR internalization, we found that it was not affected by agonists of $GABA_A$, $GABA_B$, δ-opioid, CCK and metabotropic glutamate receptor (Song et al., 2007).

7.7 Neural Pathways and Physiological Stimuli that Induce Spinal Opioid Release

7.7.1 Neural Pathways Involved in Spinal Opioid Release

It remains unclear what are the neuronal pathways that drive spinal opioid release. There are three possible sources of opioids in the spinal cord: primary afferents, dorsal horn neurons and supraspinal pathways. As discussed above, primary afferents do not contain enkephalins and are not the main source of dynorphins. Indeed, stimulating the dorsal root attached to spinal cord slices, or incubating the slices with capsaicin, did not induce any MOR internalization in dorsal horn neurons (Song and Marvizon, 2003a), indicating that opioid are not released from primary afferents or second order neurons. In contrast, chemical (Song and Marvizon, 2003b; Chen et al., 2008b) or electrical (Song and Marvizon, 2003a) stimulation of the dorsal horn produced abundant MOR internalization, which is consistent with the idea that the main sources of opioids in the dorsal horn are intrinsic neurons (Cruz and Basbaum, 1985; Standaert et al., 1986; Harlan et al., 1987; Todd and Spike, 1993). However, opioid immunoreactivity is present in numerous axons in the dorsolateral funiculus (Song and Marvizon, 2003a), an area that contains axons of supraspinal origin (Basbaum and Fields, 1979; Le Bars et al., 1987b; Wei et al., 1999). Therefore, these supraspinal axons are a possible source of spinal opioids. Another possibility is that descending axons make excitatory synapses with opioid-containing dorsal horn neurons, driving opioid release from them. The idea that opioid release in the spinal cord is driven supraspinally has been considered for some time (Basbaum et al., 1976; Basbaum and Fields, 1984; Fields et al., 1991; Mason, 1999) and is supported by indirect evidence. Thus, opioid receptor antagonists applied to the spinal cord reversed the analgesia induced by stimulation of the RVM (Zorman et al., 1982) or the PAG (Budai and Fields, 1998). However, adrenergic α_2 receptors and serotonin receptors also play a major role in mediating descending pain inhibition (Jensen and Yaksh, 1984; Budai et al., 1998). One group (Aimone et al., 1987) reported that while adrenergic and serotonin antagonists blocked the analgesia produced by stimulating the PAG or the nucleus raphe magnus, opioid antagonists did not.

7.7.2 Pain

Spinal opioid release can be evoked by noxious stimuli, both acute (Cesselin et al., 1985; Le Bars et al., 1987a, b; Cesselin et al., 1989; Bourgoin et al., 1990) and chronic (Przewlocki et al., 1986; Ballet et al., 2000). Since opioids are not released from primary afferents or second order neurons, their release by noxious stimuli may involve intraspinal circuits or a spinal cord-brainstem-spinal cord loop. Several studies used sciatic nerve stimulation to mimic noxious signals. Yaksh and Elde (1981) showed that Met-enkephalin release in the cat spinal cord was unaffected by cold block of the spinal cord, and therefore would not involve supraspinal structures. In contrast, Hutchinson et al. (1990) reported that the release of dynorphin A in the cat spinal cord was abolished by spinal transection, indicating that it does involve supraspinal modulation. Yet another group (Gear and Levine, 1995; Gear et al., 1999; Tambeli et al., 2002, 2003a, b) proposed that antinociception produced by an ascending spino-supraspinal pathway involves the nucleus accumbens.

Extensive studies on spinal opioid release evoked by noxious stimuli were conducted by the group of Cesselin, who measured Met-enkephalin in spinal cord superfusates after stimulation with different pain modalities. An important goal of those studies was to determine whether Met-enkephalin was released from the same spinal segment that received the noxious stimulus ("segmental" release), or from other spinal segments ("heterosegmental" release). The first case would be consistent with the idea that opioids are released from spinal circuits, whereas the second would indicate that opioid release is driven supraspinally. They found that whether Met-enkephalin release was segmental or heterosegmental depended on the stimulus modality: noxious mechanical stimulation produced heterosegmental release (Le Bars et al., 1987a, b), whereas noxious thermal stimulation (Cesselin et al., 1989) or subcutaneous formalin (Bourgoin et al., 1990) produced segmental Met-enkephalin release. In these studies the origin of the released enkephalin was estimated based on the position of the superfusion catheter, and therefore depends critically on the unknown extent of diffusion of peptides in the subdural space. Moreover, it is also unknown whether opioid receptors in the spinal cord are activated by Met-enkephalin or by other opioid peptides (Yaksh et al., 1983). Recently, we used MOR internalization to pinpoint the sites of opioid release in the spinal cord during noxious stimulation (Lao et al., 2008). We found that noxious mechanical stimulation of the hindpaw evoked segmental opioid release, i.e. restricted to the ipsilateral side of the spinal segments (L5 and L6) that received the stimulus. Noxious thermal stimulation of the hindpaw failed to evoke MOR internalization.

7.7.3 Stress

Besides pain, stress has been recognized as a major condition that induces opioid release. Many forms of stress produce analgesia, but this analgesia is not always mediated by opioids. Stressful stimuli have proven valuable to study pathways that produce analgesia and how they relate to particular neurophysiological states.

Studies performed during the 70s and the 80s identified many stressors that produce analgesia, including thermal challenge, restraint, hypoglycemia, social defeat and electric shock (Terman et al., 1986). Of these, electric shocks applied to the rat's paws or tail have been extensively used to study stress-induced analgesia (SIA). In these studies, opioid release was identified by the presence of analgesia that was blocked by opioid antagonists or cross-tolerant with morphine. Initially, whether SIA was opioid-mediated or not seemed to depend on subtle properties of the stimulus. Thus, the group of Liebeskind (Lewis et al., 1981; Terman et al., 1986) found that intermittent footshock produced opioid-mediated analgesia, whereas continuous footshock produced non-opioid analgesia. However, continuous footshock could also produce opioid-mediated analgesia when given for short duration or at low intensities. On the other hand, Watkins et al. (1982c) found that delivering footshock to the front paws produced opioid analgesia and to the hind paws non-opioid analgesia. The opioid form of SIA was mediated by opioid release in the spinal cord driven by a descending pathway in the dorsolateral funiculus (DLF) that originates in the nucleus raphe magnus (NRM) and the adjacent nucleus reticularis paragigantocellularis (Watkins et al., 1982b; Watkins and Mayer, 1982; Watkins et al., 1983). Opioid release was caused by stress and not by the shock stimulus itself, because classical conditioning to the footshock produced opioid-mediated analgesia through the same pathway (Watkins et al., 1982a, 1983).

Non-opioid SIA may involve descending adrenergic or serotonergic pathways, known to produce analgesia (Pertovaara, 2006). However, Watkins et al. (1984) found that norepinephrine and serotonin are not involved in opioid-independent SIA induced by hind paw footshock or by conditioning. Strangely, they do appear to be involved in opioid-dependent SIA induced by front paw footshock. Therefore, the relationship between norepinephrine, serotonin and opioid in modulating pain at the spinal cord level remains unclear. Recent studies indicate that non-opioid SIA involve endocannabinoid release in the PAG and, to a lesser extent, in the spinal cord (Hohmann et al., 2005; Suplita et al., 2006). However, this may represent a local regulation by endocannabinoids of a projection pathway, since endocannabinoids are believed to act at short distances as retrograde messengers (Hajos and Freund, 2002; Wilson and Nicoll, 2002).

More recent studies (Grahn et al., 1999; Maier and Watkins, 2005; Amat et al., 2006) have identified controllability over the stress stimulus as the key variable in SIA, and linked it to the learned helplessness paradigm. Thus, inescapable shock produces opioid-mediated analgesia, and also reduced escape responses, learning, food intake, social dominance and aggression; and increased fear, neophobia and opiate rewarding effects (Maier and Watkins, 2005). In contrast, escapable shock produced non-opioid analgesia and none of these adverse effects. Learned helplessness and opioid SIA are mediated by a neuronal circuit centered on the dorsal raphe nucleus (DRN) (Maier and Watkins, 2005), which send serotonergic projections to the PAG and activates the PAG-NRM-spinal cord analgesic pathway. The DRN also activates the basolateral amygdala and medial prefrontal cortex, and inhibits itself through 5-HT_{1A} receptors. The 5-HT neurons of the DRN are also inhibited by GABAergic neurons acting on benzodiazepine-sensitive $GABA_A$ receptors. In turn, opioid receptors inhibit the GABAergic neurons, disinhibiting the 5-HT neurons. The DRN is sensitized by uncontrollable stressors, a mechanism mediated by corticotrophin-releasing factor (CRF) type II receptors (Maier and Watkins, 2005). Conversely, previous experiences of control over stressors produce an "immunizing" effect by blocking the 5-HT cells in the DRN, a mechanism involving the ventral-medial prefrontal cortex (Amat et al., 2006).

The emerging view is that uncontrollable stress produces a fear/anxiety response involving DRN activation and spinal opioid release, whereas controllable stress induces a flight/fight response that inhibits the DRN and produces non-opioid analgesia (Maier and Watkins, 2005). The fear/anxiety response is characterized by "freezing" behavior and can lead to the detrimental effects of "learned helplessness". It is possible that freezing behavior is an adaptive response with survival value (it helps prey animals avoid detection by predators) and that it acquires its detrimental effects only in unnatural experimental conditions in which this strategy is repeatedly defeated. The fight/flight response may involve the activation of descending noradrenergic pathways that produce analgesia through α_{2A} and α_{2C} receptors (Yaksh, 1985; Fairbanks et al., 2002; Pan et al., 2002; Pertovaara, 2006), and inhibit spinal opioid release through α_{2C} receptors (Chen et al., 2008a). There is still much controversy regarding these issues and how they relate to depression, anxiety, arousal, fear and learning (Maier and Watkins, 2005; Ribeiro et al., 2005; Pfaff et al., 2007).

7.7.4 Acupuncture

Acupuncture is an ancient therapeutic method in Traditional Chinese Medicine that has been shown to be effective to treat pain and nausea (Eskinazi and Jobst, 1996; Mayer, 2000). It is most effective in its electroacupuncture version, i.e., when electric current is passed through needles inserted in acupuncture points. Some studies show that the analgesia produced by acupuncture and electroacupuncture is mediated by opioids released supraspinally (Bossut and Mayer, 1991) or in the spinal cord (Han, 2003). In particular, extensive work by the group of Ji-Sheng Han (2003) shows that two different opioid-releasing pathways are activated by electroacupuncture of different frequencies. Thus, low frequency (2 Hz) electroacupuncture activates a neural loop connecting the arcuate nucleus of the hypothalamus to the PAG, the RVM and the spinal cord, finally leading to enkephalin release in the spinal cord. In contrast, high frequency electroacupuncture activates a pathway going through the parabrachial nucleus and triggering dynorphin release in the spinal cord. Importantly, the intensity of the electrical stimulation was such that it recruited Aδ-fibers but not C-fibers. Therefore, electroacupuncture is not identical to noxious stimulation, although both stimuli evoke opioid release.

7.8 Conclusions

To summarize, the dorsal horn is especially rich in opioid peptides and opioid receptors, which play an important role in mediating the analgesia produced by opiate drugs, pain, some forms of stress and other neurophysiological conditions. However, many issues concerning the functions of opioids in the dorsal horn remain unresolved. These include:

1. Whether analgesia is produced by opioid receptors in primary afferent terminals or in dorsal horn neurons.
2. The role that opioid receptor trafficking plays in desensitization, tolerance and signaling.
3. The interaction between the different opioid receptors to modulate pain.

4. The function of atypical opioid receptors, like toll-like receptors and opioid growth factor receptors.
5. The relationship between opioids and other neurotransmitters that inhibit pain, like norepinephrine, serotonin and dopamine.
6. How descending pathways modulate the release of different opioid peptides in the dorsal horn.

References

Abbadie C, Trafton J, Liu H, Mantyh PW, Basbaum AI (1997) Inflammation increases the distribution of dorsal horn neurons that internalize the neurokinin-1 receptor in response to noxious and non-noxious stimulation. J Neurosci 17:8049–8060.

Abbadie C, Lombard MC, Besson JM, Trafton JA, Basbaum AI (2002) Mu and delta opioid receptor-like immunoreactivity in the cervical spinal cord of the rat after dorsal rhizotomy or neonatal capsaicin: an analysis of pre- and postsynaptic receptor distributions. Brain Res 930:150–162.

Adamson P, Xiang JZ, Mantzourides T, Brammer MJ, Campbell IC (1989) Presynaptic alpha 2-adrenoceptor and kappa-opiate receptor occupancy promotes closure of neuronal (N-type) calcium channels. Eur J Pharmacol 174:63–70.

Aimone LD, Yaksh TL (1989) Opioid modulation of capsaicin-evoked release of substance P from rat spinal cord in vivo. Peptides 10:1127–1131.

Aimone LD, Jones SL, Gebhart GF (1987) Stimulation-produced descending inhibition from the periaqueductal gray and nucleus raphe magnus in the rat: mediation by spinal mono-amines but not opioids. Pain 31:123–136.

Allen BJ, Rogers SD, Ghilardi JR, Menning PM, Kuskowski MA, Basbaum AI, Simone DA, Mantyh PW (1997) Noxious cutaneous thermal stimuli induce a graded release of endogenous substance P in the spinal cord: imaging peptide action in vivo. J Neurosci 17:5921–5927.

Amat J, Paul E, Zarza C, Watkins LR, Maier SF (2006) Previous experience with behavioral control over stress blocks the behavioral and dorsal raphe nucleus activating effects of later uncontrollable stress: role of the ventral medial prefrontal cortex. J Neurosci 26:13264–13272.

Aoki K, Kajiwara M, Oka T (1984) The role of bestatin-sensitive aminopeptidase, angiotensin converting enzyme and thiorphan-sensitive "enkephalinase" in the potency of enkephalins in the guinea-pig ileum. JpnJ Pharmacol 36:59–65.

Arvidsson U, Riedl M, Chakrabarti S, Lee JH, Nakano AH, Dado RJ, Loh HH, Law PY, Wessendorf MW, Elde R (1995) Distribution and targeting of a mu-opioid receptor (MOR1) in brain and spinal cord. J Neurosci 15:3328–3341.

Ballet S, Mauborgne A, Gouarderes C, Bourgoin AS, Zajac JM, Hamon M, Cesselin F (1999) The neuropeptide FF analogue, 1DME, enhances in vivo met- enkephalin release from the rat spinal cord. Neuropharmacology 38:1317–1324.

Ballet S, Mauborgne A, Hamon M, Cesselin F, Collin E (2000) Altered opioid-mediated control of the spinal release of dynorphin and met-enkephalin in polyarthritic rats. Synapse 37:262–272.

Bao L, Jin SX, Zhang C, Wang LH, Xu ZZ, Zhang FX, Wang LC, Ning FS, Cai HJ, Guan JS, Xiao HS, Xu ZQ, He C, Hokfelt T, Zhou Z, Zhang X (2003) Activation of delta opioid receptors induces receptor insertion and neuropeptide secretion. Neuron 37:121–133.

Basbaum AI, Fields HL (1979) The origin of descending pathways in the dorsolateral funiculus of the spinal cord of the cat and rat: further studies on the anatomy of pain modulation. J Comp Neurol 187:513–531.

Basbaum AI, Fields HL (1984) Endogenous pain control systems: brainstem spinal pathways and endorphin circuitry. Annu Rev Neurosci 7:309–338.

Basbaum AI, Clanton CH, Fields HL (1976) Opiate and stimulus-produced analgesia: functional anatomy of a medullospinal pathway. Proc Natl Acad Sci USA 73:4685–4688.

Basbaum AI, Cruz L, Weber E (1986) Immunoreactive dynorphin B in sacral primary afferent fibers of the cat. J Neurosci 6:127–133.

Besse D, Lombard MC, Zajac JM, Roques BP, Besson JM (1990) Pre- and postsynaptic distribution of mu, delta and kappa opioid receptors in the superficial layers of the cervical dorsal horn of the rat spinal cord. Brain Res 521:15–22.

Bewley TA, Li CH (1985) Tertiary structure in deletion analogues of human beta-endorphin: resistance to leucine aminopeptidase action. Biochemistry 24:6568–6571.

Boecker H, Sprenger T, Spilker ME, Henriksen G, Koppenhoefer M, Wagner KJ, Valet M, Berthele A, Tolle TR (2008) The Runner's High: Opioidergic Mechanisms in the Human Brain. Cereb Cortex.

Bossut DF, Mayer DJ (1991) Electroacupuncture analgesia in naive rats: effects of brainstem and spinal cord lesions, and role of pituitary-adrenal axis. Brain Res 549:52–58.

Botticelli LJ, Cox BM, Goldstein A (1981) Immunoreactive dynorphin in mammalian spinal cord and dorsal root ganglia. Proc Natl Acad Sci USA 78:7783–7786.

Bourgoin S, Le Bars D, Clot AM, Hamon M, Cesselin F (1990) Subcutaneous formalin induces a segmental release of Met-enkephalin-like material from the rat spinal cord. Pain 41:323–329.

Bras JMA, Becker C, Bourgoin S, Lombard MC, Cesselin F, Hamon M, Pohl M (2001) Met-enkephalin is preferentially transported into the peripheral processes of primary afferent fibres in both control and HSV1-driven proenkephalin a overexpressing rats. Neuroscience 103:1073–1083.

Budai D, Fields HL (1998) Endogenous opioid peptides acting at mu-opioid receptors in the dorsal horn contribute to midbrain modulation of spinal nociceptive neurons. J Neurophysiol 79:677–687.

Budai D, Harasawa I, Fields HL (1998) Midbrain periaqueductal gray (PAG) inhibits nociceptive inputs to sacral dorsal horn nociceptive neurons through alpha(2)- adrenergic receptors. J Neurophysiol 80:2244–2254.

Burbach JP, De Kloet ER (1982) Proteolysis of beta-endorphin in brain tissue. Peptides 3:451–453.

Cahill CM, Morinville A, Lee MC, Vincent JP, Collier B, Beaudet A (2001) Prolonged morphine treatment targets delta opioid receptors to neuronal plasma membranes and enhances delta-mediated antinociception. J Neurosci 21:7598–7607.

Cahill CM, Morinville A, Hoffert C, O'Donnell D, Beaudet A (2003) Up-regulation and trafficking of delta opioid receptor in a model of chronic inflammation: implications for pain control. Pain 101:199–208.

Cesselin F, Bourgoin S, Artaud F, Hamon M (1984) Basic and regulatory mechanisms of in vitro release of Met-enkephalin from the dorsal zone of the rat spinal cord. J Neurochem 43:763–774.

Cesselin F, Le Bars D, Bourgoin S, Artaud F, Gozlan H, Clot AM, Besson JM, Hamon M (1985) Spontaneous and evoked release of methionine-enkephalin-like material from the rat spinal cord in vivo. Brain Res 339:305–313.

Cesselin F, Bourgoin S, Clot AM, Hamon M, Le Bars D (1989) Segmental release of met-enkephalin-like material from the spinal-cord of rats, elicited by noxious thermal stimuli. Brain Res 484:71–77.

Chen SR, Pan HL (2006) Blocking mu opioid receptors in the spinal cord prevents the analgesic action by subsequent systemic opioids. Brain Res 1081:119–125.

Chen W, Song B, Lao L, Perez OA, Marvizon JC (2006) Intrathecal enkephalin requires inhibition of peptidases to produce analgesia and mu-opioid receptor internalization in dorsal horn neurons. Soc Neurosci Abstr 32:643.618.

Chen W, Song B, Lao L, Perez OA, Kim W, Marvizon JCG (2007) Comparing analgesia and μ-opioid receptor internalization produced by intrathecal enkephalin: requirement for peptidase inhibition. Neuropharmacology 53:664–667.

Chen W, Song B, Marvizon JC (2008a) Inhibition of opioid release in the rat spinal cord by α_{2C} adrenergic receptors. Neuropharmacology 54:944–953.

Chen W, Song B, Zhang G, Marvizon JC (2008b) Effects of veratridine and high potassium on μ-opioid receptor internalization in the rat spinal cord: stimulation of opioid release versus inhibition of internalization. J Neurosci Methods 170:285–293.

Cho HJ, Basbaum AI (1988) Increased staining of immunoreactive dynorphin cell bodies in the deafferented spinal cord of the rat. Neurosci Lett 84:125–130.

Citterio F, Corradini L, Smith RD, Bertorelli R (2000) Nociceptin attenuates opioid and gamma-aminobutyric acid(B) receptor-mediated analgesia in the mouse tail-flick assay. Neurosci Lett 292:83–86.

Collin E, Mauborgne A, Bourgoin S, Benoliel JJ, Hamon M, Cesselin F (1994) Morphine reduces the release of met-enkephalin-like material from the rat spinal cord in vivo by acting at delta opioid receptors. Neuropeptides 27:75–83.

Commons KG (2003) Translocation of presynaptic delta opioid receptors in the ventrolateral periaqueductal gray after swim stress. J Comp Neurol 464:197–207.

Costa E, Mocchetti I, Supattapone S, Snyder SH (1987) Opioid peptide biosynthesis: enzymatic selectivity and regulatory mechanisms. FASEB J 1:16–21.

Cruz L, Basbaum AI (1985) Multiple opioid peptides and the modulation of pain: immuno-histochemical analysis of dynorphin and enkephalin in the trigeminal nucleus caudalis and spinal cord of the cat. J Comp Neurol 240:331–348.

Cvejic S, Devi LA (1997) Dimerization of the delta opioid receptor: implication for a role in receptor internalization. J Biol Chem 272:26959–26964.

Diverse-Pierluissi M, Remmers AE, Neubig RR, Dunlap K (1997) Novel form of crosstalk between G protein and tyrosine kinase pathways. Proc Natl Acad Sci USA 94: 5417–5421.

Dolphin AC (2003) G protein modulation of voltage-gated calcium channels. Pharmacol Rev 55:607–627.

Duggan AW, Schaible HG, Hope PJ, Lang CW (1992) Effect of peptidase inhibition on the pattern of intraspinally released immunoreactive substance P detected with antibody microprobes. Brain Res 579:261–269.

Dyer SH, Slaughter CA, Orth K, Moomaw CR, Hersh LB (1990) Comparison of the soluble and membrane-bound forms of the puromycin-sensitive enkephalin-degrading aminopep-tidases from rat. J Neurochem 54:547–554.

Eckersell CB, Popper P, Micevych PE (1998) Estrogen-induced alteration of μ-opioid receptor immunoreactivity in the medial preoptic nucleus and medial amygdala. J Neurosci 18:3967–3976.

Eckert WA, III, McNaughton KK, Light AR (2003) Morphology and axonal arborization of rat spinal inner lamina II neurons hyperpolarized by mu-opioid-selective agonists. J Comp Neurol 458:240–256.

Eskinazi DP, Jobst KA (1996) National institutes of health office of alternative medicine-food and drug administration workshop on acupuncture. J Altern Complement Med 2:3–6.

Fairbanks CA, Stone LS, Kitto KF, Nguyen HO, Posthumus IJ, Wilcox GL (2002) alpha(2C)-adrenergic receptors mediate spinal analgesia and adrenergic-opioid synergy. J Pharmacol Exp Ther 300:282–290.

Fichna J, Janecka A, Costentin J, Do Rego JC (2007) The endomorphin system and its evolving neurophysiological role. Pharmacol Rev 59:88–123.

Fields HL, Emson PC, Leigh BK, Gilbert RF, Iversen LL (1980) Multiple opiate receptor sites on primary afferent fibres. Nature 284:351–353.

Fields HL, Heinricher MM, Mason P (1991) Neurotransmitters in nociceptive modulatory circuits. Annu Rev Neurosci 14:219–245.

Fuxe K, Agnati LF (1991) Two principal modes of electrochemical communication in the brain: volume versus wiring transmission. In: Volume Transmission in the Brain, pp 1–9. New York: Raven Press.

Gear RW, Levine JD (1995) Antinociception produced by an ascending spino-supraspinal pathway. JNeurosci 15:3154–3161.

Gear RW, Aley KO, Levine JD (1999) Pain-induced analgesia mediated by mesolimbic reward circuits. J Neurosci 19:7175–7181.

Go VL, Yaksh TL (1987) Release of substance P from the cat spinal cord. J Physiol 391:141–167.

Gomes I, Jordan BA, Gupta A, Trapaidze N, Nagy V, Devi LA (2000) Heterodimerization of mu and delta opioid receptors: a role in opiate synergy. J Neurosci 20:RC110.

Gomes I, Gupta A, Filipovska J, Szeto HH, Pintar JE, Devi LA (2004) A role for heterodimerization of mu and delta opiate receptors in enhancing morphine analgesia. Proc Natl Acad Sci USA 101:5135–5139.

Graf L, Miglecz E, Bajusz S, Szekely JI (1979) Met-enkephalin attenuates morphine tolerance in rats. Eur J Pharmacol 58:345–346.

Graf L, Paldi A, Patthy A (1985) Action of neutral metalloendopeptidase ("enkephalinase") on beta-endorphin. Neuropeptides 6:13–19.

Grahn RE, Maswood S, McQueen MB, Watkins LR, Maier SF (1999) Opioid-dependent effects of inescapable shock on escape behavior and conditioned fear responding are mediated by the dorsal raphe nucleus. Behav Brain Res 99:153–167.

Gross RA, Macdonald RL (1987) Dynorphin A selectively reduces a large transient (N-type) calcium current of mouse dorsal root ganglion neurons in cell culture. Proc Natl Acad Sci USA 84:5469–5473.

Grudt TJ, Williams JT (1993) kappa-Opioid receptors also increase potassium conductance. Proc Natl Acad Sci USA 90:11429–11432.

Grudt TJ, Williams JT (1994) mu-Opioid agonists inhibit spinal trigeminal substantia gelatinosa neurons in guinea pig and rat. J Neurosci 14:1646–1654.

Gutstein HB, Bronstein DM, Akil H (1992) Beta-endorphin processing and cellular origins in rat spinal cord. Pain 51:241–247.

Guyon A, Roques BP, Guyon F, Foucault A, Perdrisot R, Swerts JP, Schwartz JC (1979) Enkephalin degradation in mouse brain studied by a new H.P.L.C. method: further evidence for the involvement of carboxydipeptidase. Life Sci 25:1605–1611.

Hackler L, Zadina JE, Ge LJ, Kastin AJ (1997) Isolation of relatively large amounts of endomorphin-1 and endomorphin-2 from human brain cortex. Peptides 18:1635–1639.

Hajos N, Freund TF (2002) Distinct cannabinoid sensitive receptors regulate hippocampal excitation and inhibition. Chem Phys Lipids 121:73–82

Han JS (2003) Acupuncture: neuropeptide release produced by electrical stimulation of different frequencies. Trends Neurosci 26:17–22.

Harlan RE, Shivers BD, Romano GJ, Howells RD, Pfaff DW (1987) Localization of pre-proenkephalin mRNA in the rat brain and spinal cord by in situ hybridization. J Comp Neurol 258:159–184.

Harris JA, Chang PC, Drake CT (2004) Kappa opioid receptors in rat spinal cord: sex-linked distribution differences. Neuroscience 124:879–890.

Hazum E, Chang KJ, Leighton HJ, Lever OW, Jr., Cuatrecasas P (1982) Increased biological activity of dimers of oxymorphone and enkephalin: possible role of receptor crosslinking. Biochem Biophys Res Commun 104:347–353.

He L, Fong J, von Zastrow M, Whistler JL (2002) Regulation of opioid receptor trafficking and morphine tolerance by receptor oligomerization. Cell 108:271–282.

Hiranuma T, Iwao K, Kitamura K, Matsumiya T, Oka T (1997) Almost complete protection from [Met5]-enkephalin-Arg6-Gly7-Leu8 (Met-enk-RGL) hydrolysis in membrane preparations by the combination of amastatin, captopril and phosphoramidon. J Pharmacol Exp Ther 281:769–774.

Hiranuma T, Kitamura K, Taniguchi T, Kanai M, Arai Y, Iwao K, Oka T (1998) Protection against dynorphin-(1-8) hydrolysis in membrane preparations by the combination of amastatin, captopril and phosphoramidon. J Pharmacol Exp Ther 286:863–869.

Hohmann AG, Suplita RL, Bolton NM, Neely MH, Fegley D, Mangieri R, Krey JF, Walker JM, Holmes PV, Crystal JD, Duranti A, Tontini A, Mor M, Tarzia G, Piomelli D (2005) An endocannabinoid mechanism for stress-induced analgesia. Nature 435:1108–1112.

Honda M, Okutsu H, Matsuura T, Miyagi T, Yamamoto Y, Hazato T, Ono H (2001) Spinorphin, an endogenous inhibitor of enkephalin-degrading enzymes, potentiates leu-enkephalin-induced anti-allodynic and antinociceptive effects in mice. Jpn J Pharmacol 87:261–267.

Honore P, Menning PM, Rogers SD, Nichols ML, Basbaum AI, Besson JM, Mantyh PW (1999) Spinal cord substance P receptor expression and internalization in acute, short-term, and long-term inflammatory pain states. J Neurosci 19:7670–7678.

Hui KS, Gioannini T, Hui M, Simon EJ, Lajtha A (1985) An opiate receptor-associated aminopeptidase that degrades enkephalins. Neurochem Res 10:1047–1058.

Hui KS, Saito M, Hui M (1998) A novel neuron-specific aminopeptidase in rat brain synaptosomes. Its identification, purification, and characterization. J Biol Chem 273:31053–31060.

Hunt SP, Kelly JS, Emson PC, Kimmel JR, Miller RJ, Wu JY (1981) An immunohistochemical study of neuronal populations containing neuropeptides or gamma-aminobutyrate within the superficial layers of the rat dorsal horn. Neuroscience 6:1883–1898.

Hutchinson WD, Morton CR, Terenius L (1990) Dynorphin A: in vivo release in the spinal cord of the cat. Brain Res 532:299–306.

Hutchinson MR, Bland ST, Johnson KW, Rice KC, Maier SF, Watkins LR (2007) Opioid-induced glial activation: mechanisms of activation and implications for opioid analgesia, dependence, and reward. ScientificWorld J 7:98–111.

Ikeda H, Heinke B, Ruscheweyh R, Sandkuhler J (2003) Synaptic plasticity in spinal lamina I projection neurons that mediate hyperalgesia. Science 299:1237–1240.

Isaacson JS, Murphy GJ (2001) Glutamate-mediated extrasynaptic inhibition: direct coupling of NMDA receptors to Ca(2+)-activated K+ channels. Neuron 31:1027–1034.

Jensen TS, Yaksh TL (1984) Spinal monoamine and opiate systems partly mediate the antinociceptive effects produced by glutamate at brainstem sites. Brain Res 321:287–297.

Jessell TM, Iversen LL (1977) Opiate analgesics inhibit substance P release from rat trigeminal nucleus. Nature 268:549–551.

Jordan BA, Devi LA (1999) G-protein-coupled receptor heterodimerization modulates receptor function. Nature 399:697–700.

Keith DE, Murray SR, Zaki PA, Chu PC, Lissin DV, Kang L, Evans CJ, von Zastrow M (1996) Morphine activates opioid receptors without causing their rapid internalization. J Biol Chem 271:19021–19024.

Keith DE, Anton B, Murray SR, Zaki PA, Chu PC, Lissin DV, Monteillet-Agius G, Stewart PL, Evans CJ, von Zastrow M (1998) mu-Opioid receptor internalization: opiate drugs have differential effects on a conserved endocytic mechanism in vitro and in the mammalian brain. Mol Pharmacol 53:377–384.

Kemp T, Spike RC, Watt C, Todd AJ (1996) The mu-opioid receptor (MOR1) is mainly restricted to neurons that do not contain GABA or glycine in the superficial dorsal horn of the rat spinal cord. Neuroscience 75:1231–1238.

King M, Chang A, Pasternak GW (1998) Functional blockade of opioid analgesia by orphanin FQ/nociceptin. Biochem Pharmacol 55:1537–1540.

Kishioka S, Miyamoto Y, Fukunaga Y, Nishida S, Yamamoto H (1994) Effects of a mixture of peptidase inhibitors (amastatin, captopril and phosphoramidon) on Met-enkephalin-, beta-endorphin-, dynorphin-(1-13)- and electroacupuncture-induced antinociception in rats. Jpn J Pharmacol 66:337–345.

Kline RHt, Wiley RG (2008) Spinal mu-opioid receptor-expressing dorsal horn neurons: role in nociception and morphine antinociception. J Neurosci 28:904–913.

Kondo I, Marvizon JC, Song B, Salgado F, Codeluppi S, Hua XY, Yaksh TL (2005) Inhibition by spinal mu- and delta-opioid agonists of afferent-evoked substance P release. J Neurosci 25:3651–3660.

Lao L, Song B, Marvizon JCG (2003) Neurokinin release produced by capsaicin acting on the central terminals and axons of primary afferents: relationship with NMDA and GABA B receptors. Neuroscience 121:667–680.

Lao L, Song B, Chen W, Marvizon JC (2008) Noxious mechanical stimulation evokes the segmental release of opioid peptides that induce μ-opioid receptor internalization in the presence of peptidase inhibitors. Brain Res 1197:85–93.

Le Bars D, Bourgoin S, Clot AM, Hamon M, Cesselin F (1987a) Noxious mechanical stimuli increase the release of Met-enkephalin-like material heterosegmentally in the rat spinal cord. Brain Res 402:188–192.

Le Bars D, Bourgoin S, Villanueva L, Clot AM, Hamon M, Cesselin F (1987b) Involvement of the dorsolateral funiculi in the spinal release of met-enkephalin-like material triggered by heterosegmental noxious mechanical stimuli. Brain Res 412:190–195.

Lever IJ, Malcangio M (2002) CB(1) receptor antagonist SR141716A increases capsaicin-evoked release of Substance P from the adult mouse spinal cord. Br J Pharmacol 135:21–24.

Lever IJ, Bradbury EJ, Cunningham JR, Adelson DW, Jones MG, McMahon SB, Marvizon JC, Malcangio M (2001) Brain-derived neurotrophic factor is released in the dorsal horn by distinctive patterns of afferent fiber stimulation. J Neurosci 21: 4469–4477.

Lewis JW, Sherman JE, Liebeskind JC (1981) Opioid and non-opioid stress analgesia: assessment of tolerance and cross-tolerance with morphine. J Neurosci 1:358–363.

Li YW, Bayliss DA (1998) Activation of alpha 2-adrenoceptors causes inhibition of calcium channels but does not modulate inwardly-rectifying K + channels in caudal raphe neurons. Neuroscience 82:753–765.

Lisi TL, Sluka KA (2006) A new electrochemical HPLC method for analysis of enkephalins and endomorphins. J Neurosci Methods 150:74–79.

Liu NJ, Xu T, Xu C, Li CQ, Yu YX, Kang HG, Han JS (1995) Cholecystokinin octapeptide reverses mu-opioid-receptor-mediated inhibition of calcium current in rat dorsal root ganglion neurons. J Pharmacol ExpTher 275:1293–1299.

Macdonald RL, Werz MA (1986) Dynorphin A decreases voltage-dependent calcium conductance of mouse dorsal root ganglion neurones. J Physiol 377:237–249.

Maier SF, Watkins LR (2005) Stressor controllability and learned helplessness: the roles of the dorsal raphe nucleus, serotonin, and corticotropin-releasing factor. Neurosci Biobehav Rev 29:829–841.

Malcangio M, Fernandes K, Tomlinson DR (1998) NMDA receptor activation modulates evoked release of substance P from rat spinal cord. Br J Pharmacol 125:1625–1626.

Malmberg AB, Yaksh TL (1992) Isobolographic and dose-response analyses of the interaction between intrathecal mu and delta agonists: effects of naltrindole and its benzofuran analog (NTB). J Pharmacol ExpTher 263:264–275.

Mansour A, Khachaturian H, Lewis ME, Akil H, Watson SJ (1988) Anatomy of CNS opioid receptors. Trends Neurosci 11:308–314.

Mansour A, Fox CA, Akil H, Watson SJ (1995) Opioid-receptor mRNA expression in the rat CNS: anatomical and functional implications. Trends Neurosci 18:22–29.

Mantyh PW, DeMaster E, Malhotra A, Ghilardi JR, Rogers SD, Mantyh CR, Liu H, Basbaum AI, Vigna SR, Maggio JE (1995) Receptor endocytosis and dendrite reshaping in spinal neurons after somatosensory stimulation. Science 268:1629–1632.

Martin-Schild S, Gerall AA, Kastin AJ, Zadina JE (1998) Endomorphin-2 is an endogenous opioid in primary sensory afferent fibers. Peptides 19:1783–1789.

Marvizon JC, Martinez V, Grady EF, Bunnett NW, Mayer EA (1997) Neurokinin 1 receptor internalization in spinal cord slices induced by dorsal root stimulation is mediated by NMDA receptors. J Neurosci 17:8129–8136.

Marvizon JC, Grady EF, Waszak-McGee J, Mayer EA (1999) Internalization of μ-opioid receptors in rat spinal cord slices. Neuroreport 10:2329–2334.

Marvizon JC, Wang X, Matsuka Y, Neubert JK, Spigelman I (2003a) Relationship between capsaicin-evoked substance P release and neurokinin 1 receptor internalization in the rat spinal cord. Neuroscience 118:535–545.

Marvizon JCG, Wang X, Lao L, Song B (2003b) Effect of peptidases on the ability of exogenous and endogenous neurokinins to produce neurokinin 1 receptor internalization in the rat spinal cord. Br J Pharmacol 140:1389–1398.

Marvizon JC, Perez OA, Song B, Chen W, Bunnett NW, Grady EF, Todd AJ (2007) Calcitonin receptor-like receptor and receptor activity modifying protein 1 in the rat dorsal horn: localization in glutamatergic presynaptic terminals containing opioids and adrenergic α_{2C} receptors. Neuroscience 148:250–265.

Mason P (1999) Central mechanisms of pain modulation. Curr Opin Neurobiol 9:436–441.

Mauborgne A, Bourgoin S, Polienor H, Roumy M, Simonnet G, Zajac JM, Cesselin F (2001) The neuropeptide FF analogue, 1DMe, acts as a functional opioid autoreceptor antagonist in the rat spinal cord. Eur J Pharmacol 430:273–276.

Mayer DJ (2000) Biological mechanisms of acupuncture. Prog Brain Res 122:457–477.

Meunier JC, Mollereau C, Toll L, Suaudeau C, Moisand C, Alvinerie P, Butour JL, Guillemot JC, Ferrara P, Monsarrat B (1995) Isolation and structure of the endogenous agonist of opioid receptor-like ORL1 receptor. Nature 377:532–535.

Mills RH, Sohn RK, Micevych PE (2004) Estrogen-induced μ-opioid receptor internalization in the medial preoptic nucleus is mediated via neuropeptide Y–Y1 receptor activation in the arcuate nucleus of female rats. J Neurosci 24:947–955.

Minami M, Satoh M (1995) Molecular biology of the opioid receptors: structures, functions and distributions. Neurosci Res 23:121–145.

Mogil JS, Pasternak GW (2001) The molecular and behavioral pharmacology of the orphanin FQ/nociceptin peptide and receptor family. Pharmacol Rev 53:381–415.

Mogil JS, Grisel JE, Reinscheid RK, Civelli O, Belknap JK, Grandy DK (1996) Orphanin FQ is a functional anti-opioid peptide. Neuroscience 75:333–337.

Moises HC, Rusin KI, Macdonald RL (1994) Mu- and kappa-opioid receptors selectively reduce the same transient components of high-threshold calcium current in rat dorsal root ganglion sensory neurons. J Neurosci 14:5903–5916.

Morgan MM, Gold MS, Liebeskind JC, Stein C (1991) Periaqueductal gray stimulation produces a spinally mediated, opioid antinociception for the inflamed hindpaw of the rat. Brain Res 545:17–23.

Morinville A, Cahill CM, Esdaile MJ, Aibak H, Collier B, Kieffer BL, Beaudet A (2003) Regulation of delta-opioid receptor trafficking via mu-opioid receptor stimulation: evidence from mu-opioid receptor knock-out mice. J Neurosci 23: 4888–4898.

Morinville A, Cahill CM, Aibak H, Rymar VV, Pradhan A, Hoffert C, Mennicken F, Stroh T, Sadikot AF, O'Donnell D, Clarke PBS, Collier B, Henry JL, Vincent JP, Beaudet A (2004) Morphine-induced changes in δ opioid receptor trafficking are linked to somatosensory processing in the rat spinal cord. J Neurosci 24:5549–5559.

Morris BJ, Herz A (1987) Distinct distribution of opioid receptor types in rat lumbar spinal cord. Naunyn Schmiedebergs Arch Pharmacol 336:240–243.

Neal CR, Jr., Mansour A, Reinscheid R, Nothacker HP, Civelli O, Watson SJ, Jr. (1999a) Localization of orphanin FQ (nociceptin) peptide and messenger RNA in the central nervous system of the rat. J Comp Neurol 406:503–547.

Neal CR, Jr., Mansour A, Reinscheid R, Nothacker HP, Civelli O, Akil H, Watson SJ, Jr. (1999b) Opioid receptor-like (ORL1) receptor distribution in the rat central nervous system: comparison of ORL1 receptor mRNA expression with (125)I-[(14)Tyr]-orphanin FQ binding. J Comp Neurol 412:563–605.

Nishimura K, Hazato T (1993) Isolation and identification of an endogenous inhibitor of enkephalin-degrading enzymes from bovine spinal cord. Biochem Biophys Res Commun 194:713–719.

Noble F, Turcaud S, Fournie-Zaluski MC, Roques BP (1992a) Repeated systemic administration of the mixed inhibitor of enkephalin-degrading enzymes, RB101, does not induce either antinociceptive tolerance or cross-tolerance with morphine. Eur J Pharmacol 223:83–89.

Noble F, Soleilhac JM, Soroca-Lucas E, Turcaud S, Fournie-Zaluski MC, Roques BP (1992b) Inhibition of the enkephalin-metabolizing enzymes by the first systemically active mixed inhibitor prodrug RB 101 induces potent analgesic responses in mice and rats. J Pharmacol Exp Ther 261:181–190.

Noble F, Fournie-Zaluski MC, Roques BP (1993) Unlike morphine the endogenous enkephalins protected by RB101 are unable to establish a conditioned place preference in mice. Eur J Pharmacol 230:139–149.

Noble F, Smadja C, Valverde O, Maldonado R, Coric P, Turcaud S, Fournie-Zaluski MC, Roques BP (1997) Pain-suppressive effects on various nociceptive stimuli (thermal, chemical, electrical and inflammatory) of the first orally active enkephalin-metabolizing enzyme inhibitor RB 120. Pain 73:383–391.

Noble F, Banisadr G, Jardinaud F, Popovici T, Lai-Kuen R, Chen H, Bischoff L, Parsadaniantz SM, Fournie-Zaluski MC, Roques BP (2001) First discrete autoradiographic distribution of aminopeptidase N in various structures of rat brain and spinal cord using the selective iodinated inhibitor [125I]RB 129. Neuroscience 105:479–488.

Numata H, Hiranuma T, Oka T (1988) Inactivation of dynorphin-(1-8) in isolated preparations by three peptidases. Jpn J Pharmacol 47:417–423.

Nyberg F, Yaksh TL, Terenius L (1983) Opioid activity released from cat spinal cord by sciatic nerve stimulation. Life Sci 33 Suppl 1:17–20.

Nydahl KS, Skinner K, Julius D, Basbaum AI (2004) Co-localization of endomorphin-2 and substance P in primary afferent nociceptors and effects of injury: a light and electron microscopic study in the rat. Eur J Neurosci 19:1789–1799.

Oka T, Aoki K, Kajiwara M, Ishii K, Kuno Y, Hiranuma T, Matsumiya T (1986) Inactivation of [Leu5]-enkephalin in three isolated preparations: relative importance of aminopeptidase, endopeptidase-24.11 and peptidyl dipeptidase A. Nida Research Monograph 75:259–262.

Olave MJ, Maxwell DJ (2002) An investigation of neurones that possess the alpha 2C-adrenergic receptor in the rat dorsal horn. Neuroscience 115:31–40.

Olave MJ, Maxwell DJ (2003a) Axon terminals possessing the alpha 2c-adrenergic receptor in the rat dorsal horn are predominantly excitatory. Brain Res 965:269–273.

Olave MJ, Maxwell DJ (2003b) Neurokinin-1 projection cells in the rat dorsal horn receive synaptic contacts from axons that possess alpha2C-adrenergic receptors. J Neurosci 23:6837–6846.

Olave MJ, Maxwell DJ (2004) Axon terminals possessing alpha2C-adrenergic receptors densely innervate neurons in the rat lateral spinal nucleus which respond to noxious stimulation. Neuroscience 126:391–403.

Pan YZ, Li DP, Pan HL (2002) Inhibition of glutamatergic synaptic input to spinal lamina II(o) neurons by presynaptic alpha(2)-adrenergic receptors. J Neurophysiol 87:1938–1947.

Patierno S, Zellalem W, Ho A, Parsons CG, Lloyd KC, Tonini M, Sternini C (2005) N-Methyl-d-aspartate receptors mediate endogenous opioid release in enteric neurons after abdominal surgery. Gastroenterology 128:2009–2019.

Pertovaara A (2006) Noradrenergic pain modulation. Prog Neurobiol 80:53–83.

Pfaff DW, Martin EM, Ribeiro AC (2007) Relations between mechanisms of CNS arousal and mechanisms of stress. Stress 10:316–325.

Pierce TL, Grahek MD, Wessendorf MW (1998) Immunoreactivity for endomorphin-2 occurs in primary afferents in rats and monkey. Neuroreport 9:385–389.

Pohl M, Lombard MC, Bourgoin S, Carayon A, Benoliel JJ, Mauborgne A, Besson JM, Hamon M, Cesselin F (1989) Opioid control of the in vitro release of calcitonin gene-related peptide from primary afferent fibres projecting in the rat cervical cord. Neuropeptides 14:151–159.

Pohl M, Collin E, Bourgoin S, Conrath M, Benoliel JJ, Nevo I, Hamon M, Giraud P, Cesselin F (1994) Expression of preproenkephalin-A gene and presence of Met-enkephalin in dorsal root ganglia of the adult rat J Neurochem 63:1226–1234.

Przewlocki R, Lason W, Silberring J, Herz A, Przewlocka B (1986) Release of opioid peptides from the spinal cord of rats subjected to chronic pain. Nida Res Monograph 75:422–425.

Raingo J, Castiglioni AJ, Lipscombe D (2007) Alternative splicing controls G protein-dependent inhibition of N-type calcium channels in nociceptors. Nat Neurosci 10:285–292.

Raynor K, Kong H, Chen Y, Yasuda K, Yu L, Bell GI, Reisine T (1993) Pharmacological characterization of the cloned κ-, δ-, and μ-opioid receptors. Mol Pharmacol 45:330–334.

Reinscheid RK, Nothacker HP, Bourson A, Ardati A, Henningsen RA, Bunzow JR, Grandy DK, Langen H, Monsma FJ, Jr., Civelli O (1995) Orphanin FQ: a neuropeptide that activates an opioidlike G protein-coupled receptor. Science 270:792–794.

Ribeiro-da-Silva A, Pioro EP, Cuello AC (1991) Substance P- and enkephalin-like immunoreactivities are colocalized in certain neurons of the substantia gelatinosa of the rat spinal cord: an ultrastructural double-labeling study. J Neurosci 11:1068–1080.

Ribeiro SC, Kennedy SE, Smith YR, Stohler CS, Zubieta JK (2005) Interface of physical and emotional stress regulation through the endogenous opioid system and mu-opioid receptors. Prog Neuropsychopharmacol Biol Psychiatry 29:1264–1280.

Rios C, Gomes I, Devi LA (2004) Interactions between delta opioid receptors and alpha-adrenoceptors. Clin Exp Pharmacol Physiol 31:833–836.

Roques BP (2000) Novel approaches to targeting neuropeptide systems. Trends Pharmacol Sci 21:475–483.

Rougeot C, Messaoudi M, Hermitte V, Rigault AG, Blisnick T, Dugave C, Desor D, Rougeon F (2003) Sialorphin, a natural inhibitor of rat membrane-bound neutral endopeptidase that displays analgesic activity. Proc Natl Acad Sci USA 100:8549–8554.

Ruda MA, Iadarola MJ, Cohen LV, Young WS, III (1988) In situ hybridization histochemistry and immunocytochemistry reveal an increase in spinal dynorphin biosynthesis in a rat model of peripheral inflammation and hyperalgesia. Proc Natl Acad Sci USA 85:622–626.

Russell RD, Leslie JB, Su YF, Watkins WD, Chang KJ (1987) Continuous intrathecal opioid analgesia: tolerance and cross-tolerance of mu and delta spinal opioid receptors. J Pharmacol Exp Ther 240:150–158.

Sakurada T, Katsuyama S, Sakurada S, Inoue M, Tan-No K, Kisara K, Sakurada C, Ueda H, Sasaki J (1999) Nociceptin-induced scratching, biting and licking in mice: involvement of spinal NK1 receptors. Br J Pharmacol 127:1712–1718.

Scherrer G, Befort K, Contet C, Becker J, Matifas A, Kieffer BL (2004) The delta agonists DPDPE and deltorphin II recruit predominantly mu receptors to produce thermal analgesia: a parallel study of mu, delta and combinatorial opioid receptor knockout mice. Eur J Neurosci 19:2239–2248.

Schreff M, Schulz S, Wiborny D, Heollt V (1998) Immunofluorescent identification of endomorphin-2-containing nerve fibers and terminals in the rat brain and spinal cord. Neuroreport 9:1031–1034.

Schroeder JE, Fischbach PS, Zheng D, McCleskey EW (1991) Activation of mu opioid receptors inhibits transient high- and low-threshold Ca2+ currents, but spares a sustained current. Neuron 6:13–20.

Senba E, Yanaihara C, Yanaihara N, Tohyama M (1988) Co-localization of substance P and Met-enkephalin-Arg6-Gly7-Leu8 in the intraspinal neurons of the rat, with special reference to the neurons in the substantia gelatinosa. Brain Res 453:110–116.

Shane R, Wilk S, Bodnar RJ (1999) Modulation of endomorphin-2-induced analgesia by dipeptidyl peptidase IV. Brain Res 815:278–286.

Shimamura M, Hazato T, Katayama T (1983) A membrane-bound aminopeptidase isolated from monkey brain and its action on enkephalin. BiochimBiophys Acta 756:223–229.

Sinchak K, Micevych PE (2001) Progesterone blockade of estrogen activation of mu-opioid receptors regulates reproductive behavior. J Neurosci 21:5723–5729.

Soignier RD, Vaccarino AL, Fanti KA, Wilson AM, Zadina JE (2004) Analgesic tolerance and cross-tolerance to i.c.v. endomorphin-1, endomorphin-2, and morphine in mice. Neurosci Lett 366:211–214.

Song B, Marvizon JCG (2003a) Dorsal horn neurons firing at high frequency, but not primary afferents, release opioid peptides that produce μ-opioid receptor internalization in the rat spinal cord. J Neurosci 23:9171–9184.

Song B, Marvizon JC (2003b) Peptidases prevent μ-opioid receptor internalization in dorsal horn neurons by endogenously released opioids. J Neurosci 23:1847–1858.

Song B, Marvizon JCG (2005) NMDA receptors and large conductance calcium-sensitive potassium channels inhibit the release of opioid peptides that induce μ-opioid receptor internalization in the rat spinal cord. Neuroscience 136:549–562.

Song B, Chen W, Marvizon JC (2007) Inhibition of opioid release in the rat spinal cord by serotonin 5-HT$_{1A}$ receptors. Brain Res 1158:57–62.

Spike RC, Puskar Z, Sakamoto H, Stewart W, Watt C, Todd AJ (2002) MOR-1-immunoreactive neurons in the dorsal horn of the rat spinal cord: evidence for nonsynaptic innervation by substance P-containing primary afferents and for selective activation by noxious thermal stimuli. Eur J Neurosci 15:1306–1316.

Standaert DG, Watson SJ, Houghten RA, Saper CB (1986) Opioid peptide immunoreactivity in spinal and trigeminal dorsal horn neurons projecting to the parabrachial nucleus in the rat. J Neurosci 6:1220–1226.

Stone LS, Broberger C, Vulchanova L, Wilcox GL, Hokfelt T, Riedl MS, Elde R (1998) Differential distribution of alpha(2A) and alpha(2C) adrenergic receptor immunoreactivity in the rat spinal cord. J Neurosci 18:5928–5937.

Strock J, Diverse-Pierluissi MA (2004) Ca2 + channels as integrators of G protein-mediated signaling in neurons. Mol Pharmacol 66:1071–1076.

Suplita RL, 2nd, Gutierrez T, Fegley D, Piomelli D, Hohmann AG (2006) Endocannabinoids at the spinal level regulate, but do not mediate, nonopioid stress-induced analgesia. Neuropharmacology 50:372–379.

Suzuki H, Yanagisawa M, Yoshioka K, Hosoki R, Otsuka M (1997) Enzymatic inactivation of enkephalin neurotransmitters in the spinal cord of the neonatal rat. Neurosci Res 28:261–267.

Takemori AE, Portoghese PS (1993) Enkephalin antinociception in mice is mediated by delta 1- and delta 2-opioid receptors in the brain and spinal cord, respectively. Eur J Pharmacol 242:145–150.

Tambeli CH, Parada CA, Levine JD, Gear RW (2002) Inhibition of tonic spinal glutamatergic activity induces antinociception in the rat. Eur J Neurosci 16:1547–1553.

Tambeli CH, Quang P, Levine JD, Gear RW (2003a) Contribution of spinal inhibitory receptors in heterosegmental antinociception induced by noxious stimulation. Eur J Neurosci 18:2999–3006.

Tambeli CH, Young A, Levine JD, Gear RW (2003b) Contribution of spinal glutamatergic mechanisms in heterosegmental antinociception induced by noxious stimulation. Pain 106:173–179.

Tang Q, Gandhoke R, Burritt A, Hruby VJ, Porreca F, Lai J (1999) High-affinity interaction of (des-Tyrosyl)dynorphin A(2-17) with NMDA receptors. J Pharmacol Exp Ther 291:760–765.

Tang Q, Lynch RM, Porreca F, Lai J (2000) Dynorphin A elicits an increase in intracellular calcium in cultured neurons via a non-opioid, non-NMDA mechanism. J Neurophysiol 83:2610–2615.

Terman GW, Lewis JW, Liebeskind JC (1986) Two opioid forms of stress analgesia: studies of tolerance and cross-tolerance. Brain Res 368:101–106.

Tieku S, Hooper NM (1992) Inhibition of aminopeptidases N, A and W. A re-evaluation of the actions of bestatin and inhibitors of angiotensin converting enzyme. Biochem Pharmacol 44:1725–1730.

Todd AJ, Spike RC (1992) Co-localization of Met-enkephalin and somatostatin in the spinal cord of the rat. Neurosci Lett 145:71–74.

Todd AJ, Spike RC (1993) The localization of classical transmitters and neuropeptides within neurons in laminae I–III of the mammalian spinal dorsal horn. Prog Neurobiol 41:609–645.

Todd AJ, Spike RC, Russell G, Johnston HM (1992) Immunohistochemical evidence that Met-enkephalin and GABA coexist in some neurones in rat dorsal horn. Brain Res 584:149–156.

Todd AJ, McGill MM, Shehab SA (2000) Neurokinin 1 receptor expression by neurons in laminae I, III and IV of the rat spinal dorsal horn that project to the brainstem. Eur J Neurosci 12:689–700.

Trafton JA, Abbadie C, Marchand S, Mantyh PW, Basbaum AI (1999) Spinal opioid analgesia: how critical is the regulation of substance P signaling? J Neurosci 19:9642–9653.

Trafton JA, Abbadie C, Marek K, Basbaum AI (2000) Postsynaptic signaling via the mu-opioid receptor: Responses of dorsal horn neurons to exogenous opioids and noxious stimulation. J Neurosci 20:8578–8584.

Traub RJ (1996) The spinal contribution of substance P to the generation and maintenance of inflammatory hyperalgesia in the rat. Pain 67:151–161.

Traub RJ (1997) Spinal modulation of the induction of central sensitization. Brain Res 778:34–42.

Tsou K, Khachaturian H, Akil H, Watson SJ (1986) Immunocytochemical localization of pro-opiomelanocortin-derived peptides in the adult rat spinal cord. Brain Res 378:28–35.

Tuchscherer MM, Seybold VS (1989) A quantitative study of the coexistence of peptides in varicosities within the superficial laminae of the dorsal horn of the rat spinal cord. J Neurosci 9:195–205.

Vanderah TW, Gardell LR, Burgess SE, Ibrahim M, Dogrul A, Zhong CM, Zhang ET, Malan TP, Jr., Ossipov MH, Lai J, Porreca F (2000) Dynorphin promotes abnormal pain and spinal opioid antinociceptive tolerance. J Neurosci 20:7074–7079.

Waldhoer M, Fong J, Jones RM, Lunzer MM, Sharma SK, Kostenis E, Portoghese PS, Whistler JL (2005) A heterodimer-selective agonist shows in vivo relevance of G protein-coupled receptor dimers. Proc Natl Acad Sci USA 102:9050–9055.

Walwyn W, Maidment NT, Sanders M, Evans CJ, Kieffer BL, Hales TG (2005) Induction of delta opioid receptor function by up-regulation of membrane receptors in mouse primary afferent neurons. Mol Pharmacol 68:1688–1698.

Watkins LR, Mayer DJ (1982) Involvement of spinal opioid systems in footshock-induced analgesia: antagonism by naloxone is possible only before induction of analgesia. Brain Res 242:309–326.

Watkins LR, Cobelli DA, Mayer DJ (1982a) Classical conditioning of front paw and hind paw footshock induced analgesia (FSIA): naloxone reversibility and descending pathways. Brain Res 243:119–132.

Watkins LR, Cobelli DA, Mayer DJ (1982b) Opiate vs non-opiate footshock induced analgesia (FSIA): descending and intraspinal components. Brain Res 245:97–106.

Watkins LR, Cobelli DA, Faris P, Aceto MD, Mayer DJ (1982c) Opiate vs non-opiate footshock-induced analgesia (FSIA): the body region shocked is a critical factor. Brain Res 242:299–308.

Watkins LR, Young EG, Kinscheck IB, Mayer DJ (1983) The neural basis of footshock analgesia: the role of specific ventral medullary nuclei. Brain Res 276:305–315.

Watkins LR, Johannessen JN, Kinscheck IB, Mayer DJ (1984) The neurochemical basis of footshock analgesia: the role of spinal cord serotonin and norepinephrine. Brain Res 290:107–117.

Watkins LR, Wiertelak EP, Maier SF (1992) Kappa opiate receptors mediate tail-shock induced antinociception at spinal levels. Brain Res 582:1–9.

Watkins LR, Hutchinson MR, Johnston IN, Maier SF (2005) Glia: novel counter-regulators of opioid analgesia. Trends Neurosci 28:661–669.

Wei F, Dubner R, Ren K (1999) Dorsolateral funiculus-lesions unmask inhibitory or dis-facilitatory mechanisms which modulate the effects of innocuous mechanical stimulation on spinal Fos expression after inflammation. Brain Res 820:112–116.

Whistler JL, von Zastrow M (1998) Morphine-activated opioid receptors elude desensitization by beta-arrestin. Proc Natl Acad Sci USA 95:9914–9919.

Whistler JL, Chuang HH, Chu P, Jan LY, von Zastrow M (1999) Functional dissociation of mu opioid receptor signaling and endocytosis: implications for the biology of opiate tolerance and addiction. Neuron 23:737–746.

Wilson RI, Nicoll RA (2002) Endocannabinoid signaling in the brain. Science 296:678–682.

Wisner A, Dufour E, Messaoudi M, Nejdi A, Marcel A, Ungeheuer MN, Rougeot C (2006) Human Opiorphin, a natural antinociceptive modulator of opioid-dependent pathways. Proc Natl Acad Sci USA 13:13.

Yabaluri N, Medzihradsky F (1997) Down-regulation of mu-opioid receptor by full but not partial agonists is independent of G protein coupling. Mol Pharmacol 52:896–902.

Yaksh TL (1981) Spinal opiate analgesia: characteristics and principles of action. Pain 11:293–346.

Yaksh TL (1985) Pharmacology of spinal adrenergic systems which modulate spinal nociceptive processing. Pharmacol BiochemBehav 22:845–858.

Yaksh TL, Elde RP (1981) Factors governing release of methionine enkephalin-like immunoreactivity from mesencephalon and spinal cord of the cat in vivo. J Neurophysiol 46:1056–1075.

Yaksh TL, Chipkin RE (1989) Studies on the effect of SCH-34826 and thiorphan on [Met5]enkephalin levels and release in rat spinal cord. Eur J Pharmacol 167:367–373.

Yaksh TL, Jessell TM, Gamse R, Mudge AW, Leeman SE (1980) Intrathecal morphine inhibits substance P release from mammalian spinal cord in vivo. Nature 286:155–157.

Yaksh TL, Terenius L, Nyberg F, Jhamandas K, Wang JY (1983) Studies on the release by somatic stimulation from rat and cat spinal cord of active materials which displace dihydromorphine in an opiate-binding assay. Brain Res 268:119–128.

Yoshimura M, North RA (1983) Substantia gelatinosa neurones hyperpolarized in vitro by enkephalin. Nature 305:529–530.

Zadina JE, Hackler L, Ge LJ, Kastin AJ (1997) A potent and selective endogenous agonist for the μ-opiate receptor. Nature 386:499–502.

Zagon IS, Verderame MF, McLaughlin PJ (2002) The biology of the opioid growth factor receptor (OGFr). Brain Res Brain Res Rev 38:351–376.

Zhang X, Bao L, Arvidsson U, Elde R, Hokfelt T (1998) Localization and regulation of the delta-opioid receptor in dorsal root ganglia and spinal cord of the rat and monkey: evidence for association with the membrane of large dense-core vesicles. Neuroscience 82:1225–1242.

Zoli M, Agnati LF (1996) Wiring and volume transmission in the central nervous system: the concept of closed and open synapses. Prog Neurobiol 49:363–380.

Zorman G, Belcher G, Adams JE, Fields HL (1982) Lumbar intrathecal naloxone blocks analgesia produced by microstimulation of the ventromedial medulla in the rat. Brain Res 236:77–84.

Chapter 8
CGRP in Spinal Cord Pain Mechanisms

Volker Neugebauer

Abstract Calcitonin gene-related peptide (CGRP) has emerged as an important molecule at different levels of the pain neuraxis. Anatomical, neurochemical, electrophysiological and behavioral data strongly suggest that CGRP in the spinal cord enhances neurotransmission, neuronal excitability, and nocifensive behaviors in preclinical pain models. Spinal CGRP also modulates the transmission of nociceptive information to supraspinal sites, thus contributing to high integrated pain behaviors. The precise mechanism of action of CGRP is still not fully understood, in part because of the complexity of the CGRP receptor(s). There is strong evidence for postsynaptic actions in the spinal cord, but CGRP receptors appear to be localized on both pre- and post-synaptic elements and can modulate the release and action of other transmitters as well. The availability of CGRP antagonists that have been tested successfully in Phase II clinical trials for migraine headache offers an important opportunity for new and improved therapeutic strategies in certain pain states.

Abbreviations

CLR	calcitonin-like receptors
DAMGO	[D-Ala^2, NMe-Phe^4, Gly-ol^5]-enkephalin
GPCRs	G-protein coupled receptors
NO	nitric oxyde
PKA	protein kinase A
PKC	protein kinase C
RAMP1	receptor activity-modifying protein 1
RCP	receptor component protein
VGLUT2	vesicular glutamate transporter -2

V. Neugebauer (✉)
Department of Neuroscience and Cell Biology, The University of Texas Medical Branch, Galveston, TX 77555-1069, USA
e-mail: voneugeb@utmb.edu

M. Malcangio (ed.), *Synaptic Plasticity in Pain*,
DOI 10.1007/978-1-4419-0226-9_8, © Springer Science+Business Media, LLC 2009

8.1 Introduction

Non-opioid neuropeptides such as calcitonin gene-related peptide (CGRP) represent an exciting class of analgesic drug targets. They act on G-protein-coupled receptors to produce long-lasting modifications of neurotransmission and excitability and have emerged as important modulators of peripheral and central nociceptive processing (Randic 1996; Schaible 1996; Van Rossum et al. 1997; Hokfelt et al. 2000; Willis and Coggeshall 2004).

8.2 CGRP and Its Receptors

Discovered in 1981 (Amara et al. 1982) the 37-amino-acid peptide CGRP belongs to a family of structurally related peptides that also include calcitonin, amylin, adrenomedullin, internedin/adrenomedullin2, and calcitonin receptor-stimulating peptides 1–3 (Wimalawansa 1996; Van Rossum et al. 1997; Poyner et al. 2002; Chang et al. 2004; Brain and Cox 2006; Doods et al. 2007). Two CGRP isoforms exist (α-CGRP and β-CGRP). Derived from different genes they differ by only one (rat) or three (human) amino acids and show generally similar distributions, biological activities and CGRP receptor affinities, although differences may exist in some tissues and α-CGRP appears to be more abundant in sensory systems (Van Rossum et al. 1997; Poyner et al. 2002; Willis and Coggeshall 2004).

Like other members of this peptide family CGRP stimulates adenylyl cyclase, leading to cyclic AMP formation and protein kinase A (PKA) activation, through G-protein-coupled receptors (GPCRs), but other signal transduction mechanisms may be involved in various tissues (Wimalawansa 1996; Van Rossum et al. 1997; Doods et al. 2000; Poyner et al. 2002). CGRP receptors are multi-unit complexes formed by three different proteins (Poyner et al. 2002; Hay et al. 2006): the seven-transmembrane calcitonin-like receptor (CLR) (cloned by Njuki et al. 1993); the receptor activity-modifying protein 1 (RAMP1), a single transmembrane protein that forms a heterodimer with CLR and determines its ligand affinity and pharmacological profile (cloned by McLatchie et al. 1998); and a cytoplasmatic receptor component protein (RCP) that facilitates coupling of the receptor to effector systems (cloned by Luebke et al. 1996). CGRP receptors undergo agonist-induced desensitization through a mechanism that involves phosphorylation by G-protein-coupled receptor kinase 6 and protein kinase C (PKC) but not PKA (Pin and Bahr 2008).

Pharmacological evidence has been suggesting heterogeneity of the CGRP receptors for quite some time (see Hay 2007). CGRP1 receptors are defined by their high sensitivity to the peptide antagonist CGRP$_{8-37}$and to newer non-peptide antagonists BIBN4096 and MK-0974 (Doods et al. 2007); model tissue for CGRP1 receptor expression and pharmacology is the left atrium. In contrast, so-called CGRP2 receptors are characterized by their insensitivity to

CGRP$_{8-37}$ and preferential activation by linear CGRP analogues; prototypical tissue expression is in the vas deference. CGRP1 receptors are formed by the above described multi-unit complex whereas the constituents of CGRP2 receptors have not yet been identified. In fact, the "CGRP2 receptor" may not be a single receptor but a phenotype formed by a combination of multiple molecular entities. Potential candidates for CGRP2 receptors are the amylin receptor AMY1(a) (calcitonin receptor with RAMP1) and the adrenomedullin receptor AM2 (CLR with RAMP3) (Hay 2007), but unknown cofactors or even receptor splice variants could contribute as well (Poyner et al. 2002).

As a consequence, selective antagonists for CGRP1, but not CGRP2, are available at present (Poyner et al. 2002; Doods et al. 2007; Hay 2007). They include the classical peptide antagonist CGRP$_{8-37}$ and the non-peptide antagonists BIBN4096 (Olcegepant), SB-273779 and MK-0974. MK-0974 is orally active. SB-273779 is less potent than BIBN4096 and MK-0974 at the human CGRP receptor but its effects are not species-dependent (Aiyar et al. 2001). In contrast, BIBN4096 and MK-0974 have much higher affinities for human than non-primate CGRP receptors, possibly because human CLR and RAMP1 form a binding pocket with higher hydrophobicity (Mallee et al. 2002; Taylor et al. 2006).

8.3 Localization of CGRP and CGRP Receptors in the Spinal Cord

CGRP and its receptors are widely distributed in the peripheral and central nervous system (Skofitsch and Jacobowitz 1985; Yashpal et al. 1992; Quirion et al. 1992; Saper 1995; Skofitsch et al. 1995; Van Rossum et al. 1997). The localization of CGRP and its binding sites in the spinal cord has been reviewed in detail (Willis and Coggeshall 2004) and can be described as follows.

8.3.1 CGRP

The major if not exclusive source of CGRP in the spinal dorsal horn is primary afferent axons. No cells with CGRP mRNA and no descending CGRP fibers are found in the dorsal horn, but the majority of large neurons in the ventral horn, presumably motoneurons, and preganglionic autonomic neurons contain CGRP and CGRP mRNA. The removal of primary afferent input results in the essentially complete loss of CGRP in the dorsal horn. CGRP and its mRNA are expressed in more dorsal root ganglion (DRG) cells than other peptides. CGRP is found mainly in small DRG cells and unmyelinated axons (C-fibers) but also in some medium-sized and a few large DRG cells and in myelinated axons of Aδ- and even Aβ-fibers. CGRP is typically co-localized with substance P (SP); all SP-containing DRG cells express CGRP but only half of the CGRP cells express SP. CGRP also coexists with glutamate in primary afferent terminals. CGRP containing fibers and terminals are localized in laminae I, II and V of the dorsal horn as well as in the area around the central canal. In the superficial dorsal horn CGRP terminals form classical synapses with postsynaptic elements, including spinothalamic tract cells, GABAergic neurons and dynorphin containing cell bodies; they also synapse on presynaptic (predominantly non-peptidergic) central terminals.

8.3.2 CGRP Receptors

The spinal dorsal horn contains numerous CGRP binding sites and CGRP1 receptor components CLR, RAMP1 and RCP (Van Rossum et al. 1997; Ye et al. 1999; Willis and Coggeshall 2004; Cottrell et al. 2005; Ma et al. 2006; Marvizon et al. 2007). There is clear evidence for $[^{125}I]hCGRP\alpha$ binding sites in laminae I and X and a more variable species-dependent distribution in other areas of the dorsal horn, including lamina II (Yashpal et al. 1992). Dense staining for CGRP receptors was found in laminae I and II and lighter staining in deeper laminae of the dorsal horn using monoclonal antibodies (Ye et al. 1999). CGRP receptor expression was in neuronal but not glial elements. CLR, RAMP1 and RCP were also detected predominantly in laminae I and II, indicating the presence of functional CGRP1 receptor complexes (Cottrell et al. 2005; Ma et al. 2006; Marvizon et al. 2007). These findings no longer support the once hotly debated mismatch between CGRP and its receptors in the superficial dorsal horn.

There is some controversy about the synaptic localization of CGRP receptors. Monoclonal antibodies detected postsynaptic CGRP receptors in dendrites and cell bodies of superficial dorsal horn neurons, some of which were contacted by presumed primary afferent terminals, suggesting that postsynaptic CGRP receptors can participate in direct primary afferent interactions with dorsal horn neurons (Ye et al. 1999). Likewise, CLR, RCP and RAMP1 immunoreactivity was present in cell bodies of dorsal horn neurons (Cottrell et al. 2005; Ma et al. 2006; Marvizon et al. 2007). CGRP receptors and receptor components were also found presynaptically on fibers in the dorsal horn (Ye et al. 1999; Cottrell et al. 2005; Marvizon et al. 2007; but see Ma et al. 2006). Some studies showed colocalization of presynaptic CGRP receptors (and CLR) with CGRP, suggesting a possible role as autoreceptors (Ye et al. 1999; Cottrell et al. 2005). Another study, however, failed to detect a colocalization of CLR with CGRP or with primary afferent markers but showed that CLR and RAMP1 co-localized with synaptophysin and vesicular glutamate transporter-2 (VGLUT2), suggesting that these were glutamatergic presynaptic terminals (Marvizon et al. 2007). CLR and RAMP1 were also found in terminals containing opioids (enkephalin and dynorphin) and α_{2C}-adrenoreceptors (Marvizon et al. 2007).

In summary, CGRP receptors appear to be localized not only postsynaptically on dorsal horn neurons but also presynaptically on terminals where they could function as auto- or hetero-receptors.

8.4 Pain-Related Changes in Spinal CGRP Neurochemistry

Noxious stimuli and tissue inflammation increase spinal CGRP levels whereas peripheral axotomy decreases the amount of CGRP in the dorsal horn (Willis and Coggeshall 2004). Electrical stimulation of dorsal roots (Malcangio and Bowery 1996) or spinally administered capsaicin (Garry and Hargreaves 1992; Garry et al. 2000) increased the release of CGRP-like immunoreactivity in spinal dorsal horn slices. In vivo studies using antibody microprobes showed CGRP release in the superficial dorsal horn following electrical stimulation of unmyelinated, but not large myelinated, primary afferents (Morton and Hutchison 1989; Schaible et al. 1994), noxious thermal or mechanical stimulation of the skin (Morton and Hutchison 1989), and noxious, but not innocuous, mechanical stimulation of the knee joint (Schaible et al. 1994). Noxious visceral stimulation (colorectal distension, CRD) decreased the spinal content of CGRP protein, which was interpreted as increased release from afferent terminals and subsequent uptake and/or degradation (Lu et al. 2005).

A delayed increase of CGRP levels or immunoreactivity was detected in several models of inflammatory pain. CGRP release (Schaible et al. 1994) and immunoreactive staining for CGRP (Sluka and Westlund 1993a, b) in the ipsilateral dorsal horn increased several hours after the induction of a knee joint arthritis (kaolin/carrageenan model) in rodents (Fig. 8.1). CGRP immunoreactivity was more variable in the same arthritis model in monkeys and actually decreased at the 8 h time point (Sluka et al. 1992). CGRP immunoreactivity in the dorsal horn increased bilaterally in the acute (3 days), but not chronic (3 weeks), phase of a more prolonged monoarthritis model that was induced in the knee by the intraarticular injection of methylated bovine serum albumin (mBSA) after systemic sensitization of the animal to the same antigen (Mapp et al. 1993). CGRP protein levels and release in the dorsal horn were increased in rats with polyarthritis (3–4 weeks after induction with complete Freund's adjuvant, CFA) compared to normal controls (Collin et al. 1993). Bilateral increases of immunoreactive CGRP in the dorsal horn were also observed in adjuvant-induced polyarthritis after 1 and 2 months (Marlier et al. 1991; Kar et al. 1994).

Unilateral hindpaw inflammation (CFA model) increased the content of immunoreactive CGRP in the dorsal horn after several (3, 5, 8 or 14) days following an initial (12 h, 1 and 2 days) decrease (Donnerer et al. 1992; Galeazza et al. 1995; Malcangio and Bowery 1996; Mulder et al. 1997). Likewise, capsaicin-evoked release of immunoreactive CGRP increased in dorsal horn slices obtained at day 4 of a CFA-induced hindpaw inflammation (Galeazza et al. 1995).

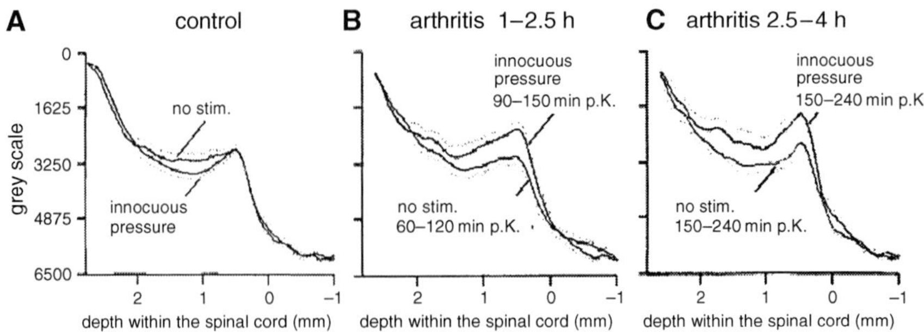

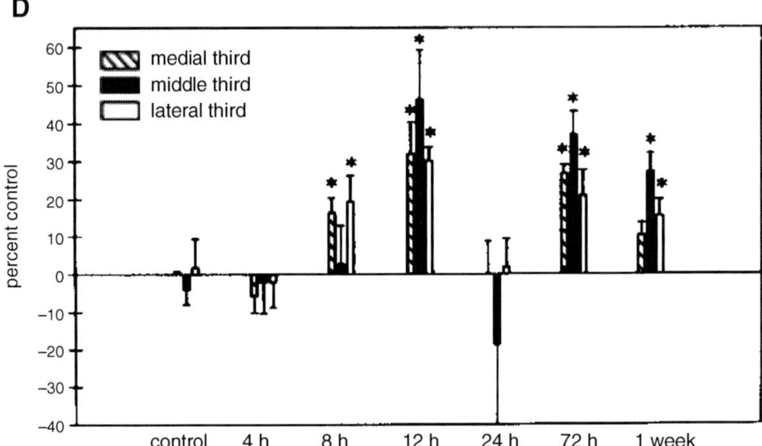

Fig. 8.1 Increased spinal CGRP release in an arthritis pain model. A, B, C Intraspinal release of CGRP before **A** and after **B, C** arthritis induction in the knee with kaolin/carrageenan was measured using microprobes coated with antibodies against CGRP. The probes were inserted into the spinal cord to a depth of 2.5 mm from the dorsal surface (located at 0 mm) for periods of 10 min. Reduced density on the grey scale indicates displacement of $[^{125}I]$CGRP binding by endogenous CGRP. **A** Averaged gray density of 40 probes present in the spinal cord during "no stimulation" and of 43 probes during innocuous pressure applied to the knee. **B** Averaged images of 20 probes present in the cord during "innocuous pressure" applied to the arthritic knee (90–150 min post-kaolin, p.K.) and of 27 "no stimulation" probes inserted between 60 and 120 min p.K. **C** Averaged images of 10 probes present in the cord during "innocuous pressure" applied to the arthritic knee (150–240 min post-kaolin) and of 13 "no stimulation" probes inserted between 150 and 240 min p. K. Pressure was applied to the knee four times for 1 min within 10 min. In the rat, 1 mm corresponds to lamina V. **D** Change of immunoreactive CGRP in the dorsal horn of the lumbar spinal cord ipsilateral to the arthritic knee (kaolin/carrageenan model). Percent changes were calculated by dividing the side-to-side difference by the contralateral (control) value. A significant increase in stain density in the ipsilateral lumbar dorsal horn compared to control and cervical (not shown) sections occurred at 8, 12, 72 h and 1 week after arthritis induction in the ipsilateral knee. Statistical significance is represented by an asterisk (*)

A, B, C: modified from Figs. 1B, 5B, D, respectively, in (Schaible et al. 1994). D: modified from Fig. 3C in (Sluka and Westlund 1993a).

However, neither basal nor electrically evoked CGRP release changed in the CFA model during the 4-week period after induction (Galeazza et al. 1995; Malcangio and Bowery 1996). Cyclophosphamide-induced cystitis (10 days) also increased CGRP immunoreactivity in the superficial dorsal horn (Vizzard 2001).

A more rapid onset of increased spinal CGRP levels and release occurred in the carrageenan and formalin models of inflammatory pain. Increased basal and capsaicin-evoked CGRP release was measured in dorsal horn slices 3 h after carrageenan-induced hindpaw inflammation (Garry and Hargreaves 1992). Increased CGRP immunoreactivity in fibers and terminals in the superficial dorsal was detected 2 h after intraplantar formalin injection (Zhang et al. 1994).

A chronic (>108 days) bilateral increase of CGRP immunoreactivity in laminae III and IV in a model of spinal cord injury was interpreted as sprouting, because CGRP is normally confined to superficial laminae I and II (Christensen and Hulsebosch 1997). Sprouting following spinal cord injury may be part of a regenerative process that can also be triggered by certain forms of peripheral nerve injury (see Willis and Coggeshall 2004). Decreased CGRP immunoreactivity was measured in the dorsal horn of neuropathic rats (2 weeks after loose ligation of the sciatic nerve, chronic constriction injury model) (Yu et al. 1996a).

In conclusion, CGRP is released in the dorsal horn in response to noxious stimuli and in different pain models, but there is substantial variability, which may reflect the different mechanisms that regulate CGRP release. A number of receptors have been implicated in the spinal release of CGRP. $GABA_A$ but not $GABA_B$ receptor activation decreased the potassium-evoked, but not spontaneous, release of CGRP from spinal dorsal horn slices (Bourgoin et al. 1992). A $GABA_B$ antagonist also had no effect on electrically evoked CGRP release. Opioid receptors can increase or decrease CGRP release. The opioid antagonist naloxone increased electrically-evoked CGRP release from spinal cord slices (Malcangio and Bowery 1996) and produced a larger increase of CGRP release in rats with CFA-induced polyarthritis than in control rats (Collin et al. 1993). Conversely, a μ-opioid receptor agonist (DAMGO) inhibited the spinal release in polyarthritic rats but not in controls (Collin et al. 1993). On the other hand, sustained morphine treatment increased the spinal release of CGRP through a Raf-1 dependent sensitization of adenylyl cyclase, increased cyclic AMP production, and PKA activation, which may play an important role in morphine hyperalgesia and analgesic tolerance (Yue et al. 2008).

CGRP release can be increased by capsaicin (Galeazza et al. 1995; Malcangio and Bowery 1996; Evans et al. 1996; Garry et al. 2000), bradykinin (Evans et al. 1996), and prostacyclin or PGI2 (Hingtgen and Vasko 1994). Bradykinin increases CGRP release through a mechanism that involves N-type, but not L- or P-type, calcium channels (Evans et al. 1996). Capsaicin-evoked CGRP release involves NMDA and non-NMDA receptors (Garry et al. 2000) as well as the production of nitric oxide (NO), which appears to act indirectly, perhaps as a retrograde messenger, but not through the elevation of cyclic GMP considered the primary mode of action of NO (Dymshitz and Vasko 1994; Garry et al. 2000).

The role of NMDA receptors in the regulation of spinal CGRP release is controversial. NMDA receptor antagonists blocked CGRP release evoked by capsaicin (Garry et al. 2000) and in the kaolin/carrageenan-induced arthritis pain model (Sluka and Westlund 1993b), but NMDA itself neither evoked release nor enhanced capsaicin-evoked release of CGRP (Nazarian et al. 2008). Finally, presynaptic α2-adrenoreceptors have been shown to inhibit CGRP release (Geppetti et al. 2005).

In contrast to the well documented pain-related changes of CGRP content and release in the spinal cord, there is relatively little information about changes in CGRP receptors. A significant decrease in binding sites for $[^{125}I]hCGRP\alpha$ was observed in adjuvant-induced polyarthritis (Kar et al. 1994). CFA-induced hindpaw inflammation produced an ipsilateral decrease of $[^{125}I]hCGRP$ binding in laminae I and II at day 4 but a bilateral increase in lamina V at days 2–8 (Galeazza et al. 1992).

8.5 Electrophysiological Effects of Spinal CGRP

The involvement of CGRP in spinal nociceptive processing is well established (Neugebauer et al. 1996; Schaible 1996; Willis and Coggeshall 2004; Sun et al. 2004a).

8.5.1 CGRP

CGRP typically has excitatory or sensitizing effects on spinal dorsal horn neurons (Miletic and Tan 1988; Ryu et al. 1988a, b; Murase et al. 1989; Biella et al. 1991; Neugebauer et al. 1996; Ebersberger et al. 2000; Sun et al. 2004a, b). Spinal application of CGRP increased background activity ("excitation") of dorsal horn neurons recorded extracellularly in anesthetized animals. The proportion of CGRP-activated cells varied substantially in different studies and was reported to be 6% (Biella et al. 1991), 25% (Neugebauer et al. 1996), 50% (Sun et al. 2004b), and 60% (Miletic and Tan 1988). Evidence from patch-clamp studies in spinal cord slices suggests that CGRP-induced activation is at least in part the result of direct cellular actions. CGRP produced an inward membrane current that depolarized dorsal horn neurons (Ryu et al. 1988a; Bird et al. 2006) and enhanced a calcium current (Murase et al. 1989). CGRP also increased spontaneous action potential firing (Ryu et al. 1988a) and neuronal excitability measured as the number of action potentials evoked in response to direct intracellular injections of depolarizing currents (Bird et al. 2006), indicating direct postsynaptic membrane effects.

In addition to direct excitation, CGRP produced sensitization of dorsal horn neurons. CGRP increased the responses of dorsal horn neurons to NMDA and AMPA (Ebersberger et al. 2000) or SP (Biella et al. 1991) applied iontophoretically next to the cell (Fig. 8.2A, B, C). CGRP also enhanced the

Fig. 8.2 Sensitization of spinal dorsal horn neurons by CGRP. A Intraspinal application of CGRP (50 nA, 5 min) by iontophoresis increased the responses of a dorsal horn neuron to iontophoretically applied NMDA (100 nA). **B** CGRP (80 nA, 5 min) enhanced the responses of another neuron to AMPA (20 nA). **C** CGRP increased the excitation of a dorsal horn neuron by substance P (SP). **D** Intraspinal application of CGRP increased the responses of a dorsal horn neuron to brief (15 s) noxious mechanical stimulation (compression) of the knee joint, ankle joint, and paw. **A, B, C, D** Extracellular single-unit recordings of wide-dynamic range (or multireceptive) neurons were made in the spinal cord of pentobarbital-anesthetized rats. Peristimulus time histograms in **A, B, C** show action potentials (spikes) per second. All compounds were administered into the spinal dorsal horn by iontophoresis
A, B: Figs. 1A and 2A in (Ebersberger et al. 2000);
C: modified from Fig. 1 in (Biella et al. 1991);
D: modified from Fig. 1A in (Neugebauer et al. 1996).

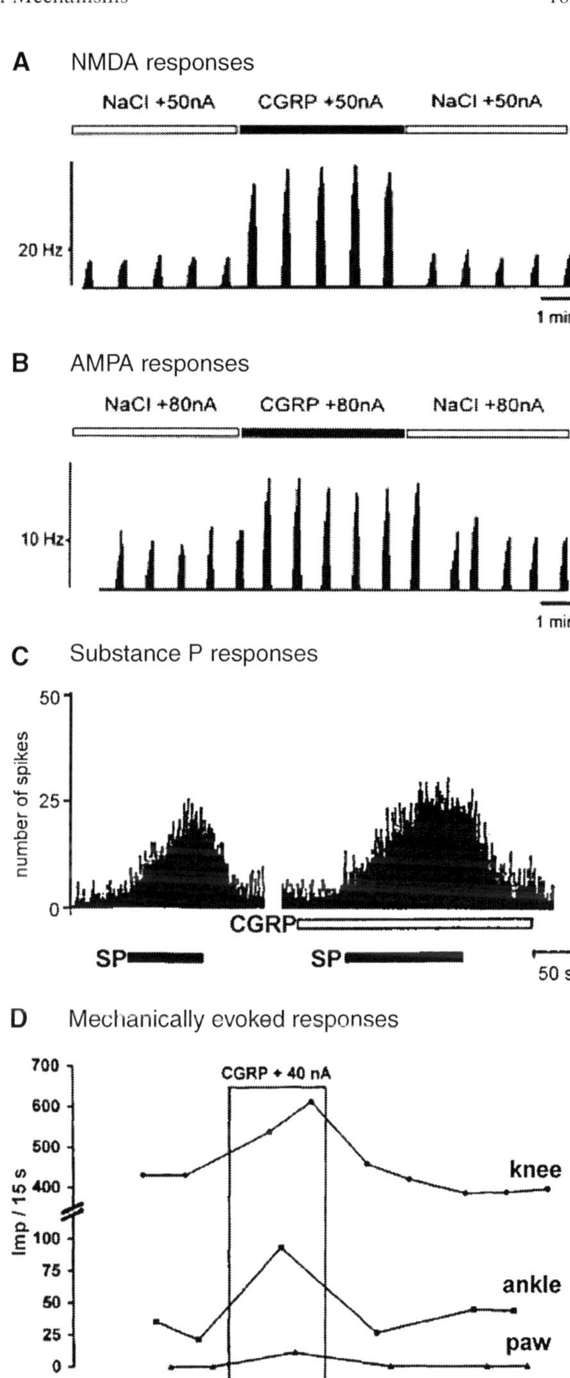

NMDA-activated inward current in dorsal horn neurons in spinal cord slices (Murase et al. 1989). The sensitizing effect of CGRP on NMDA responses is mediated through CGRP1 receptors. The interaction with AMPA receptors appears to be more complex as it was not blocked, but enhanced, by a CGRP receptor antagonist (Ebersberger et al. 2000). The enhancement of the action of SP by CGRP has been proposed to involve increased SP release (Oku et al. 1987) or inhibition of enzymatic degradation of SP (Schaible et al. 1992). CGRP increased the responses of dorsal neurons to innocuous and noxious stimulation of deep tissue such as knee and ankle joints (Fig. 8.2D) (Neugebauer et al. 1996) and to noxious but not innocuous cutaneous stimuli (Biella et al. 1991; Sun et al. 2004a, b). CGRP-induced sensitization of spinal dorsal horn neurons involves a mechanism that requires PKA and PKC (Sun et al. 2004b).

CGRP also facilitated excitatory synaptic transmission in spinal cord slices (Ryu et al. 1988b; Bird et al. 2006). Analysis of miniature excitatory postsynaptic currents (mEPSCs) showed that CGRP acted post- rather than pre-synaptically (Bird et al. 2006). Facilitatory effects of CGRP on monosynaptic EPSCs evoked in substantia gelatinosa (SG) neurons by dorsal root stimulation further suggest that CGRP can modulate direct primary afferent interactions with dorsal horn neurons (Bird et al. 2006).

8.5.2 CGRP Receptor Blockade

The facilitatory effects of CGRP on spinal dorsal horn neurons are generally blocked by a selective CGRP1 receptor antagonist ($CGRP_{8-37}$). One exception may be the modulation of AMPA responses by CGRP (see Section 8.5.1). $CGRP_{8-37}$ inhibited the responses of dorsal horn neurons to transdermal electrical stimulation of the hindpaw (Yu et al. 1999) and to brief noxious mechanical stimulation of the deep tissue (knee and ankle joints) (Neugebauer et al. 1996); in some neurons the responses to innocuous mechanical stimuli were reduced as well (Neugebauer et al. 1996). In contrast, $CGRP_{8-37}$ had little effect on the responses to brief noxious cutaneous stimuli (Sun et al. 2004a). These studies used extracellular single-unit recordings in anesthetized animals.

$CGRP_{8-37}$ consistently inhibited or prevented the central sensitization of dorsal horn neurons recorded in anesthetized animals in different pain models. Intraspinal application of $CGRP_{8-37}$ by iontophoresis during the induction of an acute arthritis in the knee (kaolin/carrageenan model) largely prevented the development of sensitization of dorsal horn neurons to noxious stimulation of peripheral deep tissue (Fig. 8.3A, B) (Neugebauer et al. 1996). This pretreatment paradigm also reduced, but did not prevent, the sensitization to innocuous stimuli and the expansion of receptive fields, a typical sign of central sensitization. When applied in the arthritis pain state 4–7 h postinduction $CGRP_{8-37}$

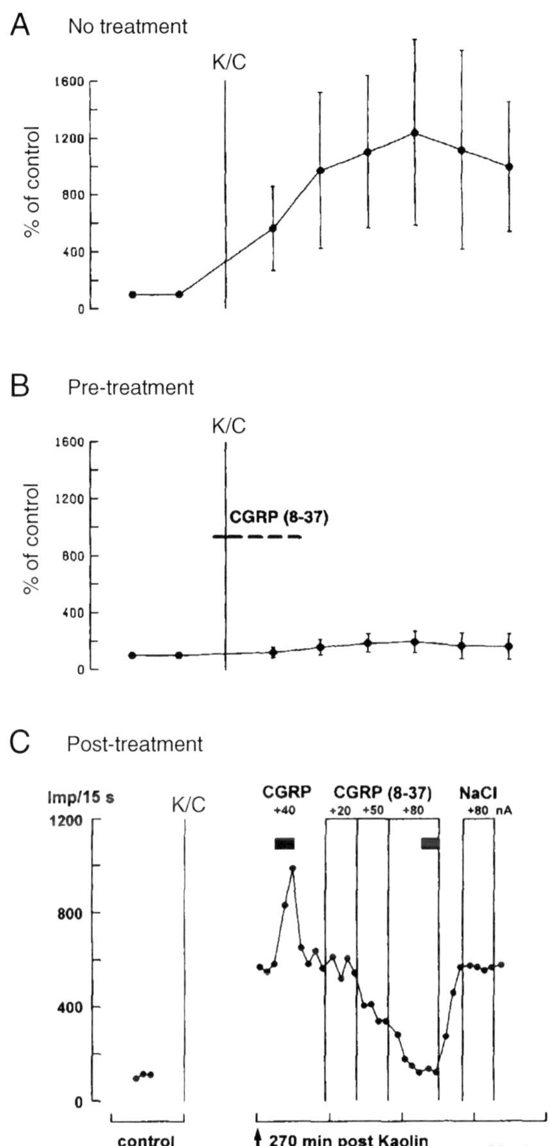

Fig. 8.3 CGRP receptor blockade prevents or inhibits pain-related sensitization of spinal dorsal horn neurons. A Responses of spinal dorsal horn neurons (n = 13) to brief (15 s) noxious mechanical stimulation (compression) of the knee increased after the induction of arthritis in the ipsilateral knee by intraarticular injections of kaolin and carrageenan (K/C). **B** Intraspinal application of a CGRP1 receptor antagonist (CGRP$_{8-37}$; 80 nA) by iontophoresis during arthritis induction largely prevented the increase of responses of dorsal horn neurons (n = 8) to noxious stimuli that is observed consistently in control neurons (see **A**). Periods of drug administration are indicated by horizontal bars. Each symbol in A and B shows the mean ± S.D. of the responses over a 1 h period. **C** Intraspinal application of CGRP$_{8-37}$ by

inhibited the enhanced responses of sensitized neurons to noxious and normally innocuous stimulation of the arthritic knee (area of primary hyperalgesia/ allodynia) and of the non-injured ankle (area of secondary hyperalgesia/ allodynia) (Fig. 8.3C) (Neugebauer et al. 1996). $CGRP_{8-37}$ reduced the background activity in less than half of the neurons and decreased the receptive field size only in about one third of the neurons. Topical application of $CGRP_{8-37}$ to the cord surface during intradermal capsaicin injection completely prevented the increase of background activity of dorsal horn neurons and their sensitization to noxious and innocuous cutaneous stimulation in the area of secondary hyperalgesia/allodynia that is typically observed in the capsaicin pain model (Sun et al. 2004a). Spinal application of $CGRP_{8-37}$ 45 min after intradermal capsaicin injection significantly inhibited the increased background activity and responses to cutaneous stimulation in the area of secondary hyperalgesia/ allodynia (Sun et al. 2004a).

These results suggest that CGRP1 receptor activation plays an important role in pain-related spinal sensitization but may not account for all aspects of central changes in certain pain models. The inability of a CGRP1 receptor antagonist to fully prevent or reverse all parameters of central sensitization in the arthritis pain model may point to the critical role of other transmitter/ modulator systems or targets downstream of (continued) CGRP receptor activation. The involvement of CGRP receptors other than CGRP1 is also a possibility (see also Chapter 12).

Despite focal spinal application of CGRP1 receptor antagonists the electrophysiological recordings in vivo could not determine the exact site of action of CGRP in the dorsal horn circuitry. A recent study demonstrated that CGRP1 receptors on SG neurons contribute to pain-related synaptic plasticity (Bird et al. 2006). Monosynaptic primary afferent input to SG neurons was inhibited by $CGRP_{8-37}$ in slices from arthritic animals (kaolin/carrageenan model) but not in slices from normal controls, suggesting pain-related endogenous activation of CGRP1 receptors at the site of primary afferent interaction with dorsal horn neurons (Fig. 8.4). The role of CGRP receptors in more chronic pain models remains to be determined.

Fig. 8.3 (continued) iontophoresis 5 h after arthritis induction with intraarticular injections of kaolin and carrageenan (K/C) inhibited the enhanced responses of a dorsal horn neuron. Numbers below drug names indicate the ejection current used for iontophoretic application. $CGRP_{8-37}$ also blocked the facilitatory effect of CGRP, suggesting that the antagonist was effective. Ejection of a comparable current through an iontophoresis barrel containing NaCl (as a control for current artifacts) had no effect. Each symbol shows the number of action potentials (impulses, imp) during brief (15 s) noxious stimulation of the knee. **A, B, C** Extracellular single-unit recordings of wide-dynamic range neurons were made in the spinal cord of pentobarbital-anesthetized rats
A, B: Modified from Fig. 4A and B; C: Fig. 6 (Neugebauer et al. 1996).

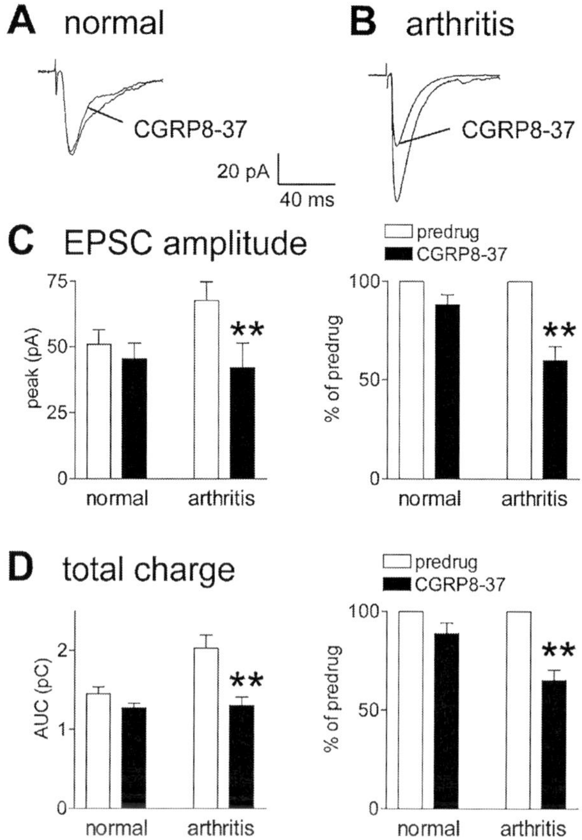

Fig. 8.4 CGRP receptor blockade inhibits pain-related synaptic plasticity in substantia gelatinosa (SG) neurons. A, B $CGRP_{8-37}$ (1 μM) inhibited monosynaptic excitatory postsynaptic currents (EPSCs) recorded in an SG neuron in a slice from an arthritic rat **B** but not in another SG neuron in a slice from a normal rat **A**. Each trace is the average of 8–10 monosynaptic EPSCs evoked by electrical stimulation of the dorsal root. **C, D** $CGRP_{8-37}$ (1 μM) significantly inhibited the EPSC peak amplitude **C**, a measure of synaptic strength, and the area under the curve (AUC, total charge, **D**) in SG neurons in slices from arthritic rats ($P < 0.01$, paired t-test, n = 5) but not in control neurons (n = 7) from normal rats. Analysis of raw data (pA, pC) is shown on the left; normalized data (% of predrug values) are shown on the right in **C** and **D**. Voltage-clamp recordings were made at −60 mV. CGRP8-37 was applied by superfusion of the slice in artificial cerebrospinal fluid (ACSF) for 10–12 min. ** $P < 0.01$ (paired t-test)
Reproduced from (Bird et al. 2006) with permission from BioMed Central.

8.5.3 Supraspinal Consequences

Lamina I neurons that receive peptidergic afferent input and have CGRP receptors (Ma et al. 2003; Braz et al. 2005; Cottrell et al. 2005; Marvizon et al. 2007) give rise to the spino-parabrachio-amygdaloid pain pathway (Gauriau and Bernard 2004). Lamina I neurons are also activated by SG neurons such as those that are sensitized through a CGRP-dependent mechanism (Bird et al. 2006). Blockade of spinal CGRP1 receptors by intrathecal application of $CGRP_{8-37}$ (1 μM, 15 min) significantly inhibited the increased responses of neurons in the central nucleus of the amygdala (CeA) in the kaolin/carrageenan-induced arthritis pain model. CeA neurons were identified as targets of the spino-parabrachio-amygdaloid pathway by orthodromic electrical stimulation in the parabrachial area. Pain-related transmission from the parabrachial area to the CeA critically involves CGRP (Han et al. 2005b). Therefore, CGRP plays an important role at the spinal and supraspinal levels of the peptidergic spino-parabrachio-amygdaloid pain pathway.

8.6 Behavioral Effects of Spinal CGRP

8.6.1 CGRP

Spinal application of CGRP typically has pro-nociceptive effects. Intrathecal CGRP produced mechanical allodynia measured as the decrease of paw withdrawal thresholds (Oku et al. 1987; Sun et al. 2003; 2004b). CGRP-induced mechanical allodynia was inhibited by a CGRP1 receptor antagonist (Sun et al. 2003) and inhibitors of PKA or PKC (Sun et al. 2004b) (Fig. 8.5A, B). In contrast, $CGRP_{8-37}$ did not antagonize the CGRP-induced increase of a nocifensive flexor reflex recorded as electromyography (EMG) activity (Xu and Wiesenfeld-Hallin 1996). Intrathecal CGRP decreased reaction times in the tail-flick test, indicating thermal hyperalgesia (Cridland and Henry 1988). Duration of the pronociceptive effects of CGRP was concentration/dose-dependent. High concentrations (0.5 μM, 10 μl; or 1 μM, 5 μl) had long-lasting (70–90 min) effects (Sun et al. 2003, 2004b) whereas the effects of lower doses (5 and 6.5 nmol, 10 μl) were short-lasting (4–15 min) (Oku et al. 1987; Cridland and Henry 1988); 30 pmol (Kawamura et al. 1989), 120 pmol (Gamse and Saria 1986) or 3.25 nmol (Cridland and Henry 1988, 1989) and even 10 nmol (Yu et al. 1994) had no effect.

Different from the pronociceptive effects of exogenous CGRP under normal conditions, endogenous CGRP does not appear to be required for normal nocifensive responses but is important in inflammatory pain. Mice deficient in the calcitonin/αCGRP gene showed normal thermo-nociception in the hotplate and paw withdrawal tests (Zhang et al. 2001). After induction of a knee joint arthritis (kaolin/carrageenan model) CGRP-knockout animals did not develop increased thermo-nociception that was observed in wildtype mice (Zhang et al. 2001). Mechano-nociception was not measured in this study.

Some evidence suggests an interaction of CGRP and SP. Pretreatment with intrathecal CGRP at a concentration that had no effect on its own increased SP-induced vocalizations to innocuous mechanical stimulation (Cridland and

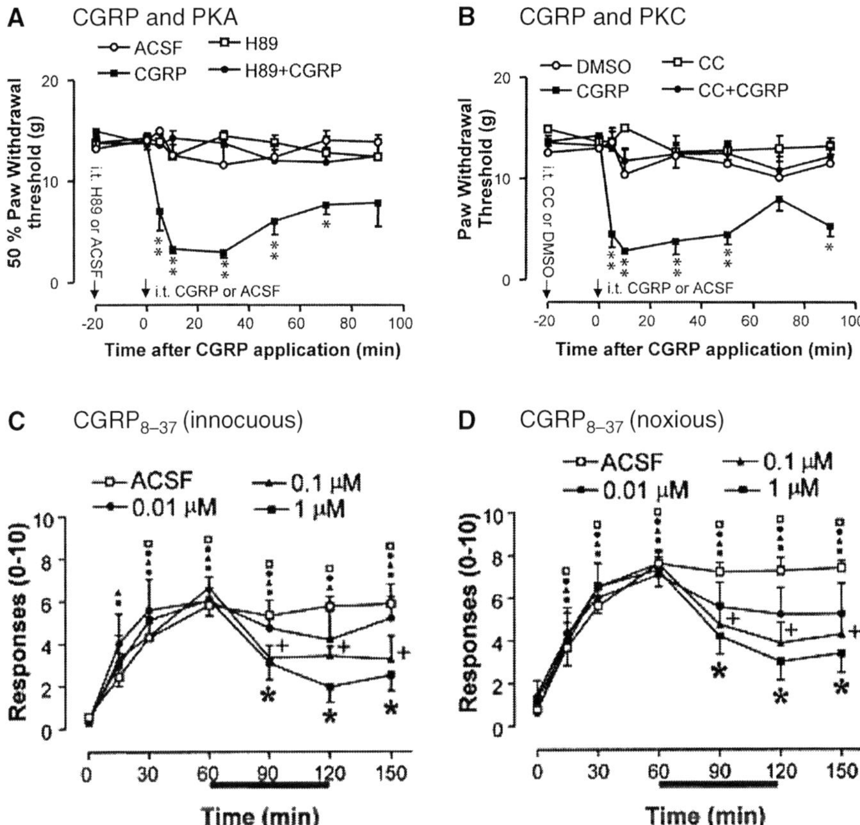

Fig. 8.5 Behavioral effects of CGRP receptor activation or blockade. A CGRP-induced mechanical allodynia in normal rats is reduced by a PKA inhibitor (H89). Intrathecal CGRP (0.5 µM, 10 µl) significantly decreased paw withdrawal thresholds. H89 (10 µM, 10 µl) inhibited the effects of CGRP. H89 or ACSF (vehicle control) was applied 20 min before CGRP. **B** CGRP-induced mechanical allodynia in normal rats is reduced by a PKC inhibitor (chelerythrine chloride, CC). Intrathecal CGRP (0.5 µM, 10 µl) significantly decreased paw withdrawal thresholds. CC (0.1 mM, 10 µl) inhibited the effects of CGRP. CC or DMSO (0.2%, vehicle control) was applied 20 min before CGRP. (A, B) Each symbol represents the means $\pm$ SE, n = 5–7 (A) or 5–10 (B). * P < 0.05, ** P < 0.01, *** P < 0.001 (compared with vehicle control at each time point; repeated-measures ANOVA followed by Newman-Keuls posttest). **C, D** A CGRP receptor antagonist (CGRP8-37) inhibits capsaicin-induced allodynia **C** and hyperalgesia **D**. Von Frey filaments were applied to the plantar surface of the hindpaw 10 times at each time point before and after intradermal capsaicin (1%, 10 µl; injected at the 0 time point). Number of paw withdrawal responses was averaged for each experimental group of animals (n = 5–7). Allodynia and hyperalgesia were determined using von Frey filaments with bending forces of 10 mN (C, innocuous stimulus) and 90 mN (D, noxious). The effect of capsaicin was inhibited by CGRP8-37 (0.01, 0.1, and 1.0 µM) administered into the dorsal horn by microdialysis (5 µl/min) from 60 to 120 min after capsaicin (thick bar). □ ● ▲ ■ P < 0.05, significant difference from baseline in each of the different lines; ⁺P < 0.05, significant difference from the 60 min time point in 0.1 µM group; *P < 0.05, significant difference from the 60 min time point in 1 µM group (Friedman repeated measures ANOVA on ranks followed by pairwise multiple comparison procedures) A, B: Figs. 1 and 2 (Sun et al. 2004b); C, D: Fig. 1 A and B (Sun et al. 2003).

Henry 1989). On the other hand, the same concentration of CGRP inhibited thermal hyperalgesia (tail-flick test) that was produced by intrathecal SP or by tail immersion in hot water (Cridland and Henry 1989). No interaction of low concentrations of CGRP with SP was detected in other studies measuring SP-induced mechanical allodynia (paw withdrawal threshold), thermal hyperalgesia (tail-flick latency), chemo-nociception (subcutaneous hypertonic saline), and aversive responses such as scratching and biting (Gamse and Saria 1986; Oku et al. 1987). However, intrathecal $CGRP_{8-37}$ (10 nmol) reversed SP-induced thermal and mechanical hyperalgesia (Yu et al. 1994).

8.6.2 CGRP Receptor Blockade

Blockade of spinal CGRP receptors inhibits nocifensive behavior in various pain models whereas mixed effects have been observed under normal conditions. Intrathecal injection of anti-CGRP antiserum increased baseline mechano-nociceptive thresholds for more than 8 h (Kuraishi et al. 1988). Likewise, intrathecal $CGRP_{8-37}$ (5 and 10 nmol but not 1 nmol) produced a prolonged increase (>60 min) of mechanical withdrawal thresholds and mechanical and thermal withdrawal latencies in normal animals (Yu et al. 1994; 1996a, b; 1998). In other studies, however, intrathecal $CGRP_{8-37}$ (2.6 and 5.2 nmol) facilitated a nocifensive flexor reflex (EMG activity) (Xu and Wiesenfeld-Hallin 1996) and intrathecal anti-CGRP antiserum or $CGRP_{8-37}$ (1 nM to 10 μM) had no effect on thermo-nociception (withdrawal latency) and mechano-nociception (withdrawal threshold or frequency) in normal animals (Kawamura et al. 1989; Bennett et al. 2000; Sun et al. 2003). The lack of antagonist effects in normal animals is in agreement with unchanged baseline nocifensive behavior of CGRP-knockout mice (see Section 8.5.1).

Blockade of spinal CGRP receptor activation consistently inhibited mechanical allodynia and mechanical and thermal hyperalgesia in models of inflammatory and neuropathic pain. Intrathecal anti-CGRP antiserum restored normal paw withdrawal latencies to radiant heat and mechanical thresholds in animals with acute inflammation (2 h after subcutaneous carrageenan) or chronic arthritis (15 days after intradermal CFA) (Kuraishi et al. 1988; Kawamura et al. 1989). The onset of drug effects was more rapid in arthritic animals compared to controls (Kuraishi et al. 1988). Administration of $CGRP_{8-37}$ (10 nM to 1 μM; 5 μl/min for 1 h) into the dorsal horn by microdialysis attenuated secondary mechanical allodynia and hyperalgesia induced by intradermal capsaicin (Fig. 8.5) (Sun et al. 2003). Pre- and post-treatment paradigms were effective in the capsaicin model but $CGRP_{8-37}$ had no effect under normal conditions (Sun et al. 2003). Intrathecal $CGRP_{8-37}$ (10 nmol but not 1 or 5 nmol) increased paw withdrawal latencies to thermal and mechanical stimuli and mechanical withdrawal thresholds in animals with hindpaw inflammation (2 h after subcutaneous carrageenan) (Yu et al. 1996b; 1998), thermal injury (4 h after hindpaw immersion in hot water) (Lofgren et al. 1997) or mononeuropathy (1 week after loose ligation of the sciatic nerve, chronic constriction injury model) (Yu et al. 1996a). $CGRP_{8-37}$ effects were less pronounced in these pain models compared to normal animals. In animals with spinal cord hemisection (28 days) intrathecal $CGRP_{8-37}$ ($1-50$ nM, 10 µl) decreased mechanical allodynia and mechanical and thermal hyperalgesia concentration-dependently (Bennett et al. 2000). $CGRP_{8-37}$ had no effect in normal animals.

A recent study (Adwanikar et al. 2007) showed that intrathecal application of the non-peptide CGRP1 receptor antagonist BIBN4096 (1 µM, 15 min) reversed the decrease of hindlimb withdrawal thresholds to mechanical stimulation of the knee (primary allodynia) in arthritic animals (kaolin/carrageenan model, $5-6$ h postinduction). The inhibitory effect of BIBN4096 on enhanced spinal reflexes was comparable to that of $CGRP_{8-37}$ (1 µM, 15 min). Neither antagonist had any effect on basal withdrawal thresholds in normal animals.

8.6.3 Supraspinal Consequences

Comparatively little attention has been paid to the effect of the spinal CGRP system on higher integrated pain behavior. A few studies mentioned that CGRP (6.5 and 100 nmol) produced spontaneous and mechanically evoked vocalizations (Cridland and Henry 1988; Bennett et al. 2000) and prolonged SP-induced vocalizations (Cridland and Henry 1989) in normal animals. Vocalizations are supraspinally organized responses that can be evoked by aversive stimuli. Audible (<16 kHz) and ultrasonic (25 ± 4 kHz) vocalizations to noxious stimuli reflect nocifensive and affective responses, respectively (Han et al. 2005a). Intrathecal administration of BIBN4096 (1 µM, 15 min) or $CGRP_{8-37}$ (1 µM, 15 min) significantly inhibited audible and ultrasonic vocalizations of arthritic animals ($5-6$ h postinduction with kaolin/carrageenan) but had no effect on the vocalizations of normal animals (Adwanikar et al. 2007). Vocalizations were also inhibited by blockade of CGRP1 receptors in the amygdala (Han et al. 2005b), suggesting that CGRP plays an important role in spino-amygdaloid mechanisms of pain behavior(s).

8.7 Concluding Remarks

Strong evidence suggests that CGRP acts as a neuromodulator if not transmitter in the spinal cord to contribute to mechanisms of pain-related plasticity and behavior. CGRP and its receptor components are present in the dorsal horn and CGRP is released in response to noxious stimuli. Exogenous CGRP mimics and CGRP receptor blockade attenuates certain aspects of spinal nociceptive plasticity and behavior. In contrast to the important role of the spinal CGRP system in various pain models its contribution to normal nociceptive processing and behavior is less clear. The mechanism(s) by which CGRP produces neural changes associated with pain behavior remain to be determined. They are likely to include second messenger–mediated modifications of postsynaptic targets such as voltage- and ligand-gated ion channels but may also involve presynaptic regulation of the release of other transmitters/neuromodulators such as amino acids and SP. The successful completion of Phase II clinical trials of CGRP1 antagonists for migraine headache (Doods et al. 2007) suggests that CGRP1 antagonists will become clinically available in the near future for the treatment of certain types and aspects of pain.

Acknowledgments The author's work is supported by NIH grants NS38261 and NS11255.

References

Adwanikar H, Ji G, Li W, Doods H, Willis WD, Neugebauer V (2007) Spinal CGRP1 receptors contribute to supraspinally organized pain behavior and pain-related sensitization of amygdala neurons. Pain 132:53–66.

Aiyar N, Daines RA, Disa J, Chambers PA, Sauermelch CF, Quiniou M, Khandoudi N, Gout B, Douglas SA, Willette RN (2001) Pharmacology of SB-273779, a nonpeptide calcitonin gene-related peptide 1 receptor antagonist. J Pharmacol Exp Ther 296:768–775.

Amara SG, Jonas V, Rosenfeld MG, Ong ES, Evans RM (1982) Alternative RNA processing in calcitonin gene expression generates mRNAs encoding different polypeptide products. Nature 298:240–244.

Bennett AD, Chastain KM, Hulsebosch CE (2000) Alleviation of mechanical and thermal allodynia by CGRP8-37 in a rodent model of chronic central pain. Pain 86:163–175.

Biella G, Panara C, Pecile A, Sotgiu ML (1991) Facilitatory role of calcitonin gene-related peptide (CGRP) on excitation induced by substance P (SP) and noxious stimuli in rat spinal dorsal horn neurons. An iontophoretic study in vivo. Brain Res 559:352–356.

Bird GC, Han JS, Fu Y, Adwanikar H, Willis WD, Neugebauer V (2006) Pain-related synaptic plasticity in spinal dorsal horn neurons: role of CGRP. Mol Pain 2:31.

Bourgoin S, Pohl M, Benoliel JJ, Mauborgne A, Collin E, Hamon M, Cesselin F (1992) gamma-Aminobutyric acid, through GABAA receptors, inhibits the potassium-stimulated release of calcitonin gene-related peptide- but not that of substance P-like material from rat spinal cord slices. Brain Res 583:344–348.

Brain SD, Cox HM (2006) Neuropeptides and their receptors: innovative science providing novel therapeutic targets. Br J Pharmacol 147 Suppl 1:S202–S211.

Braz JM, Nassar MA, Wood JN, Basbaum AI (2005) Parallel "pain" pathways arise from subpopulations of primary afferent nociceptor. Neuron 47:787–793.

Chang CL, Roh J, Hsu SY (2004) Intermedin, a novel calcitonin family peptide that exists in teleosts as well as in mammals: a comparison with other calcitonin/intermedin family peptides in vertebrates. Peptides 25:1633–1642.

Christensen MD, Hulsebosch CE (1997) Spinal cord injury and anti-NGF treatment results in changes in CGRP density and distribution in the dorsal horn in the rat. Exp Neurol 147:463–475.

Collin E, Mantelet S, Frechilla D, Pohl M, Bourgoin S, Hamon M, Cesselin F (1993) Increased in vivo release of calcitonin gene-related peptide-like material from the spinal cord in arthritic rats. Pain 54:203–211.

Cottrell GS, Roosterman D, Marvizon JC, Song B, Wick E, Pikios S, Wong H, Berthelier C, Tang Y, Sternini C, Bunnett NW, Grady EF (2005) Localization of calcitonin receptor-like receptor and receptor activity modifying protein 1 in enteric neurons, dorsal root ganglia, and the spinal cord of the rat. J Comp Neurol 490:239–255.

Cridland RA, Henry JL (1988) Effects of intrathecal administration of neuropeptides on a spinal nociceptive reflex in the rat: VIP, galanin, CGRP, TRH, somatostatin and angiotensin II. Neuropeptides 11:23–32.

Cridland RA, Henry JL (1989) Intrathecal administration of CGRP in the rat attenuates a facilitation of the tail flick reflex induced by either substance P or noxious cutaneous stimulation. Neurosci Lett 102:241–246.

Donnerer J, Schuligoi R, Stein C (1992) Increased content and transport of substance P and calcitonin gene- related peptide in sensory nerves innervating inflamed tissue: evidence for a regulatory function of nerve growth factor in vivo. Neuroscience 49:693–698.

Doods H, Arndt K, Rudolf K, Just S (2007) CGRP antagonists: unravelling the role of CGRP in migraine. Trends Pharmacol Sci 28:580–587.

Doods H, Hallermayer G, Wu D, Entzeroth M, Rudolf K, Engel W, Eberlein W (2000) Pharmacological profile of BIBN4096BS, the first selective small molecule CGRP antagonist. Br J Pharm 129:420–423.

Dymshitz J, Vasko MR (1994) Nitric oxide and cyclic guanosine 3',5'-monophosphate do not alter neuropeptide release from rat sensory neurons grown in culture. Neuroscience 62:1279–1286.

Ebersberger A, Charbel Issa P, Vanegas H, Schaible H-G (2000) Differential effects of calcitonin gene-related peptide and calcitonin gene-related peptide 8-37 upon responses to N-methyl–aspartate or (R,S)-[alpha]-amino-3-hydroxy-5-methylisoxazole-4-propionate in spinal nociceptive neurons with knee joint input in the rat. Neuroscience 99:171–178.

Evans AR, Nicol GD, Vasko MR (1996) Differential regulation of evoked peptide release by voltage-sensitive calcium channels in rat sensory neurons. Brain Res 712:265–273.

Galeazza MT, Garry MG, Yost HJ, Strait KA, Hargreaves KM, Seybold VS (1995) Plasticity in the synthesis and storage of substance P and calcitonin gene-related peptide in primary afferent neurons during peripheral inflammation. Neuroscience 66:443–458.

Galeazza MT, Stucky CL, Seybold VS (1992) Changes in [125I]hCGRP binding in rat spinal cord in an experimental model of acute, peripheral inflammation. Brain Res 591:198–208.

Gamse R, Saria A (1986) Nociceptive behavior after intrathecal injections of substance P, neurokinin A and calcitonin gene-related peptide in mice. Neurosci Lett 70:143–147.

Garry MG, Hargreaves KM (1992) Enhanced release of immunoreactive CGRP and substance P from spinal dorsal horn slices occurs during carrageenan inflammation. Brain Res 582:139–142.

Garry MG, Walton LP, Davis MA (2000) Capsaicin-evoked release of immunoreactive calcitonin gene-related peptide from the spinal cord is mediated by nitric oxide but not by cyclic GMP. Brain Res 861:208–219.

Gauriau C, Bernard J-F (2004) A comparative reappraisal of projections from the superficial laminae of the dorsal horn in the rat: the forebrain. J Comp Neurol 468:24–56.

Geppetti P, Capone JG, Trevisani M, Nicoletti P, Zagli G, Tola MR (2005) CGRP and migraine: neurogenic inflammation revisited. J Headache Pain 6:61–70.

Han JS, Bird GC, Li W, Neugebauer V (2005a) Computerized analysis of audible and ultrasonic vocalizations of rats as a standardized measure of pain-related behavior. J Neurosci Meth 141:261–269.

Han JS, Li W, Neugebauer V (2005b) Critical role of calcitonin gene-related peptide 1 receptors in the amygdala in synaptic plasticity and pain behavior. J Neurosci 25:10717–10728.

Hay DL (2007) What makes a CGRP2 receptor? Clin Exp Pharmacol Physiol 34:963–971.

Hay DL, Poyner DR, Sexton PM (2006) GPCR modulation by RAMPs. Pharmacol Ther 109:173–197.

Hingtgen CM, Vasko MR (1994) Prostacyclin enhances the evoked-release of substance P and calcitonin gene-related peptide from rat sensory neurons. Brain Res 655:51–60.

Hokfelt T, Broberger C, Xu ZQ, Sergeyev V, Ubink R, Diez M (2000) Neuropeptides – an overview. Neuropharmacology 39:1337–1356.

Kar S, Rees RG, Quirion R (1994) Altered calcitonin gene-related peptide, substance P and enkephalin immunoreactivities and receptor binding sites in the dorsal spinal cord of the polyarthritic rat. Eur J Neurosci 6:345–354.

Kawamura M, Kuraishi Y, Minami M, Satoh M (1989) Antinociceptive effect of intrathecally administered antiserum against calcitonin gene-related peptide on thermal and mechanical noxious stimuli in experimental hyperalgesic rats. Brain Res 497:199–203.

Kuraishi Y, Nanayama T, Ohno H, Minami M, Satoh M (1988) Antinociception induced in rats by intrathecal administration of antiserum against calcitonin gene-related peptide. Neurosci Lett 92:325–329.

Lofgren O, Yu L-C, Theodorsson E, Hansson P, Lundeberg T (1997) Intrathecal $CGRP_{8-37}$ results in a bilateral increase in hindpaw withdrawal latency in rats with a unilateral thermal injury. Neuropeptides 31:601–607.

Lu CL, Pasricha PJ, Hsieh JC, Lu RH, Lai CR, Wu LL, Chang FY, Lee SD (2005) Changes of the neuropeptides content and gene expression in spinal cord and dorsal root ganglion after noxious colorectal distension. Regul Pept 131:66–73.

Luebke AE, Dahl GP, Roos BA, Dickerson IM (1996) Identification of a protein that confers calcitonin gene-related peptide responsiveness to oocytes by using a cystic fibrosis transmembrane conductance regulator assay. Proc Natl Acad Sci USA 93: 3455–3460.

Ma W, Chabot JG, Quirion R (2006) A role for adrenomedullin as a pain-related peptide in the rat. Proc Natl Acad Sci USA 103:16027–16032.

Ma W, Chabot J-G, Powell KJ, Jhamandas K, Dickerson IM, Quirion R (2003) Localization and modulation of calcitonin gene-related peptide-receptor component protein-immunoreactive cells in the rat central and peripheral nervous systems. Neuroscience 120:677–694.

Malcangio M, Bowery NG (1996) Calcitonin gene-related peptide content, basal outflow and electrically-evoked release from monoarthritic rat spinal cord in vitro. Pain 66:351–358.

Mallee JJ, Salvatore CA, LeBourdelles B, Oliver KR, Longmore J, Koblan KS, Kane SA (2002) Receptor activity-modifying protein 1 determines the species selectivity of non-peptide CGRP receptor antagonists. J Biol Chem 277:14294–14298.

Mapp PI, Terenghi G, Walsh DA, Chen ST, Cruwys SC, Garrett N, Kidd BL, Polak JM, Blake DR (1993) Monoarthritis in the rat knee induces bilateral and time-dependent changes in substance P and calcitonin gene-related peptide immunoreactivity in the spinal cord. Neuroscience 57:1091–1096.

Marlier L, Poulat P, Rajaofetra N, Privat A (1991) Modifications of serotonin-, substance P- and calcitonin gene-related peptide-like immunoreactivities in the dorsal horn of the spinal cord of arthritic rats: a quantitative immunocytochemical study. Exp Brain Res 85:482–490.

Marvizon JCG, Perez OA, Song B, Chen W, Bunnett NW, Grady EF, Todd AJ (2007) Calcitonin receptor-like receptor and receptor activity modifying protein 1 in the rat dorsal horn: Localization in glutamatergic presynaptic terminals containing opioids and adrenergic [alpha]2C receptors. Neuroscience 148:250–265.

McLatchie LM, Fraser NJ, Main MJ, Wise A, Brown J, Thompson N, Solari R, Lee MG, Foord SM (1998) RAMPs regulate the transport and ligand specificity of the calcitonin-receptor-like receptor. Nature 393:333–339.

Miletic V, Tan H (1988) Iontophoretic application of calcitonin gene-related peptide produces a slow and prolonged excitation of neurons in the cat lumbar dorsal horn. Brain Res 446:169–172.

Morton CR, Hutchison WD (1989) Release of sensory neuropeptides in the spinal cord: studies with calcitonin gene-related peptide and galanin. Neuroscience 31:807–815.

Mulder H, Zhang Y, Danielsen N, Sundler F (1997) Islet amyloid polypeptide and calcitonin gene-related peptide expression are upregulated in lumbar dorsal root ganglia after unilateral adjuvant-induced inflammation in the rat paw. Brain Res Mol Brain Res 50:127–135.

Murase K, Ryu PD, Randic M (1989) Excitatory and inhibitory amino acids and peptide-induced responses in acutely isolated rat spinal dorsal horn neurons. Neurosci Lett 103:56–63.

Nazarian A, Gu G, Gracias NG, Wilkinson K, Hua XY, Vasko MR, Yaksh TL (2008) Spinal N-methyl-d-aspartate receptors and nociception-evoked release of primary afferent substance P. Neuroscience 152:119–127.

Neugebauer V, Rumenapp P, Schaible H-G (1996) Calcitonin gene-related peptide is involved in the spinal processing of mechanosensory input from the rat's knee joint and in the generation and maintenance of hyperexcitability of dorsal horn neurons during development of acute inflammation. Neuroscience 71:1095–1109.

Njuki F, Nicholl CG, Howard A, Mak JC, Barnes PJ, Girgis SI, Legon S (1993) A new calcitonin-receptor-like sequence in rat pulmonary blood vessels. Clin Sci (Lond) 85:385–388.

Oku R, Satoh M, Fujii N, Otaka A, Yajima H, Takagi H (1987) Calcitonin gene-related peptide promotes mechanical nociception by potentiating release of substance P from the spinal dorsal horn in rats. Brain Res 403:350–354.

Pin SS, Bahr BA (2008) Protein kinase C is a common component of CGRP receptor desensitization induced by distinct agonists. Eur J Pharmacol.

Poyner DR, Sexton PM, Marshall I, Smith DM, Quirion R, Born W, Muff R, Fischer JA, Foord SM (2002) International Union of Pharmacology. XXXII. The mammalian calcitonin gene-related peptides, adrenomedullin, amylin, and calcitonin receptors. Pharmacol Rev 54:233–246.

Quirion R, Van Rossum D, Dumont Y, St Pierre S, Fournier A (1992) Characterization of CGRP1 and CGRP2 receptor subtypes. Ann NY Acad Sci 657:88–105.

Randic M (1996) Plasticity of excitatory synaptic transmission in the spinal cord dorsal horn. Prog Brain Res 113:463–506.

Ryu PD, Gerber G, Murase K, Randic M (1988a) Actions of calcitonin gene-related peptide on rat spinal dorsal horn neurons. Brain Res 441:357–361.

Ryu PD, Gerber G, Murase K, Randic M (1988b) Calcitonin gene-related peptide enhances calcium current of rat dorsal root ganglion neurons and spinal excitatory synaptic transmission. Neurosci Lett 89:305–312.

Saper CB (1995) Central autonomic system. In: The rat nervous system (Paxinos G, ed), pp 107–135. San Diego, CA: Academic.

Schaible HG, Freudenberger U, Neugebauer V, Stiller RU (1994) Intraspinal release of immunoreactive calcitonin gene-related peptide during development of inflammation in the joint in vivo – a study with antibody microprobes in cat and rat. Neuroscience 62: 1293–1305.

Schaible HG, Hope PJ, Lang CW, Duggan AW (1992) Calcitonin gene-related peptide causes intraspinal spreading of substance P released by peripheral stimulation. Eur J Neurosci 4:750–757.

Schaible H-G (1996) On the role of tachykinins and calcitonin gene-related peptide in the spinal mechanisms of nociception and in the induction and maintenance of inflammation-evoked hyperexcitability in spinal cord neurons (with special reference to nociception in joints). Prog Brain Res 113:423–441.

Skofitsch G, Jacobowitz DM (1985) Calcitonin gene-related peptide: detailed immunohisto-chemical distribution in the central nervous system. Peptides 6: 721.

Skofitsch G, Wimalawansa SJ, Jacobowitz DM, Gubisch W (1995) Comparative immunohistochemical distribution of amylin-like and calcitonin gene related peptide like immunoreactivity in the rat central nervous system. Can J Physiol Pharmacol 73:945–956.

Sluka KA, Dougherty PM, Sorkin LS, Willis WD, Westlund KN (1992) Neural changes in acute arthritis in monkeys. III. Changes in substance P, calcitonin gene-related peptide and glutamate in the dorsal horn of the spinal cord. Brain Res Brain Res Rev 17:29–38.

Sluka KA, Westlund KN (1993a) Behavioral and immunohistochemical changes in an experimental arthritis model in rats. Pain 55:367–377.

Sluka KA, Westlund KN (1993b) Spinal cord amino acid release and content in an arthritis model: the effects of pretreatment with non-NMDA, NMDA, and NK1 receptor antagonists. Brain Res 627:89–103.

Sun RQ, Lawand NB, Lin Q, Willis WD (2004a) Role of calcitonin gene-related peptide in the sensitization of dorsal horn neurons to mechanical stimulation after intradermal injection of capsaicin. J Neurophysiol 92:320–326.

Sun RQ, Lawand NB, Willis WD (2003) The role of calcitonin gene-related peptide (CGRP) in the generation and maintenance of mechanical allodynia and hyperalgesia in rats after intradermal injection of capsaicin. Pain 104:201–208.

Sun RQ, Tu YJ, Lawand NB, Yan JY, Lin Q, Willis WD (2004b) Calcitonin gene-related peptide receptor activation produces PKA- and PKC-dependent mechanical hyperalgesia and central sensitization. J Neurophysiol 92:2859–2866.

Taylor CK, Smith DD, Hulce M, Abel PW (2006) Pharmacological characterization of novel alpha-Calcitonin Gene-Related Peptide (CGRP) receptor peptide antagonists that are selective for human CGRP receptors. J Pharmacol Exp Ther 319:749–757.

Van Rossum D, Hanish U-K, Quirion R (1997) Neuroanatomical localization, pharmacological characterization and functions of CGRP, related peptides and their receptors. Neurosci Biobehav Rev 21:649–678.

Vizzard MA (2001) Alterations in neuropeptide expression in lumbosacral bladder pathways following chronic cystitis. J Chem Neuroanat 21:125–138.

Willis WD, Coggeshall RE (2004) Sensory mechanisms of the spinal cord. New York: Plenum.

Wimalawansa SJ (1996) Calcitonin gene-related peptide and its receptors: molecular genetics, physiology, pathophysiology, and therapeutic potentials. Endocr Rev 17:533–585.

Xu XJ, Wiesenfeld-Hallin Z (1996) Calcitonin gene-related peptide (8-37) does not antagonize calcitonin gene-related peptide in rat spinal cord. Neurosci Lett 204:185–188.

Yashpal K, Kar S, Dennis T, Quirion R (1992) Quantitative autoradiographic distribution of calcitonin gene-related peptide (hCGRP alpha) binding sites in the rat and monkey spinal cord. J Comp Neurol 322:224–232.

Ye Z, Wimalawansa SJ, Westlund KN (1999) Receptor for calcitonin gene-related peptide: localization in the dorsal and ventral spinal cord. Neuroscience 92:1389–1397.

Yu LC, Hansson P, Lundeberg T (1994) The calcitonin gene-related peptide antagonist CGRP8-37 increases the latency to withdrawal responses in rats. Brain Res 653: 223–230.

Yu LC, Hansson P, Lundeberg T (1996a) The calcitonin gene-related peptide antagonist CGRP8-37 increases the latency to withdrawal responses bilaterally in rats with unilateral experimental mononeuropathy, an effect reversed by naloxone. Neuroscience 71:523–531.

Yu L-C, Hansson P, Brodda-Jansen G, Theodorsson E, Lundeberg T (1996b) Intrathecal CGRP$_{8-37}$-induced bilateral increase in hindpaw withdrawal latency in rats with unilateral inflammation. Br J Pharmacol 117:43–50.

Yu LC, Hansson P, Lundeberg S, Lundeberg T (1998) Effects of calcitonin gene-related peptide-(8-37) on withdrawal responses in rats with inflammation. Eur J Pharmacol 347:275–282.

Yu LC, Zheng EM, Lundeberg T (1999) Calcitonin gene-related peptide 8-37 inhibits the evoked discharge frequency of wide dynamic range neurons in dorsal horn of the spinal cord in rats. Regul Pept 83:21–24.

Yue X, Tumati S, Navratilova E, Strop D, St John PA, Vanderah TW, Roeske WR, Yamamura HI, Varga EV (2008) Sustained morphine treatment augments basal CGRP release from cultured primary sensory neurons in a Raf-1 dependent manner. Eur J Pharmacol 584:272–277.

Zhang L, Hoff AO, Wimalawansa SJ, Cote GJ, Gagel RF, Westlund KN (2001) Arthritic calcitonin/[alpha] calcitonin gene-related peptide knockout mice have reduced nociceptive hypersensitivity. Pain 89:265–273.

Zhang RX, Mi ZP, Qiao JT (1994) Changes of spinal substance P, calcitonin gene-related peptide, somatostatin, Met-enkephalin and neurotensin in rats in response to formalin-induced pain. Regul Pept 51:25–32.

Part IV
Amplification of Pain-Related Information

Chapter 9
Long-Term Potentiation in Superficial Spinal Dorsal Horn: A Pain Amplifier

Ruth Drdla and Jürgen Sandkühler

Abstract Long-term potentiation of synaptic strength (LTP) is one of the most intensively studied models of lasting signal amplification in the nervous systems. LTP has also been identified at synapses between small primary afferent Aδ- or C-fibres, many of which are nociceptive, and 2nd order neurons in superficial spinal dorsal horn. In the present chapter we review fundamental properties of spinal LTP and we describe different induction protocols including electrical nerve stimulation, acute nerve injury and noxious stimulation such as capsaicin or formalin intraplantar injections. The presently known signal transduction pathways leading to LTP in pain pathways include activation of NMDA receptor and NK1 receptors for substance P, opening of T-type voltage-gated calcium channel, release of Ca^{2+} from intracellular stores, as well as activation of PKC and CaMKII. These signalling pathways are similar to those leading to hyperalgesia. The converging and independent evidence summarized in this chapter suggests that LTP at the first synapse in pain pathways may underlie various forms of hyperalgesia following trauma, inflammation or nerve injury.

Abbreviations

CaMKII	Ca^{2+} calmodulin dependent kinase II
CREB	cAMP-response element binding protein
GABA	gamma-aminobutyric acid
GLT1	glutamate transporter 1
i.v.	intravenous
IASP	International Association for the Study of Pain
inhal.	inhalation
IP$_3$	inositol triphosphate

R. Drdla (✉)
Department of Neurophysiology, Center for Brain Research, Medical University of Vienna, Spitalgasse 4, A-1090 Vienna, Austria
e-mail: ruth.drdla@meduniwien.ac.at

M. Malcangio (ed.), *Synaptic Plasticity in Pain*,
DOI 10.1007/978-1-4419-0226-9_9, © Springer Science+Business Media, LLC 2009

LTD   long-term depression
LTP   long-term potentiation
MAPK   mitogen activated kinase
mGluR   metabotropic glutamate receptor
n.t.   not tried
NK1-R   neurokinin 1 receptor
NMDA   N-methyl-D-aspartate
NMDAR   N-methyl-D-aspartate-receptor
NO   nitric oxide
PKA   protein kinase A
PKC   protein kinase C
PLC   phospholipase C
superf.   superfusion
VDCC   voltage-dependent Ca^{2+} channel

9.1 Introduction

Acute, painful stimuli are signalled along a chain of excitation in nociceptive pathways from peripheral nociceptive nerve fibres to the brain. Hyperalgesia in course of trauma, inflammation or nerve injury may result from the amplification of nociceptive responses anywhere along the neuraxis. While sensitization of nociceptive nerve endings usually ceases when the primary cause for pain, e.g. the trauma or the inflammation has disappeared, central amplifiers may still be turned on long after healing. Long-term potentiation of synaptic strength (LTP) is one of the most intensively studied models of lasting signal amplification in the nervous systems. Here we summarize evidence that LTP in pain pathways may lead to some forms of hyperalgesia.

9.2 What is "LTP"?

Synaptic strength is defined as the magnitude of the direct postsynaptic response to a given presynaptic input. In other words synaptic strength is the size of the excitatory or inhibitory postsynaptic currents or potentials triggered by a single presynaptic action potential. Postsynaptic responses may be modulated in an activity-dependent or activity- independent fashion. For example the repetitive use of a synapse may lead to lasting increase in synaptic strength. If such increase outlasts the conditioning stimulus by at least 30 min, then it is labelled as early LTP and when it lasts for several hours or longer it is called late LTP. In principle, LTP may persist from minutes up to the lifespan of an animal or human subject. LTP can be expressed pre- and/or postsynaptically, i.e. synaptic strength can increase if the release of neurotransmitter(s) is enhanced and/or if the postsynaptic effects of the neurotransmitter(s) become stronger (Lisman and Raghavachari 2006). At excitatory synapses the postsynaptic depolarisation may or may not trigger firing of single or multiple action potentials. This depends not only on the magnitude of the postsynaptic depolarisation but among other factors also on the level of the resting membrane potential and threshold for action potential firing, i.e. on the membrane excitability of the postsynaptic neuron, the global excitatory and inhibitory input to the neuron and the extent of temporal and spatial summation. Thus, the firing rates of action potential discharges are not a measure of synaptic strength or changes such as LTP. However, action potential discharges can be used to assess downstream effects of synaptic plasticity.

LTP is a widespread phenomenon, which has been observed at numerous synapses throughout the nervous system. LTP in the hippocampus is considered a cellular model of learning and memory formation. To establish a scientific terminology, it is essential that the terms used are unambiguous and that identical phenomena are described by common phrases. For example, the long-lasting increase in synaptic strength should always be labelled "LTP" irrespective of which synapse in the nervous system is concerned. Thus, LTP at spinal synapses must not be given other denotations such as "central sensitisation".

9.3 LTP and "Central Sensitisation" Are Not Equivalent

While LTP is a clearly defined form of synaptic plasticity, "central sensitisation" is used in the literature in at least two mutually exclusive definitions. Some authors use "central sensitisation" as an umbrella term for all forms of changes within the central nervous system which ultimately lead to enhanced pain perception. Presently proposed central mechanisms of pain amplification and pain generation include changes in segmental, propriospinal and descending inhibitory and facilitatory control, paradoxical excitation by inhibitory neurotransmitters and various forms of synaptic plasticity. Thus, under this definition "central sensitization" would not define any specific mechanism but a hypothetical construct of changes within the central nervous system, ultimately leading to pathological pain.

A very different definition is used by other authors and by the task force for taxonomy of the IASP. Here, "central sensitisation" is defined as "an enhanced responsiveness of nociceptive neurons in the central nervous system to their normal afferent input". This is a very clear-cut definition with little error-proneness. The mechanisms underlying "central sensitisation" in this definition can be studied well experimentally. Nociceptive neurons in the central nervous system may, however, serve very distinct and also antagonistic functions not all of which are related to the perception of pain. Some nociceptive neurons are excitatory, others are inhibitory. Thus, LTP at synapses between nociceptive afferents and inhibitory, e.g. GABAergic interneurons in superficial spinal dorsal horn would be a mechanism of "central sensitisation" according to the definition by IASP. The enhanced response of an inhibitory nociceptive neuron to its "normal afferent input" may, however, lead to antinociception rather than pain hypersensitivity. Thus, the same phenomenon may clearly violate the former definition of "central sensitization" while nicely fulfilling the IASP criteria. Unfortunately incompatible definitions of "central sensitization" are used in the literature on pain and sometimes even within the same publication.

9.4 Methods to Assess LTP in Pain Pathways

LTP is usually measured as an increase in mono-synaptically-evoked postsynaptic currents or potentials in response to a single presynaptic action potential, either in vitro or in vivo.

Whole-cell patch-clamp recording in in vitro preparations is now the most often used technique. This enables excellent control over the composition of the intracellular fluid of the postsynaptic neurons which may be advantageous to study postsynaptic mechanisms of LTP. Perforated patch-clamp recordings or intracellular recordings with sharp electrodes can be used if dialysis of the postsynaptic neuron has to be avoided, i.e. if a diffusible mediator is involved. To evaluate LTP at the first synapses in nociceptive pathways, transverse slices with long dorsal roots attached can be prepared from lumbar spinal cord of rats or mice to study mono-synaptic, Aδ-fibre (Randic et al. 1993) or C-fibre evoked (Ikeda et al. 2003) excitatory postsynaptic potentials or currents in identified dorsal horn neurons.

Both, the induction and maintenance of LTP are context-sensitive. At spinal synapses, segmental, propriospinal and supraspinal descending systems may modulate the numerous aspects of synaptic plasticity. For example the same conditioning stimulus may cause LTP or long-term depression (LTD) of synaptic strength, depending on the level of membrane potential of the postsynaptic neurons. Thus, the polarity of synaptic plasticity is not solely determined by the pattern of the afferent input, but also by the status of the postsynaptic neuron. Systemic modulators of synaptic plasticity can, of course, only be studied in the entire animal with primary afferent nerve fibres, propriospinal and supraspinal ascending and descending connections intact. In vivo, C-fibre-evoked field potentials can be measured in superficial spinal dorsal horn e.g. in response to high intensity electrical stimulation of the sciatic nerve for up to 24 h (Liu and Sandkühler 1997). LTP of C-fibre evoked field potentials was first reported in 1995 (Liu and Sandkühler 1995). These extracellularly recorded field potentials have high thresholds ($> 5-6$ V) and long latencies ($90-150$ ms, corresponding to the conduction velocity of $< 1.2 \text{ m} \cdot \text{s}^{-1}$). They are not abolished by spinalisation or muscle relaxation, strongly suggesting that these signals reflect summation of postsynaptic, mainly mono-synaptically-evoked currents but not action potential firing (Schouenborg 1984; Liu and Sandkühler 1997). See Fig. 9.1.

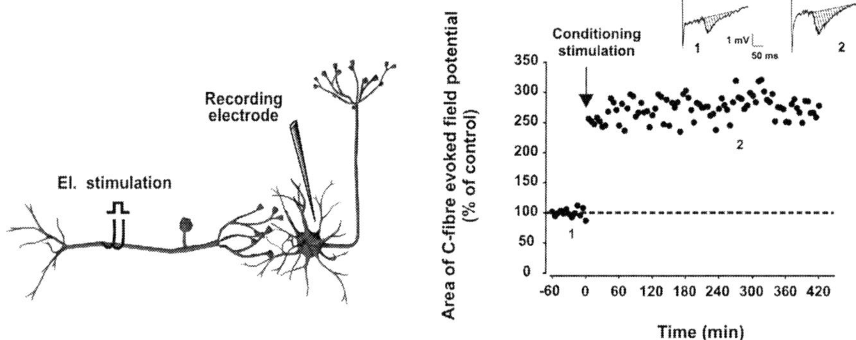

Fig. 9.1 Spinal LTP is induced upon electrical stimulation of primary afferent fibres. C-fibre evoked field potentials are recorded upon electrical stimulation of primary afferent fibres in the superficial laminae of the spinal cord dorsal horn of deeply anaesthetized adult rats. LTP is induced after electrical conditioning stimulation of the sciatic nerve at C-fibre intensity. A representative example is shown. Insets show original recordings prior to (1) or after LTP induction (2). Area of C-fibre evoked field potentials (% of control) are plotted against time (min). Modified from Drdla and Sandkühler (2008)

A typical consequence of LTP at excitatory synapses would be an increase in action potential firing of the same and perhaps also downstream neurons in response to a given stimulus. LTP-inducing conditioning stimuli have indeed been found to facilitate action potential firing of multireceptive neurons in deep dorsal horn (Afrah et al. 2002; Vikman, Duggan and Siddall 2003; Rygh et al. 2006; Haugan, Rygh and Tjølsen 2008; Pedersen and Gjerstad 2008). This long-term facilitation of action potential firing is likely due to LTP at the first synapse in the nociceptive pathway but other mechanisms must not be excluded. Action potential firing would also be enhanced if membrane excitability is increased, i.e. the thresholds for action potential firing are lowered, if inhibition is less effective or if inhibition is even reversed and becomes excitatory e.g. due to a reversal of the anion gradient in the postsynaptic neuron (Coull et al. 2003, 2005).

9.5   LTP-Inducing Protocols

The most frequently used conditioning stimulation to induce LTP at synapses in the brain is an electrical stimulus given to presynaptic nerve fibres at a high frequency (bursts of 100 Hz are most often used). LTP has been induced at synapses between primary afferent fibres and higher order neurons by high frequency stimulation both in vitro and in vivo. In spinal cord slice preparations, both, Aδ-fibre (Randic et al. 1993) and C-fibre (Ikeda et al. 2003; Ikeda et al. 2006) – evoked responses are potentiated by high frequency stimulation when postsynaptic neurons are mildly depolarized to -70 to -50 mV. The same high frequency stimulation induces LTD of Aδ-fibre-evoked responses if cells are hyperpolarized to -85 mV (Randic et al. 1993).

High frequency stimulation induces LTP selectively at C-fibre synapses with lamina I neurons that express the neurokinin 1 (NK1) receptor and send a projection to the parabrachial area. In contrast, high frequency stimulation fails to induce LTP at synapses with neurons which express the NK1 receptor and send a projection to the periaqueductal grey or at synapses with neurons that do not express the NK1 receptor and which have no identified supraspinal projection (Ikeda et al. 2003, 2006). See Fig. 9.2. Interestingly, lamina I neurons which express the NK1 receptor are indispensable for the full expression of hyperalgesia in animal models of inflammation and nerve injury (Mantyh et al. 1997; Nichols et al. 1999). This nociceptive facilitation may involve a spinal-brainstem-spinal loop (Suzuki et al. 2002).

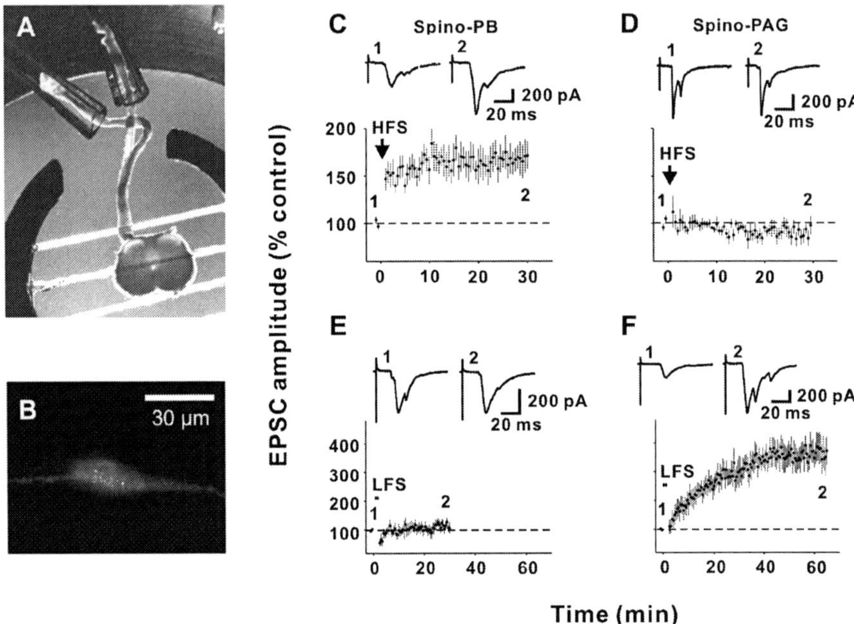

Fig. 9.2 Different forms of LTP are induced in distinct groups of lamina I projection neurons. Electrical high- or low-frequency conditioning stimulation induces LTP in lamina I neurons expressing the NK1 receptor for substance P. **(A)** Transversal slice of the spinal cord with attached dorsal root. **(B)** Projection neuron retrogradely labelled with DiI. **(C, D)** Conditioning high-frequency stimulation induces LTP in neurons with a projection to the parabrachial area (PB) but is ineffective in neurons projecting to the periaqueductal grey (PAG). **(E, F)** Conditioning low-frequency stimulation has no effect in neurons projecting to the PB, but induces LTP in spino-PAG neurons. In all electrophysiological graphs, time course of normalized amplitudes (± SEM) of monosynaptically evoked EPSCs are shown. Modified from Ikeda et al. (2003) and Ikeda et al. (2006)

High frequency stimulation at C-fibre intensity of sciatic nerve fibre afferents induces LTP of C-fibre-, but not Aß-fibre-evoked field potentials in superficial spinal dorsal horn of adult, deeply anaesthetized rats (Liu and Sandkühler 1995, 1997; Ma and Zhao 2002). Conditioning high frequency stimulation at A-fibre intensity fails to induce LTP of either A- or C-fibre-evoked field potentials in intact animals. In spinalised animals, conditioning high frequency stimulation at Aδ-fibre intensity triggers, however, LTP of C-fibre-evoked field potentials (Liu et al. 1998) likely involving a heterosynaptic induction mechanism. Likewise, in rats with a spinal nerve ligation, but not in control animals, high frequency stimulation at a low intensity (10 V, 0.5 ms pulses) induces LTP of C-fibre-evoked field potentials whereas high intensity high frequency stimulation (30 V, 0.5 ms pulses) was effective in both, control and in neuropathic animals (Xing et al. 2007). This suggests that the threshold for inducing LTP is lowered when descending inhibition is impaired or under neuropathic conditions.

High-frequency, burst-like discharges are required for LTP induction at virtually all synapses studied so far, at least if not paired with postsynaptic depolarization or during blockage of postsynaptic inhibition. Some C-fibres may discharge at these high rates (e.g. at the beginning of a noxious mechanical stimulus (Handwerker, Anton and Reeh 1987)). Low-level activity between $1-10$ imp·s^{-1} is, however, more typical for C-fibre discharges during inflammation, trauma or wound healing (Puig and Sorkin 1996). Presynaptic activity at these low frequencies is generally considered inadequate to cause a sufficiently strong rise in postsynaptic $[Ca^{2+}]_i$ for potentiation of synaptic strength. In fact, low-level presynaptic activity was either ineffective or induced synaptic LTD rather than LTP in previous studies.

Recently, we provided direct evidence that electrical stimulation of sciatic nerve fibres at low frequencies ($\sim 1-10$ Hz) and even single shocks cause substantial rise in intracellular Ca^{2+} concentration of lamina I neurons in vivo (Ikeda et al. 2006). This suggests that low level discharges in primary afferents may trigger Ca^{2+}-dependent signalling in these neurons. Indeed, we discovered a form of primary afferent-induced LTP which is triggered by conditioning stimulation at frequencies as low as 2 Hz. Importantly, this LTP induction by low frequency stimulation does not require pairing with postsynaptic depolarisation or blocking of inhibitory neurotransmitter receptors. In a spinal cord-dorsal root slice preparation conditioning electrical low frequency stimulation (2 Hz for $2-3$ min, C-fibre strength) of dorsal root afferents induces LTP selectively at C-fibre synapses with lamina I neurons that express the NK1 receptor and project to the periaqueductal grey (Ikeda et al. 2006). In the intact animal spinal dorsal neurons are under a powerful tonic inhibition arising from supraspinal, descending pathways (Basbaum and Fields 1978; Sandkühler 1996) (see Chapter 19). Descending inhibition is inevitably lost in the in vitro situation and could thereby facilitate LTP-induction. It has been shown that some stimuli evoking low level afferent input may only induce LTP when inhibitory pathways are disrupted. For example, noxious squeezing of the skin or the sciatic nerve or noxious heating of the skin has been demonstrated to induce LTP only in spinalized animals (Sandkühler and Liu 1998). C-fibre synapses between primary afferents and lamina I neurons expressing the NK1 receptor appear to be unique in that LTP can be induced by low frequency stimulation and by natural, low or high frequency, asynchronous and irregular discharge patterns in sensory nerve fibres. In animals with spinal cord and descending pathways intact, low frequency stimulation of the sciatic nerve at C-fibre intensity as well as intraplantar, subcutaneous injections of capsaicin (100 μl, 1%) or formalin (100 μl, 5%) induce LTP (Ikeda et al. 2006; Drdla and Sandkühler 2008). See Fig 9.3.

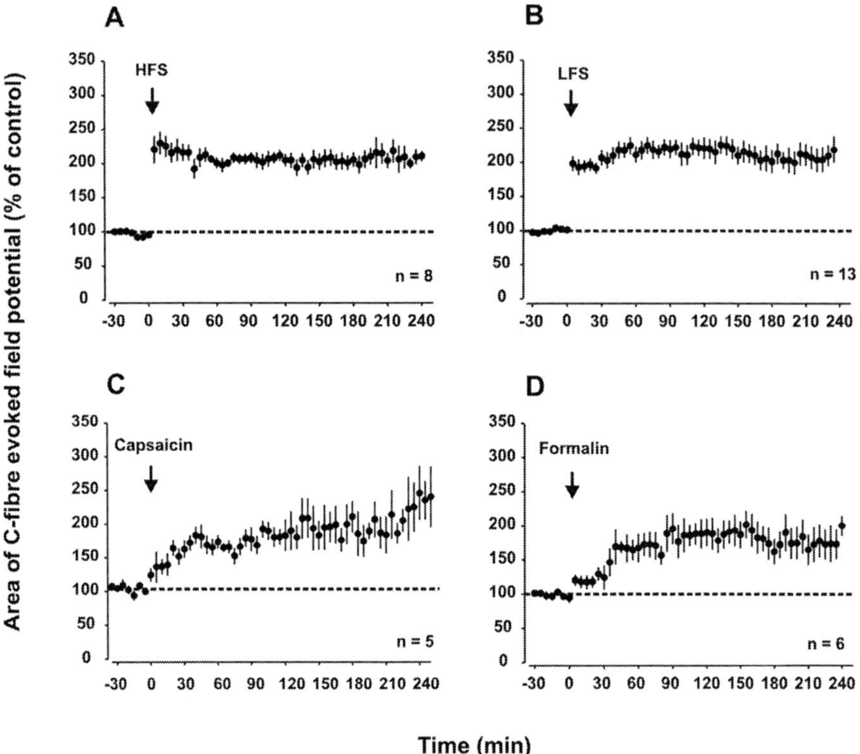

Fig. 9.3 Spinal LTP is induced by electrical conditioning stimulation as well as by natural noxious stimuli. In all graphs, mean area of C-fibre evoked field potentials recorded from superficial laminae of the spinal cord dorsal horn upon electrical stimulation of the sciatic nerve are plotted against time (± SEM). **(A, B)** LTP is induced upon electrical high- (HFS) or low-frequency stimulation (LFS, *arrow*) delivered to the sciatic nerve in deeply anaesthetized adult rats. **(C, D)** Capsaicin or formalin at time point zero (*arrow*), injected into the hindpaw of deeply anaesthetized rats, also induces slowly rising LTP of C-fibre evoked field potentials. **C** and **D** modified from Ikeda et al. (2006)

In conclusion, high frequency stimulation and low frequency stimulation may have fundamentally different effects on LTP induction at different C-fibre synapses. This finding is in line with previous reports also illustrating that the frequency of afferent barrage in C-fibres may have qualitatively different effects in spinal cord. For example, brain-derived neurotrophic factor is released from primary afferents in spinal cord slices in an activity-dependent manner by high frequency stimulation at 100 Hz but not by 1 Hz low frequency stimulation of primary afferent nerve fibres, while substance P is also released by low frequency stimulation (Lever et al. 2001).

LTP can also be induced at C-fibre synapses in the spinal cord in the absence of any presynaptic activity. In spinalised, deeply anaesthetized, adult rats, superfusion of spinal cord segments with NMDA, substance P or neurokinin A are all sufficient to induce LTP of C-fibre-evoked field potentials (Liu and Sandkühler 1998). With spinal cord and descending, including inhibitory pathways intact, spinal applications of the same substances fail, however, to induce LTP of C-fibre-evoked field potentials (Liu and Sandkühler 1998). Likewise, in rats with a spared nerve injury, but not in control animals topical application of tumour necrosis factor α to the spinal cord at the recording segments induces LTP of C-fibre-evoked field potentials (Liu et al. 2007). Spinal application of an agonist at dopamine receptors 1/5 induces a LTP of C-fibre-evoked field potentials, that develops slowly and that lasts for more than 10 h (Yang et al. 2005). Similarly, spinal application of brain-derived neurotrophic factor also induces slow onset, long-lasting LTP C-fibre-evoked field potentials (Zhou et al. 2008).

9.6 LTP at Synapses of Primary Afferent A-Fibres

Conditioning 50 Hz stimulation of sciatic nerve fibres leads to a depression of A-fibre-evoked spinal field potentials. When the $GABA_A$ receptor antagonist bicuculline is given (1 mg·kg^{-1} intraperitoneal) the same conditioning stimulus now produces LTP rather than LTD (Miletic and Miletic 2001). Similarly, 50 Hz conditioning stimulation produces short lasting potentiation followed by LTD in control animals but LTP in animals with a chronic constriction injury of sciatic nerve (Miletic and Miletic 2000). Topical application of muscimol (10 µg), a $GABA_A$ receptor agonist to spinal cord prevents tetanus-induced LTP of A-fibre-evoked field potentials in animals with a chronic constriction injury (Miletic et al. 2003). This again suggests that the polarity of synaptic plasticity is context-sensitive and not solely dominated by the type of afferent input.

9.7 Signalling Pathways of Spinal LTP

All available evidence suggests that a rise in postsynaptic $[Ca^{2+}]_i$ is the key trigger for the induction of LTP at synapses between nociceptive C-fibres and 2nd-order neurons in superficial spinal dorsal horn. We have recently demonstrated that LTP-inducing stimuli cause substantial rise in $[Ca^{2+}]_i$ in lamina I neurons not only in slice preparations, but also in intact, deeply anaesthetized animals (Ikeda et al. 2006). Sources of $[Ca^{2+}]_i$ rise are manifold and include activation of NMDA receptors upon release of glutamate likely from nociceptive nerve fibres. The NMDA receptor-mediated currents are facilitated by concomitant release of substance P binding to NK1 receptors on the same neurons, as this directly enhances NMDA receptor channel opening (Lieberman and Mody 1998) and NMDA receptor-mediated currents in lamina I neurons (Ikeda et al. 2003) and in freshly isolated dorsal horn neurons (Rusin et al. 1993). In addition, Ca^{2+} influx through voltage-sensitive Ca^{2+} channels of the T-type and release of Ca^{2+} from intracellular, ryanodine- and inositol triphosphate-receptor sensitive Ca^{2+} stores is required (Ikeda et al. 2003, 2006; Drdla and Sandkühler 2008). In AMPA receptor GluR2-deficient mice LTP induction by high frequency stimulation is enhanced, suggesting that Ca^{2+}-permeable AMPA receptors may, at least under these experimental conditions, also contribute to LTP induction (Youn et al. 2008). The downstream Ca^{2+}-dependent signalling pathways include activation of protein kinase A, protein kinase C, calcium-calmodulin-dependent protein kinase II and nitric oxide synthase (see Table 9.1, Drdla and Sandkühler 2008 and Randić, Chapter 10).

At present, there is clear evidence for a postsynaptic, Ca^{2+}-dependent form of LTP induction in spinal cord lamina I neurons (Ikeda et al. 2003, 2006). Indirect evidence suggest that in addition excitability of presynaptic terminal of primary afferents may be enhanced after LTP-inducing stimuli (Ikeda and Murase 2004).

Non-neuronal mediators such as spinal cord glial cells play a role in the succession of LTP-inducing events (Watkins and Maier 2002). The glial metabolic inhibitor fluorocitrate blocks LTP in the spinal cord dorsal horn in vivo, which is reversed by spinal administration of a NMDA receptor antagonist. Interestingly, fluorocitrate may also change the polarity of synaptic plasticity. When high frequency stimulation is given 1 h, but not 3 h after fluorocitrate LTD but no LTP of C-fibre-evoked field potentials is induced (Ma and Zhao 2002). Inhibition of the glutamate transporter 1, which is predominantly expressed by astrocytes in the spinal cord, blocks LTP and c-Fos expression in neurons, suggesting an important role for glial glutamate transporters in high-frequency stimulation-induced LTP by regulation of extracellular glutamate levels (Wang et al. 2006). When assessed with voltage-sensitive dyes the presynaptic facilitation of electrical activity in primary afferents after LTP-inducing stimuli is partially sensitive to inducible nitric oxide synthase inhibitor (AMT), a blocker of glial cell metabolisms (monofluoroacetic acid, MFA), and a metabotropic glutamate receptor group I antagonist (LY367385) (Ikeda and Murase 2004; Ikeda et al. 2007).

Importantly, the very same signal transduction pathways leading to LTP at synapses of nociceptive nerve fibres are also required for the full expression of hyperalgesia in animal models of inflammatory and neuropathic pain (Meller and Gebhart 1993; Petersen-Zeitz and Basbaum 1999; Sandkühler 2000a; Willis, Jr. 2001). See Table 9.1 for a summary of signalling elements involved in spinal LTP.

Table 9.1 Summary of signalling elements involved in spinal LTP

Target	Agonist/ Anaesthetic	LTP by HFS	LTP by LFS	In vivo	In vitro	Route	Comments	Reference(s)
NMDAR	D-AP5	Blocked	Blocked	•		Topical		Liu and Sandkühler (1995); Drdla and Sandkühler (2008)
		Blocked	Blocked		•	Superf.		Randic et al. (1993); Ikeda et al. (2003, 2006)
	MK-801	Blocked	n.t.	•		i.v.		Ikeda et al. (2006)
NK1-R	L703,606	n.t.	Blocked	•		Topical	Induces LTD	Drdla and Sandkühler (2008)
		Blocked	Blocked		•	Superf.		Ikeda et al. (2003, 2006)
	RP67580	Blocked	n.t.	•		Superf.		Liu and Sandkühler (1997)
	SR49968	Blocked	n.t.	•		Superf.		Liu and Sandkühler (1997)
mGluR (I)	(S)-4CPG	Blocked	n.t.	•		Superf. / i.v.		Azkue et al. (2003)
(II)	LY341495	No effect	n.t.	•		Superf. / i.v.		
	EGLU	No effect	n.t.	•		Superf. / i.v.		
(III)	MSOP	No effect	n.t.	•		Superf. / i.v.		
T-type VDCC	Mibefradil	n.t.	Blocked	•		Topical		Drdla and Sandkühler (2008)
	Ni^{2+}	Blocked	Blocked		•	Superf.		Ikeda et al. (2003)
$\alpha_2$adreno - R	Clonidine	Blocked	n.t.	•		Topical		Ge et al. (2006)
$GABA_A$ - R	Diazepam	Blocked	n.t.	•		Topical		Hu et al. (2006)
CaMKII	KN-62	n.t.	Blocked		•	Superf.		Ikeda et al. (2006)
	KN-93	Blocked	Blocked	•		Topical		Yang et al. (2004); Drdla and Sandkühler (2008)
	AIP	Blocked	n.t.	•		Topical		Yang et al. (2004)

Table 9.1 (continued)

Target	Agonist/ Anaesthetic	LTP by HFS	LTP by LFS	In vivo	In vitro	Route	Comments	Reference(s)
PKA	Rp-CPT-cAMPs	Blocked	n.t.	●		Topical		Yang et al. (2004)
PKC	GF109203X	n.t.	Blocked		●	Superf.		Ikeda et al. (2006)
	Chelerythrine	Blocked	Blocked	●		Topical		Yang et al. (2004); Drdla and Sandkühler (2008)
	Gö 6983	Blocked	n.t.	●		Topical		Yang et al. (2004)
PLC	U73122	n.t.	Blocked	●		Topical		Drdla and Sandkühler (2008)
		Blocked	Blocked		●	Superf.		Ikeda et al. (2003, 2006)
Ryanodine receptors	Dantrolene	n.t.	Blocked	●		Topical		Drdla and Sandkühler (2008)
IP$_3$ receptors	2-APB	Blocked			●	Superf.		Ikeda et al. (2003)
			Blocked				Induces LTD	Ikeda et al. (2006)
NO synthase	L-NMMA	n.t.	Blocked		●	Superf.		Ikeda et al. (2006)
		n.t.	Blocked	●		i.v.		
	L-NAME	Blocked	n.t.	●		Topical		Zhang et al. (2005)
MAPK/ CREB	PD98059	Blocked	n.t.	●		Superf.	Reverses established LTP	Xin et al. (2006)
Protein synthesis	Cycloheximide	No Effect	n.t.	●		Superf.	Blocks LTP-maintenance	Hu et al. (2003)
	Anisomycin	No effect	n.t.	●		Superf.		
Glia cells	Fluorocitrat	Blocked	n.t.	●		i.t.	Induces LTD	Ma and Zhao (2002)
(GLT-1)	DHK	Blocked	n.t.	●		i.i.		Wang et al. (2006)

Table 9.1 (continued)

Target	Agonist/ Anaesthetic	LTP by HFS	LTP by LFS	In vivo	In vitro	Route	Comments	Reference(s)
μ-opioid receptors	Fentanyl	Blocked	n.t.	●		i.v.		Benrath et al. (2004)
	DAMGO	Blocked	n.t.		●	Superf.		Terman et al. (2001)
Anaesthesia level	Urethane	No effect	n.t.	●		Inhal.		Benrath et al. (2004); Ikeda et al. (2006)
	Isoflurane	No effect	No effect	●				
	Sevoflurane	No effect	n.t.	●				
	Xenon (lipid solution)	Blocked	n.t.	●		i.v.		Benrath et al. (2007)

NMDAR: N-methyl-D-aspartate-receptor; NK1-R: neurokinin 1 receptor; mGluR: metabotropic glutamate receptor; VDCC: voltage-dependent Ca^{2+} channel; GABA: gamma-aminobutyric acid; CaMKII: Ca^{2+} calmodulin dependent kinase II; PKA: protein kinase A; PKC: protein kinase C; PLC: phospholipase C; IP_3: inositol triphosphate; NO: nitric oxide; MAPK: mitogen activated kinase; CREB: cAMP-response element binding protein; GLT1: glutamate transporter 1; n.t.: not tried; superf.: superfusion; inhal.: inhalation; i.v.: intravenous.

9.8 Prevention of LTP Induction

LTP induction can be prevented by means which block a Ca^{2+} rise in the postsynaptic cell. This includes clinically used μ-opioid receptor agonists. Intravenous infusion of fentanyl blocks LTP induction in vivo (Benrath et al. 2004). Similarly, LTP of spinal field potentials elicited by stimulation in the tract of Lissauer in spinal cord slices is blocked by [D-Ala^2, N-MePhe^4, Gly-ol]-enkephalin (DAMGO), a more specific agonist at these receptors (Terman et al. 2001).

Deep surgical level of anaesthesia with either urethane, isoflurane or sevoflurane is, however, insufficient to pre-empt LTP induction of C-fibre-evoked field potentials (Benrath et al. 2004). In contrast, the noble gas xenon which has NMDA receptor blocking and anaesthetic properties also prevents induction of LTP at C-fibre synapses in intact rats (Benrath et al. 2007).

9.9 Long-Term Depression and Depotentiation

Synaptic strength at primary afferents can also be depressed for a long period of time (long-term depression or LTD) by conditioning afferent stimulation (Randic et al. 1993; Sandkühler et al. 1997; Sandkühler 2000b). Both, LTP and LTD require NMDA receptor activation (Sandkühler et al. 1997). The thresholds for the induction of LTP and LTD have been suggested to be very narrowly tuned. Changes in the level of elevation of $[Ca^{2+}]_i$ might shift the threshold from LTP towards LTD (Lisman 2001). The polarity of synaptic plasticity further depends upon the magnitude (Lisman 1989), the temporal pattern (Bi and Poo 1998) and the mode of postsynaptic Ca^{2+} elevation (Nishiyama et al. 2000). These parameters are all critical for activation of distinct Ca^{2+} dependent signal transduction pathways involving protein phosphatases and kinases. Recently it has been shown that under blockade of NK1 receptors low-frequency stimulation which normally induces LTP now leads to LTD in vivo (Drdla and Sandkühler 2008), suggesting that blockade of NK1 receptors might critically change rise in $[Ca^{2+}]_i$, shifting synaptic plasticity from LTP towards LTD.

Brief, high frequency conditioning stimulation of the sciatic nerve at Aδ-fibre intensity induces LTD of C-fibre evoked field potentials in the spinal cord dorsal horn and may also reverse (depotentiate) established LTP (Liu et al. 1998). The depotentiation by Aδ-fibre stimulation is time-dependent and effective only when applied 15 or 60 min but not 3 h after LTP induction (Zhang et al. 2001).

9.10 LTP in Pain Pathways Amplifies Pain Responses

Conditioning stimuli which induce LTP at synapses of C-fibres in spinal cord also lead to long-term changes in pain-related behaviour. This has been demonstrated both, for experimental animals and for human subjects. In rats, high frequency stimulation causes thermal hyperalgesia at the ipsilateral hind paw for six days (Zhang et al. 2005). In human volunteers transcutaneous, high frequency stimulation of peptidergic nerve fibres induces a long-lasting increase in pain perception at the conditioned site (Klein et al. 2004). This perceptual correlate of spinal LTP shares time course of early LTP (Klein et al. 2007b), appears to be modality specific (Lang et al. 2007) and is blocked by ketamine suggesting that it requires the activation of NMDA receptors (Klein et al. 2007a). Enhanced sensitivity to painful stimuli is most pronounced at the conditioned site (homotopic potentiation) but also present at nearby unconditioned areas (heterotopic potentiation) (Klein et al. 2008).

9.11 Concluding Remarks

In conclusion, we suggest that LTP at synapses between primary afferent nociceptive nerve fibres and 2nd order neurons in superficial spinal dorsal horn is a potential mechanism underlying some forms of pain amplification in behaving animals and human subjects and perhaps in pain patients.

References

Afrah AW, Fiskå A, Gjerstad J et al (2002) Spinal substance P release in vivo during the induction of long-term potentiation in dorsal horn neurons. Pain 96:49–55

Azkue JJ, Liu X-G, Zimmermann M et al (2003) Induction of long-term potentiation of C fibre-evoked spinal field potentials requires recruitment of group I, but not group II/III metabotropic glutamate receptors. Pain 106:373–379

Basbaum AI, Fields HL (1978) Endogenous pain control mechanisms: review and hypothesis. Ann Neurol 4:451–462

Benrath J, Brechtel C, Martin E et al (2004) Low doses of fentanyl block central sensitization in the rat spinal cord in vivo. Anesthesiology 100:1545–1551

Benrath J, Kempf C, Georgieff M et al (2007) Xenon blocks the induction of synaptic long-term potentiation in pain pathways in the rat spinal cord in vivo. Anesth Analg 104:106–111

Bi G-Q , Poo M-M (1998) Synaptic modifications in cultured hippocampal neurons: dependence on spike timing, synaptic strength, and postsynaptic cell type. J Neurosci 18:10464–10472

Coull JAM, Beggs S, Boudreau D et al (2005) BDNF from microglia causes the shift in neuronal anion gradient underlying neuropathic pain. Nature 438:1017–1021

Coull JAM, Boudreau D, Bachand K et al (2003) Trans-synaptic shift in anion gradient in spinal lamina I neurons as a mechanism of neuropathic pain. Nature 424:938–942

Drdla R, Sandkühler J (2008) Long-term potentiation at C-fibre synapses by low-level presynaptic activity in vivo. Mol Pain 4:18

Ge Y-X, Xin W-J, Hu N-W et al (2006) Clonidine depresses LTP of C-fiber evoked field potentials in spinal dorsal horn via NO-cGMP pathway. Brain Res 1118:58–65

Handwerker HO, Anton F, Reeh PW (1987) Discharge patterns of afferent cutaneous nerve fibers from the rat's tail during prolonged noxious mechanical stimulation. Exp Brain Res 65:493–504

Haugan F, Rygh LJ, Tjølsen A (2008) Ketamine blocks enhancement of spinal long-term potentiation in chronic opioid treated rats. Acta Anaesthesiol Scand 52:681–687

Hu N-W, Zhang H-M, Hu X-D et al (2003) Protein synthesis inhibition blocks the late-phase LTP of C-fiber evoked field potentials in rat spinal dorsal horn. J Neurophysiol 89: 2354–2359

Hu X-D, Ge Y-X, Hu N-W et al (2006) Diazepam inhibits the induction and maintenance of LTP of C-fiber evoked field potentials in spinal dorsal horn of rats. Neuropharmacology 50:238–244

Ikeda H, Heinke B, Ruscheweyh R et al (2003) Synaptic plasticity in spinal lamina I projection neurons that mediate hyperalgesia. Science 299:1237–1240

Ikeda H, Murase K (2004) Glial nitric oxide-mediated long-term presynaptic facilitation revealed by optical imaging in rat spinal dorsal horn. J Neurosci 24:9888–9896

Ikeda H, Stark J, Fischer H et al (2006) Synaptic amplifier of inflammatory pain in the spinal dorsal horn. Science 312:1659–1662

Ikeda H, Tsuda M, Inoue K et al (2007) Long-term potentiation of neuronal excitation by neuron-glia interactions in the rat spinal dorsal horn. Eur J Neurosci 25: 1297–1306

Klein T, Magerl W, Hopf H-C et al (2004) Perceptual correlates of nociceptive long-term potentiation and long-term depression in humans. J Neurosci 24:964–971

Klein T, Magerl W, Nickel U et al (2007a) Effects of the NMDA-receptor antagonist ketamine on perceptual correlates of long-term potentiation within the nociceptive system. Neuropharmacology 52:655–661

Klein T, Magerl W, Treede R-D (2007b) Perceptual correlate of nociceptive long-term potentiation (LTP) in humans shares the time course of early-LTP(LTP1). J Neurophysiol 96:3551–3555

Klein T, Stahn S, Magerl W et al (2008) The role of heterosynaptic facilitation in long-term potentiation (LTP) of human pain sensation. Pain 139:507–519

Lang S, Klein T, Magerl W et al (2007) Modality-specific sensory changes in humans after the induction of long-term potentiation (LTP) in cutaneous nociceptive pathways. Pain 128:254–263

Lever IJ, Bradbury EJ, Cunningham JR et al (2001) Brain-derived neurotrophic factor is released in the dorsal horn by distinctive patterns of afferent fiber stimulation. J Neurosci 21:4469–4477

Lieberman DN, Mody I (1998) Substance P enhances NMDA channel function in hippocampal dentate gyrus granule cells. J Neurophysiol 80:113–119

Lisman J (1989) A mechanism for the Hebb and the anti-Hebb processes underlying learning and memory. Proc Natl Acad Sci USA 86:9574–9578

Lisman J, Raghavachari S (2006) A unified model of the presynaptic and postsynaptic changes during LTP at CA1 synapses. Sci STKE 2006:1–15

Lisman JE (2001) Three Ca^{2+} levels affect plasticity differently: the LTP zone, the LTD zone and no man's land. J Physiol 532:285–285

Liu X-G, Morton CR, Azkue JJ et al (1998) Long-term depression of C-fibre-evoked spinal field potentials by stimulation of primary afferent Aδ-fibres in the adult rat. Eur J Neurosci 10:3069–3075

Liu X-G, Sandkühler J (1997) Characterization of long-term potentiation of C-fiber-evoked potentials in spinal dorsal horn of adult rat: essential role of NK1 and NK2 receptors. J Neurophysiol 78:1973–1982

Liu X-G, Sandkühler J (1998) Activation of spinal N-methyl-D-aspartate or neurokinin receptors induces long-term potentiation of spinal C-fibre-evoked potentials. Neuroscience 86:1209–1216

Liu X-G, Sandkühler J (1995) Long-term potentiation of C-fiber-evoked potentials in the rat spinal dorsal horn is prevented by spinal N-methyl-D-aspartic acid receptor blockage. Neurosci Lett 191:43–46

Liu Y-L, Zhou L-J, Hu N-W et al (2007) Tumor necrosis factor-α induces long-term potentiation of C-fiber evoked field potentials in spinal dorsal horn in rats with nerve injury: the role of NF-kappa B, JNK and p38 MAPK. Neuropharmacology 52:708–715

Ma J-Y, Zhao Z-Q (2002) The involvement of glia in long-term plasticity in the spinal dorsal horn of the rat. Neuroreport 13:1781–1784

Mantyh PW, Rogers SD, Honoré P et al (1997) Inhibition of hyperalgesia by ablation of lamina I spinal neurons expressing the substance P receptor. Science 278:275–279

Meller ST, Gebhart GF (1993) Nitric oxide (NO) and nociceptive processing in the spinal cord. Pain 52:127–136

Miletic G, Draganic P, Pankratz MT et al (2003) Muscimol prevents long-lasting potentiation of dorsal horn field potentials in rats with chronic constriction injury exhibiting decreased levels of the GABA transporter GAT-1. Pain 105:347–353

Miletic G, Miletic V (2000) Long-term changes in sciatic-evoked A-fiber dorsal horn field potentials accompany loose ligation of the sciatic nerve in rats. Pain 84:353–359

Miletic G, Miletic V (2001) Contribution of GABA-A receptors to metaplasticity in the spinal dorsal horn. Pain 90:157–162

Nichols ML, Allen BJ, Rogers SD et al (1999) Transmission of chronic nociception by spinal neurons expressing the substance P receptor. Science 286:1558–1561

Nishiyama M, Hong K, Mikoshiba K et al (2000) Calcium stores regulate the polarity and input specificity of synaptic modification. Nature 408:584–588

Pedersen LM, Gjerstad J (2008) Spinal cord long-term potentiation is attenuated by the NMDA-2B receptor antagonist Ro 25-6981. Acta Physiol (Oxf) 192:421–427

Petersen-Zeitz KR, Basbaum AI (1999) Second messengers, the substantia gelatinosa and injury-induced persistent pain. Pain Suppl 6:S5–12

Puig S, Sorkin LS (1996) Formalin-evoked activity in identified primary afferent fibers: systemic lidocaine suppresses phase-2 activity. Pain 64:345–355

Randic M, Jiang MC, Cerne R (1993) Long-term potentiation and long-term depression of primary afferent neurotransmission in the rat spinal cord. J Neurosci 13: 5228–5241

Rusin KI, Jiang MC, Cerne R et al (1993) Interactions between excitatory amino acids and tachykinins in the rat spinal dorsal horn. Brain Res Bull 30:329–338

Rygh LJ, Suzuki R, Rahman W et al (2006) Local and descending circuits regulate long-term potentiation and zif268 expression in spinal neurons. Eur J Neurosci 24:761–772

Sandkühler J (1996) The organization and function of endogenous antinociceptive systems. Prog Neurobiol 50:49–81

Sandkühler J (2000a) Learning and memory in pain pathways. Pain 88:113–118

Sandkühler J (2000b) Long-lasting analgesia following TENS and acupuncture: Spinal mechanisms beyond gate control. In: Devor M, Rowbotham MC, Wiesenfeld-Hallin Z (ed) Proceedings of the 9th World Congress on Pain. IASP Press Seattle

Sandkühler J, Chen JG, Cheng G et al (1997) Low-frequency stimulation of afferent Aδ-fibers induces long-term depression at primary afferent synapses with substantia gelatinosa neurons in the rat. J Neurosci 17:6483–6491

Sandkühler J, Liu X (1998) Induction of long-term potentiation at spinal synapses by noxious stimulation or nerve injury. Eur J Neurosci 10:2476–2480

Schouenborg J (1984) Functional and topographical properties of field potentials evoked in rat dorsal horn by cutaneous C-fibre stimulation. J Physiol 356:169–192

Suzuki R, Morcuende S, Webber M et al (2002) Superficial NK1-expressing neurons control spinal excitability through activation of descending pathways. Nat Neurosci 5: 1319–1326

Terman GW, Eastman CL, Chavkin C (2001) Mu opiates inhibit long-term potentiation induction in the spinal cord slice. J Neurophysiol 85:485–494

Vikman KS, Duggan AW, Siddall PJ (2003) Increased ability to induce long-term potentiation of spinal dorsal horn neurones in monoarthritic rats. Brain Res 990:51–57

Wang Z-Y, Zhang Y-Q, Zhao Z-Q (2006) Inhibition of tetanically sciatic stimulation-induced LTP of spinal neurons and Fos expression by disrupting glutamate transporter GLT-1. Neuropharmacology 51:764–772

Watkins LR, Maier SF (2002) Beyond neurons: evidence that immune and glial cells contribute to pathological pain states. Physiol Rev 82:981–1011

Willis WD, Jr. (2001) Role of neurotransmitters in sensitization of pain responses. Ann NY Acad Sci 933:142–156

Xin W-J, Gong Q-J, Xu J-T et al (2006) Role of phosphorylation of ERK in induction and maintenance of LTP of the C-fiber evoked field potentials in spinal dorsal horn. J Neurosci Res 84:934–943

Xing G-G, Liu F-Y, Qu X-X et al (2007) Long-term synaptic plasticity in the spinal dorsal horn and its modulation by electroacupuncture in rats with neuropathic pain. Exp Neurol 208:323–332

Yang H-W, Hu X-D, Zhang H-M et al (2004) Roles of CaMKII, PKA and PKC in the induction and maintenance of LTP of C-fiber evoked field potentials in rat spinal dorsal horn. J Neurophysiol 91:1122–1133

Yang H-W, Zhou L-J, Hu N-W et al (2005) Activation of spinal D1/D5 receptors induces late-phase LTP of c-fiber evoked field potentials in rat spinal dorsal horn. J Neurophysiol 94:961–967

Youn D-H, Royle G, Kolaj M et al (2008) Enhanced LTP of primary afferent neurotransmission in AMPA receptor GluR2-deficient mice. Pain 136:158–167

Zhang H-M, Qi Y-J, Xiang X-Y et al (2001) Time-dependent plasticity of synaptic transmission produced by long-term potentiation of C-fiber evoked field potentials in rat spinal dorsal horn. Neurosci Lett 315:81–84

Zhang X-C, Zhang Y-Q, Zhao Z-Q (2005) Involvement of nitric oxide in long-term potentiation of spinal nociceptive responses in rats. Neuroreport 16:1197–1201

Zhou L-J, Zhong Y, Ren W-J et al (2008) BDNF induces late-phase LTP of C-fiber evoked field potentials in rat spinal dorsal horn. Exp Neurol 212:507–514

Chapter 10
Modulation of Long-Term Potentiation of Excitatory Synaptic Transmission in the Spinal Cord Dorsal Horn

M. Randić

Abstract There is considerable interest in understanding long-term potentiation (LTP) of glutamatergic synaptic transmission because the molecular mechanisms involved in its induction and expression are thought to be essential for learning, memory and pain. The molecular mechanisms involved in induction and expression of LTP have been widely characterized, especially in the hippocampus and have been proposed to be cellular models of learning and memory. LTP in the spinal pain pathways has been considered as one of the cellular mechanisms of post-injury pain hypersensitivity (central sensitization). Extensive evidence has indicated that changes in both the presynaptic release of glutamate and post-synaptic response to glutamate are involved in expression of LTP. This chapter attempts a brief review of some of the postsynaptic mechanisms underlying induction, expression and modulation of the high frequency stimulation-induced LTP of excitatory synaptic transmission in the superficial dorsal horn of the spinal cord. It is becoming clear that the spinal LTP, which might contribute to hyperalgesia in animal models of pain, uses multiple mechanisms involving protein phosphorylation, similar to the processes associated with hippocampal LTP. Modulation of postsynaptic AMPA and NMDA receptor function caused by phosphorylation may play an important role in the induction and expression of synaptic plasticity at dorsal horn excitatory synapses.

Abbreviations

AMPA	α-amino-3-hydroxy-5-methylisoxazole-4-propionic acid
BDNF	brain derived neurotrophic factor
CaMKII	calcium/calmodulin-dependent protein kinase II
EPSP	excitatory postsynaptic potential
ERK	extracellular signal regulated kinases
HFS	high frequency stimulation

M. Randić (✉)
Department of Biomedical Sciences, Iowa State University, Ames, IA 50011, USA
e-mail: mrandic@iastate.edu

M. Malcangio (ed.), *Synaptic Plasticity in Pain*,
DOI 10.1007/978-1-4419-0226-9_10, © Springer Science+Business Media, LLC 2009

LTD   long term depression
LTP   long term potentiation
MAPK   mitogen activated protein kinase
NMDA N-methyl-D-aspartate
PAF   primary afferent fibres
PKA   protein kinase A
PKC   protein kinase C
SDH   superficial dorsal horn
STT   spinothalamic tract

10.1 Introduction

The superficial dorsal horn of the spinal cord (SDH, laminae I/II) is the preferential site of termination of small myelinated (Aδ) and unmyelinated (C) primary afferent fibers that respond to noxious stimuli. SDH neurons include projection neurons, that conduct excitatory signals to higher brain regions for further processing, as well as local inhibitory and excitatory interneurons regulating output of projection neurons (Willis and Coggeshall, 2004). Because of the nociceptive nature of the fine afferents that make the first synaptic relay in this region of the gray matter of the spinal cord, the SDH has been considered as an important area of transmission and modulation of nociceptive information.

Ionotropic glutamate receptors are ligand-gated ion channels that are the major excitatory neurotransmitter receptors in the vertebrate central nervous system. These receptors can be subdivided on the basis of agonist pharmacology and sequence homology into three functionally distinct subclasses: AMPA (α-amino-3-hydroxy-5-methylisoxazole-4-propionic acid), kainate and NMDA (N-methyl-D-aspartate) receptors. AMPA receptors mediate the most of the fast excitatory synaptic transmission in the brain and the spinal cord (Hollmann and Heinemann, 1994; Dingledine et al., 1999). Kainate receptors contribute to the synaptic responses at excitatory synapses and can modulate presynaptic neurotransmitter release. NMDA receptors play a critical role in the modulation of excitatory synaptic transmission because of their permeability to calcium ions and ability to activate calcium-dependent signal transduction processes (Malenka and Nicoll, 1999).

Synaptic plasticity at excitatory synapses is thought to be essential for information processing in the central nervous system, and underlies complex behaviours such as learning and memory in the brain and hyperalgesia caused by tissue or nerve injury in the spinal cord. The best investigated forms of synaptic plasticity in the CNS are long-term potentiation (LTP, an increase in synaptic strength), and long-term depression (LTD, a decrease in synaptic efficacy) of excitatory synaptic transmission. The molecular mechanisms of LTP and LTD have been widely characterized (Malenka and Nicoll, 1999; Citri and Malenka,

2008), especially in the hippocampus, and have been proposed to be cellular models of learning and memory. LTP in the spinal pain pathways has been considered as one of the cellular mechanisms of post-injury pain hypersensitivity (central sensitization). In the CA1 region of the hippocampus and many regions of the brain and spinal cord, the induction of LTP and LTD is dependent on NMDA receptor activation and the resulting increase in intracellular calcium concentration (Malenka et al., 1992; Malenka and Nicoll, 1999). Ca^{2+} influx through the NMDA receptors can activate a variety of protein kinases and/or phosphatases, which in turn modulate synaptic strength. However, the essential substrates for the kinases and phosphatases that mediate changes in synaptic transmission during LTP and LTD have not been identified. Extensive evidence has indicated that changes in both the presynaptic release of glutamate and postsynaptic response to glutamate are involved in expression of LTP and LTD (Malenka and Nicoll, 1999; Nicoll, 2003). Recent studies have indicated that LTP and LTD may be expressed, by regulation of AMPA receptor function. The regulation of AMPA receptor function occurs through two distinct mechanisms: modulation of ion channel properties of the receptor and regulation of synaptic targeting of the receptor. Both of these processes are regulated by protein phosphorylation of the receptor (Song and Huganir, 2002).

10.2 NMDAR-Dependent LTP

The most wide spread form of LTP requires the activation of postsynaptic NMDARs and an increase in intracellular Ca^{2+} concentration. In agreement with the evidence obtained in the CA1 area of hippocampus (Bliss and Collingridge, 1993; Malenka and Nicoll, 1999), we demonstrated (Randić et al., 1993) that the selective competitive antagonist of the NMDAR D-AP5 has a minimal effect on basal synaptic transmission, but prevented the induction of the high-frequency stimulation (HFS)-evoked LTP at Aδ- or C-fiber synapses in the SDH (Ikeda et al., 2003, 2006), but see (Hamba et al., 2000).

It is well established that the induction of LTP in the CA1 area of hippocampus, requires synaptic activation of NMDAR during postsynaptic depolarization using different induction protocols. Experimentally this is produced in brain (Bliss and Collingridge, 1993; Malenka and Nicoll, 1999), and spinal cord (Randić et al., 1993; Ikeda et al., 2003), by applying high-frequency stimulation to the synapses, or by use of "a pairing protocol" by which the postsynaptic cell is depolarized during low frequency synaptic activation (Wei et al., 2006), or using protocols that produce spike-timing-dependent plasticity (STDP) in the brain (Dan and Poo, 2006) and as well as in the spinal cord (Youn et al., 2005; Jung et al., 2006). When the postsynaptic cell is depolarized during the induction of LTP, the Mg^{2+} block of the NMDAR channel that exists near resting membrane potential is relieved, allowing Na^+ and Ca^{2+} to enter the dendritic spine by means of the NMDAR channel. The resultant rise in $[Ca^{2+}]$ within the dendritic spine is the critical trigger for LTP. The evidence in support of the model is compelling. Specific NMDAR antagonists completely block the induction of LTP (Collingridge et al., 1983). In addition, preventing the elevation of intracellular Ca^{2+} with Ca^{2+} chelating agents blocks LTP in the brain (Lynch et al., 1983; Malenka et al., 1988) and the SDH (Rusin et al., 1992; Ikeda et al., 2003).

The regulation of NMDA receptor occurs at a number of different levels. The NMDAR is an ionotropic glutamate receptor that is both voltage- and ligand-gated (Dingledine et al., 1999). These properties enable the receptor to detect coincident synaptic input and postsynaptic depolarization, and form the basis for the involvement of the NMDAR in synaptic plasticity. NMDARs are composed of three subunit families: NR1, NR2 and NR3. Native NMDARs are composed of a heteromeric assembly of NR1 and NR2A-D subunits. All functional NMDA receptors include at least one NR1 subunit, which is required for receptor activity, whereas selective addition of NR2 subunits into the assembly enables modulation of channel kinetics (Hollmann and Heinemann, 1994). A single amino acid residue in the NR1 subunit, asparagine 528 serves as a determinant for basic properties of the NMDAR, such as high calcium permeability and voltage-dependent Mg^{2+} block (Burnashev et al., 1992). In rat dorsal horn NMDA, single cell reverse transcriptase PCR has detected the NR1 subunit and each of the four NR2 (NR2A–NR2D) subunits; the NR2B subunit being a predominant one. NMDA receptor channel modulation occurs through subunit (NR1 and NR2) phosphorylation by various intracellular protein kinases, including cyclic AMP-dependent protein kinase (PKA), calcium phospholipid-dependent protein kinase C (PKC), calcium/calmodulin-dependent protein kinase II (CaMKII), mitogen-activated protein kinases (MAPK), including extracellular signal regulated kinases (ERK), protein tyrosine kinases, as well as through dephosphorylation via the Ca^{2+}/calmodulin-dependent phosphatase calcineurin (Chen and Huang, 1992; Lieberman and Mody, 1994; Moon et al., 1994; Omkumar et al., 1996; Tingley et al., 1997). The NR1 subunit of the NMDAR undergoes a PKC-mediated phosphorylation at serine 896, as well as PKA-mediated phosphorylation at serine 890 and serine 897 (Tingley et al., 1997). PKC-mediated phosphorylation enhances channel activity (Kelso et al., 1992; Xiong et al., 1998). Neuronal responses to NMDA are potentiated by PKC, due to an increased probability of NMDAR-channel opening (Xiong et al., 1998) and a decrease in a voltage-dependent Mg^{2+} block of the NMDAR (Chen and Huang, 1992).

Experimental evidence for the existence of long-lasting activity-dependent changes in synaptic strength at excitatory synapses between primary afferent Aδ- or C-fibers and neurons in the SDH (laminae I–II) of the spinal cord was lacking until early 1990s, when it was demonstrated that brief conditioning high-intensity, high-frequency repeated trains of nociceptive input cause a long-lasting potentiation of both, Aδ- or C-fiber monosynaptically-evoked excitatory postsynaptic potentials (EPSPs)-elicited when postsynaptic neurons were depolarized (Randić et al., 1993).

A major technological advance in the investigation of synaptic plasticity in the spinal cord superficial dorsal horn was the development of the rodent (rat, mouse) spinal cord slice preparations that made LTP accessible to precise experimental analysis. To study LTP at the first synapse in the nociceptive pathways, transverse or longitudinal slices with long dorsal roots attached (8–15 mm), were prepared from lumbar spinal cord (L5–L6) of rats (18–28 day-old), or mice (63–121 day-old) to investigate monosynaptic (and polysynaptic) Aδ-fiber or C-fiber elicited EPSPs or excitatory postsynaptic currents (EPSCs) in identified SDH neurons (Randić et al., 1993; Ikeda et al., 2003). LTP has been induced at primary afferent fibre synapses with SDH neurons by a brief high-intensity, high-frequency electrical stimulation (50–100 Hz bursts for 1–3 s at Aδ or C-fiber strength) in vitro (Randić et al., 1993; Hamba et al., 2000; Ikeda et al., 2000, 2003) and in vivo (Zhang et al., 2001). LTP in wide-dynamic range deep DH neurons in vivo (Svendsen et al., 1997, 1999) and in the intermediate gray matter and ventral horn slices of the rat spinal cord has been also reported (Pockett and Figurov, 1993; Pockett, 1995).

In addition, it was more recently shown that not only HFS but also conditioning low frequency stimulation (1–10 Hz) at C-fiber strength causes LTP both in vitro and in vivo (Terman et al., 2001; Youn et al., 2005; Ikeda et al., 2006). LTP can also be induced by excitation of sensory nerve endings by subcutaneous injection of capsaicin or formalin (Ikeda et al., 2006), and by noxious skin heating and acute nerve injury (Sandkühler and Liu, 1998). In lamina I of the DH, LTP is selectively induced in nociceptive specific neurons, which express the NK1 receptor for substance P (SP) and which project to the brain (Ikeda et al., 2003, 2006). These neurons play a key role in hyperalgesia following inflammation and nerve injury (Mantyh et al., 1997).

In difference to the essential role of NMDAR in the induction of the NMDAR-dependent LTP in the CA1 area of hippocampus, induction of LTP at C-fiber synapses with lamina I DH neurons requires co-activation of multiple receptors such as NMDA receptor (Randić et al., 1993; Ikeda et al., 2003, 2006, but see Hamba et al., 2000), group I metabotropic glutamate receptors (Gerber et al., 2000; Hamba et al., 2000; Zhong et al., 2000; Park et al., 2004; Jung et al., 2006), NK1 and NK2 receptors (Rusin et al., 1992, 1993; Randić, 1996), as well as activation of L- and T-type voltage-gated Ca^{2+} channels (VGCCs, Ryu and Randić, 1990) or opening of T-type VGCCs in lamina I DH neurons (Ikeda et al., 2003, 2006). The resultant rise in $[Ca^{2+}]_i$ within the dendritic spine from all of these sources is the critical trigger for LTP. Preventing the rise in $[Ca^{2+}]_i$ in postsynaptic neurons blocks LTP (Ikeda et al., 2003). It has been recently reported that both HFS- and LFS-inducing stimuli of LTP cause a significant elevation in $[Ca^{2+}]_i$ in lamina I DH neurons not only in slice preparation but also in intact animals (Ikeda et al., 2003, 2006). Considerable evidence suggests that the increase in $[Ca^{2+}]_i$ activates Ca^{2+}-sensitive protein kinases such as CaMKII, PKC, PKA, phospholipase C, NO/cGMP, and (MAPK/ERK) pathway (Gerber et al., 1989, 2000; Cerne et al., 1992, 1993; Rusin et al., 1992, 1993; Kolaj et al., 1994; Ikeda et al., 2003; Yang et al., 2004; Zhang et al., 2005; Wei et al., 2006; Xin et al., 2006).

The important role of NMDARs in pain mechanisms in the spinal cord is well established and peripheral NMDARs are involved in nociception, as well (Fisher et al., 2000; Fundytus, 2001; Willis, 2001). NMDARs expressed in the spinal DH have been implicated in the activity-dependent plastic changes that lead to induction and maintenance of central sensitization and pathological pain (Baranauskas and Nistri, 1998; Woolf and Salter, 2000; Ji et al., 2003). NMDA receptor blockade reduces allodynia and secondary hyperalgesia (Ren et al., 1992) and blocks inflammatory (Woolf and Thompson, 1991) and neuropathic pain hypersensitivity. Spinal NMDAR activity is increased following peripheral inflammation (Guo and Huang, 2001), contributing to the increased activity of DH neurons. Intrathecal application of NMDAR agonists generates pain behaviour (Malmberg and Yaksh, 1993).

This chapter attempts a brief review of some of the postsynaptic mechanisms underlying induction, expression and modulation of the HFS-induced LTP of excitatory synaptic transmission at glutamatergic synapses in the area of the superficial dorsal horn of the spinal cord. It is becoming clear that the phenomena such as spinal LTP, which might contribute to hyperalgesia in animal models of pain, use multiple mechanisms involving protein phosphorylation, similar to the processes associated with hippocampal LTP. Modulation of postsynaptic AMPA and NMDA receptor function caused by phosphorylation may play an important role in the induction and expression of synaptic plasticity at DH excitatory synapses.

10.3 Signal Transduction Mechanisms of LTP

Studies over the past decade have progressed in mapping some of the signal transduction pathways that are involved in modulation of synaptic strength at the postsynaptic site in CA1 hippocampal neurons (Soderling and Derkach, 2000; Lisman et al., 2002; Malenka and Bear, 2004; Citri and Malenka, 2008) and to a smaller extent in the spinal dorsal horn (Woolf and Salter, 2000; Willis, 2002; Ji et al., 2003). It is clear from this work that phenomena such as LTP, use multiple mechanisms, which in many cases involve protein phosphorylation. Protein phosphorylation is a widespread mechanism for signal transduction and regulation in the nervous system, and phosphorylation of glutamate receptors by protein kinases (Gerber et al., 1989; Greengard et al., 1991; Wang et al., 1991, 1994; Cerne et al., 1992; McGlade-McCulloh et al., 1993; Kolaj et al., 1994; Tan et al., 1994; Wang and Salter, 1994; Sweatt, 2004) has been implicated in regulation of synaptic transmission and plasticity (Bliss and Collingridge, 1993). Presynaptically, protein phosphorylation causes changes in the efficiency of neurotransmitter release, and postsynaptically, phosphorylation of neurotransmitter (glutamate) receptors represents a major mechanism for the regulation of their function.

In the CA1 area of the hippocampus the combined activation, in particular of CaMKII, protein tyrosine kinases, PKA, PKC and MAPK/ERK pathway, results in phosphorylation of glutamate receptor-ion channels and the increase of postsynaptic currents. Crosstalk between these biochemical pathways can account for most features of the early-phase LTP in this brain region (Barria et al., 1997a; Nicoll, 2003). These studies have highlighted the roles in the early-phase LTP for the transient potentiation of NMDARs through tyrosine residue phosphorylation in order to enhance Ca^{2+} influx; prolonged activation of CaM-KII through autophosphorylation; inhibition of protein phosphatase 1 (PP1) through PKA phosphorylation of inhibitor 1 and prolonged potentiation of AMPA receptors by CaMKII-mediated phosphorylation. However, the specific targets and mechanisms that directly mediate changes in synaptic strength in response to neural activity still remain to be identified.

Induction of HFS-evoked LTP in the DH of the spinal cord (Gerber et al., 1989; Ikeda et al., 2003; Kolaj et al., 1994; Yang et al., 2004; Miyabe and Miletić, 2005; Zhang et al., 2005; Wei et al., 2006; Xin et al., 2006; Sandkühler, 2007), similar as in brain (Blitzer et al., 1995; Makhinson et al., 1999; Otmakhova et al., 2000; Soderling and Derkach, 2000; Kalia et al., 2004; Sweatt, 2004; Thomas and Huganir, 2004;), involves multiple mostly calcium-dependent signaling pathways, including CaMKII, PKC, PKA, MAPKs/ERK, tyrosine kinase Src. Both in the hippocampus (Malinow et al., 1989; Fukunaga et al., 1995; Otmakhov et al., 1997; Lisman et al., 2002) and in the spinal cord (Kolaj et al., 1994; Ikeda et al., 2003; Yang et al., 2004) an essential step in the LTP induction downstream of Ca^{2+} influx via NMDA receptors is activation of alpha-CaMKII. The maintenance of LTP, however, may not depend on activation of this kinase (Lisman et al., 2002).

In the hippocampus (Malenka and Nicoll, 1999; Kennedy, 2000) and the SDH (Bruggemann et al., 2000; Fang et al., 2002) CaMKII is highly expressed in the postsynaptic density of the dendritic spines, that also contain glutamate receptors. Among the proteins phosphorylated and regulated by CaMKII are the AMPA and NMDA receptors (McGlade-McCulloh et al., 1993; Raymond et al., 1993; Tan et al., 1994). This kinase has unusual property; it remains in an active state following removal of Ca^{2+} stimulus, through autophosphorylation at a threonine residue 286 in mice. The activated kinase by calcium-calmodulin undergoes a rapid autophosphorylation at Thr 286, after the triggering of LTP (Fukunaga et al., 1995; Barria et al., 1997a), which generates a constitutively active CaMKII that phosphorylates exogenous substrates. In this way a transient increase in calcium concentration in a dendritic spine can be transduced into a prolonged kinase activity that persists in the absence of raised Ca^{2+} levels until protein phosphatase 1 (PP1) dephosphorylates Thr 286 and inactivates the CaMKII. CaMKII moves into the PSD, where it binds the NR2B subunit of the NMDA receptor and transforms by it into a persistently active state. In this state it is no longer a substrate for inactivation by PP1. Once bound to NMDAR at the synapse, CAMKII can phosphorylate GluR1 AMPAR subunit on Ser 831 (Barria et al., 1997a), increasing the single-channel conductance of synaptic AMPARs (Barria et al., 1997b). Phosphorylated CaMKII also binds with NR2B subunits in the DH suggesting the presence of a similar mechanism (Fang et al., 2002; Willis, 2002). Induction of LTP in hippocampal slices results in activation of a CaMKII within 1 min, and this constitutive activity is stable for at least 1 h (Fukunaga et al., 1993). Two other protein kinase that can phosphorylate GluR1 could also be involved in its regulation. PKC can phosphorylate GluR1 (Mammen et al., 1997), and induction of LTP generates a long-lasting activation of PKC. Protein kinase A (PKA) can also phosphorylate GluR1 resulting in potentiation of the AMPAR-mediated current (Roche et al., 1996).

In the CA1 area of the hippocampus, the heteromeric AMPARs consist predominantly of GluR1 and GluR2 subunits. CaMKII can enhance whole-cell AMPA receptor-mediated current in cultured hippocampal neurons (McGlade-McCulloh et al., 1993) and phosphorylates the native AMPAR (Tan et al., 1994) and, it can phosphorylate this receptor in the CA1 region after the induction of LTP in this area (Barria et al., 1997b). CaMKII also induces trafficking and incorporation of GluR1-containing AMPARs into hippocampal synapses via RAS and activation of either ERK or phosphatidy-linositol 3-kinase pathways (Lisman et al., 2002; Malinow and Malenka, 2002; Nicoll, 2003; Sheng and Lee, 2003). The α-isoform of CaMKII is a neural-specific enzyme that has been implicated in long-lasting modifications in synaptic function. Evidence suggests that CaMKII and LTP enhance synaptic transmission in the hippocampus by the same mechanism. Moreover, post-synaptic injection of inhibitors of CaMKII, or genetic deletion of a CaMKII subunit blocks the induction of LTP (Malenka et al., 1989; Malinow et al., 1989; Silva et al., 1992; Otmakhov et al., 1997).

10.3.1 Calcium/Calmodulin-Dependent Protein Kinase II Enhances AMPA/NMDA and Synaptic Responses of Rat DH Neurons

We have demonstrated that long-lasting modulation in synaptic efficacy can be induced at primary afferent Aδ- or C-fiber synapses with neurons in the super-ficial laminae of the spinal DH by HFS of primary afferent fibers (Randić et al., 1993). Molecular mechanisms underlying the enhancement of excitatory synap-tic transmission in the SDH has still to be elucidated. Of particular interest is the role of functional modulation of ionotropic glutamate receptors (AMPA, NMDA) in synaptic transmission of nociceptive information and plasticity in the spinal DH (Randić, 1996; Gerber et al., 2000; Sandkühler, 2000, 2007). Evidence has been obtained in the CA1 area of the hippocampus that the enhancement of synaptic responses in LTP is in part due to an increased sensitivity of postsynaptic AMPA receptors (Bliss and Collingridge, 1993; Raymond et al., 1993; McGlade-McCulloh et al., 1993; Tan et al., 1994). In addition, evidence was obtained that HFS of Shaffer collaterals in the hippo-campus results in a long-lasting increase in the activity of CaMKII (Fukunaga et al., 1993). Based on the high distribution of the CaMKII in the PSDs of the SDH, we hypothesized that alpha- CaMKII contributes to the development of persistent changes in synaptic plasticity at this spinal level.

Several lines of indirect evidence support a role for this kinase in mediating the effects of Ca^{2+} on synaptic strength. A direct approach to investigating a role for CaMKII is to determine its effects on synaptic strength and LTP, when the concentration of activated kinase postsynaptically increased in DH cells. Therefore we first investigated whether the α-subunit of CaMKII can modulate the current responses of acutely isolated rat spinal DH neurons to selective agonists of AMPA and NMDA subtypes of glutamate receptors using whole-cell voltage-clamp technique. To test the involvement of CaMKII, a truncated, constitutively active form of this kinase (α-CaMKII) was directly injected into DH neurons. We demonstrated that in acutely isolated rat spinal DH neurons, the AMPA and NMDA receptors can be regulated by endogenous and exo-genous CaMKII. We found that intracellularly applied, the alpha subunit of CaMKII enhanced AMPA (Fig. 10.1C, D and E) and NMDA-mediated cur-rents (Fig. 10.1F, G and H) (Kolaj et al., 1994). Consistent with the slow rate of AMPAR phosphorylation, potentiation of current by CaMKII requires 15–30 min for maximal effect (Kolaj et al., 1994). Microcystin, a nonselective phosphatases inhibitor, also enhances AMPA and NMDA responses. In addi-tion, conventional intracellular recordings were made from L II of DH neurons in spinal cord slices to determine the effect of intracellular perfusion with the α-CaMKII on EPSPs- evoked by high intensity, HFS of primary afferent Aδ and/or C-fibers. Presence of α-CaMKII in the recording micropipette resulted in a gradual, but significant, increase in the amplitude of EPSP (Fig. 10.1A and B) (Kolaj et al., 1994). In contrast, no significant change in the PAF-evoked EPSPs occurred when the micropipette contained heat-inactivated CaMKII. Moreover, following the increase in synaptic strength by CaMKII, the HFS of primary afferent fibres failed to evoke LTP. The latter result indicates that CaMKII by itself is sufficient to increase synaptic strength and this enhance-ment seems to share the same mechanism as the enhancement seen with LTP. Excitatory synaptic transmission was enhanced by α-CaMKII, which is con-sistent with the importance of phosphorylation of the postsynaptic AMPA and NMDA receptor-ion complexes in the long-term changes in synaptic transmis-sion. Although these experiments demonstrated the involvement of CaMKII activity in the regulation of AMPAR and NMDAR function, it is not known whether this kinase acted directly by phosphorylating the AMPAR, or indir-ectly by phosphorylating other proteins that regulate AMPA and NMDA receptors function.

To summarize we demonstrated, that perfusion with the α-subunit of CaM-KII of SDH neurons mimicked two major features of LTP; it caused an increase in the amplitude of Aδ- or C-fiber-elicited EPSPs and an increase in the current responses of SDH cells to locally applied AMPA and NMDA.

Our suggestion that CaMKII plays a direct and essential role in generation of LTP following tetanic stimulation of Aδ or C-fibers is strongly supported by

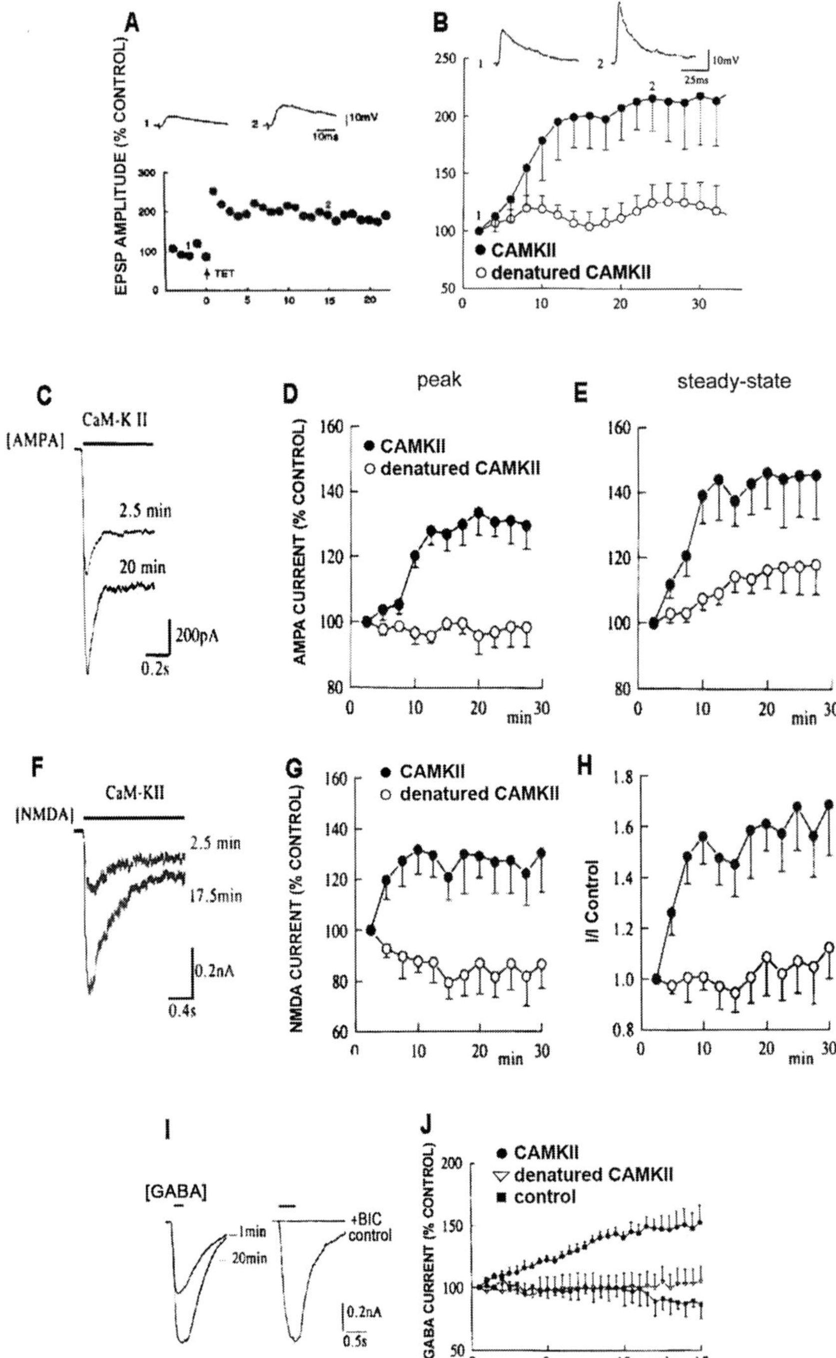

Fig. 10.1 Alpha subunit of calcium/calmodulin-dependent protein kinase II (CaMKII) enhances NMDA, AMPA and synaptic responses of rat spinal dorsal horn neurons.

our finding that excitatory synaptic transmission is enhanced, and LTP depressed (occluded) by rising the concentration of constitutively active CaM-KII within the SDH neurons (Kolaj et al., 1994). Another relevant evidence implicating CaM-KII in LTP is provided more recently by the findings that spinal application of specific inhibitors of CaM-KII (KN-93, AIP) completely blocks the induction of LTP of C-fiber-evoked potentials (Yang et al., 2004) and spinal LTP in nociceptive neurons in LI-IV of the dorsal horn (Pedersen

◀──

Fig. 10.1 (continued) (**A**) LTP of fast excitatory synaptic transmission at primary afferent synapses with neurons in the superficial dorsal horn (SDH). The graph shows the time course of LTP of excitatory postsynaptic potential (EPSP) recorded intracellularly from a SDH neuron in response to electrical stimulation (20 V, 0.1 ms) of a lumbar dorsal root. At time zero (*arrow*) dorsal root was given three tetani (at the same intensity as the test stimulus) of 1 s duration, each at 100 Hz and 10 s intervals. Above the graph are displayed individual monosynaptic EPSPs taken before (*trace 1*) and during (*trace 2*) the potentiation. (**B**) A long-lasting potentiation of excitatory synaptic transmission at primary afferent synapses with neurons in substantia gelatinosa (SG) after intracellular application of activated CaM-KII. Summarized data (mean ± SEM) showing the time courses of EPSP amplitudes expressed as percentage of control response, in the presence of activated 1 μM CaMKII (●; $n = 5$) or heat inactivated CaMKII (denatured CaMKII; ○; n–6). The traces displayed above the graph are individual monosynaptic EPSPs recorded intracellularly from a SG neuron in response to electrical stimulation of a lumbar dorsal root taken 2 and 24 min after start of recording with a solution containing activated CaMKII. (**C**, **D** and **E**): AMPA-induced currents in acutely isolated DH neurons are enhanced by activated CaMKII. (**C**) Super-imposed traces are the inward current responses evoked by 30 μM α-amino-3-hydroxy-5-methyl-4-isoxazolepropionic acid (AMPA) in acutely isolated DH neurons recorded with the microelectrode solution containing 200 nM activated CaMKII at the times indicated. (**D** and **E**) pooled data ($n = 39$) illustrating the time courses of the peak (**D**) and steady-state (**E**) component of AMPA-induced current in the presence of 200 nM of activated CaMKII (●; $n = 12$) or heat-inactivated CaMKII (denatured CaMKII; ○, $n = 7$); the holding potential was –60 mV, and the data are expressed as the percentage of the first control response. Data are presented as means ± SEM; statistical comparisons between control and experimental values (activated CaMKII vs heat-inactivated CaMKII) were determined with the use of the Student's *t*-test. (**F**) NMDA responses are enhanced by intracellular application of activated CaMKII. Superimposed traces are the inward current responses evoked by 0.1 mM NMDA (5 s) plus 50 nM glycine recorded with the microelectrode solution containing activated 200 nM CaMKII at the times indicated (**F**). (**G** and **H**) Pooled data show the time courses of the peak (**G**) and steady-state (**H**) component of NMDA-induced current in the presence of 200 nM activated CaMKII (●; $n = 14$) or heat inactivated CaMKII (denatured CaMKII; ○; $n = 6$). (**I** and **J**) γ-aminobutyric acid (GABA)-induced currents in acutely isolated DH neurons are enhanced by intracellular application of activated CaMKII included in a microelectrode solution. (**I**) (*left traces*): super-imposed traces are the inward current responses evoked by 20 μM GABA (250 ms), recorded with the CsCl filled microelectrode containing 200 nM activated CaMKII at 1 and 20 min after the rupture of the patch. (**I**) (*right traces*): show complete block of GABA-induced current by bicuculline methiodide (10 μM). (**J**) Pooled data show the time courses of the GABA-induced currents in the presence of activated CaMKII (●; $n = 7$), or heat-inactivated CaMKII (denatured CaMKII ○; $n = 4$, or control solution (■; control, $n = 4$). Adapted from Kolaj et al. (1994)

et al., 2005). It has been also reported that α-CaMKII enhances γ-aminobutyric acid, glycine and inhibitory synaptic responses (Fig. 10.1I and J) of rat hippocampal and spinal DH neurons (Wang et al., 1995; Wang and Randić, 1996).

Besides the essential role of CaMKII in LTP in the SDH, this kinase has been implicated in the activity-dependent synaptic plasticity changes that lead to induction and maintenance of central sensitization, and thereby, pathological pain (Willis, 2002). The mechanisms associated with chronic pain states are not clearly understood. However, a growing body of evidence indicates that the synaptic plasticity of DH neurons contributes to pain hypersensitivity (central sensitization) after strong noxious stimulation (Woolf and Salter, 2000; Guo et al., 2002; Willis, 2002; Ji et al., 2003).

10.4 Central Sensitization

Central sensitization is defined as an enhanced responsiveness of nociceptive neurons in the CNS to their normal afferent input (see also Drdla and Sandhkuler, Chapter 9). It is characterized by reductions in threshold and increase in responsiveness of DH neurons, as well as by enlargements of their receptive fields. Sensitization of nociceptive DH neurons, is thought to underlie the development of secondary hyperalgesia (defined as increased response to a painful stimulus) and allodynia (pain due to stimulus that normally does not evoke pain) following tissue injury (Torebjörk et al., 1992; Willis, 2002).

In *central sensitization*, responses of DH cells to stimulation of sensory receptors are enhanced without any change in the excitability of the primary afferent neurons. The present view is that central sensitization of spinothalamic tract (STT) neurons is a variety of LTP (Willis, 2002). Current evidence suggests that synaptic plasticity and central sensitization in the spinal cord share common signaling pathways. Several neurotransmitters are involved in the induction of central sensitization, including glutamate acting at AMPA, NMDA, kainate and group I metabotropic glutamate receptors (Dickenson and Sullivan, 1987; Neugebauer et al., 1993; South et al., 2003); SP acting on NK1 receptors, calcitonin gene-related peptide (CGRP) acting on CGRP1 receptor and brain-derived neurotrophic factor (BDNF) acting at tyrosine kinase B receptors (trkB and ephB receptors) (Dougherty et al., 1994; Dougherty and Willis, 1992; Mannion et al., 1999; Neugebauer et al., 1999; Thompson et al., 1999; Karim et al., 2001). The underlying mechanism for central sensitization of DH neurons, including STT cells, is change in the responsiveness of DH cells to excitatory and inhibitory amino acids released upon activation of primary afferent fibers (Willis, 2002; Ji et al., 2003). Responses to excitatory amino acids enhance and those to inhibitory amino acids decrease during central sensitization as a result from the activation of multiple signal transduction pathways. It is also known, that alteration in the responsiveness of the excitatory and inhibitory amino acid receptors is due to phosphorylation of these receptors by multiple protein kinases including CaMKII, PKC, PKA, non-receptor tyrosine kinase Src, NO/PKG and MAPK/ERK cascades (Willis, 2001, 2002; Ji et al., 2003). There is also evidence that multiple protein kinases, including CaMKII, (Fang et al., 2002) are up-regulated early in the process of central sensitization and that AMPA and NMDA receptors are phosphorylated by these protein kinases (Willis, 2002). Two mechanisms appear to contribute to increased synaptic efficacy: changes in ion channels and receptor activity due to post-translational processing, and trafficking of receptors to synaptic membrane (Woolf and Salter, 2000; Ji et al., 2003).

Several other protein kinases, PKC, PKA, protein tyrosine kinases ERK/MAPK pathway, have been implicated in the induction of LTP in the hippocampus. However, the experimental evidence supporting their role is not as strong as for CaMKII. Activation of PKA has been suggested to enhance the activity of CaMKII by decreasing protein phosphatase activity (Blitzer et al., 1998; Makhinson et al., 1999). PKC, especially the PKC isozyme PKMζ has been implicated in the maintenance of late-phase of LTP in hippocampus (Ling et al., 2002; Serrano et al., 2005; Pastalkova et al., 2006). Src tyrosine kinase has been involved in the enhancement of NMDA receptor function during induction of LTP (Kalia et al., 2004). The involvement of ERK/MAPK pathway in LTP has also lately been considered (Sweatt, 2004; Thomas and Huganir, 2004).

10.5 Role of PKC in LTP

Protein kinase C has been suggested to be involved in the induction of LTP in the hippocampus (Linden et al., 1987: Malenka et al., 1986; Hu et al., 1987; Malinow et al., 1989; Ling et al., 2002), and may also contribute to LTP in the spinal DH (Gerber et al., 1989). PKC is expressed in the DH and it has been suggested to be important in sensory signal processing including pain (Igwe and Cronwall, 2001; Willis, 2002; Yajima et al., 2003). Activation of PKC in the spinal cord causes mechanical allodynia and thermal hyperalgesia (Palecek et al., 1999). Inhibition of PKC prevents central sensitization of STT neurons produced by capsaicin (Lin et al., 1996; Willis, 2002), and potentiation of spinal NMDAR activity induced by peripheral inflammation (Guo and Huang, 2001). Peripheral noxious stimulation induces phosphorylation of the NMDA receptor NR1 subunit at the PKC-dependent site, serine-896, in spinal cord DH neurons (Brenner et al., 2004).

The finding that the spinal DH contains high levels of binding sites for phorbol esters and that PKC is present in the rat spinal DH (Worley et al., 1986; Mochly-Rosen et al., 1987) raised the possibility that PKC may play a functional role in sensory transmission and plasticity, both in the release of putative neurotransmitters and also in the signal transduction at various subclasses of glutamate receptors. Since PKC activation can be mediated directly by phorbol esters, we used these agents to examine the effects of the enzyme activation on fast and slow excitatory synaptic transmission in the spinal DH, the basal and the dorsal root stimulation-evoked release of endogenous excitatory (glutamate, aspartate) and inhibitory (GABA, glycine) amino acids, and the responsiveness of the rat DH neurons to specific agonists of various subclasses of glutamate receptors (Randić, 1996).

Using spinal cord slice preparation and intracellular recording techniques we demonstrated that 4-β-phorbol-12,13-dibutyrate and 4-β-phorbol-12,13-diacetate caused a marked and long-lasting increase (>1 h) in the amplitude and the duration of fast and slow EPSPs, evoked in DH neurons by electrical stimulation of primary afferent fibers at Aδ or C-fiber strength (Gerber et al., 1989). Furthermore, phorbol esters produced an increase in the basal and electrically-evoked release of endogenous excitatory (glutamic, aspartic) and inhibitory amino acids (glycine, GABA) (Kangrga and Randić, 1990). In the presence of TTX, phorbol esters and intracellularly injected PKC selectively enhanced, in a reversible manner, voltage, current and calcium responses of DH neurons to L-glutamate, NMDA and AMPA (Gerber et al., 1989; Rusin et al., 1992, 1993; Cerne et al., 1993). The potentiation of the NMDA response was blocked by D-APV, a specific NMDA receptor antagonist. Thus, phorbol esters appear to produce a long lasting enhancement of the excitatory synaptic transmission in the rat spinal DH slice preparation by acting both at pre- and postsynaptic sites. These results suggested that the rat spinal DH PKC may have a role in controlling the release of neurotransmitter glutamate, and may also be involved in the regulation of sensitivity of postsynaptic AMPA and NMDA receptors, the effects that may be important in strengthening of synaptic function in the spinal DH. However, the identity of endogenous substance(s) participating in these effects is less well elucidated.

In extension of our results (Gerber et al., 1989) it was reported that PKC-mediated NMDA receptor phosphorylation potentiates the channel activity (Kelso et al., 1992; Xiong et al., 1998) and neuronal responses to NMDA are enhanced by PKC, due to an increase in the probability of NMDAR channel opening (Xiong et al., 1998) and a decrease in the voltage-dependent Mg^{2+} block of the NMDA receptor channel (Chen and Huang, 1992). PKC also phosphorylates the AMPAR, thus modulating its sensitivity to glutamate (Roche et al., 1996; Barria et al., 1997b; Giese et al., 1998).

Taken together the potentiation of both AMPA and NMDA receptor (function) by PKC may contribute to the expression of LTP in the spinal DH. More recently it was demonstrated that spinal application of PKC inhibitors (chelerythrine, Gö6983) blocked the induction of C-fiber-evoked field potentials in the spinal DH and reversed spinal LTP when applied 15 min after LTP induction (Yang et al., 2004).

10.6 The Role of PKA in LTP

Evidence indicates that cyclic adenosine $3',5'$ monophosphate-dependent protein kinase (PKA) gates plasticity by modulating CaMKII activity (Blitzer et al., 1998; Soderling and Derkach, 2000), specifically the threshold for LTP induction, which is triggered by CaMKII activity, could be controlled by PKA. PKA has been implicated in the induction of LTP in the CA1 area of the hippocampus indirectly via phosphorylation of inhibitor 1, an endogenous inhibitor of protein phosphatase 1 (PP1). The function of PP1 inhibitor is primarily to prolong activation of CaMKII by autophosphorylation. This would also maintain the phosphorylation of GluR1 and facilitate LTP (Blitzer et al., 1995; Makhinson et al., 1999; Otmakhova et al., 2000). It is known that LTP induction produces a transient rise in cAMP (10–20 min) (Chetkovich and Sweatt, 1993), a transient activation of PKA (2–10 min) (Blitzer et al., 1995); and gating of CaMKII by cAMP-regulated protein phosphatase activity during LTP, but the role of this pathway in LTP remains unclear. This issue is complex, because LTP appears to involve early, intermediate and late-phases (Frey et al., 1993), which may utilize different kinase pathways, including the PKA. Inhibition of the cAMP pathway decreases the early-LTP (lasting up to 1 h after induction) at CA1 hippocampal synapses (Otmakhova et al., 2000). Moreover, when two major Ca^{2+}-dependent adenylyl cyclases (AC1 and AC8) were knocked out, the early LTP was significantly decreased.

This result strongly supports the involvement of adenylyl cyclase in the early LTP. Intermediate phase of LTP begins about an hour after induction and is also dependent on cAMP. There is general agreement that the late phase of LTP ($\sim$3 h after induction) depends on cAMP and protein synthesis (Frey et al., 1993). More recently identified mechanism contributing to synaptic plasticity is the regulation of trafficking of AMPAR at synapses (Sheng and Lee, 2003; Malinow and Malenka, 2002; Song and Huganir, 2002). AMPA receptors are tetramers assembled from any of four subunits, GluR1–GluR4 or GluRA to GluRD with variable stoichiometry (Hollmann and Heinemann, 1994; Dingledine et al., 1999). The functional role of phosphorylation of any of two sites in the GluR1 subunit, serine 831 (S831) phosphorylated by both CaMKII and PKC and serine 845 (S845) phosphorylated by PKA (Roche et al., 1996; Barria et al., 1997a) results in potentiation of AMPAR ion-channel function (Roche et al., 1996; Barria et al., 1997b; Banke et al., 2000). The state of phosphorylation at the GluR1 PKA site can control channel open time (Banke et al., 2000), and has been correlated with changes in synaptic strength.

LTP is associated with an increase in phosphorylation of GluR1 (Barria et al., 1997b) and PKA acts as a gate for synaptic plasticity directly by phosphorylating GluR1 at Ser 845 and by making AMPARs available for synaptic incorporation (Roche et al., 1996; Song and Huganir, 2002. Furthermore, PKA phosphorylation of the AMPA receptor subunits GluR1 and GluR4 directly controls the synaptic trafficking of AMPA receptors underlying plasticity (Esteban et al., 2003). Thus it is evident that PKA phosphorylation of AMPAR subunits contributes to different mechanisms underlying synaptic plasticity.

We anticipated that the SDH, might be a favorable site to study synaptic neuromodulation by the adenylate cyclase-PKA cascade. This spinal region is known to regulate synaptic strength by LTP (Randić et al., 1993), and the abundantly expressed PKA, AMPA and NMDA receptor channels in the SDH, have all been implicated in central sensitization and pathological pain (Willis, 2000; Miletić et al., 2002). The findings that PKA is present in the DH, that an adenylate cyclase activator forskolin depolarizes DH neurons and enhances their sensitivity to NMDA, the effects inhibited by H-8 a claimed antagonist of PKA (Gerber et al., 1989), as well as the fact that glutamate and kainate responses of cultured hippocampal neurons are enhanced by PKA (Greengard et al., 1991), raised the possibility that cAMP through the activation of PKA may play a functional role in the excitatory synaptic transmission and plasticity in the SDH by modulating various subclasses of glutamate receptors.

By using conventional intracellular recording in a spinal slice preparation, we have demonstrated that the elevation of intracellular concentration of cAMP by a membrane-permeable analog 8-Br cAMP, and application of phosphodiesterase inhibitor, IBMX (3-isobutyl-1-methylxanthine), enhance primary afferent stimulation-evoked monosynaptic EPSPs in LII neurons, and the responses of DH neurons to AMPA, kainate and NMDA receptors ligands. In addition, a specific peptide inhibitor of PKA (PKI_{6-22} amide), which binds with high affinity to active catalytic subunit of PKA, prevented the 8-Br cAMP-induced potentiation of NMDA-mediated current responses of DH neurons (Cerne et al., 1992).

Further studies on AMPA and NMDA receptor channels expressed in acutely isolated rat spinal DH neurons revealed that these channels are subject to neuromodulatory regulation through the adenylate-cyclase cascade. The whole-cell current response to AMPA or NMDA was enhanced by forskolin, 8-Br cAMP or intracellular perfusion of cAMP or a purified catalytic subunit of PKA, included in the pipette solution.

These results demonstrate the importance of PKA activity in the regulation of AMPAR and NMDAR function in the spinal SDH region. However, it remains unknown whether the PKA acted directly on AMPA and NMDA receptors, or indirectly by phosphorylating other synaptic proteins that regulate AMPAR and NMDAR function.

In the spinal cord, the activation of PKA contributes to both normal nociceptive processing and alterations in spinal function following injury or inflammation (Cerne et al., 1992, 1993; Malmberg et al., 1997; Lin et al., 2002; Hoeger-Bement and Sluka, 2003). Besides the involvement of PKA in LTP in the spinal DH, this kinase has been implicated in the induction and maintenance of central sensitization, and thereby, pathological pain (Randić, 1996; Miletić et al., 2002; Willis, 2002; Ji et al., 2003; Sandkühler, 2007). Evidence is as follows: hyperalgesia and allodynia can be induced by perfusion of the dorsal horn with 8-Br-cAMP; the secondary hyperalgesia and allodynia caused by intradermal injection of capsaicin are reduced by post-treatment of the spinal cord with a PKA inhibitor (Sluka and Willis, 1997); the NR1 subunit of the NMDA receptor undergoes PKA-mediated phosphorylation at serine 890 and serine 897 (Leonard and Hell, 1997; Tingley et al., 1997); the NR1 subunit of NMDARs in STT cells is phosphorylated on Serine-897 after intradermal injection of capsaicin (Zou et al., 2000); capsaicin-induced central sensitization can be blocked by PKA inhibitor (H89) (Lin et al., 2002).

PKA-dependent NMDA receptor function has been revealed in pain-related synaptic plasticity in STT neurons (Willis, 2002) and amygdala neurons (Bird et al., 2005). Indeed, intraspinal infusion of forskolin, an activator of adenylate cyclase, but not the inactive isomer (D-forskolin) caused the sensitization of the responses of STT neurons to mechanical stimulation of skin.

Altogether, these results support the view that the PKA activation, and subsequent increase in the effectiveness of ionotropic glutamate receptor function (Rusin et al., 1992, 1993; Lin et al., 2002; Willis, 2002) and/or increased neurotransmitter release contribute to central sensitization. Recent evidence provides further support that activation of PKA may play an important role in the early stages of nerve injury-elicited plasticity in the DH.

10.7 LTP in the Spinal Dorsal Horn is Blocked by Tyrosine Kinase Inhibitor

A role for protein tyrosine kinases in the induction of LTP in the hippocampus was first suggested by experiments showing that LTP is blocked by tyrosine kinase inhibitors (O'Dell et al., 1991). It was also shown that tyrosine phosphorylation of the NMDAR subunit NR2B is increased following induction of LTP in the rat dentate gyrus of hippocampus (Rostas et al., 1996), a region where LTP is blocked by tyrosine kinase inhibitors (Abe and Saito, 1993). Activation of the protein tyrosine kinase (Src) by tetanic stimulation appears to be a biochemical mechanism gating the induction of LTP in the CA1 hippocampal neurons by phosphorylating and enhancing NMDAR-mediated current (Wang and Salter, 1994; Wang et al., 1996). NMDA-mediated increase in Ca^{2+} influx into the dendritic spine triggers the biochemical pathways leading to potentiation of AMPARs-mediated current.

In the SDH, LTP requires NMDAR-dependent Ca^{2+} influx (Randić et al., 1993) and activation of serine and threonine protein kinases (Wang et al., 1991, 1994; Chen and Huang, 1992; Kelso et al., 1992). Since the function of NMDARs is regulated by tyrosine phosphorylation in spinal DH neurons (Wang and Salter, 1994; Wang et al., 1996), and Src kinase is associated with NMDA receptor-ion channel complexes in postsynaptic membrane densities at synapses (Yu et al., 1997) in the rat spinal cord, we examined the possible additional requirements of tyrosine kinase activity in the induction of LTP in the SDH (Zhong and Randić, 1997). We examined whether lavendustin A, a selective PTK inhibitor, which has high affinity for Src and Trk tyrosine kinases (Sugrue et al., 1990; Sato et al., 2003), could affect LTP. Lavendustin A selectively and completely blocked the induction of LTP when applied into the bath (Fig. 10.2B), but it had no effect on the established LTP, or normal synaptic transmission. To examine further the physiological specificity of this PTKs inhibitor we did additional control experiments. We examined lavendustin B, a structural analogue of lavendustin A, which does not block tyrosine kinase activity, and found that it failed to block LTP when applied extracellularly before tetanic stimulation.

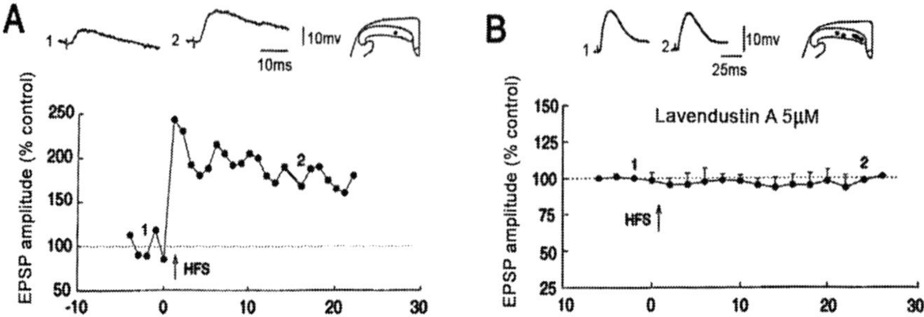

Fig. 10.2 LTP is blocked in the SDH by tyrosine kinase inhibitor. (**A**) the graph shows the time course of LTP of EPSP recorded intracellularly from a SG neuron in response to electrical stimulation of a lumbar dorsal root. At time zero (*arrow*) dorsal root was given three tetani of 1 s duration (at C-fiber strength), each at 100 Hz and 10 s intervals. Above the graph are displayed individual monosynaptic EPSPs taken before (*trace 1*) and during (*trace 2*) the potentiation. (**B**) Bath application of lavendustin A (5 μM) before tetanic stimulation blocks LTP ($n = 5$) No effect of lavendustin A applied 30 min post-tetanus was observed in 3 of these slices where lavendustin A was applied for at least 50 min. Each point in this figure represents the mean ± SEM

These data suggest that in addition to the serine/threonine protein kinases, the induction of LTP in the SDH requires tyrosine kinase activity. However, our experiments did not characterize whether protein tyrosine kinase Src or Trk B might be involved. Although the role of BDNF in the induction of LTP in the CA1 area of the hippocampus is well established (Kovalchuck et al., 2002) the role of BDNF in spinal LTP has still to be elucidated (Malcangio and Lessmann, 2003). BDNF (acting at trk B receptors)-mediated GluR1 phosphorylation potentially regulates synaptic plasticity postsynaptically through NR2B subunit of the NMDAR (Kovalchuck et al., 2002).

Prolonged increase in tyrosine phosphorylation of the NR2B subunit of the NMDAR in the spinal cord during the development and maintenance of inflammatory hyperalgesia has been reported (Guo et al., 2002). In the latter case, the findings correlate in vivo the NMDAR tyrosine phosphorylation with the development and maintenance of inflammatory hyperalgesia and suggest that signal transduction upstream to NR2B tyrosine phosphorylation involves G-protein coupled receptors (NK1, group I mGluRs) and PKC and Src family protein tyrosine kinases (Guo et al., 2002).

10.8 Modulation of Primary Afferent Neurotransmission by Tachykinins Acting at Presynaptic and Postsynaptic Sites

The role of tachykinins in long-term modulation of excitatory synaptic transmission related to pain is now well established (for reviews see: Petersen-Zeitz and Basbaum, 1999; Woolf and Salter, 2000; Ji et al., 2003) (see Hua and Yaksh, Chapter 6). Tachykinins produce their effects in nervous tissue through three molecularly distinct receptor subtypes. The neurokinin receptors NK-1, NK-2 and NK-3, are preferentially activated by the endogenous peptides, substance P (SP), neurokinin A (NKA) and neurokinin B (NKB), respectively. SP and NK-1 bind to the NK1 receptor, which is heavily expressed by projection neurons of the SDH (Todd et al., 2002) (see Todd, Chapter 1).

SP and NKA are synthesized in small diameter primary afferent neurons which project to areas of the spinal cord involved in the processing of nociceptive information including laminae I, II, and V of the spinal DH (Hökfelt et al., 1975). SP and NKA are released in the spinal DH in response to a variety of noxious stimuli (Yaksh et al., 1980; Duggan et al., 1988). SP causes a slow depolarization in DH neurons and an increased response to C-fiber input (Murase and Randić, 1984). SP selectively excites nociceptive DH neurons, potentiates excitatory synaptic transmission in SDH neurons (Randić, 1996), increases cytosolic free calcium concentration via mobilization of intracellular stores (Heath et al., 1994), potentiates low- and high-voltage-activated Ca^{2+} conductance (Ryu and Randić, 1990), stimulates phosphoinositide turnover, cyclic AMP formation and activates ERK in these neurons (Wei et al., 2006). The effects of SP are preferentially mediated by NK-1 receptors, and the loss of DH neurons expressing these receptors prevents the development of hyperalgesia (Mantyh et al., 1997). SP is functionally involved in the slow excitatory synaptic transmission evoked upon high intensity repetitive stimulation of PAFs (Randić, 1996; Galik et al., 2008), and hyperalgesic states (Thompson et al., 1994). In nociceptive lamina 1 DH neurons, NMDA and NK-1 receptor-activated multiple signal transduction pathways and activation of low-threshold voltage-gated calcium channels synergistically facilitate PAF-elicited LTP at synapses from nociceptive fibers (Ikeda et al., 2006). Moreover, the genetic deletion of the NK1 receptors blocked the expression of the HFS-elicited LTP in adult mice (Fig. 10.3H). However, the specific contribution of SP to the development of LTP in the spinal cord has yet to be determined.

An interaction of tachykinins and glutamate in central nociceptive processing is suggested by several findings. Tachykinins and glutamate co-exist at the central terminals of C- and some Aδ- primary afferent fibers (DeBiasi and Rustioni, 1988). SP enhances glutamate-induced currents in acutely isolated spinal DH neurons (Randić et al., 1990), and potentiation of STT neuron response to chemical and mechanical stimuli (non-noxious and noxious) following combined application of NMDA and SP was observed (Willis, 2002).

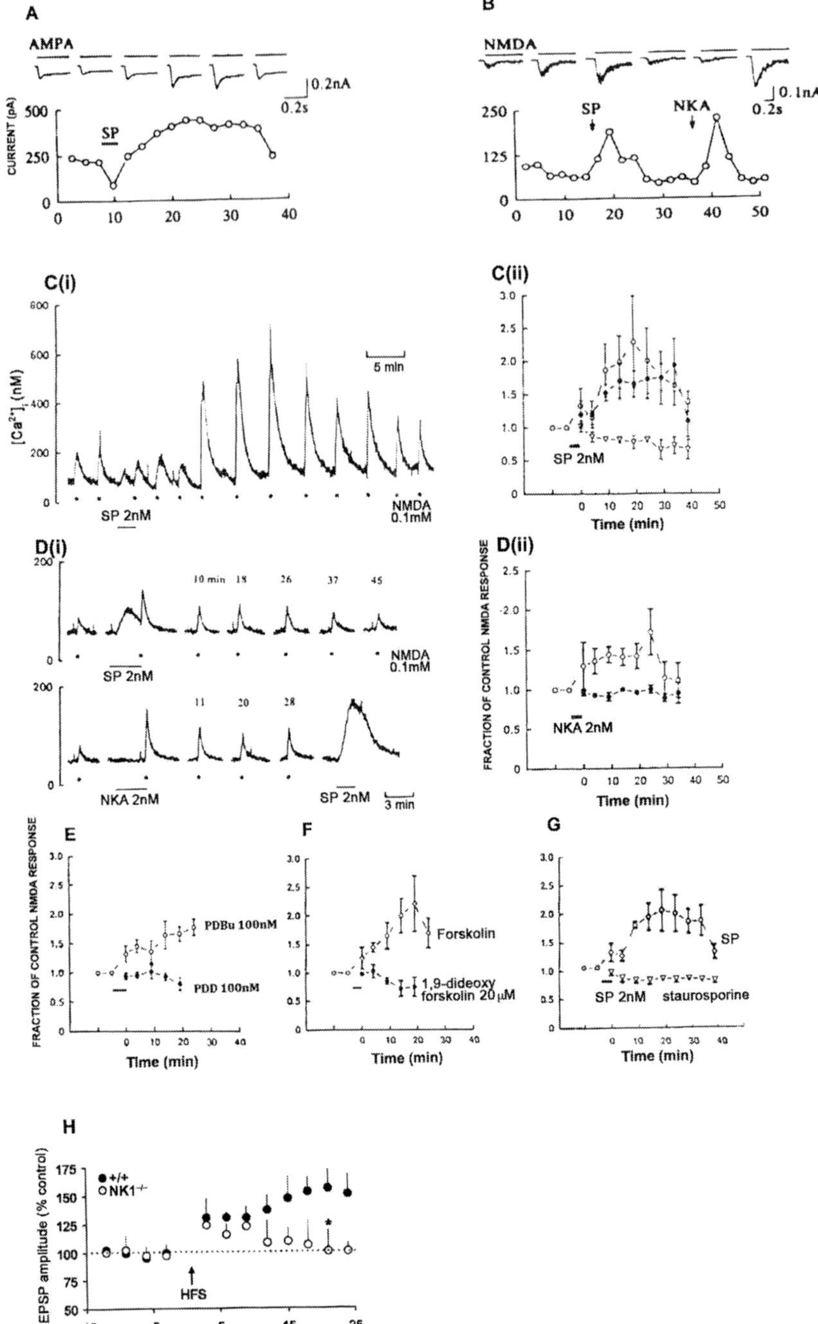

Fig. 10.3 Tachykinins potentiate NMDA and AMPA responses in acutely isolated neurons from the spinal dorsal horn. (**A** and **B**) AMPA and NMDA-induced current responses in two

Fig. 10.3 (continued) Cs^+-loaded DH neurons are potentiated by substance P (SP) and neurokinin A (NKA). (**A**) Traces show inward current responses evoked by 10 μM AMPA recorded at 2.5 min intervals from a SDH neuron held at −60 mV before, during and after superfusion of the cell with SP (1 nM, 2.5 min). Graph shows a time-course of changes in the peak amplitude of the AMPA-induced current recorded before, during and after the superfusion with SP in the same DH neuron, TTX $(5 \times 10^{-7}$ M) was present throughout. (**B**) Traces show inward current responses evoked by 0.1 mM NMDA recorded at 2.5 min intervals from a DH neuron held at −60 mV and during 10 s application of 0.1 mM NMDA plus 2 nM SP or 0.1 mM NMDA plus 2 nM NKA. Time course of changes in the peak amplitude of the initial transient component of NMDA-induced current recorded before, during and after SP + NMDA or NKA + NMDA co-administration in the same neurons is shown by the graph. (**C** and **D**) Tachykinin effects on NMDA (100 μM)-dependent increases in $[Ca^{2+}]_i$ in single DH neurons. (**C(i)**) The effects of SP (2 nM) applied for a 2-min period after control applications of NMDA as denoted by the *closed circles* (15-s applications). In this particular cell, SP alone produced a small increase in the $[Ca^{2+}]_i$. (**C(ii)**) Time course of data from the 17 cells in which the effects of SP were examined on NMDA-dependent increase in $[Ca^{2+}]_i$. Responding cells were defined as >20% increase above control responses for at least two NMDA applications after the addition of SP. The NMDA response after removal of the SP was not included in this analysis. For this and the following time courses, the control response was the mean value of the two NMDA-dependent increase in $[Ca^{2+}]_i$ responses immediately before the addition of SP. Data for each time point are the pooled data for a 5-min bin. The *open circles* represent the data from four cells in which SP potentiated NMDA responses and also increased the basal $[Ca^{2+}]_i$. The *filled circles* represent data in which SP potentiated the NMDA responses, but alone did not increase the basal $[Ca^{2+}]_i$. The *open triangles* represent the time course of NMDA responses for nine cells in which SP failed to potentiate the increase in $[Ca^{2+}]_i$ produced by NMDA. The *vertical bars* represent the standard error of the mean value of the increases (SEM) unless the data points are the mean of three observations, in which case the *vertical bar* represents the standard deviation. (**D(i)**) The potentiation of NMDA responses by SP (2 nM) and NKA (2 nM) in a single cell. The numbers above the NMDA applications denote the time after application of the tachykinin to the neuron. This experiment also shows that the cell still responds to SP with an increase in the $[Ca^{2+}]_i$ after the NKA failed to produce such a response. (**D(ii)**) The courses of data from the 12 cells in which the effects of NKA (2 nM) were examined on the NMDA-dependent increase in the $[Ca^{2+}]_i$. The *open circle* represent data from six cells in which NKA potentiated NMDA responses and the *filled circles* represent the time course of NMDA responses for nine cells in which NKA (2 nM) failed to potentiate the increase in $[Ca^{2+}]_i$ produced by NMDA. (**E**) Pooled data illustrating the time course of the effects of 4β-phorbol-12,13-dibutyrate (PDBu) (100 nM) in the three cells in which a potentiation of the NMDA-induced increase in $[Ca^{2+}]_i$ was observed (*open circles*). Also shown is the data (*filled circles*) for the effects of 4a-phorbol 12,13-didecanoate (PDD: 100 nM) a phorbol ester that does not activate PKC (data from three cells). The *vertical bars* represent the standard error of the mean value of the increases (SEM) unless data point is the mean of three values in which case the *vertical bar* represents the standard deviation. (**F**) Forskolin (20 μM) potentiated the responses to NMDA in a DH neuron. Pooled data illustrating the time course of the effects of forskolin (20 μM) in the four of six cells in which a potentiation was observed (*open circles*). Also shown (*filled circles*) are the data for the effects of 1,9-dideoxyforskolin (20 μM) an inactive form of forskolin (data from four cells). The *vertical bars* represent the standard error of the mean value of the increases (SEM) unless data point is the mean of three values in which case the *vertical bar* represents the standard deviation. (**G**) Staurosporine (200 nM) prevented the potentiating effect of SP (2 nM) on NMDA (100 μM) responses in a DH neuron. The effect of SP (2 nM) (*open circles*) for eight of the 17 cells in which a potentiation of the NMDA was observed. The time course of the response in pooled data from 5-min bins of data is shown. The *open*

The modulatory actions of tachykinins appeared to involve both the presynaptic and postsynaptic sites. We provided a more direct evidence for this view by showing that the action of glutamate in the DH is enhanced presynaptically and postsynaptically by SP and calcitonin gene-related peptide (CGRP), another neuropeptide present in small primary afferent fibers (Kangrga and Randić, 1990; Randić et al., 1990; Rusin et al., 1992, 1993). When the PAFs were repetitively stimulated at C-fiber strength, SP and CGRP potentiated the basal and PAF-stimulation evoked release of glutamate from spinal cord slices (Kangrga and Randić, 1990). Neonatal capsaicin treatment prevented the PAF stimulation-evoked release of glutamate and the SP- induced increase in the basal release of glutamate. However, the origins of the EAA that are released, and the mechanisms underlying the enhancement of the release of glutamate and aspartate by tachykinins and CGRP, remained to be elucidated. The finding that glutamate and SP coexist in primary afferent terminals in the SDH (DeBiasi and Rustini, 1988), coupled with the potentiation of the basal and PAF-evoked efflux of glutamate by SP and NKA (Kangrga and Randić, 1990), provide evidence for a role of tachykinins in the regulation of glutamate release.

It would seem that excitatory co-release of amino acids and tachykinins could serve to interact cooperatively to result in a potentiation of depolarizing action at postsynaptic sites on DH neurons. These pre- and postsynaptic mechanisms of action of tachykinins and CGRP, are likely to have important physiological implications for strengthening the synaptic connections in the spinal DH, and nociception.

10.8.1 Modulation of NMDA Responses in Acutely Isolated Rat Dorsal Horn Neurons by Tachykinins

The functional role of tachykinins in the rat spinal DH could be related, at least in part, to an interaction with glutamate receptor(s) at the postsynaptic site. We observed that glutamate-activated conductance in rat spinal DH neurons was enhanced by SP (Randić et al., 1990), and we used acutely isolated DH neurons under whole-cell voltage clamp conditions to analyse the postsynaptic interactions between neurokinin and glutamate receptors

Fig. 10.3 (continued) *triangles* show the data from seven cells in which SP was applied during staurosporine application. Adapted from Rusin et al. (1993). (**H**) Primary afferent fibers-evoked LTP requires the activation of NK1 receptor. Summary graphs (mean ± SEM) showing the magnitude and the time course of LTP induced by high frequency stimulation (*arrow*; at time zero), consisting of a burst of 100 pulses (30–35 V/0.5 ms) at 100 Hz, repeated three times at 10 s intervals, in wild type ($+/+$, *filled circles*; $n = 4$ slices from 4 mice) and NK1 knockout (NK1 $-/-$; *open circles*; $n = 4$ slices from 4 mice). Sample traces are the average of four to six consecutive EPSPs

(Murase et al., 1989). Direct membrane effects of glutamate, NMDA and AMPA receptors activation was measured as an inward current generated by exogenously local pressure-applied NMDA or AMPA to acutely isolated DH neurons (Murase et al., 1989).SP and NKA both caused a long-lasting enhancement (> 1 h) of glutamate-, NMDA- and AMPA-induced currents recorded in a subset of acutely isolated neurons from the rat DH (Randić et al., 1990; Rusin et al., 1992) (Fig. 10.3A and B). Infusion of a calcium chelating agent (1,2-bis(2-aminophenoxy)-ethane-N,N,N',N'-tetra-acetic acid (BAPTA) in the patch pipette blocked the NMDAR potentiation by SP, leading to the suggestion that Ca^{2+} may mediate the potentiation. To further elucidate the mechanism underlying this enhancement of NMDAR function, we measured the effects of tachykinins and glutamate receptor agonists on $[Ca^{2+}]_i$, in these cells. SP, but not NKA, increased $[Ca^{2+}]_i$ in a subpopulation of neurons, in agreement with previous studies. The increase in $[Ca^{2+}]_i$ was found to be due to Ca^{2+} influx through voltage-sensitive Ca^{2+} channels. SP and NKA also potentiated the increase in $[Ca^{2+}]_i$ produced by NMDA, but not by AMPA, kainate or 50 mM K^+. However, the results of this study clearly dissociate the ability of tachykinins to increase $[Ca^{2+}]_i$ from their ability to enhance NMDA effects (Rusin et al., 1993). Thus both, SP (Fig. 10.3C(i) and C(ii)) and NKA (Fig. 10.3D(i) and D(ii)) enhanced NMDA effects, whereas only SP increased $[Ca^{2+}]_i$. Furthermore, SP enhanced NMDA effects in cells regardless an increase in $[Ca^{2+}]_i$ suggesting that basal $[Ca^{2+}]_i$ allows effective G-protein function and operation of signal transduction pathways in general (Rusin et al., 1992) and Ca^{2+} influx is not required for the SP-induced enhancement of NMDA receptor function either in hippocampal granule cells (Lieberman and Mody, 1998) or spinal DH neurons (Rusin et al., 1993).

10.8.2 Possible Cellular and Molecular Mechanisms of the SP Enhancement of NMDA Response

The exact molecular mechanisms underlying the enhancement of NMDA receptor-activated conductance by tachykinins in the DH have yet to be elucidated. The possibility that SP directly modifies kinetic properties of single NMDA channels in acutely isolated DH neurons has not been examined in our study (Rusin et al., 1992, 1993). However, SP was found to produce a robust enhancement of single NMDA channel function through a readily diffusible, but yet unidentified intracellular message in acutely isolated adult rat hippocampal dentate gyrus granule cells (Libermann and Mody, 1998). Calcium influx or activation of protein kinase C were not required for the SP-induced increase in NMDA channel open durations. Although the NK1 receptor found in DH neurons appears to be coupled to phospholipase C hydrolysis and activation of PKC, the NK1 receptor present in dentate granular cells may be coupled to a different second-messenger system, since neither activators nor antagonists of PKC or PKA have any effect on the mean open duration of NMDA channels (Liebermann and Mody, 1998). Another way in which the activation of SP receptors may modify glutamate and NMDAR-activated conductance of DH neurons is indirectly through the regulation of intracellular second messenger mechanisms.

The intracellular pathway linking SP receptor activation to changes in NMDA responsiveness may involve PKC, since perfusion of rat spinal cord slices with phorbol esters enhanced the depolarizing responses to NMDA and L-glutamate in DH neurons (Gerber et al., 1989). In agreement with this finding, it was reported that phorbol esters, enhance the amplitude of NMDA current in oocytes (Kelso et al., 1992), and CA1 hippocampal neurons. Consistent with these results we have demonstrated that active phorbol esters potentiate the NMDA-dependent increases in $[Ca^{2+}]_i$ in DH neurons (Fig. 10.3E). Furthermore, the PKC (and PKA) inhibitors staurosporine (Fig. 10.3G) and H-7 prevented the SP-induced potentiation of the NMDA-mediated current and calcium response (Gerber et al., 1989; Rusin et al., 1992, 1993). Previous studies have demonstrated that SP can activate PKC, and that heteromeric NMDA receptors constructed in vitro from cloned subunits can be directly potentiated by PKC, and this effect can be blocked by staurosporine. Exactly how the function of the NMDA receptor is altered is not completely clear. PKC has two effects on the NMDA receptor-activated channels: it increases the probability of single NMDA-receptor channel openings (Xiong et al., 1998) and it reduces the affinity of Mg^{2+} to NMDA channels (Chen and Huang, 1992). The PKC-mediated phosphorylation has been suggested to result in a reduction of the Mg-dependent blockade of NMDA receptors (Chen and Huang, 1992) via direct or indirect mechanisms (Zheng et al., 1999; Liao et al., 2001).

The possibility that the activation of PKA may also be involved in the regulation of the activity of NMDA receptors is supported by several findings. It has been reported that NMDA-induced currents of acutely isolated DH neurons, and NMDA-induced increases in $[Ca^{2+}]_i$, are modulated by the activity of PKA (Rusin et al., 1992, 1993). The finding that NMDA-evoked current was potentiated by treatment of acutely isolated DH neurons with forskolin (Fig. 10.3F), or 8-Br cAMP, as well as by intracellular perfusion with cAMP or a catalytic subunit of PKA (Rusin et al., 1992; Cerne et al., 1993), and that the enhancement of NMDA response was prevented by H8, a claimed antagonist of PKA, is consistent with the results obtained in DH neurons using the spinal slice preparation (Gerber et al., 1989; Cerne et al., 1992). In agreement with our results, it has been recently shown (Bird et al., 2005) that pain-related synaptic plasticity in the amygdala is accompanied by PKA-mediated enhanced NMDA receptor function and increased phosphorylation of NR1 subunit of the NMDAR. Moreover, a conditional deletion of the NR1 subunit of the NMDA receptor in adult spinal cord DH (lamina II) reduces current responses mediated by exogenously applied NMDA (South et al., 2003).

Our studies also indicate that the activity of NMDAR expressed in acutely isolated neurons from the SDH can be modulated by activated CaMKII (Kolaj et al., 1994). The NMDA receptors contain consensus phosphorylation sites for CaMKII in NR1 and NR2 subunits (Hollmann and Heinemann, 1994). In addition there is evidence that NMDA receptors are directly phosphorylated by CaMKII. These results suggest the possibility that at least three second messenger systems may be involved in the regulation of the activity of neuronal NMDA receptors.

Besides PKC, PKA and CaMKII, the extracellular-signal regulated kinase (ERK) might also participate in the interactions of tachykinins with glutamate. The ERK kinase, one of the major subfamilies of MAPKs, is a serine-threonine kinase critical for the transduction of signals from cell surface receptors to the nucleus (Impey et al., 1998). The ERK cascades are suggested to contribute to excitatory synaptic plasticity in the CNS, including the spinal DH (Ji et al., 2003; Sweatt, 2004). However, many of their upstream signaling pathways remain to be identified. It has been recently reported that activation of spinal glutamate (NMDA, AMPA and mGluRs) and NK-1 receptors via increases in postsynaptic Ca^{2+} levels is important for activation of Ca^{2+}-dependent signaling cascades, including ERK in DH neurons (Ji et al., 2003; Wei et al., 2006), similar as in brain (Impey et al., 1998). Activation of NMDA receptors phosphorylates ERK in hipppocampal neurons, possibly through protein kinase A (PKA) and protein kinase C (PKC) (Impey et al., 1998;), and in the spinal cord (Ji et al., 2003; Hu and Gereau IV, 2003; Kawasaki et al., 2004). It was also recently reported that calcium/calmodulin-stimulated adenylyl cyclases (AC1 and AC8) contribute to activation of ERK in spinal DH neurons in adult rats and mice (Wei et al., 2006). AC1 and AC8 couple NMDA receptor-induced cytosolic Ca^2 increase to cAMP signaling pathways (Chetkovich and Sweatt, 1993).

10.9 Enhanced LTP of Primary Afferent Neurotransmission in AMPA Receptor GluR2-Deficient Mice

Although pharmacological blockade of NMDARs generally prevents LTP induction in most brain and SDH neurons, high frequency (100 Hz) tetanic stimulation can induce LTP independently of these receptors in the brain. It has been shown that Ca^{2+}-permeable AMPA-receptors mediate LTP in interneurons in the amygdala (Mahanty and Sah, 1998). This form of LTP requires an elevation of postsynaptic Ca^{2+} for its induction.

Ca^{2+}-permeable-AMPA receptors (AMPARs) are expressed in the superficial dorsal horn (SDH), laminae (I/II) of the spinal cord, the area involved in transmission and modulation of sensory information, including nociception (Engelman et al., 1999; Hartmann et al., 2004). A possible role of Ca^{2+}-permeable-AMPARs in synaptic strengthening has been suggested in postnatal DH cultures (Gu et al.,

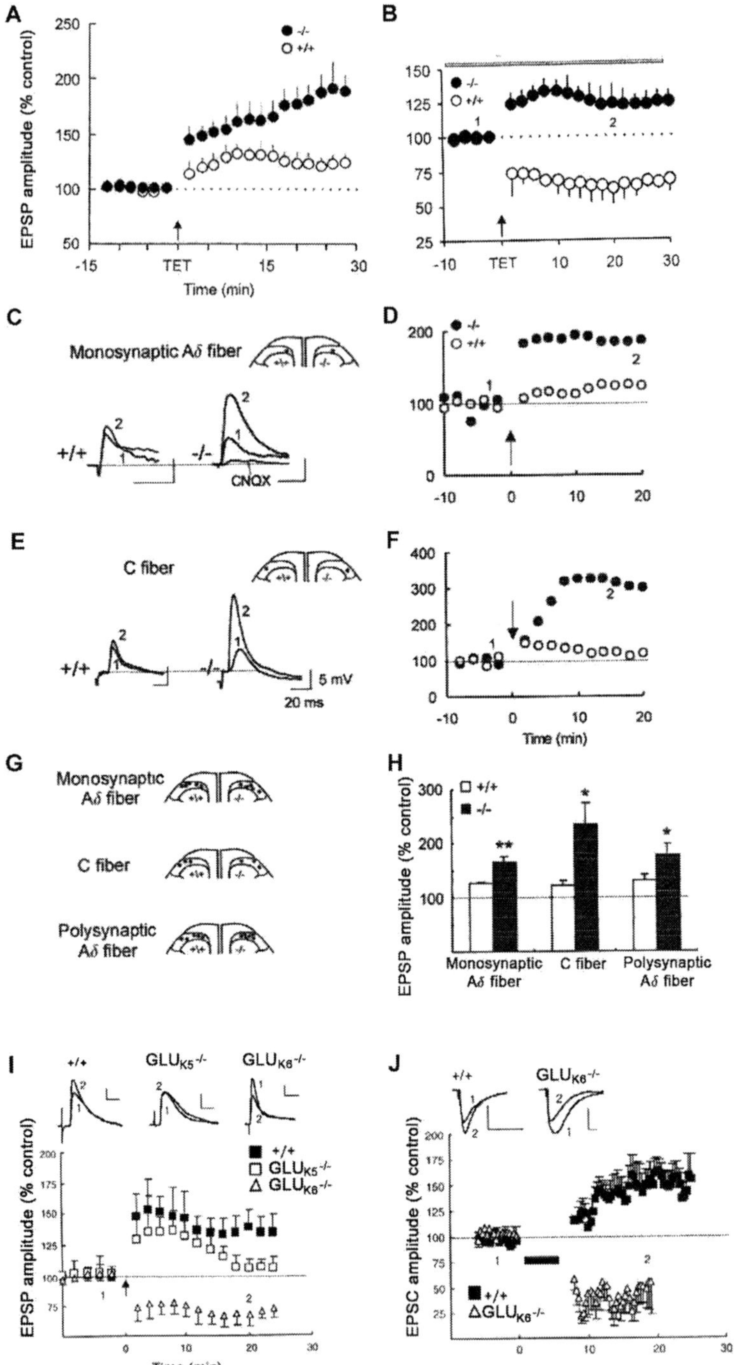

Fig. 10.4 Enhanced LTP of primary afferent neurotransmission in AMPA receptor GluR2-deficient mice. (**A**) Primary afferents – evoked EPSPs from $+/+$ and GluR2$^{-/-}$ SDH neurons

1996), but their role in the long-lasting activity-dependent plasticity of primary afferent neurotransmission in the adult mouse SDH has not been investigated. In the recent study the role of Ca^{2+}-permeable-AMPARs in the regulation of long-lasting synaptic plasticity, specifically LTP and LTD in the SDH, was investigated using mice deficient in AMPAR GluR2 subunit. We have shown (Youn et al., 2008) that the GluR2 mutants exhibited no changes in passive membrane properties, but a significant increase in rectification of excitatory postsynaptic currents, the finding suggesting increased expression of Ca^{2+}-permeable-AMPARs. In the absence of GluR2, high-frequency stimulation of small-diameter primary afferent fibers induced LTP that is enhanced and non-saturating in the SDH (Fig. 10.4A) at both primary afferent Aδ- or C-fiber monosynaptic (Fig. 10.4C, D, E and F) and polysynaptic pathways, whereas neuronal excitability and paired-pulse depression were normal (Youn et al., 2008). The LTP can be induced in the presence of NMDA receptor antagonist D-AP5 (Fig. 10.4B), and L-type Ca^{2+} channel blockers, suggesting that Ca^{2+}-permeable AMPARs are sufficient to induce LTP in the SDH neurons of adult mouse spinal cord. In contrast, the induction of HFS-LTD is reduced in the SDH of GluR2 mutants.

◀──

Fig. 10.4 (continued) were intracellularly recorded in the presence of 10 μM bicuculline methiodide and 2 μM strychnine to block $GABA_A$ and glycine receptors activity. High frequency stimulation (100 Hz for 1 s, delivered 3 times at 0.1 Hz, at Aδ- or C-fiber strength) was applied (*TET, arrow*) to induce LTP. Summary graphs demonstrate the magnitude and the time course of LTP induced in $+/+$ (o; 12 neurons, 10 slices, 9 mice) and $GluR2^{-/-}$ mice (●; 17 neurons, 13 slices, 12 mice). (**B**) LTP in the presence of NMDA receptor blockade in $GluR2^{-/-}$ mice. The summarized time course graphs from SDH neurons intracellularly recorded in the presence of the NMDA receptor antagonist D-AP5 (50–100 μM) demonstrates LTP in $GluR^{-/-}$ mice (●; 11 neurons, 7 slices, 7 mice) and long-term depression (LTD) in $+/+$ mice (o; 10 neurons, 7 slices, 7 mice). (**C–H**) Enhanced LTP in $GluR2^{-/-}$ mice SDH neurons receiving different synaptic inputs. Superimposed traces (*left column*) and time course graphs (*right column*) demonstrate the enhanced LTP of monosynaptic Aδ-fiber (**C** and **D**) and C-fiber (**E** and **F**)-evoked EPSPs, induced by HFS stimulation (*arrows*). Numbers on the graphs (**D** and **F**) indicate the corresponding time of the sampled traces. CNQX, a non-NMDAR antagonist, almost completely blocked the enhanced monosynaptic Aδ-fiber-evoked EPSP from GluR2 $^{-/-}$ mice. (**C**) Insets show the location of SDH neurons recorded. Histogram (**H**) compares the magnitude of LTP of EPSPs-evoked from monosynaptic and polysynaptic primary afferent Aδ-fiber, and C-fiber between $+/+$ and $GluR2^{-/-}$ SDH neurons (cell locations in **G**) ($p^* < 0.05$; $** p < 0.01$). (**I** and **J**) Primary afferent fibers-evoked LTP requires the activation of glutamate receptor 6 (Glu_{K6}). (**I**) Superimposed traces above the graphs are averages of 3–5 primary-afferent-fibers-evoked monosynaptic EPSPs in SG neurons from wild-type ($+/+$), Glu_{K5} knockout ($Glu_{K5}^{-/-}$) and $Glu_{K6}^{-/-}$ mice, taken before (1) and after (2) HFS. Calibration: x axis, 20 ms; y axis, 2 mV. A graph shows magnitude and time course of LTP in SG neurons ($n=6$) of wild-type mice and $Glu_{K5}^{-/-}$ ($n=4$) mice, and LTD in $Glu_{K6}^{-/-}$ ($n=7$) mice, induced by tetanic stimulation (100 Hz, 3 s) of primary afferents. (**J**) A graph shows that spike-timing stimulation paradigm (200 depolarizing current pulses of 2 nA/2 ms at 1 Hz, indicated by a *solid bar* on the graph, induced LTP of monosynaptic Aδ-fiber-evoked excitatory postsynaptic currents (EPSCs) in all tested SG cells of wild-type mice (3 cells, 3 mice), but instead only LTD in $Glu_{K6}^{-/-}$ mice (3 cells, 3 mice). Superimposed traces above the graph are Aδ fiber-evoked monosynaptic EPSCs in wild-type and $Glu_{K6}^{-/-}$ mice, taken before and after the stimulation period. Calibration: x axis, 20 ms; y axis, 20 pA. Adapted from Youn et al. (2008)

These results suggest an important role for AMPAR GluR2 subunit in regulating synaptic plasticity with potential relevance for long-lasting hypersensitivity in pathological states. Blockade of Ca^{2+}-permeable AMPA receptors in the rat spinal cord diminishes the development of hyperalgesia and allodynia associated with peripheral injury (Jones and Sorkin, 2004).

10.10 Concluding Remarks

Spinal neuron LTP following activation of primary afferent fibres, which might contribute to hyperalgesia in animal models of pain, uses multiple mechanisms involving protein phosphorylation, similar to the processes associated with hippocampal LTP. Modulation of postsynaptic AMPA and NMDA receptor function caused by phosphorylation may play an important role in the induction and expression of synaptic plasticity at dorsal horn excitatory synapses.

References

Abe K, Saito H (1993) Tyrosine kinase inhibitors, herbimycin A and lavendustin A, block formation of long-term potentiation in the rat dentate gyrus in vivo. Brain Res 621: 167–170

Banke TG, Bowie D, Lee H, et al. (2000) Control of GluR1 AMPA receptor function by cAMP-dependent protein kinase. J Neurosci 20: 89–102

Baranauskas G, Nistri A (1998) Sensitization of pain pathways in the spinal cord: cellular mechanisms. Prog Neurobiol 54: 349–365

Barria A, Derkach V, Soderling T (1997a) Identification of the Ca^{2+}/calmodulin-dependent protein kinase II regulatory phosphorylation site in the α-amino-3-hydroxy-5-methyl-4-isoxazole-propionate-type glutamate receptor. J Biol Chem 272: 32727–32730

Barria A, Muller D, Derkach V (1997b) Regulatory phosphorylation of AMPA-type glutamate receptors by CaM-KII during long-term potentiation. Science 276: 2042–2045

Bird GC, Lash LL, Han JS, et al. (2005) Protein kinase A-dependent enhanced NMDA receptor function in pain-related synaptic plasticity in rat amygdala neurones. J Physiol 564: 907–921

Bliss TV, Collingridge GL (1993) A synaptic model of memory: long-term potentiation in the hippocampus. Nature 361: 31–39

Blitzer RD, Wong T, Nouranifar R, et al. (1995) Postsynaptic cAMP pathway gates early LTP in hippocampal CA1 region. Neuron 15: 1403–1414

Blitzer RD, Connor JH, Brown GP, et al. (1998) Gating of CaMKII by cAMP-regulated phosphatase activity during LTP. Science 280: 1940–1942

Brenner GJ, Ji RR, Shaffer S, et al. (2004) Peripheral noxious stimulation induces phosphorylation of the NMDA receptor NR1 subunit at the PKC-dependent site, serine 896, in spinal cord dorsal horn neurons. Eur J Neurosci 20: 375–384

Bruggemann I, Schultz S, Wibomy D, et al. (2000) Colocalization of the mu-opoid receptor and calcium/calmodulin-dependent kinase II in distinct pain-processing brain regions. J Neurophysiol 80: 2797–2800

Burnashev N, Schoepfer R, Monyer H, et al. (1992) Control by asparagine residues of calcium permeability and magnesium blockade in the NMDA receptor. Science 257: 1415–1419

Cerne R, Jiang M, Randić M (1992) Cyclic adenosine 3′,5 ′-monophosphate potentiates excitatory amino acid and synaptic responses of rat spinal dorsal horn neurons. Brain Res 596: 111–123

Cerne R, Rusin KI, Randić M (1993) Enhancement of the N-methyl-D-aspartate response in spinal dorsal horn neurons by cAMP-dependent protein kinase. Neurosci Lett 161: 124–128

Chen L, Huang LYM (1992) Protein kinase C reduces Mg^{2+} block of NMDA receptor channels as a mechanism of modulation. Nature 356: 521–523

Chetkovich DM, Sweatt JD (1993) NMDA receptor activation increases cyclic AMP in area CA1 of the hippocampus via calcium/calmodulin stimulation of adenylyl cyclase. J Neurochem 61: 1933–1942

Citri A, Malenka RC (2008) Synaptic plasticity: multiple forms, functions, and mechanisms. Neuropsychopharmacol 33: 18–41

Collingridge GL, Kehl SJ, McLennan H (1983) Excitatory amino acids in synaptic transmission in the Schaffer collateral-commissural pathway of the rat hippocampus. J Physiol 334: 33–46

Dan Y, Poo MM (2006) Spike timing-dependent plasticity: from synapse to perception. Physiol Rev 86: 1033–1048

DeBiasi S, Rustioni A (1988) Glutamate and substance P coexist in primary afferent terminals in the superficial laminae of spinal cord. Proc Natl Acad Sci USA 85: 7820–7824

Dickenson AH, Sullivan AF (1987) Evidence for a role of the NMDA receptor in the frequency dependent potentiation of deep rat dorsal horn nociceptive neurons following C-fibre stimulation. Neuropharmacol 26: 1235–1238

Dingledine R, Borges K, Bowie D, et al. (1999) The glutamate receptor ion channels. Pharmacol Rev 51: 7–61

Dougherty PM, Willis WD (1992) Enhanced responses of spinothalamic tract neurons to excitatory amino acids accompany capsaicin-induced sensitization in the monkey. J Neurosci 12: 883–894

Dougherty PM, Palecek J, Paleckova V, et al. (1994) Neurokinin 1 and 2 antagonists attenuate the responses and NK1 antagonists prevent the sensitization of primate spinothalamic tract neurons after intradermal capsaicin. J Neurophysiol 72: 1464–1475

Duggan AW, Hendry IA, Morton CR, et al. (1988) Cutaneous stimuli releasing immunoreactive SP in the dorsal horn of the cat. Brain Res 451: 261–273

Engelman HS, Allen TB, MacDermott AB (1999) The distribution of neurons expressing calcium-permeable AMPA receptors in the superficial laminae of the spinal cord dorsal horn. J Neurosci 19: 2081–2089

Esteban JA, Shi SH, Wilson C, et al. (2003) PKA phosphorylation of AMPA receptor subunits controls synaptic trafficking underlying plasticity. Nat Neurosci 6: 136–143

Fang L, Wu J, Lin Q, et al. (2002) Calcium-calmodulin-dependent protein kinase II contributes to spinal cord central sensitization. J Neurosci 22: 4196–4204

Fisher K, Coderre TJ, Hagen NA (2000) Targeting the N-methyl-D-aspartate receptor for chronic pain management. Preclinical animal studies, recent clinical experience and future research directions. J Pain Symptom Manage 20: 358–373

Frey U, Huang YY, Kandel ER (1993) Effects of cAMP simulate a late stage of LTP in hippocampal CA1 neurons. Science 260: 1661–1664

Fukunaga K, Miller D, Miyamoto E, et al. (1995) Increased phosphorylation of Ca^{2+}/ calmodulin-dependent protein kinase II and its endogenous substrates in the induction of long-term potentiation. J Biol Chem 270: 6119–6124

Fukunaga K, Stoppini L, Miyamoto E, et al. (1993) Long-term potentiation is associated with an increased activity of Ca^{2+}/calmodulin-dependent protein kinase II. J Biol Chem 268: 7863–7867

Fundytus ME (2001) Glutamate receptors and nociception. Implications for the drug treatment of pain. CNS Drugs 15: 29–58

Galik J, Youn DH, Kolaj M, et al. (2008) Involvement of group I metabotropic glutamate receptors and glutamate transporters in the slow excitatory synaptic transmission in the spinal cord dorsal horn. Neurosci 154: 1372–1387

Gerber G, Kangrga I, Ryu PD, et al. (1989) Multiple effects of phorbol esters in rat spinal dorsal horn. J Neurosci 9: 3606–3617

Gerber G, Youn DH, Hsu CH, et al. (2000) Spinal dorsal horn synaptic plasticity: involvement of group I metabotropic glutamate receptors. In: Sandkühler J, Bromm B, Gebhart GF (eds) Nervous system plasticity and chronic pain. Prog Brain Res vol. 129, Elsevier, Amsterdam, pp. 115–134

Giese KP, Fedorov NB, Filipkowski RK, et al. (1998) Autophosphorylation at Thr286 of the α-calcium/calmodulin kinase II in LTP and learning. Science 279: 870–873

Greengard P, Jen J, Nairn AC, et al. (1991) Enhancement of the glutamate response by cAMP-dependent protein kinase in hippocampal neurons. Science 253: 1135–1138

Gu JG, Albuquerque C, Lee CJ, et al. (1996) Synaptic strengthening through activation of Ca^{2+} permeable AMPA receptors. Nature 381: 793–796

Guo H, Huang LY (2001) Alteration in the voltage dependence of NMDA receptor channels in rat dorsal horn neurones following peripheral inflammation. J. Physiol Lond 537: 115–123

Guo W, Zou S, Guan Y, et al. (2002) Tyrosine phosphorylation of the NR2B subunit of the NMDA receptor in the spinal cord during the development and maintenance of inflammatory hyperalgesia. J Neurosci 22: 6208–6217

Hamba M, Onodera K, Takahashi T (2000) Long-term potentiation of primary afferent neurotransmission at trigeminal synapses of juvenile rats. Eur J Neurosci 12: 1128–1134

Hartmann B, Ahmadi S, Heppenstall PA, et al. (2004) The AMPA receptor subunits GluR-A and GluR-B reciprocally modulate spinal synaptic plasticity and inflammatory pain. Neuron 44: 637–650

Heath MJ, Womack MD, MacDermott AB (1994) Substance P elevates intracellular calcium in both neurons and glial cells from the dorsal horn of the spinal cord. J Neurophysiol 72: 1192–1198

Hoeger-Bement MK, Sluka KA (2003) Phosphorylation of CREB and mechanical hyperalgesia is reversed by blockade of the cAMP pathway in a time-dependent manner after repeated intramuscular acid injections. J Neurosci 23: 5437–5445

Hökfelt T, Kellerth JO, Nilsson G, et al. (1975) Substance P: localization in the central nervous system and in some primary sensory neurons. Science 190: 889–890

Hollmann M, Heinemann S (1994) Cloned glutamate receptors. Annu Rev Neurosci 17: 31–108

Hu GY, Hvalby O, Walaas SI, et al. (1987) Protein kinase C injection into hippocampal pyramidal cells elicits features of long-term potentiation. Nature 328: 426–429

Hu HJ, Gereau IV RW (2003) ERK integrates PKA and PKC signaling in superficial dorsal horn neurons. II. Modulation of neuronal excitability. J Neurophysiol 90: 1680–1688

Igwe OJ, Cronwall BM (2001) Hyperalgesia induced by peripheral inflammation is mediated Neurosci 154: 1372–1387 by protein kinase C beta II enzyme in the rat spinal cord. Neurosci 104: 875–890

Ikeda H, Asai T, Murase K (2000) Robust changes of afferent-induced excitation in the rat spinal dorsal horn after conditioning high-frequency stimulation. J. Neurophysiol 83: 2412–2420

Ikeda H, Heinke B, Ruscheweyh R, et al. (2003) Synaptic plasticity in spinal lamina I projection neurons that mediate hyperalgesia. Science 299: 1237–1240

Ikeda H, Stark J, Fischer H, et al. (2006) Synaptic amplifier of inflammatory pain in the spinal dorsal horn. Science 312: 1659–1662

Impey S, Obrietan K, Wong ST, et al. (1998) Cross talk between ERK and PKA is required for Ca^{2+} stimulation of CREB-dependent transcription and ERK nuclear translocation. Neuron 21: 869–883

Ji RR, Kohno T, Moore KA, et al. (2003) Central sensitization and LTP: do pain and memory share similar mechanisms. Trends Neurosci. 26: 696–705

Jones TL, Sorkin LS (2004) Calcium-permeable alpha-amino-3-hydroxy-5-methyl 4-isoxazole propionic acid/kainate receptors mediate development, but not maintenance, of secondary allodynia evoked by first-degree burn in the rat. J Pharmacol Exp Ther 310: 223–229

Jung SJ, Kim SJ, Park YK, et al. (2006) Group I mGluR regulates the polarity of spike-timing dependent plasticity in substantia gelatinosa neurons. Biochem Biophys Res Comm 347: 509–516

Kalia LV, Gingrich JR, Salter MW (2004) Src in synaptic transmission and plasticity. Oncogene 23: 8007–8016

Kangrga, I. Randić, M (1990) Tachykinins and calcitonin gene-related peptide enhance release of endogenous glutamate and aspartate from the rat spinal dorsal horn slice. J Neurosci 10: 2026–2038

Karim F, Wang CC, Gereau IV RW (2001) Metabotropic glutamate receptor subtypes 1 and 5 are activators of extracellular signal-regulated kinase signaling required for inflammatory pain in mice. J Neurosci. 21: 3771–3779

Kawasaki Y, Kohno T, Zhuang ZY, et al. (2004) Ionotropic and metabotropic receptors, protein kinase A, protein kinase C, and Src contribute to C-fiber-induced ERK activation and cAMP response element-binding protein phosphorylation in dorsal horn neurons, leading to central sensitization. J Neurosci 24: 8310–8321

Kelso SR, Nelson TE, Leonard JP (1992) Protein kinase C-mediated enhancement of NMDA currents by metabotropic glutamate receptors in *xenopus* oocytes. J. Physiol 449: 705–718

Kennedy MB (2000) Signal-processing machines at the postsynaptic density. Science 290: 750–754

Kolaj M, Cerne R, Cheng G, et al. (1994) Alpha subunit of calcium/calmodulin-dependent protein kinase enhances excitatory amino acid and synaptic responses of rat spinal dorsal horn neurons. J Neurophysiol 72: 2525–2531

Kovalchuck Y, Hany E, Kafitz Kw, et al. (2002) Postsynaptic induction of BNDF-mediated long-term potentiation. Science 295: 1729–1734

Leonard AS, Hell JH (1997) Cyclic AMP-dependent protein kinase and protein kinase C phosphorylate N-methyl-D-aspartate receptor at different sites. J Biol Chem 272: 12107–12115

Liao GY, Wagner DA, Hsu MH, et al. (2001) Evidence for direct protein kinase C-mediated modulation of N-methyl-D-aspartate receptor current. Mol Pharmacol 59: 960–964

Lieberman DN, Mody I (1994) Regulation of NMDA channel function by endogenous Ca^{2+}-dependent phosphatase. Nature 369: 235–239

Lieberman DN, Mody I (1998) Substance P enhances NMDA channel function in hippocampal dentate gyrus granule cells. J Neurophysiol 80: 113–119

Lin Q, Peng YB, Willis WD (1996) Possible role of protein kinase C in the sensitization of primate spinothalamic neurons. J Neurosci 16: 3026–3034

Lin Q, Wu J, Willis WD (2002) Effects of protein kinase A activation on the responses of primate spinothalamic neurons to mechanical stimuli. J. Neurophysiol 88: 214–221

Linden DJ, Shen FS, Marakami K, et al. (1987) Enhancement of long-term potentiation by cis-unsaturated fatty acid: relation to protein kinase C and phospholipase Mζ. J Neurosci 7: 3783–3792

Ling DS, Benardo LS. Serrano PA, et al. (2002) Protein kinase Mζ is necessary and sufficient for LTP maintenance. Nat Neurosci 5: 295–296

Lisman J, Schulman H, Cline H (2002) The molecular basis of CaMKII function in synaptic and behavioural memory. Nature Rev Neurosci 3: 175–190

Lynch G, Larson J, Kelso S, et al. (1983) Intracellular injections of EGTA block induction of hippocampal long-term potentiation. Nature 305: 719–721

Mahanty NK, Sah P (1998) Calcium-permeable AMPA receptors mediate long-term potentiation in interneurons in the amygdala, Nature 394: 683–687

Makhinson M, Chotiner JK, Watson JB, et al. (1999) Adenylyl cyclase activation modulates activity-dependent changes in synaptic strength and Ca^{2+}/calmodulin-dependent kinase II autophosphorylation. J Neurosci 19: 2500–2510

Malcangio M, Lessmann V (2003) A common thread for pain and memory synapses? Brain-derived neurotrophic factor and trk B receptors. Trends Pharmacol Sci 24: 116–121

Malenka RC, Madison DV, Nicoll RE (1986) Potentiation of synaptic transmission in the hippocampus by phorbol esters. Nature 321: 175–177

Malenka RC, Kauer JA, Zucker RS, et al. (1988) Postsynaptic calcium is sufficient for potentiation of hippocampal synaptic transmission. Science 242: 81–84

Malenka RC, Kauer JA, Perkel DJ, et al. (1989) An essential role for postsynaptic calmodulin and protein kinase activity in long-term potentiation. Nature 340: 554–557

Malenka RC, Lancaster B, Zucker RS (1992) Temporal limits on the rise in postsynaptic calcium required for the induction of long-term potentiation. Neuron 9: 121–128

Malenka RC, Nicoll RA (1999) Long-term potentiation – a decade of progress. Science 285: 1870–1874

Malenka RC, Bear MF (2004) LTP and LTD: an embarrassment of riches. Neuron 44: 5–21

Malinow R, Malenka RC (2002) AMPA receptor trafficking and synaptic plasticity. Annu Rev Neurosci 25: 103–126

Malinow R, Schulman H, Tsien RW (1989) Inhibition of postsynaptic PKC or CaMKII blocks induction but not expression of LTP. Science 245: 862–866

Malmberg AB, Yaksh TL (1993) Spinal nitric oxide synthesis inhibition blocks NMDA-induced thermal hyperalgesia and produces antinociception in the formalin test in rats. Pain 54: 291–300

Malmberg AB, Brandon EP, Idzerda RL, et al. (1997) Diminished inflammation and nociceptive pain with preservation of neuropathic pain in mice with a targeted mutation of the type I regulatory subunit of cAMP-dependent protein kinase. J Neurosci 17: 7462–7470

Mammen AL, Kameyama K, Roche KW (1997) Phosphorylation of the alpha-amino-3-hydroxy-5-methylisoxazole-4-propionic acid receptor GluR1 subunit by calcium/calmodulin-dependent kinase II. J Biol Chem 272: 32528–32533

Mannion RJ, Costigan M, Decosterd I, et al. (1999) Neurotrophins: peripherally and centrally acting modulators of tactile stimulus-induced inflammatory pain hypersensitivity. Proc Natl Acad Sci USA 96: 9385–9390

Mantyh PW, Rogers SD, Honore P, et al. (1997) Inhibition of hyperalgesia by ablation of Lamina I spinal neurons expressing the SP receptor. Science 278: 275–279

McGlade-McCulloh E, Yamamoto H, Tan SE, et al. (1993) Phosphorylation and regulation of glutamate receptors by calcium/calmodulin-dependent protein kinase II. Nature 362: 640–642

Miletić G, Pankratz MT, Miletić V (2002) Increases in phosphorylation of cyclic AMP response element binding protein (CREB) and decreases in the content of calcineurin accompany thermal hyperalgesia following chronic constriction injury in rats. Pain 99: 493–500

Miyabe T, Miletić V (2005) Multiple kinase pathways mediate the early sciatic-ligation-associated activation of CREB in the rat spinal dorsal horn. Neurosci Lett 381: 80–85

Mochly-Rosen D, Basbaum AI, Koshland DE Jr. (1987) Distinct cellular and regional localization of immunoreactive protein kinase C in rat brain. Proc Natl Acad Sci USA 84: 4660–4664

Moon IS, Apperson ML, Kennedy MB (1994) The major tyrosine-phosphorylated protein in the postsynaptic density fraction is N-methyl-D-aspartate receptor subunit 2B. Proc Natl Acad Sci USA 91: 3954–3958

Murase K, Randić M (1984) Actions of substance P on rat spinal dorsal horn neurones. J Physiol 346: 203–217

Murase K, Ryu PD, Randić M (1989) Excitatory and inhibitory amino acids and peptide-induced responses in acutely isolated rat spinal dorsal horn neurons. Neurosci Lett 103: 56–63

Neugebauer V, Chen PS, Willis WD (1999) Role of metabotropic glutamate receptor subtype mGluR1 in brief nociception and central sensitization of primate STT cells. J Neurophysiol 82: 272–282

Neugebauer V, Lucke T, Schaible HG (1993) N-methyl-D-aspartate (NMDA) and non-NMDA receptor antagonists block the hyperexcitability of dorsal horn neurons during development of acute arthritis in rat's knee joint. J Neurophysiol 70: 1365–1377

Nicoll RA (2003) Expression mechanisms underlying long-term potentiation; a postsynaptic view. Phil Trans R Soc Lond B 358: 721–726

O'Dell TJ, Kandel ER, Grant SGN (1991) Long-term potentiation in the hippocampus is blocked by tyrosine kinase inhibitors. Nature 353: 558–560

Omkumar RV, Kiely MJ, Rosenstein AJ, et al. (1996) Identification of a phosphorylation site for calcium/calmodulin-dependent protein kinase II in the NR2B subunit of the N-methyl-D-aspartate receptor. J Biol Chem 271: 31670–31678

Otmakhov N, Griffith L, Lisman JE (1997) Postsynaptic inhibitors of calcium/calmodulin-dependent protein kinase type II block induction but not maintenance of pairing-induced long-term potentiation. J Neurosci 17: 5357–5365

Otmakhova NA, Otmakhov N, Mortenson LH, et al. (2000) Inhibition of the cAMP pathway decreases early long-term potentiation at CA1 hippocampal synapses. J Neurosci 20: 4446–4451

Palecek J, Paleckova V, Willis WD (1999) The effect of phorbol esters on spinal cord amino acid concentrations in responsiveness of rats to mechanical and thermal stimuli. Pain 80: 597–605

Park YK, Galik J, Ryu PD, et al. (2004) Activation of presynaptic group I metabotropic glutamate receptors enhances glutamate release in the rat spinal cord substantia gelatinosa. Neurosci Lett 361: 220–224

Pastalkova E, Serrano P, Pinkhasova D, et al. (2006) Storage of spatial information by the maintenance mechanism of LTP. Science 313: 1141–1144

Pedersen LM, Lien GF, Bollerud I, et al. (2005) Induction of long-term potentiation in single nociceptive dorsal horn neurons is blocked by the CaMKII inhibitor AIP. Brain Res 1041: 66–71

Petersen-Zeitz KR, Basbaum AI (1999) Second messengers, the substantia gelatinosa and injury-induced persistent pain. Pain Suppl. 6: S5–S12

Pockett S, Figurov A (1993) Long-term potentiation and depression in the ventral horn of the rat spinal cord in vitro. NeuroReport 4: 97–99

Pockett S (1995) Long-term potentiation and depression in the intermediate gray matter of rat spinal cord in vitro. Neurosci 67: 791–798

Randić M, Hećimović H, Ryu PD (1990) Substance P modulates glutamate induced currents in acutely isolated rat spinal dorsal horn neurons. Neurosci Lett 117: 74–80

Randić M, Jiang MC, Cerne R (1993) Long-term potentiation and long-term depression of primary afferent neurotransmission in the rat spinal cord. J Neurosci 13: 5228–5241

Randić M (1996) Plasticity of excitatory synaptic transmission in the spinal cord dorsal horn. In: Kumazawa T, Kruger L, Mizumura K (eds) The polymodal receptor: a gateway to pathological pain. Prog Brain Res, vol 113. Elsevier, Amsterdam, pp. 463–506

Raymond LA, Blackstone CD, Huganir RL (1993) Phosphorylation of amino acid neuro-transmitter receptors in synaptic plasticity. Trends Neurosci 16: 147–153

Ren K, Hylden JLK, Williams GM, et al. (1992) The effects of a non-competitive NMDA receptor antagonist, MK-801, on behavioural hyperalgesia and dorsal horn neuronal activity in rats with unilateral inflammation. Pain 50: 331–344

Roche KW, O'Brien RJ, Mammen AL, et al. (1996) Characterization of multiple phosphorylation sites in the AMPA receptor GluR1 subunit. Neuron 16: 1179–1188

Rostas JA, Brent VA, Voss K, et al. (1996) Enhanced tyrosine phosphorylation of the 2B subunit of the N-methyl-D-aspartate receptor in long-term potentiation. Proc Natl Acad Sci USA 93: 10452–10456

Rusin KI, Ryu PD, Randić M (1992) Modulation of excitatory amino acid responses in rat dorsal horn neurons by tachykinins. J Neurophysiol 68: 265–286

Rusin KI, Bleakman D, Chard PS, et al. (1993) Tachykinins potentiate N-methyl-D-aspartate responses in acutely isolated neurons from the dorsal horn. J. Neurochem 60: 952–960

Ryu PD, Randić M (1990) Low- and high-voltage activated calcium currents in rat spinal dorsal horn neurons. J Neurophysiol 63: 273–285

Sandkühler J, Liu XG (1998) Induction of long-term potentiation at spinal synapses by noxious stimulation or nerve injury. Eur J Neurosci 10: 2476–2480

Sandkühler J (2000) Learning and memory in pain pathways. Pain 88: 113–118

Sandkühler J (2007) Understanding LTP in pain pathways. Mol Pain 3: 9

Sato E, Takano Y, Kuno Y, et al. (2003) Involvement of spinal tyrosine kinase in inflammatory and N-methyl-D-aspartate-induced hyperalgesia in rats. Eur J Pharmacol 468: 191–198

Serrano P, Yao Y, Sactor TC (2005) Persistent phosphorylation by protein kinase M zeta maintains late-phase long-term potentiation. J Neurosci 25: 1979–1984

Sheng M, Lee H (2003) AMPA receptor trafficking and synaptic plasticity: major unanswered questions. Neurosci Res 46: 127–134

Silva AJ, Stevens CF, Tonegawa S, et al. (1992) Deficient hippocampal long-term potentiation in α-calcium-calmodulin kinase II mutant mice. Science 257: 201–206

Sluka KA, Willis WD (1997) The effects of G protein and protein kinase inhibitors on the behavioural responses of rat to intradermal injection of capsaicin. Pain 71: 165–178

Soderling TR, Derkach VA (2000) Postsynaptic protein phosphorylation and LTP. Trends Neurosci 23: 75–80

Song I, Huganir RL (2002) Regulation of AMPA receptors during synaptic plasticity. Trends Neurosci 25: 578–588

South SM, Kohno T, Kaspar BK, et al. (2003) A conditional deletion of the NR1 subunit of the NMDA receptor in adult spinal cord dorsal horn reduces NMDA currents and injury-induced pain. J Neurosci 23: 5031–5040

Sugrue MM, Brugge JS, Marshak DR, et al. (1990) Immunocytochemical localization of the neuron-specific form of the c-src gene product, $pp60^{c\text{-src}\,(+)}$ in rat brain. J Neurosci 10: 2513–2527

Svendsen F, Tjølsen A, Hole, K (1997) LTP of spinal A-β and C-fibre evoked responses after electrical sciatic nerve stimulation. NeuroReport 8: 3427–3430

Svendsen F, Tjølsen A, Rygh LJ, et al. (1999) Expression of long-term potentiation in single wide-dynamic range neurons in the rat is sensitive to blockade of glutamate receptors. Neurosci Lett 259: 25–28

Sweatt JD (2004) Mitogen-activated protein kinases in synaptic plasticity and memory. Curr Opin Neurobiol 14: 311–317

Tan SE, Wenthold RJ, Soderling TR (1994) Phosphorylation of AMPA-type glutamate receptors by calcium/calmodulin-dependent protein kinase II and protein kinase C in cultured hippocampal neurons. J Neurosci 14: 1123–1129

Terman GW, Eastman CL, Chavkin C (2001) Mu opiates inhibits long-term potentiation induction in the spinal cord slice. J Neurophysiol 85: 485–494

Thomas GM, Huganir RL (2004) MAPK cascade signaling and synaptic plasticity. Nat Rev Neurosci 5: 173–183

Thompson SW, Dray A, Urbán L, et al. (1994) Injury-induced plasticity of spinal reflex activity: NK1 neurokinin receptor activation and enhanced A- and C-fiber mediated responses in the rat spinal cord in vitro. J Neurosci 14: 3672–3687

Thompson SW, Bennett DL, Kerr BJ, et al. (1999) Brain-derived neurotrophic factor is an endogenous modulator of nociceptive responses in the spinal cord. Proc Natl Acad Sci USA 96: 7714–7718

Tingley WG, Ehlers MD, Kameyama K, et al. (1997) Characterization of protein kinase A and protein kinase C phosphorylation of N-methyl-D-aspartate receptor NR1 subunit using phosphorylation site-specific antibodies. J Biol Chem 272: 5157–5166

Todd AJ, Puskar Z, Spike RC et al. (2002) Projection neurons in laminal of rat spinal cord with the neurokinin 1 receptor are selectively innervated by substance P-containing afferents and respond to noxious stimulation. J Neurosci 22: 4103–4113

Torebjörk HE, Lundberg LE, LaMotte RH (1992) Central changes in processing of mechanoreceptive input in capsaicin-induced secondary hyperalgesia in humans. J Physiol 448: 765–780

Urbán L, Randić M (1984) Slow excitatory transmission in rat dorsal horn: possible mediation by peptides. Brain Res 290: 336–341

Wang LY, Salter MW, MacDonald JF (1991) Regulation of kainate receptors by cAMP-dependent protein kinase and phosphatases. Science 253: 1132–1135

Wang LY, Dudek EM, Browning MD, et al. (1994) Modulation of AMPA/kainate receptors in cultured murine hippocampal neurones by protein kinase C. J Physiol 475: 431–437

Wang YT, Salter MW (1994) Regulation of NMDA receptors by tyrosine kinases and phosphatases. Nature 369: 233–235

Wang RA, Cheng G, Kolaj M, et al. (1995) α-Subunit of calcium/calmodulin-dependent protein kinase II enhances γ-aminobutyric acid and inhibitory synaptic responses of rat neurons in vitro. J Neurophysiol 73: 2099–2106

Wang RA, Randić M (1996) Alpha subunit of CaMKII increases glycine currents in acutely isolated rat spinal neurons. J Neurophysiol 75: 2651–2653

Wang YT, Yu YM, Salter MW (1996) Ca(2+)-independent reduction of N-methyl-D-aspartate channel activity by protein tyrosine phosphatase. Proc Natl Acad Sci USA 93: 1721–1725

Wei F, Vadakkan KI, Toyoda H, et al. (2006) Calcium calmodulin-stimulated adenylyl cyclases contribute to activation of extracellular signal-regulated kinase in spinal dorsal horn neurons in adult rats and mice. J Neurosci 26: 851–861

Willis WD (2001) Mechanisms of central sensitization of nociceptive dorsal horn neurons. In: Patterson MM, Grau JW (eds) Spinal cord plasticity. Alterations in reflex function, Kluwer Academic, New York, pp 127–161

Willis WD (2002) Long-term potentiation in spinothalamic neurons. Brain Res Rev 40: 202–214

Willis WD, Coggeshall RE (2004) Sensory mechanisms of the spinal cord. Primary afferent neurons and the spinal dorsal horn. 3rd ed. Kluwer/Academic/Plenum, New York, pp 1–560

Woolf CJ, Thompson SW (1991) The induction and maintenance of central sensitization is dependent on N-methyl-D-aspartatic acid receptor activation: implications for the treatment of post-injury pain hypersensitivity states. Pain 44: 293–299

Woolf CJ, Salter MW (2000) Neuronal plasticity: increasing the gain in pain. Science 288: 1765–1768

Worley PF, Baraban JM, De Souza EB, et al. (1986) Mapping second messenger systems in the brain: differential localization of adenylate cyclase and protein kinase C. Proc Natl Acad Sci USA 83: 4053–4057

Xin WJ, Gong QJ, Xu JT (2006) Role of phosphorylation of ERK in induction and maintenance of LTP of the C-fiber evoked field potentials in spinal dorsal horn, J Neurosci Res 84: 934–943

Xiong ZG, Raouf R, Lu WY, et al. (1998) Regulation of N-methyl-D-aspartate receptor function by constitutively active protein kinase C. Mol Pharmacol 54: 1055–1063

Yajima Y, Narita M, Shimamura M, et al. (2003) Differential involvement of spinal protein kinase C and protein kinase A in neuropathic and inflammatory pain in vivo. Brain Res 992: 288–293

Yaksh TL, Jessell TM, Gamse R, et al. (1980) Intrathecal morphine inhibits substance P release from mammalian spinal cord in vivo. Nature 286: 155–157

Yang HW, Hu XD, Zhang HM, et al. (2004) Roles of CaMKII, PKA, and PKC in the induction and maintenance of LTP of C-fiber-evoked field potentials in rat spinal dorsal horn. J Neurophysiol 91: 1122–1133

Youn DH, Voitenko N, Gerber G, et al. (2005) Altered long-term synaptic plasticity and kainate induced Ca^{2+} transients in the substantia gelatinosa neurons in Glu_{K6}-deficient mice. Mol Brain Res 142: 9–18

Youn DH, Royle G, Kolaj M, et al. (2008) Enhanced LTP of primary afferent neurotransmission in AMPA receptor GluR2-deficient mice. Pain 136: 158–167

Yu XM, Askalan R, Keil II GJ, et al. (1997) NMDA channel regulation by channel-associated protein tyrosine kinase Src. Science 275; 674–678

Zhang HM, Qi YJ, Xiang XY, et al. (2001) Time-dependent plasticity of synaptic transmission produced by long-term potentiation of C-fiber evoked field potentials in rat spinal dorsal horn. Neurosci Lett 315: 81–84

Zhang XC, Zhang YQ, Zhao ZQ, et al. (2005) Involvement of nitric oxide in long-term potentiation of spinal nociceptive responses in rats. Neuroreport 16: 1197–1201

Zheng X, Zhang L, Wang AP, et al. (1999) Protein kinase C potentiation of N-methyl-D-aspartate receptor activity is not mediated by phosphorylation of N-methyl-D-aspartate receptor subunit. Proc Natl Acad Sci USA 96: 15262–15267

Zhong J, Randić M (1997) Long-term potentiation in the spinal cord dorsal horn is blocked by tyrosine kinase inhibitor. Neurosci Abstr 23: 98

Zhong J, Gerber G, Kojić L, et al. (2000) Dual modulation of excitatory synaptic transmission by agonists of group I metabotropic glutamate receptors in the rat spinal dorsal horn. Brain Res 887: 359–377

Zou X, Lin Q, Willis WD (2000) Enhanced phosphorylation of NMDA receptor 1 subunits in spinal cord dorsal horn and spinothalamic tract neurons after intradermal injection of capsaicin in rats. J Neurosci 20: 6989–6997

Chapter 11
Windup in the Spinal Cord

Stephen W.N. Thompson

Abstract One of the most fundamental features of nociception is the sensitization of neuronal responsiveness that manifests as an increase in action potential activity to repeated nociceptive input. This sensitization can occur at all levels of the neuraxis including the primary afferent nociceptor, neurons within the dorsal and ventral horn of the spinal cord as well as supraspinal neurons. Because of its overall importance as a model for synaptic plasticity, sensitization of response of post-synaptic neurons has been widely studied. In the context of nociception, sensitization of the response of post-synaptic neurons in the mammalian spinal cord has received intensive attention. Sensitization of neuronal responsiveness takes many forms, one of the simplest that can be observed in both dorsal horn interneurons and ventral horn motoneurons is known as 'windup' (WU). Windup refers to the incrementing discharge of action potentials that may be recorded from post-synaptic dorsal or ventral horn neurons in response to the application of repetitive, low frequency, short duration (electrical) stimuli applied to an afferent input. In this chapter, I will describe the windup phenomenon and define its relationship with central sensitization.

Abbreviations

BAPTA	1,2-bis(2-aminophenoxy)ethane-N,N,N-tetraacetic acid
BDNF	brain derived neurotrophic factor
CGRP	calcitonin gene related peptide
DCVs	dense core vesicles
EPSPs	excitatory post synaptic potentials
NK1, NK2, NK3	neurokinin-1,2,3
NMDA	N-methyl-D-aspartate

S.W.N. Thompson (✉)
Biomedical Science, University of Plymouth, Plymouth, PL4 8AA, UK
e-mail: stephen.thompson@plymouth.ac.uk

M. Malcangio (ed.), *Synaptic Plasticity in Pain*,
DOI 10.1007/978-1-4419-0226-9_11, © Springer Science+Business Media, LLC 2009

NT4/5	neurotrophin 4/5
PKC	protein kinase C
TrkB	tropomyosin receptor kinase B
WU	windup

11.1 Introduction

In the mid nineteen sixties, Lorne Mendell and Patrick Wall, using extracellular recordings from the axons of cat spinocervical cells first described the 'windup' (WU) phenomenon as the incrementing response to repeated low frequency stimulation of an afferent modality which in the intact animal evoke sensations of pain, i.e. C-fibre nociceptors (Mendell and Wall, 1965; Mendell, 1966). They described three of most fundamental properties of windup. First, it was a postsynaptic phenomenon. Second, that in intact, uninjured animals it required the activation of unmyelinated afferent fibres (C-fibres) and third, that it could occur at low stimulation frequencies (0.3–2.0 Hz). Subsequently windup has been observed directly from dorsal horn neurons (Wagman and Price, 1969; Schouenborg and Sjolund, 1983; Zhang et al., 1991) from locus coeruleus, trigeminal and thalamic neurons (Hirata and Aston-Jones, 1994; Chung et al., 1979; Kawakita et al., 1993). It has been observed, either directly or indirectly in rats, cats, monkeys, humans, turtles and snails. The parameters of electrical stimulation and the frequency dependency of windup have been well described in previous reviews (Baranauskas and Nistri, 1998; Herrero et al., 2000).

To summarize, in the intact state, windup is threshold dependant; it is only observed following activation of afferent C-fibres. Windup is also frequency dependant; it is only triggered at frequencies of activation of afferent C-fibres above 0.3 Hz. It is maximal at 1–2 Hz and above frequencies of 20 Hz habituation, or wind-down of the response is seen (Schouenborg, 1984). It is important at this point to make the distinction between the parameters that induce windup under normal conditions and those parameters that are required following peripheral injury. Thus in situations of spinal cord hyperexcitability, windup may be induced by lower frequencies of stimulation and by different modalities of afferent activation (Thompson et al., 1994).

A further important distinction that must be made is that between those parameters of afferent activity that *induce* windup and those parameters that *maintain* windup. This was shown by Li and colleagues using low frequency repetitive electrical stimulation of cutaneous C-fibres (0.5 Hz) to evoke WU in dorsal horn WDR neurons of rats (Li et al., 1999). These investigators showed that once WU occurred, the enhanced responses of these WDR neurons could be maintained for long periods (100 s) using peripheral electrical stimuli delivered at much lower frequencies that by themselves do not evoke windup

(0.1 Hz). This phenomenon has been referred to as 'windup-maintenance' (WU-M) (Staud et al., 2004) and will be discussed in context of the relationship between windup to central sensitization.

11.2 Windup and Central Sensitization

Central sensitization refers to the increased excitability of neurons within the spinal cord that is induced following a peripheral damaging stimulus or brief electrical activation of nociceptors (Woolf, 1983). In terms of animal behaviour this is manifest as a heightened sensitivity of the area of injury, a spread of hypersensitivity to regions of uninjured tissue (secondary hypersensitivity) and the generation of pain by low threshold afferent fibres (A-beta) that do not normally evoke pain (allodynia) (Torebjork et al., 1992). At the cellular level this directly correlates to changes in the receptive field properties of individual dorsal horn neurons. Hence following brief electrical activation of nociceptors or peripheral injury, spinal neurons display reduced electrical thresholds for activation, an increase in responsiveness to their adequate stimulus, an increase in the size of their receptive fields and the recruitment of novels inputs (Cook et al., 1987). Because those stimuli, capable of producing central sensitization were also those capable of inducing windup (Wall and Woolf, 1986; Woolf and Wall, 1986a) and because of their similar sensitivity to receptor blockers (Woolf and Thompson, 1991; Woolf and Wall, 1986b; Xu et al., 1991), a causal relationship between the two was proposed. Central to understanding the relationship between windup and central sensitization was the ability to analyse, in careful detail, the synaptic potentials generated by A and C-fibre afferents in dorsal horn neurons in the in vitro isolated rat spinal cord preparation (King and Thompson, 1989; Sivilotti et al., 1993; Thompson et al., 1990, 1992, 1993; Woolf et al., 1988). Under normal circumstances, C-fibres generate very long duration excitatory postsynaptic potentials, A fibres do not. These C-fibre-evoked synaptic events may last up to 20 s following a single synchronous C-fibre input (Thompson et al., 1990). Low frequency repetitive activation of C-fibre input induces a temporal summation of these very long duration EPSPs that results in a cumulative depolarisation of the membrane potential of the postsynaptic neuron (Thompson et al., 1990). The high quality intracellular recordings available using this preparation showed for the first time the subthreshold membrane potential changes underlying the accelerating action potential discharge of windup. With intracellular recording it could be seen that following repetitive high threshold afferent stimulation, some neurons displayed postsynaptic cumulative depolarisation associated with action potential windup. Other neurons showed significant cumulative depolarisation without action potential windup and yet others had neither cumulative depolarisation nor action potential windup (Thompson et al., 1993). In a further detailed analysis of properties of the cumulative depolarisation required for windup, it could be also shown that some neurons in which a cumulative depolarisation could not be evoked by repetitive C-fibre activity at resting membrane potential, did so in response to the same stimulus when depolarised by a few millivolts by intracellular current injection (Sivilotti et al., 1993). This study also showed is that it is not the absolute amplitude of the cumulative depolarisation which determines the ability of a neuron to display windup, rather the rate at which the cumulative depolarisation increases between consecutive stimuli (Sivilotti et al., 1993). Other studies have further analysed the relationship between cumulative depolarisation, action potential windup and central sensitization. In an in vitro spinal cord preparation, repetitive stimulation of C-fibre afferents (the conditioning stimuli), heterosynaptically enhances post-synaptic responses to low-intensity stimulation of unconditioned inputs (the test stimuli) (Thompson et al., 1993). This heterosynaptic facilitation has several features in common with central sensitization of spinal neurons that contribute to postinjury pain hypersensitivity. This study analysed the relationship between the size of the cumulative depolarisation following C-fibre conditioning stimuli, the presence or absence of homosynaptic facilitation (windup) induced by the conditioning stimulus and the degree of heterosynaptic facilitation induced in response to delivery of a test stimulus to the same neuron. Results showed that a spinal cord neuron was just as likely to display heterosynaptic facilitation to a test stimulus in the absence of homosynaptic facilitation (action potential windup) than if the conditioning stimulus did produce homosynaptic action potential windup. The crucial factor however was the underlying cumulative depolarisation. A significant difference was observed therefore between the amplitude of the cumulative depolarisation evoked by the conditioning stimulus in those neurons that did show heterosynaptic windup (9.1±3.1 mV) compared to those neurons that did not show heterosynaptic facilitation (3.3±0.5 mV).

It is apparent therefore that whilst action potential windup discharge may be indicative of a neuron that can show central sensitization, it is intuitive that many more neurons, whilst not demonstrating action potential windup following repetitive activity, have the *potential* to show altered postsynaptic responses to further stimuli because of persistent changes in postsynaptic membrane potential.

Hence, a fundamental dissociation between action potential windup and central sensitization. Yes, because the synaptic processes that produce windup and central sensitization are similar, it is correct to infer the likelihood that windup may lead to central sensitization. The absence however of action potential windup does not mean that central sensitization cannot occur. Instead there is a reservoir of potential sensitization determined by previous synaptic inputs.

11.2.1 *Pharmacology of Windup: Glutamate and Neuropeptides*

What is the nature of the synaptic history that allows heterosynaptic facilitation to manifest itself? Since it is considered that cellular events leading to action potential windup also produce some of the classical features of central sensitization it is useful to examine those cellular mechanisms that link short term changes in synaptic excitability to longer term changes in spinal cord excitability. Since the discovery in 1987 by two different groups of the sensitivity of action potential windup to blockage of the N-methyl-D-aspartate (NMDA) subtype of glutamate receptors (Davies and Lodge, 1987; Dickenson and Sullivan, 1987), the pharmacology of windup has been intensively studied (see Herrero et al., 2000 for review).

Fundamental to activity-dependant synaptic plasticity in the mammalian nervous system is the activation of the NMDA class of glutamate receptor. The original model explaining the involvement of NMDA receptors in the generation of action potential windup in the spinal cord depended upon the blockade of the NMDA receptor at normal resting membrane potentials by magnesium ions and the voltage dependant relief of this block at more depolarised levels. The model suggested that since the very long EPSPs generated by C-fibre stimulation had both glutamate and tachykinin receptor-mediated components (Urban et al., 1994), temporal summation of these EPSPs generated the depolarisation required for relief of the NMDA receptor block permitting flow of Na^+ and Ca^{2+} ions into the cell in a positive feed-forward mechanism. NK1 and NK3 but not the NK2 class of tachykinin receptor are thought to underlie slow substance P and neurokinin-B evoked slow potentials respectively (see Herrero et al., 2000; Barbieri and Nistri, 2001). In addition to this direct biophysical activation of the NMDA receptor, a secondary amplification mechanism depends upon subsequent entry of Ca^{2+} ions via the NMDA ionophore (Chen and Huang, 1992). The original model suggested that the increase in intracellular Ca^{2+} would activate protein kinase C (PKC) which would in turn directly phosphorylate serine/threonine residues of the NMDA receptor which in turn would generate a greater current through these receptors at resting membrane potentials (Chen and Huang, 1992). More recent evidence has pointed to a tyrosine phosphorylation-dependent enhancement of NMDA receptor activity via pathways involving PKC, Src, CAKβ/Pyk2 and Fyn (Köhr and Seeburg, 1996; Suzuki and Okumura-Noji, 1995; Wang and Salter, 1994; Yu et al., 1997). Recently a predominant role of NR2B subunit containing NMDA receptors has been ascribed in windup formation in wide dynamic range neurons in the rat spinal cord (Kovacs et al., 2004).

11.2.2 Pharmacology of Windup: Neurotrophins

In addition to glutamate and neuropeptides contained within C-fibre primary afferent terminals in the dorsal horn of the spinal cord, several other neuromodulators are known to be expressed in the spinal dorsal horn, are associated with primary afferent terminals and are known to be upregulated and released following peripheral injury or primary afferent fibre activation. The neurotrophin, brain-derived neurotrophic factor (BDNF) is present in a subpopulation of small diameter, presumed nociceptive, cell bodies within rat DRG and central terminals within the spinal cord (Mannion et al., 1999; Michael et al., 1997). Initial ultrastructural studies localized BDNF to dense core vesicles (DCVs) (Michael et al., 1997) where it co-localises with substance P and calcitonin gene related peptide (CGRP) in central terminals of nociceptive afferents. Recent work has confirmed this using high resolution ultrastructural immunocytochemistry and demonstrated that BDNF is solely packaged in DCVs in both central and peripheral neurons (Salio et al., 2007). BDNF also co-localizes with the synaptic marker synaptotagmin (Fawcett et al., 1997). BDNF is released into the dorsal horn following electrical activation of C-fibres or capsaicin activation of these fibres (Lever et al., 2001). BDNF activity is mediated by its high-affinity receptor, the tropomyosin receptor kinase B (trkB), which also recognizes NT-4/5 (Kaplan and Miller, 1997; Kaplan and Stephens, 1994). In the adult CNS, a full-length trkB (fl-trkB) receptor and two truncated receptor forms (tr-trkB) are found generated by alternative splicing of the trkB mRNA (Barbacid, 1994; Klein et al., 1990; Middlemas et al., 1991). A recent ultrastructural study has described fl-trkB localization at synapses between first and second order sensory neurons in spinal lamina II and has shown both pre- and post-synaptic distribution of fl-trkB (Salio et al., 2005) (see Chapter 5). Because of the considerable evidence indicating activation of the BDNF/trkB system in various injury models (Kerr et al., 999; Mannion et al., 1999; Lever et al., 2003; Pezet et al., 2002a,b; Slack et al., 2004, 2005), BDNF has been suggested to function as a neuromodulator of synaptic transmission (Michael et al., 1997; Kerr et al., 1999; Thompson et al., 1999; Mannion et al., 1999). The crucial question with regard to a potential role of BDNF in windup, is the time course over which BDNF exerts its effects and the mechanism(s) by which it does so. In several areas of the CNS, including the spinal cord, hippocampus, dentate gyrus, visual cortex and insular cortex, BDNF can trigger a long-lasting increase in synaptic efficacy (BDNF-LTP) (Escobar et al., 2003; Kang and Schuman, 1995; Messaoudi et al., 2002; Zhou et al., 2008). The time course of this persistent increase in synaptic efficacy (minutes/hours), is far greater than that associated with spinal cord windup (seconds/minutes). In addition, the characteristics of the afferent barrage required to induce BDNF-LTP may not be related to the type of physiological sensory input observed in vivo, even under conditions of inflammation. Furthermore, the cellular signalling mechanisms known to underpin these prolonged changes (BDNF-induced protein synthesis) may not be consistent with short-term increases in synaptic efficacy. What then is the evidence that BDNF facilitates synaptic transmission from primary afferent neurons to spinal cord neurons within a time frame consistent with the generation of windup? One of the first indications was evidence that acute, electrically-evoked C-fibre responses in an in vitro spinal cord preparation could be significantly enhanced by superfusion of the spinal cord with BDNF (Kerr et al., 1999). In this study, A-fibre evoked responses were not affected by BDNF and the facilitation was reversed by the BDNF scavenging antibody trkB-IgG, all-be-it under conditions in which levels of endogenous BDNF were enhanced. Furthermore, these authors also demonstrated a direct enhancement of NMDA evoked synaptic responses by BDNF suggesting a mechanism of BDNF facilitation via NMDA receptor facilitation. The role of endogenous BDNF was subsequently confirmed, also utilising an in vitro approach, using BDNF-deficient neonatal mice (Heppenstall and Lewin, 2001). These mice exhibited reduced ventral root potentials (VRPs) compared to controls. Importantly this study showed that windup was also decreased in these mice. Detailed analysis of the effect of neurotrophins upon acute synaptic potentials in the mammalian spinal cord has revealed complex interactions, dependent upon age, stimulation intensity and afferent source (Arvanian and Mendell, 2001; Garraway et al., 2003). Up to the first postnatal week, brief exposure to BDNF produces short-lasting post-synaptic facilitation followed by long-lasting presynaptic inhibition of dorsal root evoked EPSPs in motoneurons (Arvanian and Mendell, 2001). In contrast, synaptic currents evoked in lamina II dorsal horn neurons by dorsal root stimulation and recorded from animals of a similar age, revealed a significant facilitation by BDNF (Garraway et al., 2003). Whilst these synaptic recordings were produced at repetitive stimulation rates (0.017 Hz) well below those that result in windup, significant among the findings were first that the synaptic facilitation required functional postsynaptic NMDA receptors and second, experiments using the intracellular calcium chelator BAPTA indicated that increased intracellular calcium was necessary to the facilitatory effect of BDNF. It thus appears that under normal conditions at low rates of afferent stimulation frequency, neurotrophins potentially contribute a neuromodulatory influence upon long duration postsynaptic events in the spinal dorsal horn.

11.2.3 Pharmacology of Windup: Non-Synaptic Component

In addition to glutamatergic, neurokinin and neurotrophin contributions and interactions to the generation of windup, other studies have shown that a further non-synaptic component may be involved. Blockade of windup by NMDA and tachykinin receptor antagonists is only partial (Thompson et al., 1990, 1993; Davies and Lodge, 1987). The possibility that intrinsic properties of the postsynaptic membrane underlie some of the characteristics of windup have been addressed in both mammals and lower vertebrates. Neurons in the spinal dorsal horn of turtles support action potential windup in response to low frequency, high intensity dorsal root stimulation (Russo and Hounsgaard, 1994, 1996). Furthermore, this action potential windup may be evoked without the need for presynaptic stimulation but rather following intracellular stimulation with short duration repetitive depolarising pulses (Russo and Hounsgaard, 1994). This depolarisation-induced action potential windup is inhibited by the L-type calcium channel blocker nifedipine and facilitated by the L-type calcium channel opener BayK. Early studies in the mammalian dorsal horn, have shown low and high threshold activated calcium currents in dorsal horn neurons (Murase and Randić, 1983; Ryu and Randic, 1990). Further analysis demonstrated that mammalian dorsal horn neurons display regenerative plateau potentials that are generated by depolarising current pulses which produce a gradual increase in action potential discharge that are sensitive to dihydropyridines or enhanced when Ca^{2+} was substituted with Ba^{2+} (Morisset and Nagy, 1998, 1999). The question is whether or not these calcium-dependant plateau potentials are involved in mammalian windup or not? One report using extracellular recording of the windup of ventral root potentials following C-fibre stimulation did not show any effects of nifedipine upon windup (Herrero et al., 2000). In contrast, studies using intracellular recordings from rat dorsal horn neurons have shown that in the presence of NMDA receptor blockers there is an absolute correlation between windup and the production of calcium-sensitive plateau potentials (Morisset and Nagy, 1999, 2000). Furthermore it was shown that a homosynaptic sensitization of dorsal horn neurons, similar to that evoked by synaptically driven windup could be produced by repetitive intracellular stimulation with depolarising current pulses (Morisset and Nagy, 2000). Windup therefore is strongly associated with calcium influx into spinal neurons, either via the NMDA ionophore itself, voltage dependant calcium channels or release from intracellular stores.

Being strongly associated with calcium influx therefore the cumulative depolarisation of windup is a strong candidate to be a link between short term homosynaptic facilitation and long term central sensitization.

11.3 Concluding Remarks

Under normal circumstances the cumulative depolarisation evoked by 0.5–2.0 Hz C-fibre stimulation outlasts the period of afferent activity for a maximum of 10–20 s (Thompson et al., 1990). The duration of action potential windup associated with this cumulative depolarisation is variable but normally does not last longer than the period of cumulative depolarisation (Thompson et al., 1990). Because of the relatively brief time course of events of the C-fibre evoked cumulative depolarisation and the associated action potential windup, the phenomenon of windup cannot be directly equated to the longer duration events of central sensitization and hyperalgesia. However as has been mentioned, stimuli that lead to windup also lead to several of the characteristics of central sensitization and that windup may therefore be an early initiator of central sensitization (Cook et al., 1987). Electrophysiological recordings of extracellular responses of single dorsal horn neurons has shown that windup-inducing stimuli applied to the receptive field resulted in expansion of receptive fields (Li et al., 1999). Receptive field expansion, along with a decrease in response threshold, increased response to suprathreshold stimuli and the recruitment of previously ineffective inputs are electrophysiological characteristics of central sensitization (Cook et al.,

1987; Woolf and King, 1990) and correlates of secondary hypersensitivity, hyperalgesia, and allodynia respectively. Whilst the study of Li et al., showed that windup was insufficient to produce sensitization to A-fibre input, it does show that the stimuli that produce windup can also produce some of the characteristics of more prolonged forms of central plasticity, namely central sensitization. A further extremely interesting feature of this study was that, once windup occurred, the enhanced responses of these neurons could be maintained for very long periods using further peripheral conditioning stimuli delivered at very low frequencies which by themselves were incapable of producing windup (Li et al., 1999) This has been referred to as windup-maintenance (WU-M) (Staud et al., 2004). This is an important finding, since it shows that WU-M may reflect mechanisms related to the early phase of central sensitization. Crucially in these experiments, WU-M in dorsal horn neurons was not related to prolonged action potential after discharges (Li et al., 1999). More likely WU-M relies upon cellular events occurring during and after nociceptive stimulation and result in amplification of previously ineffective synapses via the maintenance of subthreshold cumulative membrane depolarisations. Evidence again comes from the very high quality recordings obtainable from intracellular recordings from dorsal horn neurons. Intracellular recordings from dorsal horn neurons in vivo have shown that the excitatory, action-potential evoking, receptive field of these cells is surrounded by a subliminal region capable of generating only subthreshold responses (Woolf and King, 1987). Injury, inflammation or repetitive electrically evoked C-fibre activity (windup) produce sensitization of the surrounding subliminal area whereby stimulation of the corresponding receptive field area now evokes action potentials instead of subthreshold responses. This change from subthreshold to suprathreshold is a result of an afferent induced cumulative membrane potential that brings neurons to threshold in response to previously subthreshold input. The activation of voltage-dependant plateau potentials and the likelihood that cumulative depolarisations are maintained by subsequent low frequency nociceptive inputs may provide the basis for windup maintenance. It is apparent therefore that there is a continuum of events leading from cumulative summation of slow C-fibre evoked EPSPs to the maintenance of subthreshold changes in dorsal horn neurons to later stages that involve intracellular calcium-dependant signalling cascades that require activation of kinases, ion channel phosphorylation, transcription factor activation and novel gene expression and which form the basis of central sensitization and hyperalgesia (Woolf and Salter, 2000).

References

Arvanian, V.L., Mendell, L.M. (2001) Acute modulation of synaptic transmission to motoneurons by BDNF in the neonatal rat spinal cord. *Eur J Neurosci.* (11):1800–1808.
Baranauskas, G., Nistri, A. (1998) Sensitization of pain pathways in the spinal cord: cellular mechanisms. *Prog Neurobiol.* 54:349–365.

Barbacid, M. (1994) The Trk family of neurotrophin receptors. *J Neurobiol.* (11):1386–1403.

Barbieri, M., Nistri, A (2001) Depression of windup of spinal neurons in the neonatal rat spinal cord in vitro by an NK3 tachykinin receptor antagonist. *J Neurophysiol.* 85(4):1502–1511.

Chen, L., Huang, L.Y.M. (1992) Protein kinase C reduces Mg block of NMDA-receptor channels as a mechanism of modulation. *Nature.* 356:521–523.

Chung, J.M., Kenshalo Jr., D.R., Gerhart, K.D., Willis Jr., W.D. (1979) Excitation of primate spinothalamic neurones by cutaneous C-fibre volleys. *J Neurophysiol.* 42:1354–1369.

Cook, A.J., Woolf, C.J., Wall, P.D., McMahon, S.B. (1987) Dynamic receptive field plasticity in rat spinal dorsal horn following C-primary afferent input. *Nature.* 325:151–153.

Davies, S.N., Lodge, D., (1987) Evidence for the involvement of *N*-methylaspartate receptors in 'wind-up' of class 2 neurones in the dorsal horn of the rat. *Brain Res.* 424:402–406.

Dickenson, A.H., Sullivan, A.F. (1987) Evidence for a role of the NMDA receptor in the frequency dependent potentiation of deep rat dorsal horn nociceptive neurones following C-fibre stimulation. *Neuropharmacol.* 26;1235–1238.

Escobar, M.L., Figueroa-Guzmán, Y., Gómez-Palacio-Schjetnan, A. (2003) In vivo insular cortex LTP induced by brain-derived neurotrophic factor. *Brain Res.* November 21;991(1–2):274–279.

Fawcett, J.P., Aloyz, R., McLean, J.H., Pareek, S., Miller, F.D., McPherson, P.S., Murphy, R.A. (1997) Detection of brain-derived neurotrophic factor in a vesicular fraction of brain synaptosomes. *J Biol Chem.* April 4;272(14):8837–8840.

Garraway, S.M., Petruska, J.C., Mendell, L.M. (2003) BDNF sensitizes the response of lamina II neurons to high threshold primary afferent inputs. *Eur J Neurosci.* (9):2467–2476.

Heppenstall, P.A., Lewin, G.R. (2001) BDNF but not NT-4 is required for normal flexion reflex plasticity and function. *Proc Natl Acad Sci USA.* 98(14):8107–8112.

Herrero, J.F., Laird, J.M.A., Lopez-Garcia, J.A. (2000) Wind-up of spinal cord neurones and pain sensation: much ado about something? *Prog Neurobiol.* 61:169–203.

Hirata, H., Aston-Jones, G. (1994) A novel long-latency response of locus coeruleus neurons to noxious stimuli: mediation by peripheral C-fibers. *J Neurophysiol.* 71:1752–1761.

Kang, H., Schuman, E.M. (1995) Long-lasting neurotrophin-induced enhancement of synaptic transmission in the adult hippocampus. *Science.* 267(5204):1658–1662.

Kaplan, D.R., Miller, F.D. (1997) Signal transduction by the neurotrophin receptors. *Curr Opin Cell Biol.* (2):213–221.

Kaplan, D.R., Stephens, R.M. (1994) Neurotrophin signal transduction by the Trk receptor. *J Neurobiol.* (11):1404–1417.

Kawakita, K., Dostrovsky, J.O., Tang, J.S., Chiang, C.Y. (1993) Responses of neurons on the rat thalamic submedius to cutaneous, muscle and visceral nociceptive stimuli. *Pain.* 55(3):327–338.

Kerr, B., Trivedi, P.M., Dassan, P., Bradbury, E., French, J., Shelton, D.B., Bennett, D.B.M, McMahon, S.B., Thompson, S.W.N. (1999) Brain derived neurotrophic factor modulates nociceptive sensory inputs and NMDA-evoked responses in the rat spinal cord. *J Neurosci.* 19(12):5138–5148.

King, A.E., Thompson, S.W.N. (1989) Characterization of deep dorsal horn neurones in the rat spinal cord in vitro; synaptic and excitatory amino acid induced excitations. *Comp Biochem Physiol.* 93:171–176.

Klein, R., Conway, D., Parada, L.F., Barbacid, M. (1990) The trkB tyrosine protein kinase gene codes for a second neurogenic receptor that lacks the catalytic kinase domain. *Cell.* 61(4):647–656.

Köhr, G., Seeburg, P.H. (1996) Subtype-specific regulation of recombinant NMDA receptor-channels by protein tyrosine kinases of the Src family. *J Physiol.* 492:445–452.

Kovacs, G., Kocsis, P., Tarnawa, I., Horvath, C., Szombathelyi, Z., Farkas, S. (2004) NR2B containing NMDA receptor dependent windup of single spinal neurons *Neuropharmacology.* 46:23–30.

Lever, I.J., Bradbury, E.J., Cunningham, J.R., Adelson, D.W., Jones, M.G., McMahon, S.B., Marvizón, J.C., Malcangio, M. (2001) Brain-derived neurotrophic factor is released in the dorsal horn by distinctive patterns of afferent fiber stimulation. *J Neurosci.* June 15;21(12):4469–4477.

Lever, I., Cunningham, J., Grist, J., Yip, P.K., Malcangio, M. (2003) Release of BDNF and GABA in the dorsal horn of neuropathic rats. *Eur J Neurosci.* September; 18(5):1169–1174.

Li, J., Simone, D.A., Larson, A.A. (1999) Windup leads to characteristics of central sensitisation. *Pain.* 79:75–82.

Mannion, R.J., Costigan, M., Decosterd, I., Amaya, F., Ma, Q.P., Holstege, J.C., Ji, R.R., Acheson, A., Lindsay, R.M., Wilkinson, G.A., Woolf, C.J. (1999) Neurotrophins: peripherally and centrally acting modulators of tactile stimulus-induced inflammatory pain hypersensitivity. *Proc Natl Acad Sci USA.* 96(16):9385–9390.

Mendell, L.M. (1966) Physiological properties of unmyelinated fiber projection to the spinal cord. *Exp Neurol.* 16(3):316–332.

Mendell, L.M., Wall, P.D. (1965) Response of single dorsal horn cells to peripheral cutaneous unmyelynated fibres. *Nature.* 206:97–99.

Messaoudi, E., Ying, S.W., Kanhema, T., Croll, S.D., Bramham, C.R. (2002) Brain-derived neurotrophic factor triggers transcription-dependent, late phase long-term potentiation in vivo. *J Neurosci.* 22(17):7453–7461.

Michael, G.J., Averill, S., Nitkunan, A., Rattray, M., Bennett, D.L., Yan, Q., Priestley, J.V. (1997) Nerve growth factor treatment increases brain-derived neurotrophic factor selectively in TrkA-expressing dorsal root ganglion cells and in their central terminations within the spinal cord. *J Neurosci.* 17(21):8476–8490.

Middlemas, D.S., Lindberg, R.A., Hunter, T. (1991) trkB, a neural receptor protein-tyrosine kinase: evidence for a full-length and two truncated receptors. *Mol Cell Biol.* (1):143–153.

Morisset, V., Nagy, F. (1998) Nociceptive integration in the rat spinal cord: role of non-linear membrane properties of deep dorsal horn neurons. *Eur J Neurosci.* (12):3642–3652.

Morisset, V., Nagy, F. (1999) Ionic basis for plateau potentials in deep dorsal horn neurons of the rat spinal cord. *J Neurosci.* 19(17):7309–7316.

Morisset, V., Nagy, F. (2000) Plateau potential-dependent windup of the response to primary afferent stimuli in rat dorsal horn neurons. *Eur J Neurosci.* (9):3087–3095.

Murase, K., Randic, M. (1983) Electrophysiological properties of rat spinal dorsal horn neurones in vitro: calcium-dependent action potentials. *J Physiol.* January; 334:141–153.

Pezet, S., Cunningham, J., Patel, J., Grist, J., Gavazzi, I., Lever, I.J., Malcangio, M. (2002b) BDNF modulates sensory neuron synaptic activity by a facilitation of GABA transmission in the dorsal horn. *Mol Cell Neurosci.* (1):51–62.

Pezet, S., Lever, I.J., Malcangio, M., Perkinton, M., Thompson, S.W.N., Williams, R.J., McMahon, S.B. (2002a) Noxious stimulation induces TrkB receptor and down-stream ERK phosphorylation in the spinal dorsal horn. *Mol Cell Neurosci.* 21(4):684–695.

Russo, R.E., Hounsgaard, J. (1994) Short-term plasticity in turtle dorsal horn neurons mediated by L-type Ca^{2+} channels. *Neuroscience.* 61:191–197.

Russo, R.E., Hounsgaard, J. (1996) Plateau-generating neurones in the dorsal horn in an in vitro preparation of the turtle spinal cord. *J Physiol.* 493:39–54.

Ryu, P.D., Randic, M. (1990) Low- and high-voltage-activated calcium currents in rat spinal dorsal horn neurons. *J Neurophysiol.* 63(2):273–285.

Salio, C., Averill, S., Priestley, J.V., Merighi, A. (2007) Costorage of BDNF and neuropeptides within individual dense-core vesicles in central and peripheral neurons. *Dev Neurobiol.* 67(3):326–338.

Salio, C., Lossi, L., Ferrini, F., Merighi, A. (2005) Ultrastructural evidence for a pre- and postsynaptic localization of full-length trkB receptors in substantia gelatinosa (lamina II) of rat and mouse spinal cord. *Eur J Neurosci.* (8):1951–1966.

Schouenborg, J. (1984) Functional and topographical properties of field potentials evoked in rat dorsal horn by cutaneous C-fiber stimulation. *J Physiol.* 356:169–192.

Schouenborg, J., Sjolund, B.H. (1983) Activity evoked by A- and C-afferent fibers in rat dorsal horn neurons and its relation to a fexion refex. *J Neurophysiol.* 50:1108–1121.

Sivilotti, L.G., Thompson, S.W.N., Woolf, C.J. (1993) Rate of rise of the cumulative depolarization evoked by repetitive stimulation of small-caliber afferents is a predictor of action potential windup in rat spinal neurons in vitro. *J Neurophysiol.* 69:1621–1631.

Slack, S.E., Grist, J., Mac, Q., McMahon, S.B., Pezet, S. (2005) TrkB expression and phospho-ERK activation by brain-derived neurotrophic factor in rat spinothalamic tract neurons. *J Comp Neurol.* 489(1):59–68.

Slack, S., Pezet, S., McMahon, S.B., Thompson, S.W.N., Malcagnio, M. (2004) Brain-derived neurotrophic factor induces NMDA receptor subunit one phosphorylation via ERK and PKC in the spinal cord. *Eur J Neurosci.* 20:1769–1778.

Staud, R., Price, D.D., Robinson, M.E., Mauderli, A.P., Vierck, C.J. (2004) Maintenance of windup of second pain requires less frequent stimulation in fibromyalgia patients compared to normal controls. *Pain.* 110:689–696.

Suzuki, T., Okumura-Noji, K. (1995) NMDA receptor subunits epsilon 1 (NR2A) and epsilon 2 (NR2B) are substrates for Fyn in the postsynaptic density fraction isolated from the rat brain. *Biochem Biophys Res Commun.* 216(2):582–588.

Thompson, S.W.N., Bennett, D.L.H., Kerr, B.J., Bradbury, E.J., McMahon, S.B. (1999) BDNF is an endogenous modulator of nociceptive responses in the spinal cord. *Proc Natl Acad Sci USA.* 96:7714–7718.

Thompson, S.W.N., Dray, A., Urban, L. (1994) Injury-induced plasticity of spinal reflex activity: NK1 neurokinin receptor activation and enhanced A- and C-fibre mediated responses in rat spinal cord in vitro. *J Neurosci.* 14:3672–3687.

Thompson, S.W.N., Gerber, G., Sivilotti, L.G., Woolf, C.J. (1992) Long duration ventral root potentials in the neonatal rat spinal cord in vitro; the effects of ionotropic and metabotropic excitatory amino acid receptor antagonists. *Brain Res.* 595:87–97.

Thompson, S.W.N., King, A.E., Woolf, C.J. (1990) Activity-dependent changed in rat ventral horn neurones in vitro summation of prolonged afferent evoked postsynaptic depolarizations produce a D-APV sensitive windup. *Eur J Neurosci.* 2:638–649.

Thompson, S.W.N., Woolf, C.J., Sivilotti, L.G. (1993) Small caliber afferent inputs produce a heterosynaptic facilitation of the synaptic responses evoked by primary afferent A-fibers in the neonatal rat spinal cord in vitro. *J Neurophysiol.* 69:2116–2128.

Torebjork, H.E., Lundberg, L.E.R., LaMotte, R.H. (1992) Central changes in processing of mechanoreceptive input in capsaicin-induced secondary hyperalgesia in humans. *J Physiol (Lond).* 448;765–780.

Urban, L., Thompson, S.W.N., Dray, A. (1994) Modulation of spinal excitability: co-operation between neurokinin and excitatory amino acid neurotransmitters. *Trends Neurosci.* 17:432–438.

Wagman, I.H., Price, D.D. (1969) Responses of dorsal horn cells of *M. mulatta* to cutaneous and sural nerve A and C fibre stimuli. *J Neurophysiol.* 32:803–817.

Wall, P.D., Woolf, C.J. (1986) The brief and the prolonged facilitatory effects of unmyelinated afferent input on the rat spinal cord are independently influenced by peripheral nerve section. *Neuroscience.* 17:1199–1205.

Wang, Y.T., Salter, M.W. (1994) Regulation of NMDA receptors by tyrosine kinases and phosphatases. *Nature.* 369(6477):233–235.

Woolf, C.J. (1983) Evidence for a central component of post-injury pain hypersensitivity. *Nature.* 306:686–688.

Woolf, C.J., King, A.E. (1987) Physiology and morphology of multireceptive neurones with C-afferent fibre inputs in the deep dorsal horn of the rat lumbar spinal cord. *J Neurophysiol.* 58:460–479.

Woolf, C.J., King, A.E. (1990) Dynamic alterations in the cutaneous mechanoreceptive fields of dorsal horn neurons in the rat spinal cord. *J Neurosci.* 10(8):2717–2726.

Woolf CJ, Salter MW. (2000) Neuronal plasticity: increasing the gain in pain. *Science.* 288(5472):1765–1769.

Woolf, C.J., Thompson, S.W.N. (1991) The induction and maintainance of central sensitization is dependent on *N*-methyl-D-aspartic acid receptor activation; implications for the treatment of post-injury pain hypersensitivity states. *Pain.* 44:293–299.

Woolf, C.J., Thompson, S.W.N., King, A.E. (1988) Prolonged primary afferent induced alterations in dorsal horn neurones, an intracellular analysis in vivo and in vitro. *J Physiol (Paris).* 83:255–266.

Woolf, C.J., Wall, P.D. (1986a) The relative effectiveness of C primary afferent fibres of different origins in evoking a prolonged facilitation of the fexor reflex in the rat. *J Neurosci.* 6:1433–1443.

Woolf, C.J., Wall, P.D. (1986b) Morphine sensitive and morphine-insensitive actions of C-fibre input on the rat spinal cord. *Neurosci Letts.* 64;221–225.

Xu, X.-J., Maggi, C.A., Wiesenfeld-Hallin, Z. (1991) On the role of NK-2 tachykinin receptors in the mediation of spinal reflex excitability in the rat. *Neuroscience.* 44:483–490.

Yu, X.M., Askalan, R., Keil 2nd, G.J., Salter, M.W. (1997) NMDA channel regulation by channel-associated protein tyrosine kinase Src. *Science.* 275(5300):674–678.

Zhang, D., Owens, C.M., Willis, W.D. (1991) Intracellular study of electrophysiological features of primate spinothalamic tract neurons and their responses to afferent inputs. *J Neurophysiol.* 65:1554–1566.

Zhou, L.J., Zhong, Y., Ren, W.J., Li, Y.Y., Zhang, T., Liu, X.G. (2008) BDNF induces late-phase LTP of C-fiber evoked field potentials in rat spinal dorsal horn. *Exp Neurol.* 212(2):507–514.

Part V

Mechanisms and Targets for Chronic Pain

Chapter 12
Pain from the Arthritic Joint

Hans-Georg Schaible and Andrea Ebersberger

Abstract Nociceptive input from the joint is processed in spinal cord neurons which are either only activated by mechanical stimulation of the joint and other deep tissue, e.g. adjacent muscles, or in neurons which receive convergent inputs from joint, muscles and skin. Neurons with joint input show pronounced hyperexcitability during development of joint inflammation (enhanced responses to mechanical stimulation of the inflamed joint as well as to healthy adjacent deep structures, reduction of mechanical threshold in high threshold neurons and expansion of the receptive field). This state of hyperexcitability is maintained during persistent inflammation. The neurons are under strong control of descending inhibition which increases at least during the acute phase of inflammation. Both the induction of inflammation-induced spinal hyperexcitability and its maintenance are dependent on glutamate, substance P, neurokinin A, and CGRP. Spinal prostaglandin E_2 supports the induction of spinal hyperexcitability. By contrast, spinal prostaglandin D_2 rather attenuates spinal hyperexcitability during established inflammation.

Abbreviations

AMPA	alpha-amino-3-hydroxy-5-methyl-4-isoxazoleproprionic acid
CGRP	calcitonin gene-related peptide
COX	cyclooxygenase
DNIC	diffuse noxious inhibitory control
FCA	Freund's complete adjuvant
IkB	IkappaB protein
IKK	IκB Kinase
K/C	kaolin/carrageenan

H.-G. Schaible (✉)
Department of Physiology, Friedrich-Schiller-University of Jena, Teichgraben 8,
D-07740 Jena, Germany
e-mail: hans-georg.schaible@mti.uni-jena.de

M. Malcangio (ed.), *Synaptic Plasticity in Pain*,
DOI 10.1007/978-1-4419-0226-9_12, © Springer Science+Business Media, LLC 2009

NF-κB   nuclear factor-κB
NMDA   *N*-methyl-D-aspartate
nNOS, eNOS, iNOS   neuronal, endothelial, inducible nitric oxide synthase
NS   nociceptive specific (neuron)
PG   prostaglandin
WDR   wide dynamic range (neuron)

12.1 Pain Sensations in the Joint

The major conscious sensation in deep tissue such as joint and muscle is pain.
In a normal joint pain is most commonly elicited by twisting or hitting the joint.
In awake humans direct stimulation of fibrous structures in the joint with
innocuous mechanical stimuli evoke pressure sensations. Pain is elicited when
noxious mechanical, thermal and chemical stimuli are applied to the fibrous
structures such as ligaments and fibrous cartilage (Kellgren and Samuel, 1950;
Lewis, 1942). No pain is elicited by stimulation of cartilage, and stimulation of
normal synovial tissue rarely evokes pain (Kellgren and Samuel, 1950).

Joint inflammation is characterized by hyperalgesia and persistent pain at
rest which is usually dull and badly localized (Kellgren, 1939; Kellgren and
Samuel, 1950; Lewis, 1938, 1942). In the state of hyperalgesia the application of
noxious stimuli causes stronger pain than normal, and pain is even evoked by
innocuous mechanical stimuli such as movements in the working range and
gentle to moderate pressure. The severity of pain is the result of peripheral (see
Section 12.2.2) and central sensitization (see Section 12.3.4). Pain associated
with degenerative osteoarthritis shows similarities and differences to arthritic
pain. It may increase when the joint is being loaded, but it may also be reduced
during walking. Some patients with osteoarthritis suffer from severe pain during rest at night when the joint is immobile.

12.2 The Nociceptive Input from the Joint

12.2.1 Innervation of Joints

A typical joint nerve contains thickly or thick myelinated Aβ fibres, thinly
or thin myelinated Aδ fibres, and a high proportion ($\sim$ 80%) of unmyelinated C fibres. The latter are either sensory afferents or sympathetic effer-
ents (each $\sim$ 50%). Articular Aβ fibres terminate as corpuscular endings of
the Ruffini-, Golgi- and Pacini-type in fibrous capsule, articular ligaments,
menisci and adjacent periosteum. Articular Aδ and C fibres terminate as
free nerve endings in the fibrous capsule, adipose tissue, ligaments, menisci,
periosteum, and synovial layer. The cartilage is not innervated (Schaible and

Grubb, 1993). A large proportion of joint afferents are peptidergic containing substance P, CGRP, and somatostatin. Neurokinin A, galanin, enkephalins, and neuropeptide Y have also been identified. Neuropeptides influence the inflammatory process in the periphery and modify spinal processing of joint input (see Section 12.4.3).

12.2.2 Response Properties of Joint Afferents and Peripheral Sensitization

The activity of joint afferents has been mainly recorded in nerves supplying cat and rat knee joints and rat ankle joint. Most Aβ fibres are either strongly or weakly activated by innocuous stimuli. In the medial articular nerve of the cat knee more than 50% of Aδ fibres and about 70% of sensory C fibres with detectable receptive fields have been classified as high threshold units whereas the remaining units respond to innocuous stimuli. In addition, the joint nerves contain a large proportion of silent nociceptors most of which are C fibres. They do not respond to innocuous and noxious mechanical stimuli of the normal joint (Schaible and Grubb, 1993).

During development of arthritis the mechanosensitivity in joint afferents can increase. Some low threshold Aβ fibres show transiently increased responses to joint movements in the initial hours of inflammation. They do not develop resting discharges. Many low threshold Aδ and C fibres exhibit increased responses to movements in the working range. Most strikingly, a large proportion of the high threshold afferents are sensitized such that they respond to movements in the working range of the joint. Many units develop ongoing discharges in the resting position (Coggeshall et al., 1983; Grigg et al., 1986; Schaible and Schmidt, 1985, 1988). Furthermore, initially mechanoinsensitive afferents (silent nociceptors) become mechanosensitive and thus are recruited for the encoding of noxious events during inflammation (Grigg et al., 1986; Schaible and Schmidt, 1988). Increased mechanosensitivity is also found during chronic forms of arthritis, suggesting that mechanical sensitization is an important neuronal basis for persistent hyperalgesia of the inflamed joint (Grubb et al., 1991; Guilbaud et al., 1985).

12.2.3 Spinal Termination of Joint Afferents

Articular nerves supplying the knee or elbow joint of rat, cat and monkey enter the spinal cord via several dorsal roots and project to several spinal segments. Thus the afferents influence sensory neurons and reflex pathways in several spinal segments. Staining of whole nerves with horseradish peroxidase has yielded somewhat contradictory results concerning intraspinal termination of the afferents. Some studies show dense projections of joint afferents to lamina I and to deep laminae IV, V, and VI (and VII). However, other studies have described projections mainly into lamina II and III (Schaible and Grubb, 1993).

12.3 Spinal Cord Neurons with Joint Input

12.3.1 Receptive Fields, Thresholds, Response Properties

Neurons with input from the joint are located in the superficial and deep dorsal horns and in particular in cat and monkey also in the ventral horn. This distribution matches the spinal termination of joint afferents. Within the grey matter, neurons with input from nociceptors differ from neurons with input from low threshold proprioreceptors which are involved in processing of sensory information arising from muscle spindles. The latter neurons often show regular and sometimes high frequency discharges that are modulated by the leg position. The following paragraphs will only deal with neurons involved in nociception.

In both cat and rat, mechanonociceptive inputs from the joint are either processed in dorsal horn neurons that respond to mechanical stimulation of both deep tissue and the skin (Fig. 12.1A) or in neurons that respond only to

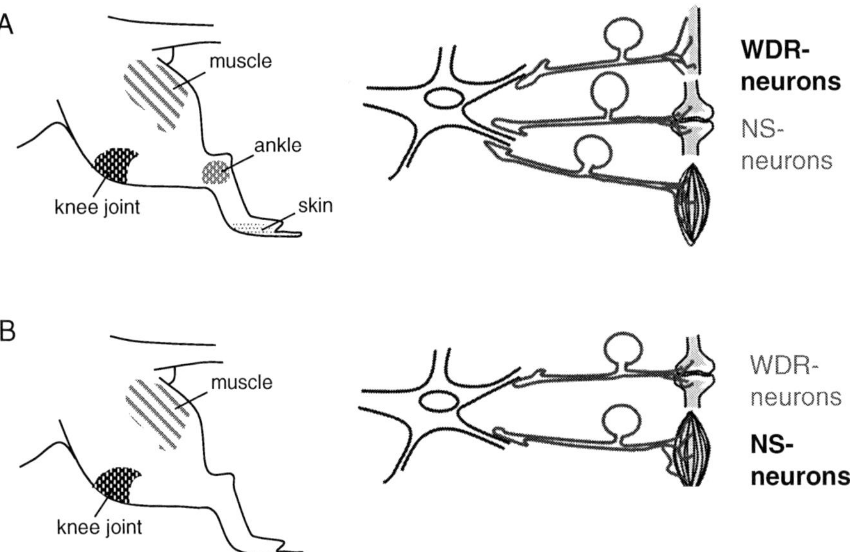

Fig. 12.1 Typical receptive fields of spinal cord neurons with input from the knee joint. (**A**) Neuron with convergent inputs from skin (on the foot), joints (knee and ankle) and muscle. Most neurons of this type are wide dynamic range (WDR) neurons, few are nociceptive-specific (NS) neurons. (**B**) Neuron with input from deep tissue only (knee joint and muscle). Most of these neurons are NS neurons

mechanical stimulation of deep tissue (Fig. 12.1B). As show in Fig. 12.1, receptive fields of single sensory neurons are usually not restricted to one joint but include other deep tissue (another joint, muscles). If present, the cutaneous receptive field is often located more distally than receptive fields in deep tissue (see Fig. 12.1A). Some neurons have bilateral receptive fields (c.f. Schaible and Grubb, 1993).

Concerning mechanical thresholds, neurons are either nociceptive-specific (NS) or wide-dynamic-range (WDR) neurons. Nociceptive-specific neurons respond only to intense pressure and/or to painful movements such as forceful supination and pronation. These stimuli elicit pain. WDR neurons respond to both innocuous pressure and noxious pressure, encoding stimulus intensity by the frequency of action potentials. They may be weakly activated by movements in the working range but show much stronger responses to painful movements. The typical discharge pattern of a WDR neuron is shown in Fig. 12.2A. The typical stimulation sites are displayed in Fig. 12.2B. By and large, NS neurons have smaller receptive fields restricted to deep tissue in joint and muscle

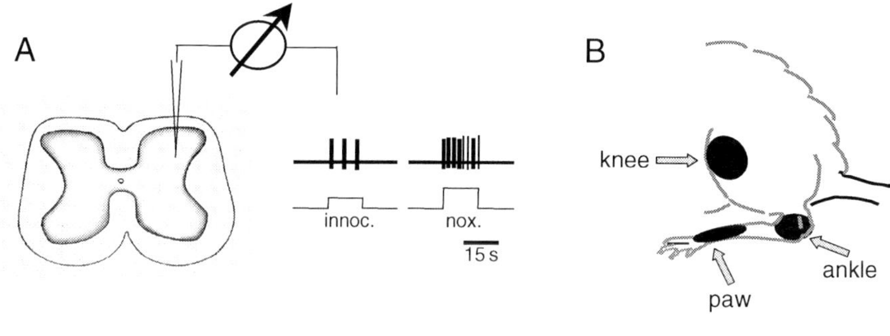

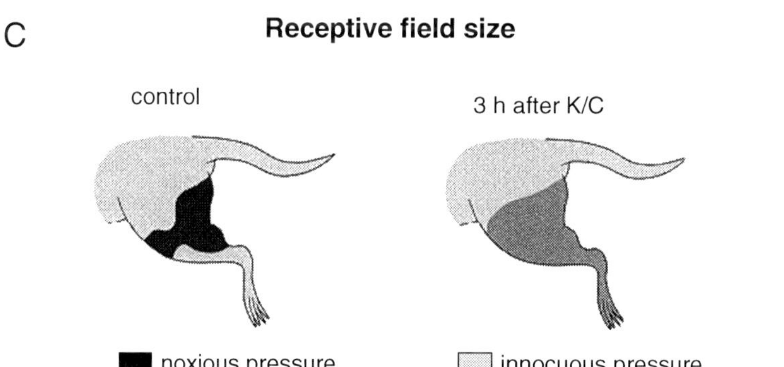

Fig. 12.2 Recordings from spinal cord neurons with joint input. (**A**) Typical response pattern (extracellularly recorded action potentials) of a WDR neuron in situ to innocuous and noxious pressure. (**B**) Typical mechanical stimulation sites during recording from neurons with input from the knee. (**C**) Expansion of the receptive field towards the paw of a NS neuron during development of inflammation. K/C: kaolin/carrageenan injection into the knee joint

(see Fig. 12.1B), and they do not have a receptive field in the skin whereas most neurons with convergent inputs from skin and deep tissue (Fig. 12.1A) are WDR neurons (Schaible, 2006a; Schaible and Grubb, 1993).

12.3.2 Projections of Spinal Cord Neurons with Joint Input

Neurons with input from joints project to different supraspinal sites (cerebellum, spinocervical nucleus, thalamus, reticular formation) or to spinal interneurons and motoneurons (Schaible, 2006a; Schaible and Grubb, 1993). In the cat neurons were identified in the ventral horn which belong to the spinoreticular tract and are predominantly or exclusively excited by noxious deep tissue stimulation (Fields et al., 1977; Meyers and Snow, 1982).

Segmental projections are important for the generation of motor and sympathetic reflexes. Noxious stimulation of the joint evokes nociceptive withdrawal reflexes (Schaible and Grubb, 1993; Woolf and Wall, 1986). During acute chemical stimulation or inflammation of the joint spinal motor reflexes are enhanced (Ferrell et al., 1998; He et al., 1988; Woolf and Wall, 1986). Articular dysfunction and ligamentous strain may cause in fact muscle spasms (Mense, 1997). However, the reflex pattern of some γ-motoneurons changes in the course of inflammation from excitatory to inhibitory (He et al., 1988). This may create a new motoric balance allowing the leg with the inflamed knee joint to be held in a position where the nociceptive input is kept to a minimum.

12.3.3 Inhibition by Descending and Heterotopic Inhibitory Systems

Spinal cord neurons with joint and muscle input are tonically inhibited by descending systems (Cervero et al., 1991; Yu and Mense, 1990). The interruption of descending inhibition lowers the excitation threshold of spinal cord neurons for mechanical stimulation, substantially increases their responses to suprathreshold stimulation and the size of their receptive fields, and causes (increased) ongoing discharges. Descending inhibition seems to be much stronger for the responses to deep input than for the responses to cutaneous input (Yu and Mense, 1990). Neurons with joint input are also inhibited by heterotopic noxious stimuli, thus underlying diffuse noxious inhibitory control (DNIC) (LeBars and Villanueva, 1988).

12.3.4 Inflammation-Evoked Hyperexcitability of Spinal Cord Neurons with Joint Input

Experimental models are used to study neuronal mechanisms underlying inflammatory hyperalgesia. Acute arthritis can be induced by the intraarticular injections of crystals such as urate and kaolin or carrageenan. The injection of kaolin and carrageenan (K/C) into the joint produces edema and granulocytic infiltration within 1–3 h with a plateau after 4–6 h. Animals show limping of the affected leg and enhanced sensitivity to pressure onto the joint. The injection of Freund's complete adjuvant (FCA) into a single joint can cause monoarthritis which lasts 2–4 weeks. In this model limping or guarding of the leg and enhanced sensitivity to pressure onto the joint develop within a day, reach a peak within 3 days and are maintained up to several weeks. The injection of at a high dose of FCA into the tail base or lymph node causes polyarthritis (Schaible and Grubb, 1993). More recently, other models such as collagen-induced polyarthritis (Inglis et al., 2007) and antigen-induced monoarthritis (Boettger et al., 2008; Segond von Banchet et al., 2000) are also being used.

During the development of kaolin/carrageean- (K/C-) induced inflammation in the knee joint, both WDR and NS neurons with knee input develop a state of hyperexcitability towards mechanical stimuli within 1–3 h (Dougherty et al., 1992; Neugebauer and Schaible, 1990; Neugebauer et al., 1993; Schaible et al., 1987). WDR neurons show enhanced responses to innocuous and noxious mechanical stimuli applied to the inflamed joint, NS neurons exhibit stronger responses to noxious mechanical stimuli and a reduction in their mechanical threshold such that the application of innocuous stimuli to the inflamed joint is sufficient to excite the neurons. In addition both types of neurons develop enhanced responses to mechanical stimuli applied to adjacent and remote healthy regions such as muscles of the lower limb or the ankle joint. The total receptive field can expand. Figure 12.2C shows the typical expansion of receptive field of a neuron with knee input to the ankle and the paw and the mechanical threshold which is lowered to the innocuous range.

Increased responses to stimuli applied to the inflamed joint result most likely from the enhanced synaptic input from sensitized joint afferents (see Section 12.2.2). However, enhanced responses to stimulation of ankle and paw must result from a spinal mechanism because these regions are not inflamed. Presumably synapses of afferents from the border of the original receptive field are too weak to activate the neuron under normal conditions but, when the neuron is rendered hyperexcitable, these previously ineffective synapses elicit suprathreshold effects. The increased responses to stimulation of the inflamed area are thought to be the neuronal mechanism of primary hyperalgesia whereas the increased responses to stimuli applied to healthy tissue are thought to underly secondary hyperalgesia adjacent to and remote from inflamed tissue.

The spinal "functional connection" between knee and paw and the change of synaptic effectiveness during inflammation has also been shown in recordings of field potentials in the spinal cord. Electrical stimulation of the posterior articular nerve of the knee joint can evoke field potentials in lumbar spinal segments. The elicited N2 and N3 waves (generated by synaptic activation of dorsal horn neurons by thinly myelinated afferents) gradually increases after induction of a local inflammation in the paw by capsaicin (Rudomin and Hernández, 2008).

Central sensitization persists in rats with chronic unilateral arthritis at the ankle (Grubb et al., 1993). During chronic FCA-induced inflammation in the knee joint, secondary hyperalgesia at the ankle lasts several weeks, and this hypersensitivity is associated with enhanced responses of spinal cord neurons to A and C fibre inputs (Martindale et al., 2007). During chronic polyarthritis spinal cord neurons have more sensitive and expanded cutanoeus receptive fields (Menétrey and Besson, 1982).

Persistent inflammation-evoked spinal activation has also been shown by c-Fos labeling. During acute urate crystal-induced ankle inflammation as well as during chronic FCA-induced paw inflammation (Menétrey et al., 1989) and FCA-induced polyarthritis (Abbadie and Besson, 1992), numerous neurons express c-Fos in lamina I, in the deep dorsal and also in the ventral horn of several segments. During chronic inflammation c-Fos expression remains elevated over weeks until recovery occurs, and, at this stage, labeling of c-Fos is mainly seen in the deep dorsal horn and only marginally in the superficial dorsal horn.

Deep input seems to be particularly able to induce long term changes in the nociceptive system. The stimulation of primary afferents in deep tissue (muscle and joint) evokes more prolonged facilitation of a nociceptive flexor reflex than stimulation of cutaneous afferents (Woolf and Wall, 1986), and capsaicin injection into deep tissue elicits a much more prolonged hyperalgesia than injection of capsaicin into the skin (Sluka, 2002). However, spinal sensitization is dampened by inhibitory influences. Tonic descending inhibition (Cervero et al., 1991; Danziger et al., 1999; Schaible et al., 1991) as well as DNIC (Calvino et al., 1987; Danziger et al., 2001) are increased during acute inflammation but normalized in the chronic stage of inflammation (Danziger et al., 1999, 2001).

Human studies support the concept of central sensitization. In humans, the areas of referred pain can be mapped upon noxious stimulation at restricted sites. When a noxious chemical stimulus, e.g. 6% NaCl, is applied to a muscle, pain is felt in a large area far beyond the stimulation site. Such areas are significantly larger during pathological conditions such as osteoarthritis (Bajaj et al., 2001). The enlargement of painful areas corresponds to the expansion of receptive fields of spinal cord neurons suggesting that the described neuronal changes in the spinal cord are the relevant mechanism of secondary hyperalgesia induced by noxious stimulation of deep tissue (Arendt-Nielsen et al., 2000).

12.4 Molecular Mechanisms of Synaptic Excitation and Spinal Hyperexcitability

12.4.1 General Principles

Transmitters which mediate the synaptic transmission of joint input (excitatory amino acids, excitatory neuropeptides) are also involved in the generation of inflammation-evoked spinal hyperexcitability. Peripheral nociceptive fibres play a key role in triggering the process of spinal sensitization because after sensitization fibres release larger amounts of excitatory amino acids and neuropeptides (see below). In addition, mediators which do not significantly contribute to normal nociception (prostaglandins) come into play. Whether a reduction of inhibition as in neuropathic pain states (c.f. Schaible, 2006b) plays a role is unknown. Furthermore, according to histochemical studies glial cells may be involved in enhanced neuronal excitability during joint inflammation (Inglis et al., 2007; Sun et al., 2007).

12.4.2 Excitatory Amino Acids (Glutamate)

Glutamate is the major transmitter in the synaptic activation of spinal cord neurons with joint input. The ionophoretic application of antagonists at AMPA/kainate (non-NMDA) receptors close to neurons with joint input reduces the responses to both innocuous and noxious pressure applied to the joint whereas the application of NMDA receptor antagonists reduces only the responses to noxious pressure. Thus, in our hands, ionotropic non-NMDA receptors mediate synaptic transmission of both low and high threshold joint afferents whereas NMDA receptors are only activated during noxious stimulation (Neugebauer et al., 1993).

During acute joint inflammation the intraspinal release of glutamate is enhanced (Sluka and Westlund, 1992; Sorkin et al., 1992). The ionophoretic application of antagonists at AMPA/kainate and NMDA receptors to spinal cord neurons as well as systemic application of NMDA receptor antagonists prevents the development of inflammation-evoked spinal hyperexcitability (Neugebauer et al., 1993). Importantly, antagonists at both receptor types reduce neuronal responses also once inflammation is established, even in models of chronic inflammation (Neugebauer et al., 1993, 1994a). Thus ionotropic glutamate receptors play a key role both in the generation and the maintenance of inflammation-evoked spinal hyperexcitability. In addition, metabotropic glutamate receptors contribute to the generation and maintenance of spinal hyperexcitability (Neugebauer et al., 1994b).

NMDA receptors are involved in the regulation of spinal NOS isoforms. During monoarthritis the expression of nNOS, iNOS, and eNOS in the dorsal horn is increased; ketamine reduces nNOS expression and increases iNOS and eNOS expression (Infante et al., 2007). The functional consequences of these changes are yet to be determined.

12.4.3 Neuropeptides

In numerous joint afferents substance P, neurokinin A and CGRP are coexpressed with glutamate. Noxious, but not innocuous compression of the normal joint enhances the intraspinal release of these peptides above baseline. The pattern of release changes during inflammation. During acute arthritis neuropeptides are released by innocuous intensity stimulation of the joint thus creating a change of the neurochemical environment in the spinal cord (Hope et al., 1990; Schaible et al., 1990, 1994). As a further indicator of spinal release of substance P during arthritis, movements of an arthritic joint have been found to induce internalization of the neurokinin 1 receptor (Sharif Naeini et al., 2005). During monoarthritis the expression of substance P and its (neurokinin 1) receptor increase in the superficial dorsal horn (Sharif Naeini et al., 2005).

Excitatory neuropeptides facilitate the responses of spinal cord neurons and they "open"synaptic pathways such that more neurons respond to stimulation (Mense, 1997). A short ionophoretic application of substance P and CGRP to spinal cord neurons can cause reversible increases of ongoing discharge and response to mechanical stimulation. Vice versa, spinal application of antagonists at neurokinin 1, neurokinin 2 and CGRP receptors attenuate the response of spinal cord neurons to noxious pressure applied to the normal joint (Neugebauer et al., 1995, 1996a,b), consistent with the enhanced release of the related peptides upon noxious stimulation.

Ionophoretic application of antagonists at neurokinin 1, neurokinin 2 and CGRP receptors attenuates the development of inflammation-evoked hyperexcitability. An example is shown in Fig. 12.3A (see also Neugebauer, Chapter 8).

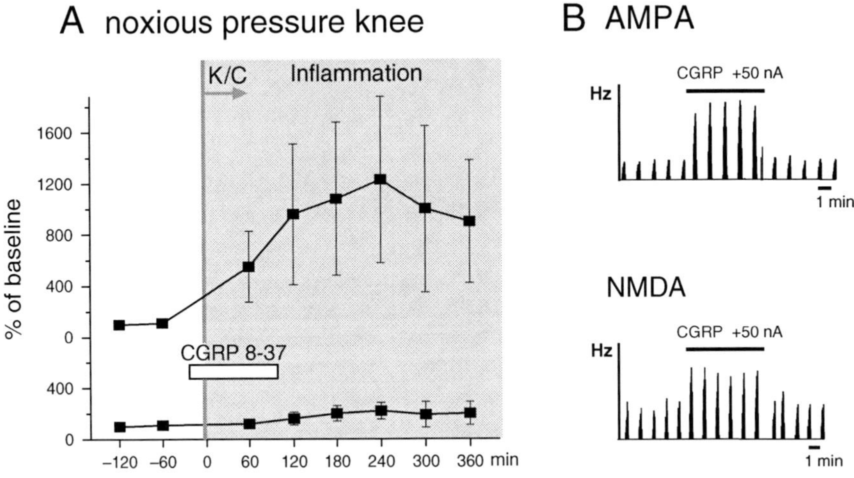

Fig. 12.3 Recordings from spinal cord neurons showing the role of CGRP. (**A**) Attenuation of inflammation-evoked spinal hyperexcitability by the CGRP1 receptor antagonist CGRP8-37. The *upper curve* shows the average increase of the responses to noxious pressure onto the knee in untreated neurons ($n = 13$). The *lower curve* displays the average increase of responses to noxious pressure onto the knee when CGRP8-37 was administered ionophoretically (at 80 nA) to the recorded neurons ($n = 8$). Values show mean $\pm$ SD, and the preinflammatory baseline was set 100%. (**B**) Effect of CGRP on responses of neurons to AMPA and NMDA. AMPA pulses (20 nA, 10 s each) and NMDA pulses (100 nA, 10 s each) were given before, during and after coapplication of CGRP. **A** is reproduced from Neugebauer et al. (1996b), **B** from Ebersberger et al. (2000)

Neurons in control rats show pronounced increase of their responses to noxious pressure onto the knee after injection of kaolin and carrageenan (K/C) into the knee. In another group of neurons the CGRP receptor antagonist CGRP8-37 was ionophoretically applied close to the neurons in the initial phase of inflammation. In these experiments the responses to noxious pressure increased only marginally. The antagonists at these receptors also reduce established hyperexcitability, however, their effect is less pronounced than that of antagonists at glutamate receptors (Neugebauer et al., 1995, 1996a,b). Probably, the activation of these peptide receptors enhances the sensitivity of glutamatergic synaptic transmission (Ebersberger et al., 2000). Ionophoretic application of CGRP increased the responses of the neurons to AMPA and NMDA pulses (Fig. 12.3B).

12.4.4 Spinal Prostaglandins

Spinal prostaglandins (PGs) are synthetized in DRG neurons and in the spinal cord by cyclooxygenases (COX) 1 and 2. PGE_2 receptors are located on primary afferent neurons and on spinal cord neurons indicating that PGs act presynaptically (influencing the release of synaptic mediators) and postsynaptically (influencing excitability) (Vanegas and Schaible, 2001). During inflammation in the joint, release of PGE_2 within the dorsal and ventral horn is significantly enhanced (Ebersberger et al., 1999; Yang et al., 1996). This is likely to result from an upregulation of spinal COX-2 which is already increased at 3 h after induction of knee joint inflammation (Fig. 12.4A).

The application of PGE_2 to the spinal cord surface facilitates the responses of spinal cord neurons to mechanical stimulation of the normal joint, similarly to peripheral inflammation. The application of the COX inhibitor indomethacin to the spinal cord before and during the development of inflammation significantly attenuates the generation of hyperexcitability. Figure 12.4B shows large increase of the responses of control neurons (vehicle on the spinal cord) to noxious pressure onto the knee during development of inflammation, and much smaller increase of responses in neurons of rats in which indomethacin had been applied to the spinal cord. Thus spinal PGs are involved in the generation of inflammation-evoked spinal hyperexcitability (Vasquez et al., 2001). However, spinal application of indomethacin does not reduce enhanced responses of spinal cord neurons to mechanical stimulation of the knee joint when knee inflammation and spinal hyperexcitability are established for several hours; indomethacin decreases the responses of spinal cord neurons only after the subsequent systemic application (Fig. 12.4C) (Vasquez et al., 2001). Because topical application of indomethacin to the spinal cord reduces the release of PGE_2 under these conditions (Fig. 12.4D), we believe that the continuous presence of PGE_2 is not required for the maintenance of inflammation-evoked spinal hyperexcitability.

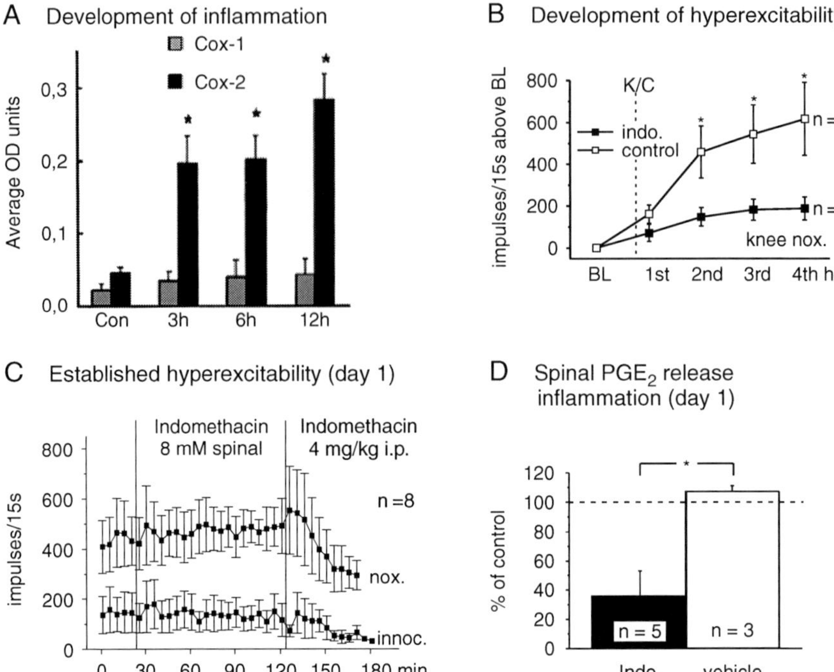

Fig. 12.4 Effect of the cyclooxygenase inhibitor indomethacin on responses of spinal cord neurons with knee input. (**A**) Upregulation of spinal cyclooxygenase-2 during development of K/C-induced inflammation in the knee. (**B**) Hourly increase of responses to noxious pressure onto the knee during development of knee inflammation. *Open squares*: neurons in control experiments, *filled squares*: neurons in rats treated with spinal indomethacin. The preinflammatory baseline (BL) is set to 0. (**C**) Responses of 8 neurons with input from the inflamed knee joint before and after spinal administration of indomethacin and after i.p. application of indomethacin. Each symbol shows averaged responses at intervals of 5 min. (**D**) Spinal release of PGE_2 from the spinal cord of rats which received either vehicle or indomethacin to the spinal cord. Three samples were collected before drug application, three after drug application. **A** is reproduced from Ebersberger et al. (1999), **B** and **C** from Vasquez et al. (2001)

Further support for a differential role of PGE_2 in the generation and maintenance of spinal hyperexcitability is provided by experiments on the spinal effect of EP receptor agonists (Bär et al., 2004) and on the inhibition of the transcription factor NFκB in the spinal cord (Ebersberger et al., 2006). The enhancement of responses of spinal cord neurons to mechanical stimulation of the normal knee joint by spinal PGE_2 is mimicked by the spinal application of agonists at the EP1 receptor (which enhances calcium influx in neurons), and by agonists at the EP2 and EP4 receptors (which activate G_s proteins and adenylylcyclases). However, once inflammation and spinal hyperexcitability are established, only the EP1 receptor agonist can further increase responses to mechanical stimulation of the inflamed knee whereas the EP2 and the EP4 agonists do not influence neuronal responses. On the other hand, spinal application of an agonist at the EP3 receptor (most isoforms are coupled to G_i proteins and reduce cAMP levels) has no influence on neuronal responses when the joint is normal but reduces the responses to mechanical stimulation of the inflamed knee (Bär et al., 2004). Thus, the status of the spinal cord determines which EP receptor agonist causes a spinal effect, and the level of cAMP could be an important molecular factor.

The activation of COX-2 depends on the activation of the transcription factor nuclear factor-κB (NF-κB). In unstimulated tissue NF-κB is bound in the cytoplasma to IκBα and IκBβ which prevent it from entering the nucleus. After stimulation, IκB kinase (IKK) phosphorylates IκB and causes its degradation, thus allowing the unbound NF-κB to enter the nucleus. Hence IKK inhibitors reduce NF-κB-mediated effects (Barnes and Karin, 1997; Chen et al., 2003). Spinal application of a specific IKK inhibitor before and after injection of kaolin and carrageenan into the knee totally prevents spinal hyperexcitability during developing joint inflammation. However, during established inflammation the IKK inhibitor cannot reduce the response to mechanical stimulation of the inflamed knee within 2.5 h after spinal administration (Ebersberger et al., 2006). The pattern of effect of the IKK inhibitor is similar to that of indomethacin (see above), and because NF-κB inhibitors prevent the upregulation of spinal cyclooxygenases (Lee et al., 2004; Tegeder et al., 2004) these data collectively suggest spinal PGE_2 is mainly important for the generation of inflammation-evoked spinal hyperexcitability but not for its maintenance.

The other major prostaglandin in the central nervous system including the spinal cord is PGD_2 (Willingale et al., 1997). Topical application of PGD_2 to the spinal cord at a high dose can causes sensitization of spinal cord neurons for mechanical stimulation of the normal joint similarly to PGE_2. This effect may result from synaptic facilitation due to an increase of the spinal release of substance P and CGRP from primary afferent neurons (Andreeva and Rang, 1993; Jenkins et al., 2001; Nakae et al., 2005). However, during joint inflammation PGD_2 dose-dependently reduces responses of spinal cord neurons to stimulation of the inflamed knee joint. This effect is mimicked by spinal application of the DP1 receptor agonist BW245C and, vice versa, spinal application of an antagonist at the DP1 receptor (BWA868C) increased responses to stimulation of the inflamed knee indicating that endogenous PGD_2 reduces neuronal discharge (Fig. 12.5A). These data indicate that under inflammatory conditions PGD_2 rather counteracts the facilitatory effect of PGE_2 (Telleria-Diaz et al., 2008). Indeed, PGD_2 can reduce the enhanced discharges evoked by PGE_2. Spinal application of PGE_2 at a high dose causes a persistent increase of the response to noxious pressure onto the knee (Fig. 12.5B, upper panel). The application of PGD_2 after PGE_2 partially reverses this effect (Fig. 12.5B, lower panel). Reduction of spinal hyperexcitability by PGD_2 may be a neuroprotective effect (Grill et al., 2008), via DP1 receptors (Liang et al., 2005). It may be caused by activation of DP1 receptors on inhibitory spinal interneurons (Minami et al., 1997).

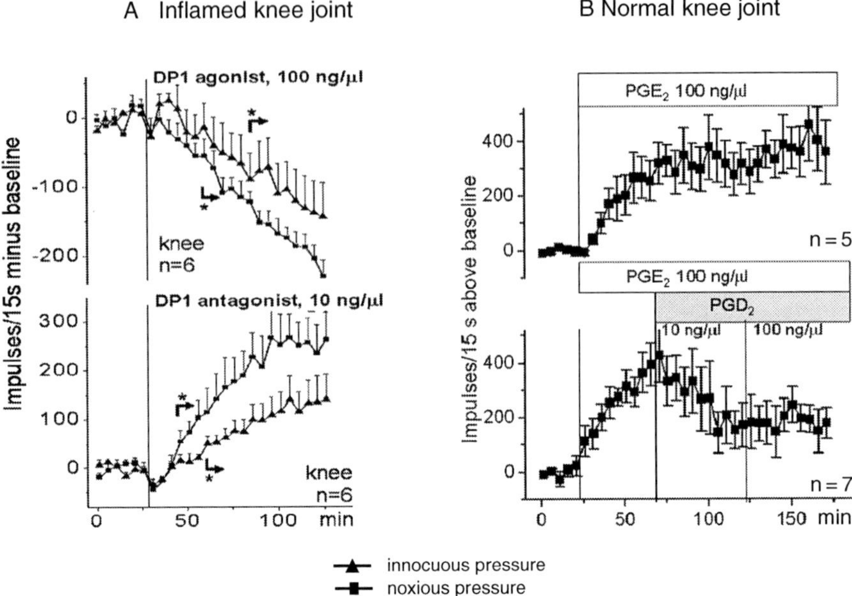

Fig. 12.5 (**A**) Effects of the DP1 receptor agonist BW245C and of the DP1 receptor antagonist BWA868C on the responses of spinal cord neurons in rats with acute knee joint inflammation. Graphs show changes of responses to innocuous and noxious pressure applied to the inflamed knee. The predrug baseline was set to 0, *vertical lines* show the time of drug application. *Asterisks* indicate the first interval of 15 min after drug application in which the values were significantly different from baseline which was set to 0 ($p < 0.05$, Wilcoxon matched paired signed rank test). (**B**) Effects of the coadministration of PGE₂ and PGD₂ on responses to mechanical stimulation of the normal knee joint. Increase of responses to noxious stimulation of the knee by PGE₂ alone (*top*) and reduction of increased responses by coapplication of PGD₂ at two doses (*bottom*). The *baseline* was set to 0. Reproduced from Telleria-Diaz et al. (2008)

12.5 Conclusions

The understanding of neuronal mechanisms of pain from arthritis is growing. Quite clearly, however, there is a bias towards experiments on acute arthritis. Much less is known about the neuronal mechanisms of chronic arthritic pain. Is chronic pain just a prolonged form of acute pain or does the quality of nociception change? This important question should be addressed in further research because many forms of joint pain are chronic. It will also be important to work out whether pain mechanisms during inflammation and osteoarthritis are similar or whether there are significant differences.

References

Abbadie C, Besson J-M (1992) C-fos expression in rat lumbar spinal cord during the development of adjuvant-induced arthritis. Neuroscience 48:985–993.

Andreeva L, Rang HP (1993) Effect of bradykinin and prostaglandins on the release of calcitonin gene-related peptide-like immunoreactivity from the rat spinal cord in vitro. Br J Pharmacol 108:185–190.

Arendt-Nielsen L, Laursen RJ, Drewes AM (2000) Referred pain as an indicator for neural plasticity. In: J Sandkühler, B Bromm, GF Gebhart (Eds) Progress in Brain Research, vol. 129. Elsevier, Amsterdam, pp. 343–356.

Bajaj P, Bajaj P, Graven-Nielsen T, Arendt-Nielsen L (2001) Osteoarthritis and its association with muscle hyperalgesia: an experimental controlled study. Pain 93:107–114.

Bär K-J, Natura G, Telleria-Diaz A, Teschner P, Vogel R, Vasquez E, Schaible H-G, Ebersberger A (2004) Changes in the effect of spinal prostaglandin E_2 during inflammation – Prostaglandin E (EP1–EP4) receptors in spinal nociceptive processing of input from the normal or inflamed knee joint. J Neurosci 24:642–651.

Barnes PJ, Karin M (1997) Nuclear factor-κB – a pivotal transcription factor in chronic inflammatory diseases. New Engl J Med 336:1066–1071.

Boettger MK, Hensellek S, Richter F, Gajda M, Stöckigt R, Segond von Banchet G, Bräuer R, Schaible H-G (2008) Antinociceptive effects of TNF-α neutralization in a rat model of antigen-induced arthritis. Evidence for a neuronal target. Arthritis Rheum 58:2368–2378.

Calvino B, Villanueva L, LeBars D (1987) Dorsal horn (convergent) neurones in the intact anaesthetized arthritic rat. II. Heterotopic inhibitory influences. Pain 31:359–379.

Cervero F, Schaible H-G, Schmidt RF (1991) Tonic descending inhibition of spinal cord neurones driven by joint afferents in normal cats and in cats with an inflamed knee joint. Exp Brain Res 83:675–678.

Chen L-W, Egan L, Li Z-W, Greten FR, Kagnoff MF, Karin M (2003) The two faces of IKK and NF-κB inhibition: prevention of systemic inflammation but increased local injury following intestinal ischemia-reperfusion. Nat Med 9:575–581.

Coggeshall RE, Hong KAP, Langford LA, Schaible H-G, Schmidt RF (1983) Discharge characteristics of fine medial articular afferents at rest and during passive movements of inflamed knee joints. Brain Res 272:185–188.

Danziger N, Weil-Fugazza J, LeBars D, Bouhassira D (1999) Alteration of descending modulation of nociception during the course of monoarthritis in the rat. J Neurosci 19:2394–2400.

Danziger N, Weil-Fugazza J, LeBars D, Bouhassira D (2001) Stage-dependent changes in the modulation of nociceptive neuronal activity during the course of inflammation. Eur J Neurosci 13:230–240.

Dougherty PM, Sluka KA, Sorkin LS, Westlund KN, Willis WD (1992) Neural changes in the acute arthritis in monkeys. I. Parallel enhancement of responses of spinothalamic tract neurons to mechanical stimulation and excitatory amino acids. Brain Res Rev 17:1–13.

Ebersberger A, Buchmann M, Ritzeler O, Michaelis M, Schaible H-G (2006) The role of spinal nuclear factor-κB in spinal hyperexcitability. NeuroReport 17:1615–1618.

Ebersberger A, Charbel Issa P, Vanegas H, Schaible H-G (2000) Differential effects of CGRP and CGRP8-37 upon responses to NMDA and AMPA in spinal nociceptive neurons with knee input in the rat. Neuroscience 99:171–178.

Ebersberger A, Grubb BD, Willingale HL, Gardiner NJ, Nebe J, Schaible H-G (1999) The intraspinal release of prostaglandin E_2 in a model of acute arthritis is accompanied by an upregulation of cyclooxygenase-2 in the rat spinal cord. Neuroscience 93:775–781.

Ferrell WR, Wood L, Baxendale RH (1998) The effect of acute joint inflammation on flexion reflex excitability in the decerebrate, low spinal cat. Quart J Exp Physiol 373:353–365.

Fields HL, Clanton CH, Anderson SD (1977) Somatosensory properties of spinoreticular neurons in the cat. Brain Res 120:49–66.

Grigg P, Schaible H-G, Schmidt RF (1986) Mechanical sensitivity of group III and IV afferents from posterior articular nerve in normal and inflamed cat knee. J Neurophysiol 55:635–643.

Grill M, Heinemann A, Hoefler G, Peskar BA, Schuligoi R (2008) Effect of endotoxin treatment on the expression and localization of spinal cyclooxygenase, prostaglandin synthases, and PGD$_2$ receptors. J Neurochem 104:1345–1357.

Grubb BD, Birrell J, McQueen DS, Iggo A (1991) The role of PGE$_2$ in the sensitization of mechanoreceptors in normal and inflamed ankle joints of the rat. Exp Brain Res 84: 383–392.

Grubb BD, Stiller RU, Schaible H-G (1993) Dynamic changes in the receptive field properties of spinal cord neurons with ankle input in rats with unilateral adjuvant-induced inflammation in the ankle region. Exp Brain Res 92:441–452.

Guilbaud G, Iggo A, Tegner R (1985) Sensory receptors in ankle joint capsules of normal and arthritic rats. Exp Brain Res 58:29–40.

He X, Proske U, Schaible H-G, Schmidt RF (1988) Acute inflammation of the knee joint in the cat alters responses of flexor motoneurones to leg movements. J Neurophysiol 59: 326–340.

Hope PJ, Jarrott B, Schaible H-G, Clarke RW, Duggan AW (1990) Release and spread of immunoreactive neurokinin A in the cat spinal cord in a model of acute arthritis. Brain Res 533:292–299.

Infante C, Diaz M, Hernández A, Constandil L, Pelissier T (2007) Expression of nitric oxide synthase isoforms in the dorsal horn of monoarthritic rats: effects of competitive and uncompetitive N-methyl-D-aspartate antagonists. Arth Res Ther 9:R53.

Inglis JJ, Notley CA, Essex D, Wilson AW, Feldmann M, Anand P, Williams R (2007) Collagen-induced arthritis as a model of hyperalgesia. Functional and cellular analysis of the analgesic actions of tumor necrosis factor blockade. Arthritis Rheum 56:4015–4023.

Jenkins DW, Feniuk W, Humphrey PP (2001) Characterization of the prostanoid receptor types involved in mediating calcitonin gene-related peptide release from cultured rat trigeminal neurones. Br J Pharmacol 134:1296–1302.

Kellgren JH (1939) Some painful joint condition and their relation to osteoarthritis. Clin Sci 4:193–205.

Kellgren JH, Samuel EP (1950) The sensitivity and innervation of the articular capsule. J Bone Joint Surg 32-B:84–91.

LeBars D, Villanueva L (1998) Electrophysiological evidence for the activation of descending inhibitory controls by nociceptive afferent pathways. In: HL Fields, J-M Besson (Eds) Progress in Brain Research, vol. 77. Elsevier, Amsterdam, pp. 275–299.

Lee KM, Kang BS, Lee HL, Son SJ, Hwang SH, Kim DS, Park J-S, Cho H-J (2004) Spinal NF-κB activation induces COX-2 upregulation and contributes to inflammatory pain hypersensitivity. Eur J Neurosci 19:3375–3381.

Lewis T (1938) Suggestions relating to the study of somatic pain. Br Med J 1:321–325.

Lewis T (1942) Pain. MacMillan, London.

Liang X, Wu L, Hand T, Andreasson K (2005) Prostaglandin D$_2$ mediates neuronal protection via the DP1 receptor. J Neurochem 92:477–486.

Martindale JC, Wilson AW Reeve AJ, Chessell IP, Headley PM (2007) Chronic secondary hypersensitivity of dorsal horn neurones following inflammation of the knee joint. Pain 133:79–86.

Menétrey D, Besson J-M (1982) Electrophysiological characteristics of dorsal horn cells in rats with cutaneous inflammation resulting from chronic arthritis. Pain 13:343–364.

Menétrey D, Gannon A, Levine JD, Basbaum AI (1989) Expression of c-fos protein in interneurons and projection neurons of the rat spinal cord in response to noxious somatic, articular, and visceral stimulation. J Comp Neurol 285:177–195.

Mense S (1997) Pathophysiologic basis of muscle pain syndromes. Myofasc Pain – Update in diagnosis and treatment. Phys Med Rehabil Clin N Am 8:23–53.

Meyers DER, Snow PJ (1982) The responses to somatic stimuli of deep spinothalamic tract cells in the lumbar spinal cord of the cat. J Physiol 329:355–371.

Minami T, Okuda-Ashitaka E, Nishizawa M, Mori H, Ito S (1997) Inhibition of nociceptin-induced allodynia in conscious mice by prostaglandin D_2. Br J Pharmacol 122:605–610.

Nakae K, Hayashi F, Hayashi M, Yamamoto N, Iino T, Yoshikawa S, Gupta J (2005) Functional role of prostacyclin receptor in rat dorsal root ganglion neurons. Neurosci Lett 388:132–137.

Neugebauer V, Lücke T, Grubb BD, Schaible H-G (1994a) The involvement of N-methyl-D-aspartate (NMDA) and non-NMDA receptors in the responsiveness of rat spinal neurons with input from the chronically inflamed ankle. Neurosci Lett 170:237–240.

Neugebauer V, Lücke T, Schaible H-G (1993) N-methyl-D-aspartate (NMDA) and non-NMDA receptor antagonists block the hyperexcitability of dorsal horn neurones during development of acute arthritis in rat's knee joint. J Neurophysiol 70:1365–1377.

Neugebauer V, Lücke T, Schaible H-G (1994b) Requirement of metabotropic glutamate receptors for the generation of inflammation-evoked hyperexcitability in rat spinal cord neurons. Eur J Neurosci 6:1179–1186.

Neugebauer V, Rümenapp P, Schaible H-G (1996a) The role of spinal neurokinin-2 receptors in the processing of nociceptive information from the joint and in the generation and maintenance of inflammation-evoked hyperexcitability of dorsal horn neurons in the rat. Eur J Neurosci 8:249–260.

Neugebauer V, Rümenapp P, Schaible H-G (1996b) Calcitonin gene-related peptide is involved in the generation and maintenance of hyperexcitability of dorsal horn neurons observed during development of acute inflammation in rat's knee joint. Neuroscience 71:1095–1109.

Neugebauer V, Schaible H-G (1990) Evidence for a central component in the sensitization of spinal neurons with joint input during development of acute arthritis in cat's knee. J Neurophysiol 64:299–311.

Neugebauer V, Weiretter F, Schaible H-G (1995) The involvement of substance P and neurokinin-1 receptors in the hyperexcitability of dorsal horn neurons during development of acute arthritis in rat's knee joint. J Neurophysiol 73:1574–1583.

Rudomin P, Hernández E (2008) Changes in synaptic effectiveness of myelinated joint afferents during capsaicin-induced inflammation of the footpad in the anaesthetized cat. Exp Brain Res 187:71–84.

Schaible H-G (2006a) Basic mechanisms of deep somatic pain. In: McMahon SB, Koltzenburg M (Eds) Wall and Melzack's Textbook of Pain, 5th edn. Elsevier, Churchill, Livingston, pp. 621–633.

Schaible H-G (2006b) Peripheral and central mechanisms of pain generation. In: Stein C (Ed) Handbook of Experimental Pharmacalogy, vol. 177. Springer-Verlag, Berlin, Heidelberg, pp. 4–28.

Schaible H-G, Freudenberger U, Neugebauer V, Stiller U (1994) Intraspinal release of immunoreactive calcitonin gene-related peptide during development of inflammation in the joint in vivo – a study with antibody microprobes in cat and rat. Neuroscience 62:1293–1305.

Schaible H-G, Grubb BD (1993) Afferent and spinal mechanisms of joint pain. Pain 55:5–54.

Schaible H-G, Jarrott B, Hope PJ, Duggan AW (1990) Release of immunoreactive substance P in the cat spinal cord during development of acute arthritis in cat's knee: A study with antibody bearing microprobes. Brain Res 529:214–223.

Schaible H-G, Neugebauer V, Cervero F, Schmidt RF (1991) Changes in tonic descending inhibition of spinal neurons with articular input during the development of acute arthritis in the cat. J Neurophysiol 66:1021–1032.

Schaible H-G, Schmidt RF (1985) Effects of an experimental arthritis on the sensory properties of fine articular afferent units. J Neurophysiol 54:1109–1122.

Schaible H-G, Schmidt RF (1988) Time course of mechanosensitivity changes in articular afferents during a developing experimental arthritis. J Neurophysiol 60:2180–2195.

Schaible H-G, Schmidt RF, Willis WD (1987) Enhancement of the responses of ascending tract cells in the cat spinal cord by acute inflammation of the knee joint. Exp Brain Res 66:489–499.

Segond von Banchet G, Petrow PK, Bräuer R, Schaible H-G (2000) Monoarticular antigen-induced arthritis leads to pronounced bilateral upregulation of the expression of neurokinin 1 and bradykinin 2 receptors in dorsal root ganglion neurones of rats. Arthritis Res 2: 424–427.

Sharif Naeini R, Cahill CM, Ribeiro-da-Silva A, Ménard HA, Henry JL (2005) Remodelling of spinal nociceptive mechanisms in an animal model of monoarthritis. Eur J Neurosci 22:2005–2015.

Sluka KA (2002) Stimulation of deep somatic tissue with capsaicin produces long-lasting mechanical allodynia and heat hypoalgesia that depends on early activation of the cAMP pathway. J Neurosci 22:5687–5693.

Sluka KA, Westlund K (1992) An experimental arthritis in rats: dorsal horn aspartate and glutamate increases. Neurosci Lett 145:141–144.

Sorkin LS, Westlund KN, Sluka KA, Dougherty PH, Willis WD (1992) Neural changes in acute arthritis in monkeys. IV: time course of amino acid release into the lumbar dorsal horn. Brain Res Rev 17:39–50.

Sun S, Cao H, Han M, Li TT, Pan HL, Zhao ZQ, Zhang YQ (2007) New evidence for the involvement of spinal fraktalkine receptor in pain facilitation and spinal glial activation in rat model of monoarthritis. Pain 129:64–75.

Tegeder I, Niederberger E, Schmidt R, Kunz S, Gühring H, Ritzeler O, Michaelis M, Geisslinger G (2004) Specific inhibition of IκB kinase reduces hyperalgesia in inflammatory and neuropathic pain models in rats. J Neurosci 24:1637–1645.

Telleria-Diaz A, Ebersberger A, Vasquez E, Schache F, Kahlenbach J, Schaible H-G (2008) Different effects of spinally applied prostaglandin D_2 (PGD_2) on responses of dorsal horn neurons with knee input in normal rats and in rats with acute knee inflammation. Neuroscience 156:184–192.

Vanegas H, Schaible H-G (2001) Prostaglandins and cyclooxygenases in the spinal cord. Prog Neurobiol 64:327–363.

Vasquez E, Bär K-J, Ebersberger A, Klein B, Vanegas H, Schaible H-G (2001) Spinal prostaglandins are involved in the development but not the maintenance of inflammation-induced spinal hyperexcitability. J Neurosci 21:9001–9008.

Willingale HL, Gardiner NJ, McLymont N, Giblett S, Grubb BD (1997) Prostanoids synthesized by cyclo-oxygenase isoforms in rat spinal cord and their contribution to the development of neuronal hyperexcitability. Br J Pharmacol 122:1593–1604.

Woolf CJ, Wall PD (1986) Relative effectiveness of C primary afferent fibres of different origins in evoking a prolonged facilitation of the flexor reflex in the rat. J Neurosci 6:1433–1442.

Yang LC, Marsala M, Yaksh TL (1996) Characterization of time course of spinal amino acids, citrulline and PGE_2 release after carrageenan/kaolin-induced knee inflammation: a chronic microdialysis study. Pain 67:345–354.

Yu X-M, Mense S (1990) Response properties and descending control of rat dorsal horn neurons with deep receptive fields. Neuroscience 39:823–831.

Chapter 13
Spinal Mechanisms of Visceral Pain and Hyperalgesia

Fernando Cervero and Jennifer M.A. Laird

Abstract Visceral pain is the most frequent form of clinically relevant pain. The study of its mechanisms is therefore immediately relevant to human pain conditions but it also offers a unique insight into the generation of hyperalgesic states. All forms of visceral pain generate enhancements of pain sensitivity in locations remote from the originating injury, a process known as "referred hyperalgesia" that is equivalent to the secondary hyperalgesia that develops following a somatic injury. In some cases, referred hyperalgesia can be the only manifestation of an altered pain state in the absence of an apparent injury or dysfunction of an internal organ. Referred hyperalgesia, like secondary hyperalgesia, is the expression of an alteration of sensory processing in the CNS and analysis of the molecular targets implicated in its generation can shed light on the general mechanisms of pain hypersensitivity. In this chapter the spinal cord mechanisms implicated in the generation of visceral hyperalgesia are discussed with reference to an animal model of referred visceral hyperalgesia and to some of the potential molecular mediators of visceral hyperalgesic states.

Abbreviations

AMPA	alpha-amino-3-hydroxy-5-methyl-4-isoxazole propionic acid
CaMKII	Ca(2 +)/calmodulin-dependent protein kinase II
CNS	central nervous system
DRG	dorsal root ganglia
DRRs	dorsal root reflexes
ERK	extracellular signal-regulated kinases
GABA	gamma-amino-butyric-acid
GluR	glutamate receptor
LTP	long-term potentiation

F. Cervero (✉)
Anaesthesia Research Unit, McGill University, 3655 Promenade Sir William Osler, Montreal, Quebec, Canada H3G 1Y6
e-mail: fernando.cervero@mcgill.ca

M. Malcangio (ed.), *Synaptic Plasticity in Pain*,
DOI 10.1007/978-1-4419-0226-9_13, © Springer Science+Business Media, LLC 2009

MAP   mitogen-activated protein kinases
mRNA   messenger ribo-nucleic acid
NKCC1   Na + -K + -2Cl- co-transporter
NMDA N-methyl-D-aspartate
PAD   primary afferent depolarization
TrpV1   transient receptor potential vanilloid receptor type 1

13.1 Introduction

Pain is a dynamic sensation and the generation of hyperalgesic states is the most characteristic expression of such dynamism. A sustained painful stimulus leads to an enhanced sensory state whereby innocuous stimulation is now felt as painful and painful sensations are amplified. These sensory alterations, grouped under the general umbrella term of hyperalgesic states, are the most prominent feature of chronic pain and are mediated by plastic changes in the way both the peripheral and central nervous systems process injury related information, particularly in the presence of inflammation or nerve injury (Cervero and Laird, 2004).

Hyperalgesic states can be triggered by the activation of nociceptive systems from both somatic and visceral tissues and it is likely that some of the mechanisms that mediate hyperalgesia are common to all forms of hypersensitivity to pain. However there are also profound differences in the way the nervous system processes sensory information from the skin, muscles and joints – the somatic structures – or from the internal organs. These differences relate to the paucity of the sensory innervation of the viscera compared to that of the skin, to the functional properties of the sensory receptors that innervate internal organs, to the central nervous system organization of the pathways that mediate visceral pain and to the motor and autonomic actions that are triggered by visceral noxious stimulation. It is often these unique features of the sensory processing of visceral signals that give pain of internal origin its distinct clinical characteristics, which can be quite different from those of somatic pain (Cervero and Laird, 1999).

It is well known that peripheral nociceptors from both somatic and visceral organs are capable of altering their functional characteristics after an injury or inflammation. This property, known as sensitization, not only explains hyperalgesic states that develop at the site of injury but also drives the more complex alterations in pain perception in areas adjacent or remote from the injury, or even in the absence of a primary injury, as for example in some neuropathic pain states, that require the participation of CNS mechanisms. The first opportunity for alteration of the normal neural processing is in the dorsal horn of the spinal cord – and its trigeminal equivalent in the brain stem – where primary afferent nociceptors make synaptic contact with second order neurons. Much attention has been given over the last 30 or 40 years to the dorsal horn relay of the nociceptive pathway as a potential candidate for the generation of changes leading to a hyperalgesic state. The term "central sensitization" has been used to qualify these changes by contrast with those taking place in the periphery. However, the extent to which peripheral and central sensitization are homologous concepts and even the meaning of the term central sensitization have been recently questioned (Sandkhuler, 2007) (Drdla and Sandkhuler, Chapter 9). It is very likely that supraspinal mechanisms contribute equally or even to a greater extent to the final generation of hyperalgesic states.

In this chapter we review the spinal mechanisms of visceral hyperalgesia, with an emphasis on the molecular targets that have been recently found to play a role in the generation of visceral hyperalgesic states. The study of visceral pain and hyperalgesia is not only of immediate clinical relevance – visceral pain is the most frequent form of pain seen in the clinic – but, because of the special characteristics of visceral pain, can also offer new insights into general pain mechanisms that can be more difficult to explore using somatic pain models. For instance, one of the most striking features of hyperalgesic states triggered by visceral injury or inflammation is the appearance of areas of secondary hyperalgesia remote from the originating stimulus and referred to the surface of the body. This separates clearly the primary (i.e. peripheral or at the site of injury) and the secondary (i.e. centrally mediated) components of the hyperalgesic state and allows differential analysis of the mechanisms that mediate either form of hyperalgesia. We have recently described an animal model of acute visceral pain and referred visceral hyperalgesia in mice that offers a very useful separation, in both the location and the time course, of the primary and secondary components of the hyperalgesic state (Laird et al., 2001). We will briefly describe this model later in this chapter.

Another important advantage of studying the mechanisms of visceral pain and hyperalgesia in the spinal cord relay is due to the different distribution and termination of sensory afferents in the spinal cord from the skin and from internal organs. Primary afferents innervating somatic tissues constitute more than 90% of all the afferents that project to the spinal cord and do so in somatotopically tightly organized and packed bundles whose distribution in the dorsal horn is very restricted (Cervero and Connell, 1984; Cervero et al., 1984). On the other hand visceral afferents make 10% or less of the afferent projection to the cord, terminate over a large number of spinal segments and diverge extensively while activating many spinal cord neurons. These anatomical and functional properties, which explain the poorly localized and diffuse nature of visceral pain, present clear advantages for the experimenter seeking to identify the molecular mechanisms triggered by noxious stimulation, as the effects of such stimuli are projected over a larger area of the cord and their amplification result in a much larger signal, particularly when one considers the relatively low number of primary afferents involved.

13.2 An Animal Model to Address Visceral Pain and Hyperalgesia

Ideally, an animal model of visceral pain should take into account the special characteristics and clinical features of the visceral pain conditions observed in humans. Some of the most widely used animal models of visceral pain measure simple behavioural responses to an acute painful stimulus applied to internal organs. Despite their popularity, these models may not be the most appropriate or predictive models of clinically relevant conditions. Animal models such as the writhing test or the colorectal distension test address only very short-lived acute reactions to a brief noxious stimulation of internal organs of the type that it is seldom the cause of prolonged visceral pain in humans. Other, more sophisticated models, attempt to reproduce a mechanism that participates in the processing of painful stimuli from internal organs and the triggering of more prolonged pain states. These mechanism-based models have proved very useful in dissecting the fundamental features of visceral pain processing.

Secondary hyperalgesia is the result of an alteration in the central processing of impulses from low-threshold mechanoreceptors, such that, these impulses are able to activate nociceptive neurons, thus evoking pain. This alteration is initially triggered and later maintained by the enhanced afferent discharges from the primary hyperalgesic area at the site of injury. In the case of visceral organs, secondary hyperalgesia is referred to the surface of the body. The primary focus is located in an internal organ, where nociceptors are sensitized by the originating stimulus and send enhanced discharges to the CNS that in turn trigger and maintain a secondary hyperalgesic area referred to the surface of the body.

We have developed an animal model in mice aimed at reproducing the mechanisms that generate referred hyperalgesia (Laird et al., 2001). It is based on the generation of an intense discharge in a group of visceral nociceptors that in turn trigger the central alteration responsible for secondary hyperalgesia. A convenient way to trigger such discharge is by the application of the TrpV1 receptor ligand, capsaicin. The model is based in the intracolonic application of capsaicin in mice. This procedure evokes two dose-dependent behavioural reactions: (i) an acute visceral pain response that last for 15–20 min and that is due to the activation of nociceptors in the colon by capsaicin and (ii) a long lasting referred hyperalgesic state in the abdominal and pelvic regions that is the expression of the altered central processing induced by the incoming afferent volley from the colonic nociceptors (Fig. 13.1). The referred hyperalgesia can be assessed behaviourally and lasts for more than 24 h following a single application of capsaicin. The essence of this model is to approach the dynamic component of visceral pain that is not necessarily time-locked to the duration of the initiating stimulus. Moreover, the model offers a long time window that permits substantial pharmacological studies.

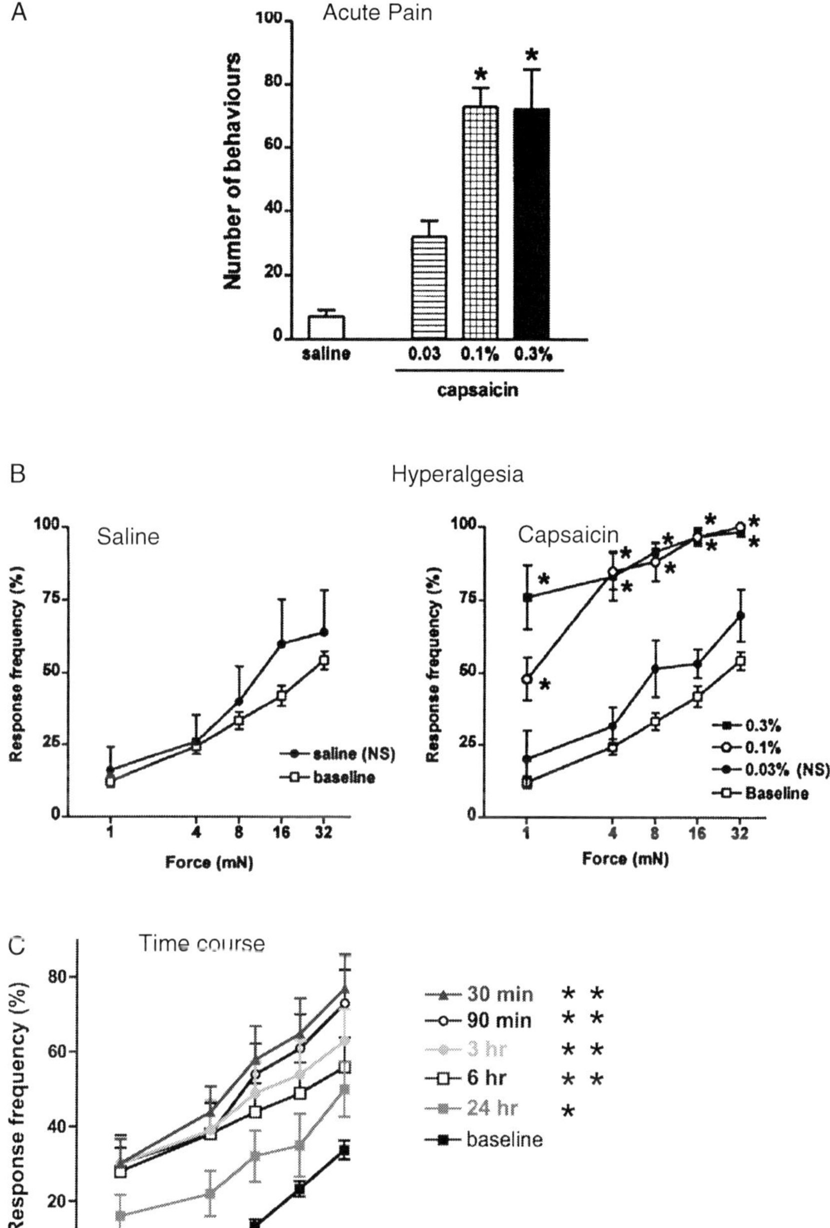

Fig. 13.1 (continued)

13.3 Spinal Cord Mechanisms of Visceral Hypersensitivity

We have examined three molecular mechanisms that may underlie visceral referred hyperalgesia in the first few minutes of the initiation of hyperalgesia, the development of hyperalgesia over minutes to hours and the maintenance phase over hours to days.

Over several years, we have developed the hypothesis that tactile allodynia is the result of altered processing at the level of the first synaptic relay in the CNS where afferent Aβ-fibres gain access to the nociceptive pathway through a pre-synaptic link with the terminals of A∂- and C afferent fibres (Cervero and Laird, 1996; Cervero et al., 2003; Price et al., 2005). Under normal circumstances activation of low-threshold afferents causes presynaptic inhibition of nociceptive afferents. The circuit involves activation of AMPA receptors on GABA-ergic spinal interneurones, which in turn release GABA onto nociceptive afferent terminals, depolarizing the nociceptive terminals which leads to a reduction of transmitter release by these terminals, and thus a reduction of afferent transmission into the spinal dorsal horn. This is the basis of the "gate-control theory". Inflammation and other forms of peripheral injury have been shown to enhance PAD to the point that the depolarization evokes action potentials in the primary afferent terminals (Willis, 1999). These discharges, known as Dorsal Root Reflexes (DRRs) can be detected antidromically (Rees et al., 1994, 1995; Lin et al., 2000) but can also cause excitation of second order neurons in the spinal cord (Cervero and Laird, 1996; Garcia-Nicas et al., 2006). In this way PAD, which is normally an inhibitory process, can be transformed into an excitatory one if the afferent depolarization is large enough to evoke spikes on the afferent terminals (Cervero and Laird, 1996; Garcia-Nicas et al., 2006).

This mechanism may also apply in visceral pain states and could explain the very rapid induction of referred visceral hyperalgesia which occurs within minutes of visceral stimuli. Using the visceral pain model described above, we examined the effects of intracolonic capsaicin on two key components of this proposed circuitry, firstly the chloride co-transporter that maintains the higher

◄───

Fig. 13.1 (continued) Model of visceral pain and referred hyperalgesia in mice. (A) Acute behavioral reactions (licking of abdomen, stretching, abdominal retractions) evoked by intracolonic instillation of saline or capsaicin. Data are shown as mean±SEM of the number of behaviors observed in the 20 min post-administration. (B) Referred abdominal hyperalgesia measured as responses to mechanical stimulation of the abdomen with von Frey hairs of five intensities. Data are shown as mean percent response frequency ±SEM before (baseline) and 20 min after intracolonic instillation of saline or of three concentrations of capsaicin. (C) Time course of the referred hyperalgesia measured as responses to mechanical stimulation of the hindpaws with von Frey hairs of 5 intensities. Data are shown at various time points after intracolonic instillation of 0.1% capsaicin. In all figures (*) indicates groups that were significantly different from saline-treated mice at $p<0.05$; and (**) at $p<0.001$. Data from Laird et al. (2001) and Galan et al. (2003)

intracellular chloride concentration of primary afferents, such that GABA-A receptor activation evokes an efflux of chloride and thus depolarization of the membrane and secondly, the AMPA receptor expressed on the GABA-ergic interneurons.

13.3.1 Chloride Co-Transporters and Visceral Hyperalgesic States

The Na+-K+-2Cl- co-transporter isoform 1 (NKCC1) is a member of the cation-dependent Cl- transporter family whose main function is to move chloride ions into the cell using the energy of the sodium gradient created by the Na-K-ATPase pump (Hass and Forbush, 1998; Alvarez-Leefmans et al., 1998). Cation chloride co-transporters contribute to the maintenance and homeostasis of cell volume (Hoffmann and Dunham, 1995) and, in the nervous system, play a critical role in the regulation and control of neuronal excitability. In particular, the NKCC1 co-transporter is responsible for the intracellular accumulation of chloride in hippocampal, dorsal root ganglia (DRG), and sympathetic ganglion neurons (Alvarez-Leefmans et al., 1998; Ballanyi and Grafe, 1985; Hara et al., 1992; Misgeld et al., 1986). The function of the NKCC1 co-transporter is tightly regulated by phosphorylation of Thr184 and Thr189 in the intracellular N-terminal domain and this has been demonstrated to be a mechanism for the enhancement of the co-transporter's activity (Darman and Forbush, 2002). In neurons, AMPA and group-1 metabotropic glutamate receptor activation have been shown to increase the activity of the NKCC1 co-transporter via a Ca2+/CaM-kinase-II dependent mechanism (Schomberg et al., 2001).

Recently, several studies have suggested a role for the NKCC1 co-transporter in nociceptive processing and in the generation and maintenance of hyperalgesic states. Disruption of the gene encoding NKCC1 causes an impaired behavioral response to the hot plate test, demonstrated by longer response latencies in the NKCC1 knock-out mice (Sung et al., 2000). We have shown that these NKCC1 knock-out mice have an increased behavioral threshold to noxious heat and a reduction in the stroking hyperalgesia (touch-evoked pain or allodynia) evoked by intradermal capsaicin injections (Laird et al., 2003). NKCC1 mRNA is expressed predominately in small and medium diameter primary afferent neurons (Price et al., 2006). The NKCC1 population of sensory neurons includes a large proportion that express TrpV1 and thus are presumably capsaicin sensitive nociceptors.

Using the intracolonic capsaicin model, we have found that capsaicin treatment evokes a fast and transient phosphorylation of the NKCC1 co-transporter in spinal cord tissues 10 min after the stimulus that returns towards basal levels 90 min after treatment (Galan and Cervero, 2005) (Fig. 13.2). There is relatively little NKCC1 mRNA expression in the superficial dorsal horn, suggesting that the NKCC1 protein there is likely expressed on both spinal neurons and the

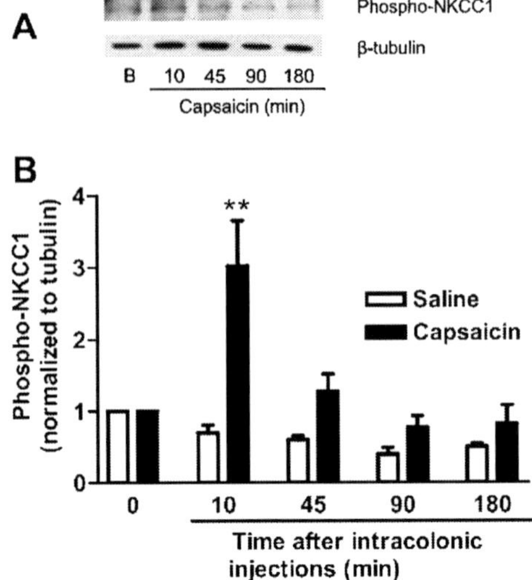

Fig. 13.2 Intracolonic capsaicin induces NKCC1 co-transporter phosphorylation. Representative Western blot from membrane protein extracts showing the time course of NKCC1 phosphorylation. Lane marked "B" in the immunoblot contains control (Basal) tissue. Membranes were re-incubated with b-tubulin as a loading control. The graph below shows the quantification of phospho-NKCC1 normalized to b-tubulin values after intracolonic capsaicin ($n=6$, per time point). *Asterisks* indicate those groups that were significantly different from control levels (**$p<0.01$). Data from Galan and Cervero (2005)

membrane of primary afferent terminals (see Price et al., 2006 for further discussion). Therefore, the phosphorylation of NKCC1 could result in increased activity in afferent terminals, which would result in enhanced PAD. Activation of NKCC1 would also increase the excitability of spinal neurons by changing their intracellular chloride concentration.

The activation of spinal NKCC1 in visceral hyperalgesia seems likely mediated via a Ca2 + /CaM-kinase-II mechanism, as described in other brain areas. Spinal CaMKII can be activated by painful stimuli (intradermal injection of capsaicin into the hind paw) and a CaMKII inhibitor blocks the resulting hyperalgesia (Fang et al., 2002). Likewise, we found that mice treated with intracolonic capsaicin showed a rapid activation of α-CaMKII with a peak 5.2-fold increase over control levels. The development of hyperalgesia was blocked by local spinal (intrathecal) application of the CaMKII inhibitor, KN-93, given before the painful visceral stimulus (Galan et al., 2004).

Painful visceral stimuli can also evoke a translocation of NKCC1 co-transporter to the membrane, which would be expected to further alter the excitability of primary afferent terminals and spinal cord neurons (Galan and Cervero, 2005). Intracolonic capsaicin induces *in vivo* trafficking of the NKCC1 co-transporter from the cytosol to the plasma membrane in the spinal cord of adult mice. This was demonstrated by both subcellular fractionation of proteins and cross-linking of membrane proteins. We detected a 40% depletion of NKCC1 in the cytosolic fraction in parallel with a 50% increase of NKCC1 in the membrane fraction 90 and 180 min after intracolonic capsaicin (Galan and Cervero, 2005). These times correspond to the plateau phase of the referred mechanical hyperalgesia evoked in the abdominal region by the visceral stimulus (Laird et al., 2001; Galan et al., 2003, 2004).

The time-courses of the NKCC1 phosphorylation and the NKCC1 trafficking suggest that these two events could contribute to the initial development and the subsequent maintenance of the capsaicin-induced hyperalgesia, respectively. If activation and membrane mobilization of NKCC1 occurred in the spinal cord terminals of primary afferents, this would produce an increase in the intracellular concentration of chloride and an enhanced GABA-mediated PAD. Another potential mechanism would be that an increase in PAD would directly activate voltage-dependent calcium channels on primary afferent terminals to increase tonic glutamate release onto nociceptors (Jang et al., 2002). However, alterations of chloride homeostasis in spinal cord neurons could also contribute to the generation of hyperalgesia. For instance, it has been shown that the reduction in the expression of postsynaptic KCC2 in spinal lamina I neurons underlies neuropathic pain following peripheral nerve injury (Coull et al., 2003).

13.3.2 *Visceral Hypersensitivity and AMPA Trafficking in the Spinal Cord*

Central sensitization is the result of synaptic plasticity and is a memory trace of previous painful stimuli (for review see Sandkühler, 2000, Ji et al., 2003) (DrDla and Sandkhuler, Chapter 9; Randic, Chapter 10). One of the best-characterized examples of synaptic plasticity in the mammalian nervous system is activity-dependent long-term potentiation (LTP), whereby brief, intense conditioning stimuli produce prolonged enhancement of responses to subsequent stimuli. LTP can be evoked by electrical stimulation in the population of superficial spinal cord neurons that mediate hyperalgesia (Ikeda et al., 2003). It can also be induced in intact, anaesthetized rats by natural pain-producing stimuli (Sandkühler and Liu, 1998) and an LTP-like enhancement in pain sensitivity can be provoked in human subjects by high frequency electrical stimulation of the skin (Klein et al., 2004). Thus it is likely that an LTP-like process underlies the generation of a hyperalgesic state (Sandkühler, 2000, Ji et al., 2003).

Potentiation of spinal nociceptive transmission by synaptic delivery of AMPA receptors, via an NMDA receptor- and $Ca(2+)$/calmodulin-dependent protein kinase II (CaMKII)-dependent pathway, has been proposed to underlie certain forms of hyperalgesia. Trafficking of the AMPA subclass of glutamate receptors from the cytosol into the post-synaptic membrane plays a critical role in LTP in hippocampal slices in culture (Malinow and Malenka, 2002). In the hippocampus, the insertion of GluR1 subunits of the AMPA receptor into the synapse is induced by LTP and is a tightly regulated process, whereas GluR2/3 subunits appear to be constitutively cycled in and out the plasma membrane to maintain normal transmission (Malinow and Malenka, 2002).

To examine AMPA-receptor trafficking in spinal neurons after natural activation of the pain pathway in adult animals in vivo, we compared control mice with mice in which visceral pain and hyperalgesia was induced by intracolonic capsaicin. Subcellular fractionation of proteins from the lumbrosacral spinal cord was followed by immunoblotting with specific antibodies against GluR1 and GluR2/3 AMPA-R subunits. There was a pronounced increase in the abundance of the AMPA-R subunit GluR1 in the synaptosomal membrane fraction with a peak 3.7-fold increase 180 min after treatment and a corresponding decrease in the levels in the cytosolic fraction. In contrast to the pronounced and prolonged effects of the painful visceral stimulus on GluR1 distribution, capsaicin treatment had no effect on the intracellular distribution of GluR2/3 in spinal tissue (Fig. 13.3A).

We found that referred hyperalgesia was inhibited by pre-treatment with Brefeldin A applied topically to the spinal cord, whilst it did not have analgesic effects in naïve animals. Brefeldin A is a fungal metabolite used as an antibiotic that specifically and reversibly blocks protein transport from the endoplasmic reticulum to the Golgi apparatus thereby inhibiting exocytosis (Fujiwara et al., 1988) and also has the advantage of low toxicity; the LD50 in mice by intraperitoneal injection is greater than 200 mg/kg (Harri et al., 1963). Further, Broutman and Baudry (2001) showed that Brefeldin A does not affect synaptic transmission in the hippocampus, at concentrations that both prevented recruitment of GluR1 subunits and the induction of LTP. Therefore the inhibition of hyperalgesia we observed seems unlikely to be due to direct toxicity or inhibition of synaptic transmission. Brefeldin A at the doses used in the behavioral experiment clearly inhibited the trafficking of GluR1 subunits to the membrane, thus GluR1 recruitment may contribute to the development and expression of hyperalgesia (Fig. 13.3B). However, Brefeldin A treatment also may have inhibited the exocytosis of other glutamate receptor subunits or other classes of proteins to the cell surface that are required for the behavioral expression of hyperalgesia.

Larsson and Broman (2008) found using neuronal tracing and postembedding immunogold labeling that acute inflammatory hyperalgesia is associated with an elevated density of GluR1-containing AMPA receptors, as well as an increased synaptic ratio of GluR1 to GluR2/3 subunits, at synapses in the superficial dorsal horn. They also reported increased and decreased levels of activated CaMKII at synapses formed by peptidergic and nonpeptidergic nociceptive fibers, respectively, suggesting that the observed redistribution of AMPA receptor subunits does not depend on postsynaptic CaMKII activity. In contrast, we observed that the GluR1 trafficking in spinal neurons in vivo after visceral stimuli followed the rules established for AMPA-receptor trafficking in hippocampal neurons. The signaling pathways that drive the insertion of GluR1 subunits into the plasma membrane during LTP in vitro include a

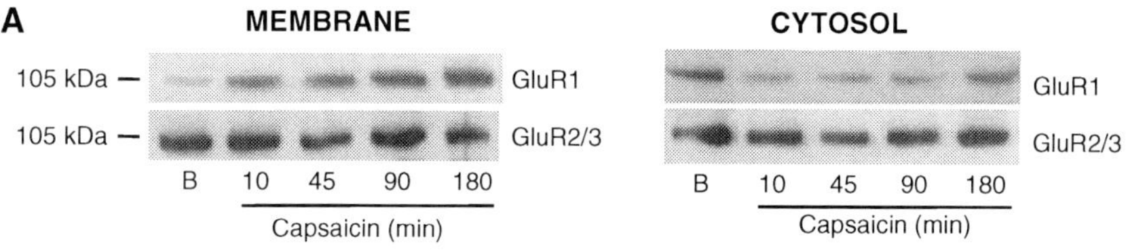

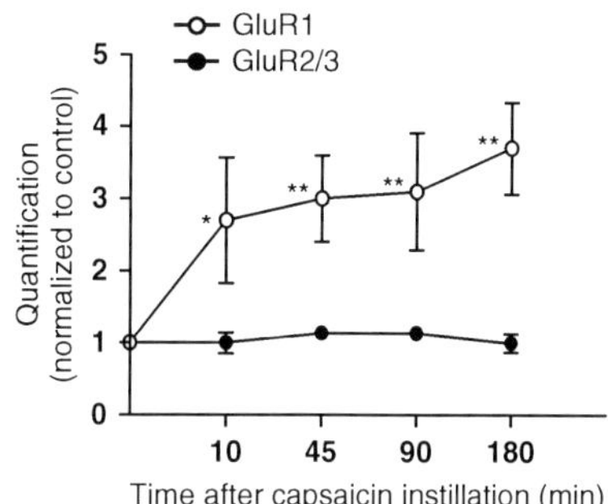

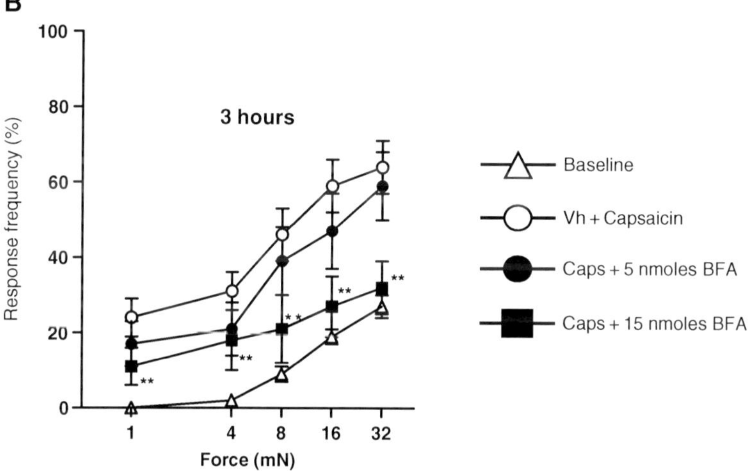

Fig. 13.3 AMPA trafficking induced by a visceral noxious stimulus. (**A**) Time course of subcellular distribution of GluR1 and GluR2/3 in the plasma membrane fraction and in the cytosolic fraction. Lanes marked "B" in the immunoblots contain control (basal) tissue. The graph below shows quantification of GluR1 and GluR2/3 normalized to levels detected in control spinal tissue, which has the arbitrary value of "1", run on the same gel, $n = 4$, per time point. *Asterisks* indicate those groups that were significantly different from control levels (*$p < 0.05$ and **$p < 0.01$). (**B**) Responses to mechanical stimulation of the abdomen with von Frey hairs of five intensities. Data are shown as mean percent response frequency ($\pm$ SEM) before (*baseline*) and at various time points after intracolonic instillation of 0.1% capsaicin or two concentrations of brefeldin A. *Asterisks* indicate responses that were significantly different from baseline ($p < 0.01$). Data from Galan et al. (2004)

requirement for calcium-calmodulin kinase II (CaMKII) activity (Malinow and Malenka, 2002, Contractor and Heinemann, 2002, Derkach et al., 2007). In our studies, as described above, intracolonic capsaicin produced an activation of CaMKII. Furthermore, animals pre-treated with intrathecal vehicle showed the expected plasma membrane enrichment of GluR1 90 min after the visceral stimulus, whereas those pre-treated with the CaMKII inhibitor, KN-93, showed no evidence of GluR1 accumulation in the plasma membrane fraction. Likewise, capsaicin-treated mice showed a depletion of GluR1 from the cytosol, that was prevented in mice pre-treated with the CaMKII inhibitor. These findings indicate that the painful visceral stimulus drives synaptic delivery of GluR1-containing receptors in a CaMKII-dependent manner. Similarly, Pezet et al. (2008) showed that somatic inflammatory pain induced by formalin injection produced an increase in CaMKII phosphorylation and trafficking of GluR1 subunits but not GluR2/3 subunits to the membrane.

The AMPA subunit GluR1 is found in most but not all glutamatergic synapses in the superficial dorsal horn (Polgar et al., 2008) whereas inhibitory spinal interneurons, expressing glycine and GABA, all express GluR1 subunits (Spike et al., 1998). Thus AMPA trafficking may not only enhance excitatory synapses, driving an LTP-type process, but may increase GABA release and thus PAD via strengthening of inputs onto the inhibitory interneurons driving the PAD.

13.3.3 Role of Intracellular Signalling Kinases

Visceral hyperalgesia is not only induced very rapidly, but also lasts for a long time after the initiating stimulus. For example, patients suffering from renal colic caused by a kidney stone exhibit pronounced referred hyperalgesia in pain-free periods of several weeks between bouts of colic pain (Giamberadino, 1999). In the visceral pain model we describe above, referred hyperalgesia persists for more than 24 h after the initiating capsaicin application. We have examined the role of extracellular signal-regulated kinases (ERKs) in persistent referred hyperalgesia, in part because they play an established role in learning, memory and synaptic plasticity in other brain regions (for reviews, see Impey et al., 1999; Mazzucchelli and Brambilla, 2000). Further, we found that spinal ERK1/2 activation evoked by paw inflammation is also correlated with persistent secondary hyperalgesia (Galan et al., 2002).

Mitogen-activated protein (MAP) kinases, also known as ERKs, are members of a family of serine/threonine protein kinases that mediate intracellular signal transduction in response to a variety of stimuli. Upon activation by upstream MAP kinase kinases, ERKs translocate into the nucleus, phosphorylate transcription factors and regulate the transcription of related genes. Recent evidence suggests a role for ERKs in nociceptive processing in the spinal cord in models of somatic pain. ERK1 and ERK2 are expressed in the spinal cord and are rapidly activated in rodent dorsal horn neurons by acute noxious stimuli like formalin or capsaicin (Thomas and Hunt, 1993; Ji et al., 1999; Karim et al., 2001). An upstream inhibitor of ERK signaling, PD 98059, reduces acute pain in the formalin model, suggesting a role for ERKs in acute nociceptive processing by a non-transcriptional mechanism. In addition, ERK may also have a role in persistent hyperalgesia. Persistent inflammatory hyperalgesia can be induced by paw inflammation with carrageenan or Freund's adjuvant. Spinal ERK is activated by these stimuli (Galan et al., 2002; Ji et al., 2002) and inflammatory hyperalgesia can be prevented or reduced by ERK inhibitors (Sammons et al., 2000; Ji et al., 2002). Spinal ERK1/2 activation evoked by paw inflammation is also correlated with secondary hyperalgesia (Galan et al., 2002).

We have investigated the role of ERK1/2 in spinal processing of visceral pain and particularly, in the development of referred visceral hyperalgesia (Galan et al., 2003). We used the model described above (Laird et al., 2001) to induce referred hyperalgesia in mice by the instillation of capsaicin into the colon. In this model we detected an activation of ERK1/2 in the lumbosacral spinal cord extracted 30 min after colonic instillation of capsaicin or mustard oil (Fig. 13.4A). ERK1/2 activation was measured by western blotting using a specific antibody. ERK activation correlated well with the expression of hyperalgesia and it was specifically localized to the lumbosacral spinal region where colonic afferents have been shown to terminate in the mouse (Payette et al., 1987). We saw no evidence of ERK1/2 activation in the thoracic and cervical cord after intracolonic stimulation indicating that ERK1/2 activation after noxious stimulation of the terminal colon was specifically localized to the lumbosacral spinal cord (Fig. 13.4A).

Previous immunohistochemical experiments have suggested a possible nuclear localization of the activated ERK1/2 (Ji et al., 1999; Karim et al., 2001). To test whether intracolonic capsaicin and mustard oil induced translocation of ERK1/2 from the cytosol into the nucleus upon activation, we tried subcellular fractionation of protein from the lumbosacral spinal cord. Capsaicin treated-mice showed no significant cytosolic ERK1/2 activation when compared to the basal control level. However intracolonic capsaicin produced a clear accumulation of phosphorylated ERK1/2 in the nuclear fraction with a peak of 2.9 and 3.0-fold respectively at 90 min. Nuclear ERK1/2 activation was still significantly greater than basal levels at 180 min post-treatment.

Furthermore, we tested the involvement of ERK activation in the expression of visceral hyperalgesia by treating animals with the ERK inhibitor U0126. This produced a dose-dependant inhibition of referred hyperalgesia but not of the spontaneous pain response. Importantly, the ERK inhibitor had no effect on pain-related behavior until 3 h after capsaicin treatment, even though it was given before capsaicin, suggesting modulation of transcription rather than post-translational events, particularly since the effect lasted up to 24 h post-capsaicin (Fig. 13.4B). Activation of the ERK pathway in spinal neurons contributes to transcriptional regulation of the tachykinin NK-1 receptor (Ji et al., 2002),

A

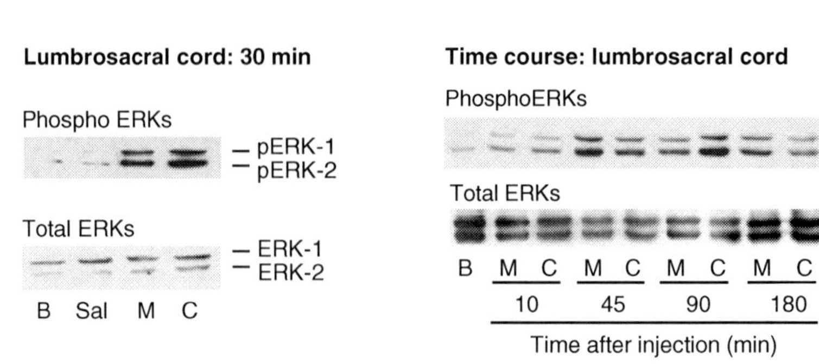

B

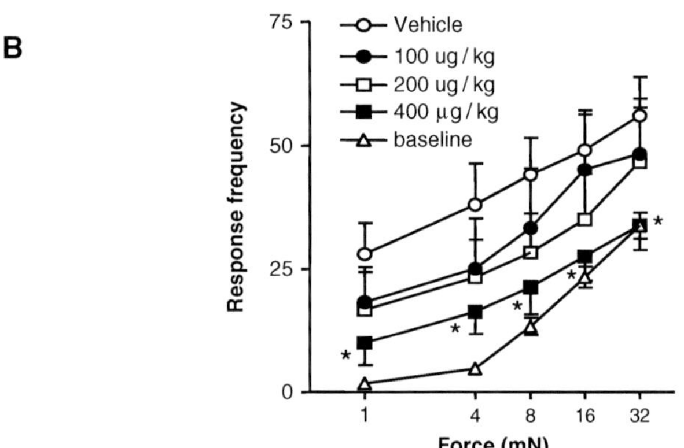

Fig. 13.4 Role of ERKs in visceral hyperalgesia. (**A**) (*left*): Western blots of lumbosacral spinal tissue 30 min after intracolonic instillation of saline (Sal), mustard oil (M) or capsaicin (C). In all cases, the 2 bands shown represent ERK-1 (p44) and ERK-2 (p42). (**A**) (*right*): Time course of pERK1/2 activation in lumbrosacral segments after mustard oil or capsaicin instillation. Lanes marked "B" contain control (basal) spinal tissue. As loading control, blots were also probed with an antibody against total ERK1/2. (**B**) Behavioral responses to mechanical stimulation of the abdomen with von Frey hairs of five intensities 6 h after intracolonic instillation of 0.1% capsaicin. U0126 (100, 200 or 400 µg/kg i.v.) or vehicle was given systemically 10 min before the tests. Data are shown as mean percent response frequency (±SEM). The *baseline* data are shown on each graph for comparison and *asterisks* indicate responses that were significantly different from those of the vehicle-treated group ($p < 0.05$). Data from Galan et al. (2003)

important in excitatory nociceptive processing. Likewise, increases in phospho-ERK are also associated with increased GABA release from spinal cord slices (Pezet et al., 2002), which may contribute to increased PAD.

ERK activation is required for transcriptional regulation associated with plasticity in other brain regions, for example, long-term-potentiation (LTP)-dependent transcription in the dentate gyrus of adult rats in vivo (Davis et al.,

2002). Our results with subcellular fractionation showed a significant translocation of phosphorylated ERK1/2 from cytoplasm into the nucleus starting at 45 min, whereas there was no significant increase of phosphorylated ERK1/2 in the cytoplasmic fraction for at least 3 h. Thus, these data strongly support a role for ERK signaling in the maintenance rather than the induction phase of secondary hyperalgesia via transcriptional regulation, whereas it does not seem to participate in other features of the pain behavior such as acute pain and primary hyperalgesia.

13.4 Concluding Remarks

Overall, the evidence we have accumulated using a model of referred visceral hyperalgesia induced by brief application of intracolonic capsaicin suggests a sequential time course of molecular events in the spinal cord, all of which are associated with the development of hyperalgesia and in several instances we have direct or indirect evidence they are required for the expression of the hyperalgesia. The first events, occurring within a few minutes of capsaicin application, and before behavioral expression of hyperalgesia can be measured, are phosphorylation of CaMKII and NKCC1 and the translocation of the AMPA receptor subunit, GluR1 to the plasma membrane.

It seems likely that NKCC1 phosphorylation is downstream of the CaMKII activation, and we have shown that GluR1 translocation is dependent on CaMKII activation since it is blocked by pre-emptive treatment with a CaMKII inhibitor. NKCC1 phosphorylation shows a peak at 10 min and has declined markedly at 45 min after capsaicin application. In contrast, the translocation of GluR1 to the membrane is sustained for up to 3 h after intracolonic capsaicin. However, although NKCC1 phosphorylation is short-lived, NKCC1 is also trafficked to the membrane, with a significant increase in the membrane fraction measurable from 90 min post-capsaicin, suggesting that increased chloride transport activity may initially be sustained by activation of the existing pool of NKCC1 and then later on by increasing the size of the pool of co-transporter molecules in the membrane. The later, maintenance stage of hyperalgesia is associated with activation of ERKs and their translocation to the nucleus. Treatment with an ERK inhibitor prevents the maintenance of hyperalgesia from 3 to 24 h after capsaicin application.

We have been able to take advantage of the unique features of visceral pain with the clear anatomical separation of the primary and secondary (referred) hyperalgesia to unambiguously ascribe these changes to secondary hyperalgesia produced by changes in central processing. Further, the changes observed are mostly rather large. For example, we detected a substantial depletion of GluR1 in the cytosolic fraction ($\sim$40% reduction) after a painful visceral stimulus, even though we extracted the whole lumbrosacral spinal cord. This suggests that a large proportion of the total cytosolic GluR1 present in this region of the spinal cord is mobilized by the activating stimulus used, which at first sight is surprising, even when one considers that GluR1 subunits are localized preferentially in the pain-processing areas of the spinal cord. This demonstrates the value of using visceral pain stimuli to maximize the chances of observing a clear effect. As mentioned above, cutaneous afferent input to the spinal cord is tightly somatotopically organized in limited areas, which would likely also restrict the spatial extent of any GluR1 trafficking associated with plasticity after a natural stimulus. In contrast, spinal processing of visceral information is widely distributed and depends to a greater extent on amplification within the CNS.

We conclude that the mechanisms we have described here are particularly important in visceral nociceptive processing in order to set in motion the CNS signal amplification that is such a prominent feature of visceral pain processing.

References

Alvarez-Leefmans FJ, Nani A, Marquez S (1998) Chloride transport, osmotic balance, and presynaptic inhibition. In: Presynaptic Inhibition and Neural Control (Rudomin P, Romo R, Mendell LM), pp. 50–79. New York: Oxford University Press.

Ballanyi K, Grafe P (1985) An intracellular analysis of g-aminobutiric-acid-associated ion movements in rat sympathetic neurons. J Physiol. (Lond) 365:41–58.

Broutman G, Baudry M (2001) Involvement of the secretory pathway for AMPA receptors in NMDA-induced potentiation in hippocampus. J Neurosci. 21(1):27–34.

Cervero F, Connell LA (1984) Distribution of somatic and visceral primary afferent fibres within the thoracic spinal cord of the cat. J Comp Neurol. 230:88–98.

Cervero F, Connell LA, Lawson SN (1984). Somatic and visceral primary afferents in the lower thoracic dorsal root ganglia of the cat. J Comp Neurol. 228:422–431.

Cervero F, Laird JMA (1996) Mechanisms of touch-evoked pain (allodynia): a new model. Pain. 68(1):13–23.

Cervero F, Laird JMA (1999) Visceral pain. Lancet. 353(9170):2145–2148.

Cervero F, Laird JMA (2004) Referred visceral hyperalgesia: from sensations to molecular mechanisms. In: Hyperalgesia: Molecular Mechanisms and Clinical Implications (Handwerker HO, Brune K), pp. 229–250. Seattle, WA: IASP Press

Cervero F, Laird JM, Garcia-Nicas E (2003) Secondary hyperalgesia and presynaptic inhibition: an update. Eur J Pain. 7(4):345–351.

Contractor A, Heinemann SF (2002) Glutamate receptor trafficking in synaptic plasticity. Sci STKE. 2002:RE14.

Coull JA, Boudreau D, Bachand K, Prescott SA, Nault F, Sik A, De Koninck P, De Koninck Y (2003) Trans-synaptic shift in anion gradient in spinal lamina I neurons as a mechanism of neuropathic pain. Nature. 424(6951):938–942.

Darman RB, Forbush B (2002) A regulatory locus of phosphorylation in the N terminus of the Na-K-Cl cotransporter, NKCC1. J Biol Chem. 277(40):37542–37550.

Davis S, Vanhoutte P, Pagès C, Caboche J, Laroche S (2002) The MAPK/ERK cascade targets both Elk-1 and cAMP response element-binding protein to control long-term potentiation-dependent gene expression in the dentate gyrus in vivo. J Neurosci. 20: 4563–4572.

Derkach VA, Oh MC, Guire ES, Soderling TR (2007) Regulatory mechanisms of AMPA receptors in synaptic plasticity. Nat Rev Neurosci. February; 8(2):101–113. Review.

Fang L, Wu J, Lin Q, Willis WD (2002) Calcium-calmodulin-dependent protein kinase II contributes to spinal cord central sensitization. J Neurosci. 22(10):4196–4204.

Fujiwara T, Oda K, Yokota S, Takatsuki A, Ikehara Y (1988) Brefeldin A causes disassembly of the Golgi complex and accumulation of secretory proteins in the endoplasmic reticulum. J Biol Chem. 263:18545–18552.

Galan A, Cervero F (2005) Painful stimuli induce in vivo phosphorylation and membrane mobilization of mouse spinal cord NKCC1 co-transporter. Neuroscience. 133:245–252.

Galan A, Cervero F, Laird JM (2003) Extracellular signaling-regulated kinase-1 and -2 (ERK 1/2) mediate referred hyperalgesia in a murine model of visceral pain. Mol Brain Res. 116:126–134.

Galan A, Laird JMA, Cervero F (2004) In vivo recruitment by painful stimuli of AMPA receptor subunits to the plasma membrane of spinal cord neurons. Pain. 112:315–323.

Galan A, Lopez-Garcia JA, Cervero F, Laird JMA (2002) Activation of spinal extracellular signaling-regulated kinase-1 and-2 by intraplantar carrageenan in rodents. Neurosci Lett. 322:37–40.

Garcia-Nicas E, Laird JMA, Cervero F (2006) GABAA receptor blockade reverses the injury-induced sensitization of Nociceptor Specific (NS) neurons in the spinal dorsal horn of the rat. J Neurophysiol. 96:661–670.

Giamberadino MA (1999) Recent and forgotten aspects of visceral pain. Eur J Pain. 3:77–92.

Hara M, Inoue M, Yasukura T, Ohhishi S, Mikami Y, Inagaki C (1992) Uneven distribution of intracellular Cl- in rat hippocampal neurons. Neurosci Lett. 143:135–138.

Harri E, Loeffier W, Sigg HP, Stahelin H, Tamm H (1963) Uber die Isolierung neuer Stoffwechselproduckte aus Penicillium brefeldianum DODGE. Helv Chim Acta. 46: 1235–1243.

Hass M, Forbush B III (1998) The Na-K-Cl cotransporters. J Bioenerg Bioeng. 30:161–172.

Hoffmann EK, Dunham PB (1995) Membrane mechanisms and intracellular signalling in cell volume regulation. Int Rev Cytol. 161:173–262.

Ikeda H, Heinke B, Ruscheweyh R, Sandkuhler J (2003) Synaptic plasticity in spinal lamina I projection neurons that mediate hyperalgesia. Science. 299:1237–40.

Impey S, Obrietan K, Storm DR (1999) Making new connections: role of ERK/MAK kinase signalling in neuronal plasticity. Neuron. 23:11–14.

Jang IS, Jeong HJ, Katsurabayashi S, Akaike N (2002) Functional roles of presynaptic GABAA receptors on glycinergic nerve terminals in the rat spinal cord. J Physiol. (Lond) 541:423–434.

Ji R, Baba H, Brenner GJ, Woolf CJ (1999) Nociceptive-specific activation of ERK in spinal neurons contributes to pain hypersensitivity. Nat Neurosci. 2:1114–1119.

Ji R, Befort K, Brenner G, Woolf CJ (2002) ERK MAP kinase activation in superficial spinal cord neurons induces prodynorphin and NK-1 upregulation and contributes to persistent inflammatory pain hypersensitivity. J Neurosci. 22:478–485.

Ji RR, Kohno T, Moore KA, Woolf CJ (2003) Central sensitization and LTP: do pain and memory share similar mechanisms? Trends Neurosci. 26:696–705.

Karim F, Wang C, Gereau RW (2001) Metabotropic glutamate receptor subtypes 1 and 5 are activators of extracellular signal-regulated kinase signaling required for inflammatory pain in mice. J Neurosci. 21:3771–3779.

Klein T, Magerl W, Hopf HC, Sandkühler J, Treede RD (2004) Perceptual correlates of nociceptive long-term potentiation and long-term depression in humans. J Neurosci. 24: 964–971.

Laird JMA, Garcia-Nicas E, Delpire E, Cervero F (2003) Presynaptic inhibition and spinal pain processing: a possible role of the NKCC1 cation-chloride co-transporter in hyperalgesia in mice. Neurosci Lett. 361(1–3):200–203.

Laird JM, Martinez-Caro L, Garcia-Nicas E, Cervero F (2001) A new model of visceral pain and referred hyperalgesia in the mouse. Pain. 92(3):335–342.

Larsson M, Broman J (2008) Translocation of GluR1-containing AMPA receptors to a spinal nociceptive synapse during acute noxious stimulation. J Neurosci. 28(28):7084–7090.

Lin Q, Zou X, Willis WD (2000) Adelta and C primary afferents convey dorsal root reflexes after intradermal injection of capsaicin in rats. J Neurophysiol. 84:2695–2698.

Malinow R, Malenka RC. (2002) AMPA receptor trafficking and synaptic plasticity. Annu Rev Neurosci. 25:103–126.

Mazzucchelli C, Brambilla R (2000) Ras-related and MAPK signaling in neuronal plasticity and memory formation. Cell Mol Life Sci. 57:604–611.

Misgeld U, Deisz RA, Dodt HU, Lux HD (1986) The role of chloride transport in postsynaptic inhibition of hippocampal neurons. Science. 232:1413–1415.

Payette RF, Tennyson VM, Pham TD, Mawe GM, Pomeranz HD, Rothman TP, Gershon MD (1987) Origin and morphology of nerve fibers in the aganglionic colon of the lethal spotted (ls/ls) mutant mouse. J Comp Neurol. 257:237–252.

Pezet S, Malcangio M, Lever IJ, Perkinton MS, Thompson SW, et al. (2002) Noxious stimulation induces Trk receptor and downstream ERK phosphorylation in spinal dorsal horn. Mol Cell Neurosci. 21:684–695.

Pezet S, Marchand F, D'Mello R, Grist J, Clark AK, Malcangio M, Dickenson AH, Williams RJ, McMahon SB (2008) Phosphatidylinositol 3-kinase is a key mediator of central sensitization in painful inflammatory conditions. J Neurosci. April 16;28(16):4261–4270.

Polgar E, Watanabe M, Hartmann B, Grant SG, Todd AJ (2008) Expression of AMPA receptor subunits at synapses in laminae I–III of the rodent spinal dorsal horn. Mol Pain. January 23;4:5.

Price TJ, Cervero F, de Koninck Y (2005) Role of cation-chloride-cotransporters (CCC) in pain and hyperalgesia. Curr Topics Med Chem. 5:547–555.

Price TJ, Hargreaves KM, Cervero F (2006) Protein expression and mRNA cellular distribution of the NKCC1 co-transporter in the dorsal root and trigeminal ganglia of the rat. Brain Res. 27:146–158.

Rees H, Sluka KA, Westlund KN, Willis WD (1994) Do dorsal root reflexes augment peripheral inflammation? Neuroreport. 5:821–824.

Rees H, Sluka KA, Westlund KN, Willis WD (1995) The role of glutamate and GABA receptors in the generation of dorsal root reflexes by acute arthritis in the anaesthetized rat. J Physiol. 484:437–445.

Sammons MJ, Raval P, Davey PT, Rogers D, Parsons AA, Bingham S (2000) Carrageenan-induced thermal hyperalgesia in the mouse: role of nerve growth factor and the mitogen-activated protein kinase pathway. Brain Res. 876:48–54.

Sandkhuler J (2007) Understanding LTP in pain pathways. Mol Pain. April 3;3:9.

Sandkühler J (2000) Learning and memory in pain pathways. Pain. 88:113–118.

Sandkühler J, Liu X (1998) Induction of long-term potentiation at spinal synapses by noxious stimulation or nerve injury. Eur J Neurosci. 10:2476–2480.

Schomberg SL, Su G, Haworth RA, Sun D (2001) Stimulation of Na-K-2Cl cotransporter in neurons by activation of non-NMDA ionotropic receptor and group-I mGluRs. J Neurophysiol. 85(6):2563–2575.

Spike RC, Kerr R, Maxwell DJ, Todd AJ (1998) GluR1 and GluR2/3 subunits of the AMPA-type glutamate receptor are associated with particular types of neurone in laminae I–III of the spinal dorsal horn of the rat. Eur J Neurosci. 10:324–333.

Sung KW, Kirby M, McDonald MP, Lovinger DM, Delpire E (2000) Abnormal GABAA receptor-mediated currents in dorsal root ganglion neurons isolated from Na-K-2Cl cotransporter null mice. J Neurosci. 20(20):7531–7538.

Thomas KL, Hunt SP (1993) The regional distribution of extracellularly regulated kinase-1 and -2 messenger RNA in the adult rat central nervous system. Neuroscience. 56:741–757.

Willis WD Jr (1999) Dorsal root potentials and dorsal root reflexes: a double-edged sword. Exp Brain Res. 124(4):395–421.

Chapter 14
Descending Modulation of Pain

Lucy Bee and Anthony Dickenson

Abstract Brainstem structures engage descending facilitatory and inhibitory neurones to potentiate or suppress the passage of sensory inputs from spinal loci to the brain. The final output for this bidirectional control is the rostral ventromedial medulla (RVM), which shapes sensory processing via relays between the spinal cord and brain, ultimately influencing pain perception via On and Off cells. In this chapter we look at how the balance between these cells' output can be reversibly and transiently altered so that nociceptive signals are enhanced or suppressed. Moreover, we look at how the descending modulatory system may become maladaptive and durably altered so that pain outlasts its biological usefulness and cause; following nerve injury for example, increased descending facilitatory output may establish an unrelenting feed-forward compensatory circuit between the periphery, spinal cord and brain to support chronic pain. We discuss how descending serotonergic action at spinal receptors is a major player in this top-down pain mechanism, and also look at the contribution of the descending noradrenergic system. Finally, we suggest that activity in the spino-bulbo-spinal circuit influences analgesic drug efficacy, and speculate on the contribution of aberrant supraspinal function to centrally-based pains such as fibromyalgia syndrome (FMS), as well as opioid-induced hyperalgesia.

Abbreviations

AMPA	α-amino-3-hydroxy-5-methyl-4-isoxazoleproprionate
CFA	Complete Freunds Adjuvant
CIBP	Cancer-Induced Bone Pain
CWMP	chronic widespread muscle pain
5,7-DHT	5,7-dihydroxytryptamine

A. Dickenson (✉)
Neuroscience, Physiology and Pharmacology, University College London, London
WC1E 6BT, UK
e-mail: anthony.dickenson@ucl.ac.uk

M. Malcangio (ed.), *Synaptic Plasticity in Pain*,
DOI 10.1007/978-1-4419-0226-9_14, © Springer Science+Business Media, LLC 2009

DLF   dorsolateral funiculus
DLPT   dorso-lateral pontine tegmentum
*f*MRI   functional magnetic resonance imaging
FMS   fibromyalgia syndrome
GABA   γ-amino butyric acid
GAD   glutamic acid decarboxylase
GBP   gabapentin
5HT   5-hydroxytryptamine, also known as serotonin
IASP   International Association for the Study of Pain
LTP   long term potentiation
MOR   μ-opioid receptor
NA   noradrenaline
NC   nucleus cuneiformis
NK   neurokinin
NMDA N-methyl-D-aspartate
NNT   number needed to treat
NRM   nucleus raphé magnus
PAG   periaqueductal gray
PGB   pregabalin
RVM   rostral ventromedial medulla
SNRI   serotonin and noradrenaline reuptake inhibitor
SPA   stimulation-produced analgesia
SSRI   selective serotonin reuptake inhibitor
TCA   tricyclic antidepressants

14.1 Introduction

Pain is a conscious experience that consists of sensory-discriminative, cognitive-evaluative and affective-emotional components. Recognition of the multi-dimensional nature of both the detection and perception of pain is central to the International Association for the Study of Pain (IASP)'s definition of this phenomenon as 'an unpleasant sensory and emotional experience associated with actual or potential tissue damage, or described in terms of such damage' (Merskey, 1994). Stemming from this definition is the acknowledgement that there is no fixed, dependable relationship between the intensity of a potentially harmful, or 'noxious' stimulus and perceived pain intensity. Instead this relationship is governed, to different extents, by individual variation (including genetics), previous damage, and the condition of tissue and nerves. In this context, it is clear that pain is not simply a series of afferent relays between the periphery and the brain, but a network where alterations, modulations, and plasticities can shift the gain and emphasis of sensory messages in a complex manner. Furthermore, the relationship between noxious input and painful output is influenced by demographics (e.g. age and gender), the environmental framework in which the stimulus is received, as well as the psychological and emotional states of the recipient. With so many contributory factors, it is at once easy to predict the problem of pain that challenges physicians and scientists alike, presenting millions of people worldwide with a largely unmet clinical need; management of chronic pain is complex and requires an understanding that multiple mechanisms may be responsible for single symptoms and single mechanisms may be responsible for multiple symptoms in both the central and peripheral nervous systems.

Transient and acute pain serve as useful biological warning functions that protect the integrity and survival of an organism. However, when pain outlasts its usefulness, existing beyond the time it takes for injuries to heal (or possibly existing in the absence of any identifiable 'organic' cause), incurring secondary symptoms or co-morbidities such as anxiety and depression, it becomes problematic and can greatly decrease function and quality of life. This progression towards pain deemed as 'chronic' means that pain becomes a syndrome in its own right rather than a by-product of some other process, and it is in this respect that it should be attended to and treated. Moreover, not only are co-morbidities a problem for the patient *per se*, but sleep problems, anxiety and depression can enhance nociception, since the affective and sensory components of pain interact.

The nervous system's capacity for re-organisation is on the one hand advantageous since pain behaviours can be subordinated to more pressing and urgent needs of the animal, including survival and attending to innate or learned stressors, yet it can alternatively present a problem if the system becomes maladaptive and permits long-lasting pain in the absence of injury. Such abnormalities may result from changes at peripheral, spinal, and/or supraspinal levels; with respect to the former, a multiplicity of mechanisms that include nociceptor alterations, ion channel dysregulation and prolonged exposure of the neuronal environment to agents such as growth factors and inflammatory mediators can lower nociceptive thresholds and enhance their responsiveness to peripheral stimuli, whilst re-wiring of primary afferent fibres can lead to an expansion of peripheral receptive fields. These changes promote noxious transmission from a more spatially and sensory diverse range of stimuli, consequently giving rise to phenomena such as allodynia (pain due to a stimulus that does not usually provoke pain) and hyperalgesia (an exaggerated response to an already noxious stimulus). Peripheral changes may be acute and reversible, leading to a dynamic and somewhat transient enhancement of the responses of central neurones, or alternatively the changes may set a precedent for protein phosphorylation and alterations in gene expression, giving rise to long-lasting increases in synaptic efficacy in the dorsal horn. On this, it is well established that mechanisms of central spinal sensitisation contribute to the enhancement of messages that ascend to higher centres (D'Mello and Dickenson, 2008). This enhancement is further underscored by spinal changes that may include recruitment and activation of microglia and loss of segmental spinal inhibitory mechanisms. Moreover, the balance between facilitation and inhibition can be tipped in favour of the former due to alterations in descending controls that project to the spinal cord from supraspinal areas, to further impact and potentiate painful outcome.

Sensory information from the periphery synapses in the dorsal horn of the spinal cord and is transferred up to supraspinal areas via functionally separate and anatomically parallel pathways, which, together with their targets, eventually decode the nociceptive signal into an elaborate pain experience. These pathways, including the spino-parabrachial and spino-thalamic pathways, relay through areas that attach emotional and contextual meaning to the nociceptive signal, in addition to areas that align homeostatic processes such as blood pressure and heart rate to the signal's exaction. Moreover, activity in these areas determines the signal that passes back down to the spinal cord to alter the volume of nociceptive processing therein, since there is significant feedback regulation of sensory signals in the dorsal horn by means of a spino-bulbo-spinal loop; this circuit originates in lamina I of the spinal cord, projects to supraspinal areas and finally passes back down through midbrain and brainstem areas to spinal and trigeminal neurones to either inhibit pain transmission via pain suppression pathways (Basbaum and Fields, 1978), or facilitate pain transmission via positive feed-forward pathways (Suzuki et al., 2002).

14.2 Descending Modulatory Control and the Placebo Effect

This so-called 'descending modulatory control' provides a neural substrate through which the brain, under the direction of forebrain motivational, cognitive and affective systems, can influence nociception and pain, not only with respect to their processing and interpretation, but also with respect to their treatment. The brain's ability to influence pain is the principle that underlies the placebo effect, which in this context refers to the capacity of an inert substance to produce analgesia with no pharmacological basis to do so. The success of placebo crucially depends on the subject's conscious perception of the 'therapeutic intervention', with improvement attributable to expectation of relief. *f*MRI scans have identified common neural mechanisms shared by placebo analgesia and opioid analgesia, and activity in these areas can be suppressed in both circumstances by the μ-opioid receptor (MOR) antagonist naloxone (Benedetti et al., 1999). Furthermore, tests and imaging studies have shown that the analgesic placebo effect is partly mediated by descending systems that use endogenous opioids as neuromodulators (Petrovic et al., 2002), whilst co-variant activity between different brainstem areas during placebo analgesia further emphasises the role of this phylogenetically old part of the brain in pain control.

14.3 Top-Down Modulation of Spinal Processing
   from the Brainstem

A major component of the final common output, or 'efferent limb' of the descending modulatory system is the rostral ventromedial medulla (RVM) which has the serotonin-rich nucleus raphe magnus (NRM) at its core and sits at the base of the brain close to the ponto-medullary junction, ideally positioned to filter neuronal signals that descend to the spinal cord. The RVM receives direct input from the midbrain periaqueductal gray (PAG) and the nucleus cuneiformis (NC), and indirect input from the prefrontal cortex, the amygdala and the anterior cingulate cortex. It is also reciprocally innervated by the dorsolateral pontine tegmentum (DLPT), a brainstem area that clusters together cell bodies from noradrenergic neurones that can directly and indirectly inhibit dorsal horn neuronal activity. RVM descending projections travel in the dorsolateral funiculus (DLF) and branch bilaterally in the dorsal horn (Holstege and Kuypers, 1982; Jones and Light, 1990), whereupon synapses are made with the terminals of primary afferent neurones, ascending tract neurones, intrinsic interneurones and the terminals of other descending neurones. Anterograde labels and antidromic stimulation have identified intense RVM innervation of laminae I, II and V (Basbaum, 1981; Skagerberg et al., 1985), which are territories occupied by the central terminals of nociceptive neurones. The RVM does not however exclusively serve nociception since its neurones also descend to circumcanular lamina X to modulate parasympathetic and sympathetic outflow, affecting autonomic targets such as the heart, blood vessels and the adrenal medulla in circumstances unrelated to pain. Ultimately however, the RVM's modulation of nociception may trump other functions and supersede autonomic responses, causing them to react in line with the noxious stimulus (for example by increasing heart rate and blood pressure).

Early conclusions regarding the role of supraspinal areas in pain control came from seminal studies based on stimulation-produced analgesia (SPA), since electrically stimulating the PAG caused rats to show no overt signs of distress to an otherwise nociceptive procedure (Reynolds, 1969). It was further shown that electrically stimulating the PAG can block intractable pain in humans (Hosobuchi et al., 1977), and that SPA can also be triggered downstream of the PAG in the RVM through descending pathways (Fields and Basbaum, 1978). Over the years this was extended to show that similar sites could support opioid analgesia, and that within the brainstem serotonin and noradrenaline were implicated in both processes. Later it became clear that there are descending noradrenergic pathways from the brainstem that modulate spinal activity with a relatively universal inhibitory action, mediated through spinal α_2-adrenoceptors (Millan, 2002). By contrast, the multitude of 5HT receptors suggested that this monoamine might play roles other than simple inhibition of function. Indeed, there is now a great deal of evidence to suggest that the maintenance of chronic pain states, whether the result of nerve trauma or inflammation, is dependant to a large degree on inappropriate activation of descending pathways from the brainstem (Urban et al., 1999; Porreca et al., 2001; Monconduit et al., 2002; Ren and Dubner, 2002), and so moving on from early emphases on descending inhibitions, it is now clear that descending facilitatory pathways are required for the full expression of chronic pain (Burgess, 2002). Ergo, whilst the RVM may increase its inhibitory output and serve a protective role in certain circumstances and situations pertaining to survival (i.e. stress-induced analgesia), it may alternatively increase its facilitatory drive such that pain transmission increases (for example during neuropathic pain). Moreover, the RVM may become maladaptive and permit long-lasting abnormal pain. In this way, acute pain which has a probable limited duration and identifiable temporal and causal relationship to injury or disease may progress to chronic pain and outlast the time it takes for the injury to heal. Indeed, although the transition from acute to chronic pain is clearly multifactorial, brainstem mechanisms that are themselves driven by afferent activity as well as central cognitive events such as depression, fear and anxiety are likely important candidates in these processes.

14.4 RVM Output Neurones

Much of the research on supraspinal control of nociceptive processing has focussed on the RVM, an area that contains subsets of neurochemically distinct projection neurones, some of which are serotonergic (Fields et al., 1983a, Gao and Mason, 2001). Electrical stimulation of the RVM results in inhibition or facilitation of spinal nociception depending upon the intensity of the stimulus (Urban et al., 1999; Fields, 2004), with low intensity stimulation tending to produce facilitation, and high intensity stimulation producing inhibition. Likewise, at low intra-RVM doses of glutamate, spinal nociception is facilitated, whilst at higher doses spinal nociception is suppressed (Zhuo et al., 2002). The bi-directional output from the RVM is a function of On and Off cells which can respectively increase and decrease dorsal horn activity (Fields et al., 1983a). The responses of On and Off cells to peripheral noxious stimuli are inversely predictive of their responses to systemic or local mu-opioid administration (Fields et al., 1983b, Barbaro et al., 1986); thus On cells, which show a burst of activity immediately prior to a reflex response to noxious heat are inhibited by morphine, whilst Off cells, which are thought to produce tonic inhibition that is turned off by pain, show a pause in firing before the same withdrawal response and are (indirectly) activated by mu-opioids (Fig. 14.1). Findings of an increased On cell discharge following prolonged noxious stimulation led to speculation that these cells enhance nociception through activation of descending facilitatory pathways. Moreover, evidence suggests that activation of On cells is critical for the hypersensitivities associated with a range of pain states (Porreca et al., 2001; Neubert et al., 2004; Vera-Portocarrero et al., 2006).

On the basis of early RVM electrophysiological recordings, it was originally thought that the activities of On and Off cells were biphasic and antagonistic. However, this thinking has been revised and it is now accepted that both groups of cells can be active at the same time, which means that the *net output* projecting upon the spinal cord depends on the variable weight of the excitatory and inhibitory influences from the brainstem. Indeed, in the presence of colonic inflammation, microinjecting NMDA receptor antagonists into the RVM attenuates behavioural hypersensitivities, whilst AMPA or kainite receptor

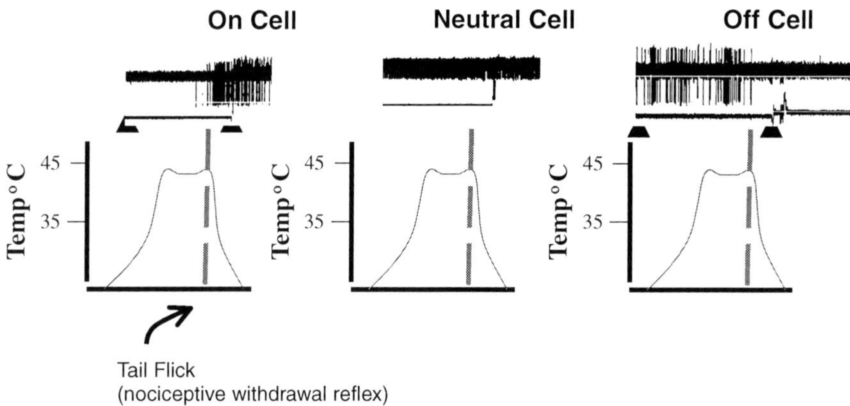

Fig. 14.1 The typical firing patterns of RVM On, Neutral and Off cells just prior to nociceptive withdrawal of the tail in response to noxious heat. On cells accelerate firing, neutral cells show no discernible change in activity, and Off cells show a pause in activity immediately before the reflex response

antagonists enhance hypersensitivities, which highlights that in the presence of inflammatory pain, inhibitory and facilitatory neurones are simultaneously active (Coutinho et al., 1998). In addition, it has been shown that there is an increase in both the On cell discharge and number of Off cells recorded from in arthritic rats (Montagne-Clavel and Oliveras, 1994), which undermines the 'mutually exclusive' concept and the associated theory that On cells function as inhibitory interneurones within the RVM to suppress Off cell activity. Recent work in which simultaneous recordings from On and Off cells are made during heat-evoked paw withdrawal in lightly anaesthetised rats showed that Off cells typically cease firing *before* the On cell reflex-related burst in firing (Cleary et al., 2008). This contradicts the idea that On cells may mediate the Off cell pause and instead points to an independent role of these facilitatory neurones in pain modulation.

Long-term plasticity in the RVM associated with chronic pain may be partly attributable to changes in the activity of neutral cells, a class of RVM neurones that show little or no consistent change in firing pattern related to the nocifensive withdrawal response. It is possible that these cells switch responsiveness to become On cell-like in defined circumstances (Urban et al., 1999). In the context of hindpaw inflammation, this transition is said to last for at least 24 h, a timescale that correlates with temporal changes in RVM activity following injury. Additionally, chronic morphine has been reported to increase the number of On cells in the RVM at the expense of neutral cells (Meng and Harasawa, 2007). It has therefore been proposed that chronic morphine sensitises a subpopulation of neutral cells to noxious stimulation, which may well account, at least in part, for the clinical and laboratory phenomenon of opioid-induced hyperalgesia, or 'paradoxical pain' (Mao et al., 1995; Celerier et al., 2001; Ossipov et al., 2005).

14.5 The RVM and Opioid Analgesia

The RVM's central role in opioid analgesia was established following observations that anti-nociceptive doses of MOR agonists in the RVM could elicit antinociception (Yaksh et al., 1976; Dickenson et al., 1979), and that inactivating this brainstem area could attenuate the anti-nociceptive actions of systemically administered morphine (Proudfit, 1980). RVM neurones express multiple opioid receptor subtypes, with functionally distinct neurones bearing different expression profiles (Marinelli et al., 2002). MOR agonists directly hyperpolarise and inhibit facilitatory neurones (putative On cells), and reduce excitatory post-synaptic currents in approximately half of spinally projecting neurones (Heinricher et al., 1992, Marinelli et al., 2002). Equivalent doses of this same agonist can inhibit RVM inhibitory post-synaptic currents and activate Off cells. Given that the direct cellular actions of morphine are mediated through a G-protein coupled inhibitory receptor, and also that MOR and the GABA synthesising enzyme GAD double-label in a subset of RVM neurones, it is highly likely that morphine-induced activation of Off cells occurs via disinhibition. On the other hand it is thought that inhibitory actions of opioids on On cells occur via more direct synaptic connections (Fig. 14.2). Consistent with these suggestions are findings that enkephalin-containing terminals in the RVM directly appose the axons of intracellularly labelled On cells but not Off cells (Mason et al., 1992). Furthermore, slice recordings have shown that more than three quarters of spinally-projecting RVM neurones directly respond to MOR agonists (these responsive neurones may fully or partially represent the population of On cells) whilst an even greater proportion of 'unidentified' RVM neurones (i.e. not specifically spinally-projecting) also respond to MOR stimulation (these neurones may represent GABAergic neurones that in turn synapse with Off cells) (Marinelli et al., 2002). Thus for clarity, evidence seems to suggest that brainstem-mediated morphine analgesia occurs through the direct inhibition of On cells (that otherwise facilitate bulbospinal neurones) and through the indirect facilitation of Off cells which leads to inhibition.

14.6 RVM Neurones and Serotonin

The heterogeneity of RVM neurones extends beyond opioid receptors and responses to noxious stimuli, to their electrophysiological and pharmacological properties (Mason, 1997; Marinelli et al., 2004; Zhang et al., 2006). In particular, different opioid-responding neurones stain differently for tryptophan hydroxylase (TPH), a cellular marker of serotonin content (Marinelli et al., 2002). Early ideas suggested that RVM-mediated morphine analgesia occurred via the actions of serotonin (sourced from the RVM) on inhibitory 5HT receptors in the dorsal horn of the spinal cord (Wigdor and Wilcox, 1987). However, cellular recordings from NRM neurones have since shown that

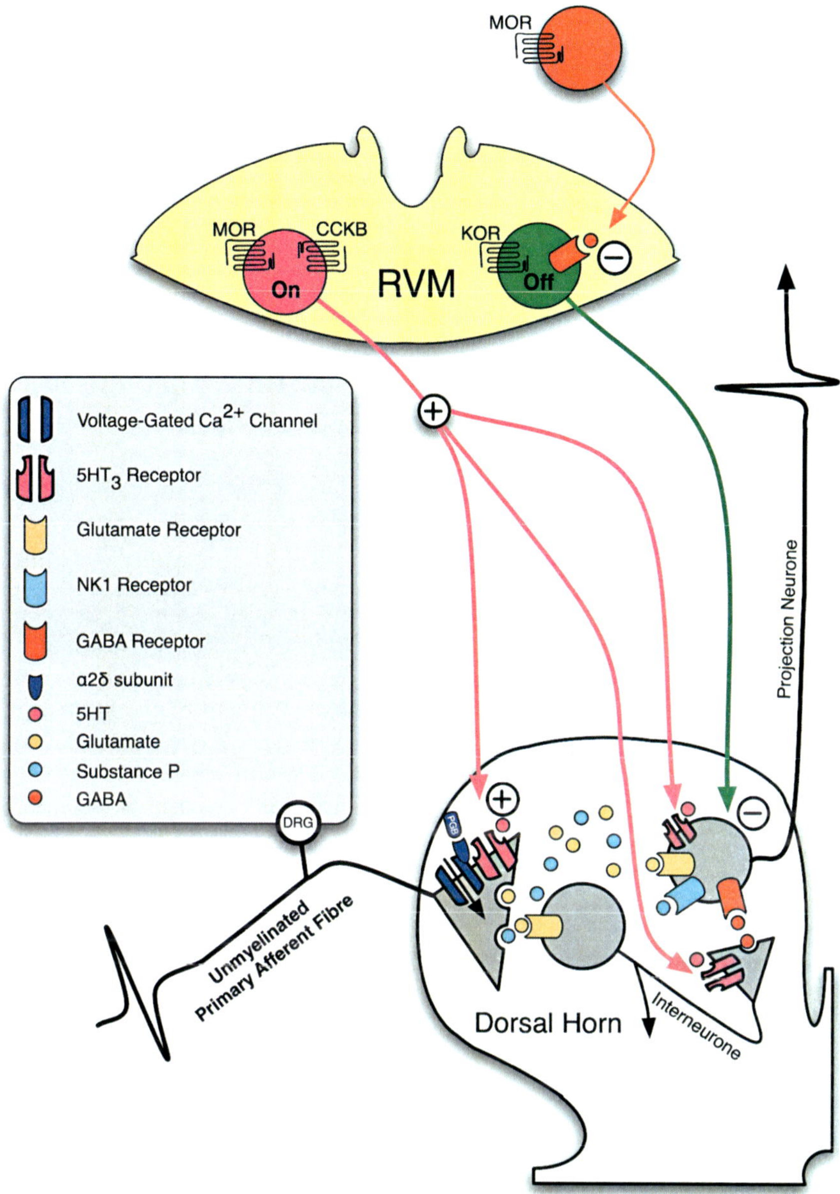

Fig. 14.2 Neuronal connections between the periphery, spinal cord and supraspinal areas. The RVM gives rise to facilitatory On cells and inhibitory Off cells that can modulate the processing of sensory information in the dorsal horn of the spinal cord. RVM Off cells receive an extrinsic GABAergic input that is sensitive to mu-opioid agonists (hence Off cells are disinhibited by MOR agonists) and their firing inhibits spinal cord activity. On cells on the other hand are directly inhibited by mu-opioid receptor agonists

activation of serotonergic neurones is not necessary for morphine analgesia (Gao et al., 1998; Arvidsson et al., 1995). Notwithstanding observations that systemic morphine increases the concentration of 5HT metabolites in the RVM and spinal cord (Rivot et al., 1988; Rivot et al., 1989; Matos et al., 1992), and that intrathecal 5HT receptor antagonists can modulate the consequent analgesia (Wigdor and Wilcox, 1987), it may be the case that morphine affects other RVM-driven processes, such as behavioural state, which may themselves alter serotonergic tone. Thus, a morphine-evoked increase in 5HT release may be indirect and secondary to primary opioid effects (Gao et al., 1998).

14.7 Different 5HT Receptors Mediate the Differential Effects of Spinal Serotonin

Like On and Off cells, the terminals of descending serotonergic axons are well placed in superficial dorsal horn laminae to modulate nociception. The capacity of 5HT to decrease or increase spinal neuronal responses is attributable to functionally opposing receptor subtypes (Yaksh and Wilson, 1979; Hylden and Wilcox, 1983). This bi-directionality is apparent in mice lacking serotonergic neurones (Zhao et al., 2007), since although responses to mechanical stimuli are enhanced, responses to inflammation are reduced as a consequence of this ablation. Interestingly, there may be a sub-modality specificity of the brainstem's responses to, and regulation of, sensory stimuli, with a preference for modulating mechanical sensory input, since inactivating the RVM or transecting the spinal cord selectively reduces mechanical – but not thermal-hypersensitivities in nerve-injured rats (Sung et al., 1998), a finding which is mirrored by the selective inhibition of mechanical-evoked responses in naïve animals by spinal ondansetron, a $5HT_3$ receptor antagonist, which likely blocks the spinal consequence of a descending serotonergic facilitatory drive (Suzuki et al., 2004). Furthermore, noxious mechanical stimulation results in increased extracellular 5HT in the RVM that is paralleled by increases in heart rate and blood pressure, yet thermal stimulation does not influence brainstem monoamine concentration despite affecting cardiovascular function (Karlsson et al., 2006). The pro-nociceptive actions of spinal 5HT have been well described and are mediated, at least in part, through excitatory $5HT_{2A}$ and the aforementioned $5HT_3$ receptors (Green et al., 2000; Zeitz et al., 2002; Sasaki et al., 2006). However, consistent inhibitory effects of $5HT_1$ receptor activation have been reported in animals, and $5HT_{1B/D}$ receptors have established roles in migraine treatment. In the case of non-cranial pains, anti-nociceptive actions of descending 5HT are mediated predominantly by inhibitory $5HT_{1A}$ receptors (Millan, 2002).

There have been reports of *hypo*algesia induced by $5HT_3$ receptor stimulation on acute pain scores (Glaum and Anderson, 1988; Glaum et al., 1990), which may be due to facilitation of GABA release from interneurones in the dorsal horn. However, $5HT_3$ receptor-mediated facilitation of spinal activity prevails in more enduring models of pain, with effects attributed to a direct pre-synaptic facilitation of excitatory neurotransmitters from the central terminals of small-diameter primary afferent neurones. Given the significant analgesia that is exerted by tropisetron and ondansetron within laboratory and clinical settings (Crisp et al., 1991; McCleane et al., 2003), it is feasible that in the chronic state, up-regulated descending serotonergic facilitations from the brainstem (Suzuki et al., 2004) mask parallel descending serotonergic inhibitions that may prevail during the acute phase of nociception (possibly before the brainstem has been sufficiently primed by aberrant peripheral activity to increase its On cell output). Indeed, after spinal cord injury, spinal 5HT (which is exclusively sourced from the brainstem) transiently reduces behavioural hypersensitivities, yet the longer lasting effect of increased 5HT content in spinal segments contributing to allodynic dermatomes is $5HT_3$ receptor-mediated facilitation (Oatway et al., 2004). Consistent with the hypothesis that serotonergic facilitations assume an enhanced role during chronic pain, LTP-like phenomena in deep dorsal horn neurones are partly regulated by $5HT_3$ receptor-mediated descending facilitatory controls (Rygh et al., 2006).

The nature of the link between serotonergic neurones and brainstem On and Off cells is ill-defined and remains the subject of continued investigation (Braz and Basbaum, 2008). Evidence suggests that there are parallel and interconnected serotonergic and non-serotonergic descending pathways arising from the RVM (Braz and Basbaum, 2008), and that intrinsic serotonergic neurones are highly involved with, and critical integrators of, the output from the RVM. Whilst precise correlations should not be assumed, many spinally-projecting μ-opioid responsive 'secondary' neurones, which appear similar to On cells *in vivo* (Fields et al., 1983b, 1991; Heinricher et al., 1994), use 5HT as a neurotransmitter (Marinelli et al., 2002). This suggests that some, but not necessarily all descending On cells may use 5HT as a neurotransmitter, and in reverse that some, but not necessarily all descending 5HT-containing neurones are On cells, hence the observation that brainstem serotonergic neurones identified by antidromic conduction velocities show excitatory responses to noxious sensory stimuli (Wessendorf and Anderson, 1983).

In many ways the RVM is seen as a relay between higher brain areas and the spinal cord and can be influenced by supraspinal areas involved in pain processing, such as the PAG, the anterior cingulate and the insular cortex. The functional consequence of these connections is that the brain can make appropriate adjustments to nociceptive processing informed by environmental contingencies such as threat or recovery from injury (Johansen and Fields, 2004). Thus it is likely that the efficacy profile of analgesic agents that have a central site of action, including anti-depressants and anti-convulsants, will be variable and dependent on the individual, and in terms of their actions on noradrenaline and 5HT will lead to different functional effects on pain.

14.8 The Spino-Bulbo-Spinal Loop

Over the past six years, a general hypothesis concerning the regulation of spinal excitability and the control of chronic pain has been proposed, building upon previous work on both ascending and descending pathways (Hunt, 2000; Suzuki et al., 2002; Suzuki et al., 2004). Essentially, starting from earlier observations, a key role for ascending pathways that are derived from a small group of neurones has been described. These neurones sit almost exclusively within laminae I and III of the dorsal horn and express the Substance P receptor NK1 (Cheunsuang et al., 2002) and are needed for both wind-up and LTP of spinal neurones following activation of nociceptive sensory afferents (Ikeda et al., 2003; Rygh et al., 2006). As described above, these neurones project upon the brainstem, particularly within areas that subsequently send information to the limbic system and somatosensory cortex (Gauriau and Bernard, 2004) Both neuropathic and inflammatory pain states are attenuated when this pathway is lesioned when a saporin-Substance P toxic conjugate, which specifically ablates neurones expressing the NK1 receptor, is applied intrathecally (Nichols et al., 1999). The lamina I–III/NK1 pathway is essential for the generation of both wind-up and LTP in deep dorsal horn neurones since both events were lost following this ablation, in addition to the coding of peripheral stimuli and properties of deep dorsal horn neurones in neuropathic rats. By contrast, these neurones had increased responses to innocuous stimuli and enlarged peripheral receptive fields (Suzuki et al., 2002, 2004). The link between this spinal bulbo-spinal loop and the descending serotonergic pathway came from observations that many of the effects of this ablation could be reproduced in part by either ablating descending serotonergic pathways, or by antagonising excitatory $5HT_3$ receptors in the spinal cord with ondansetron (Suzuki et al., 2002). Moreover, full activation of RVM serotonergic neurones did not occur as usual following lamina I–III/NK1 ablation. More recently it has been shown that loss of descending serotonergic pathways following intrathecal administration of 5,7-DHT, a neurotoxin that depletes cells of 5HT without influencing other neural and non-neuronal systems (Duan and Sawynok, 1987), severely attenuates the maintenance phases of both neuropathic and inflammatory pain states (Rahman et al., 2006; Svensson et al., 2006).

14.9 Anti-Depressants and Anti-Convulsants for the Treatment of Chronic Pain

The role of anti-depressants and anti-convulsants in the therapy of chronic pain is clear from clinical studies and meta-analyses of published data where the numbers needed to treat for these classes of drug with respect to neuropathic pain are approximately 1 in 3–5 (i.e. $NNT^1 = 3-5$). The initial use of compounds in these classes was pragmatic and not based on any clear mechanistic basis. Indeed, part of the basis for the use of antidepressants in pain treatment was the recognition of depression as a treatable co-morbidity in many pain states (Meyer-Rosberg et al., 2001). However, although pain and depression utilise similar neurochemical substrates, and the diffusely organised limbic and sensory areas of the brain inter-connect, anti-depressants confer analgesic efficacy in the absence of depressive symptoms; it is clear that mood changes are independent of the ability of these drugs to reduce pain and that the necessary doses and time-courses to achieve efficacy in each setting often differ. Over the years the use of animal models of neuropathic pain has enabled a basis for the use of anti-depressants and anti-convulsants to be determined, yet their modes of action are very different which lends credence to the concept that multiple co-existing mechanisms are at play in chronic pain states that generate common symptoms and common treatments (Hansson and Dickenson, 2005). Anti-convulsants have a rational basis of action in that epilepsy and pain share the common characteristic of arising from excessive activity in neuronal circuits. Within this class, the main drugs are excitability blockers such as carbamazepine and lamotrigine, in addition to the 'alpha-2 delta ligands' gabapentin and pregabalin. Although the main action of these so-called 'gabapentinoids' are through spinal calcium channels (Gee et al., 1996; Luo et al., 2001), their state-dependent actions require descending serotonergic and noradrenergic pathways (Suzuki et al., 2005; Tanabe et al., 2005).

The analgesic actions of anti-depressants are similar in that they all increase the availability of monoamines, however their differential effects on the synaptic levels of these modulatory transmitters varies, from TCA/ SNRI effects on *both* noradrenaline and serotonin, through to the SSRIs which are selective for serotonin. Anti-depressant agents are often prescribed in the first-line treatment of many chronic pain states, in particular those of a neuropathic origin; a meta-analysis of several studies concluded that 30% of patients with neuropathic pain showed a 50% improvement in pain intensity following the use of anti-depressants (Sindrup and Jensen, 2000). TCAs head the anti-depressant efficacy table, achieving an NNT of approximately 2–3, with SNRIs lagging slightly behind with a NNT of 4–5, and SSRIs trailing still further with a NNT of approximately 7. Some data on SSRI efficacy question their analgesic capacity (Max et al., 1992), whilst other data report equivocal responses with efficacy described as 'moderate' at best, or 'clinically insufficient' (Saarto and Wiffen, 2005; Sindrup et al., 2005). The relative roles of the two monoamines in nociception may be gleaned from the clear analgesic effects of SNRIs such as duloxetine

---

[1] NNT is a statistical term that allows the effectiveness of therapeutics to be compared. It refers to the number of patients that need to be treated to prevent one adverse outcome. Hence, with respect to pain relief, the lower the number, the more effective the analgesic.

compared to the lesser effects of SSRIs such as fluoxetine in patients with neuropathic pain, suggesting a need for noradrenaline-mediated inhibitions in the drug effect.

14.10 Noradrenergic Inhibitory Pathways from the Brainstem

There is a large body of evidence showing that descending noradrenergic pathways exert inhibitory influences onto the spinal cord, and pharmacological data suggests that this is mediated primarily via activation of α_2-adrenoceptors (Howe et al., 1983; Yaksh, 1985; Proudfit, 1988). These receptors are present in high density within superficial laminae of the dorsal horn and are located on post-synaptic dorsal horn cells and primary afferent terminals (Nicholas et al., 1993; Shi et al., 1999; Nicholson et al., 2005). Immunohistochemical studies have shown that NK1 receptor expressing neurones in the superficial dorsal horn receive contacts from glutamatergic cells that have axons expressing α_{2C}-adrenoceptors (probably primary afferent neurones) (Olave and Maxwell, 2003). Thus, activation of these adrenoceptors with specific agonists results in a reduction in noxious-evoked neuronal activity and pain behaviours. Moreover, plasticity in descending modulatory systems during pathological pain can also include alteration in inhibitory, as well as facilitatory, output from the brainstem (Ren and Ruda, 1996; Green et al., 1998; Suzuki et al., 2002).

In a study of the effects of atipamezole, a selective α_2-adrenoceptor antagonist, it was shown through targeted ablation techniques that superficial dorsal horn cells expressing NK1 receptors are necessary for the activation of descending inhibitory pathways (Rahman et al., 2007). Thus the same spinal cells that drive descending $5HT_3$ receptor-mediated facilitations also drive descending α_2-adrenoceptor-mediated inhibitions from the brainstem. However, the former must dominate since ablation of the lamina I–III/NK1 cells led to a net *reduction* in evoked neuronal responses in the spinal cord, an effect that is enhanced even further in favour of $5HT_3$ receptor-mediated facilitations in pathophysiological states. Given the increased behavioural responses of rats to mechanical stimuli after nerve injury, it is not surprising, but still intriguing, that the shift to this abnormal state occurs in the direction of increased spinal $5HT_3$ receptor-mediated facilitations, and moreover that this shift is matched by reduced α_2-adrenoceptor-mediated inhibitory controls. Noradrenergic inhibitions mediated through α_2-adrenoceptors would therefore seem a key contributor to anti-depressant effectiveness in pain, since 5HT-mediated mechanisms can variably enhance and inhibit pain. This principle is upheld in mutant mice that lack central serotonergic neurones, since the SSRI fluoxetine failed to affect behavioural scores of inflammatory pain in these mice, yet the SNRI duloxetine remained effective, albeit at a lower analgesic efficacy (Zhao et al., 2007). The outcome of anti-depressant mediated increases in noradrenaline and

5HT, based on the factors considered above, could include restoration of reduced α_2-adrenoceptor inhibitory controls and a potential increase in $5HT_1$ receptor-mediated inhibitions. The latter effect may however be offset by increased synaptic 5HT driving increased $5HT_3$ receptor-mediated facilitations at spinal levels. However, it may also be the case that due to tight regulation of serotonergic transmission, increases in synaptic 5HT lead to an autoreceptor-mediated inhibition of neurotransmission, which would consequently reduce 5HT content within the synapse, and thus reduce $5HT_3$ receptor-mediated facilitations.

14.11 Descending Facilitations Influence Treatment Outcome and are Active in Different Models of Chronic Pain

In the case of nerve injury, it would appear that altered monoamine pathways, namely a net decrease in α_2-adrenoceptor-mediated inhibitions and an increase in $5HT_3$ receptor-mediated facilitations, favour hypersensitivity. Moreover, gabapentin (GBP) and pregabalin (PGB), drugs that are used in the mainline treatment of neuropathic pain, depend on activity in these pathways for their state-dependent efficacy (Suzuki et al., 2005), since in addition to their key spinal site of action (Hwang and Yaksh, 1997; Kaneko et al., 2000; Shimoyama et al., 2000), a supraspinal component exists (Petroff et al., 1996). With respect to the noradrenergic system, gabapentin is thought to act within the locus coeruleus in the brainstem to evoke descending inhibitions that culminate on spinal α_2-adrenoceptors (Tanabe et al., 2005). In terms of the serotonergic system, the injury-dependent enhancement of descending facilitatory controls acting on spinal $5HT_3$ receptors are necessary for the analgesic actions of GBP and PGB in neuropathic animals. This dependence may be a function of the close proximity of presynaptic voltage-gated Ca^{2+} channels and $5HT_3$ receptors on the central terminals of primary afferent neurones (Suzuki et al., 2005). Thus in rats that had lost a significant proportion of their lamina I/III projection neurones (following intrathecal delivery of saporin-Substance P), and therefore had a reduced spino-bulbo-spinal drive terminating on $5HT_3$ receptors, nerve ligation surgery did not cause typical behavioural hypersensitivities, whilst electrophysiological recordings revealed reduced neuronal responses in the dorsal horn. Furthermore, in these animals GBP was completely without effect, which is similar to the finding that PGB was ineffective in rats with brainstem injections of the toxic conjugate dermorphin-saporin, which specifically lesions cells in the RVM expressing the μ-opioid receptor (MOR) (see Fig. 14.3) (Bee and Dickenson, 2008).

In these set of experiments, behavioural, electrophysiological and pharmacological techniques were combined to show that the supraspinal facilitatory drive is essential for neuronal processing of supra-threshold stimuli in normal and neuropathic states, and that descending facilitatory neurones maintain

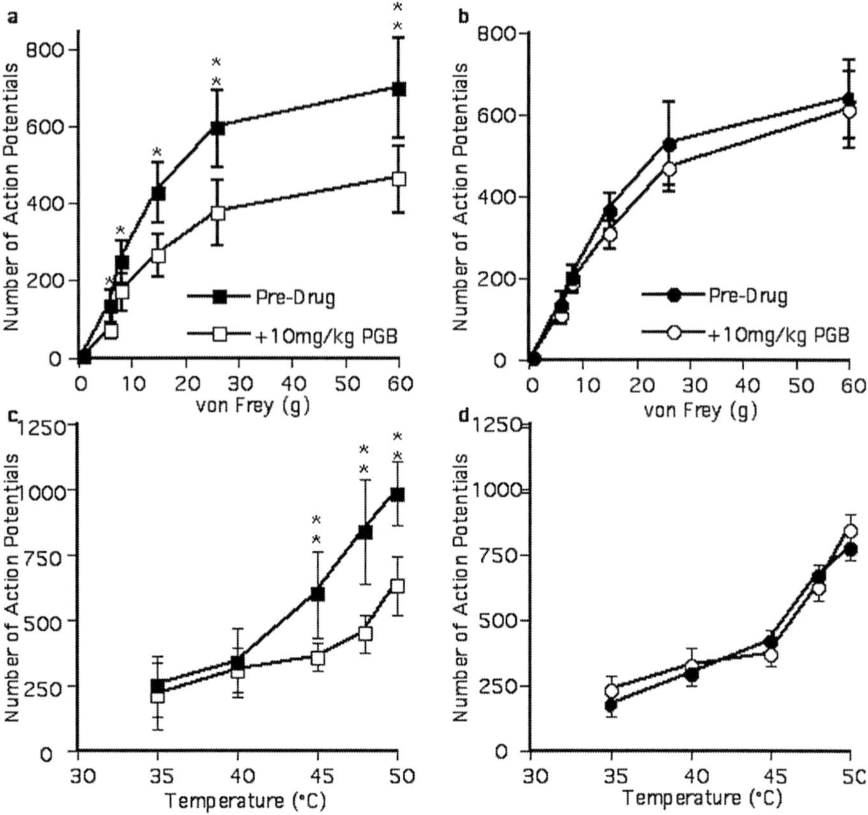

Fig. 14.3 Spinal neuronal responses to hindpaw mechanical (**a,b**) and thermal (**c,d**) stimulation in neuropathic rats that either had or had not their descending facilitatory drive intact (*left* and *right* hand panels respectively) are shown both before (*filled squares/circles*) and after (*open squares/circles*) the systemic administration of 10 mg/kg pregabalin. This drug reduced spinal neuronal responses to many of the applied stimuli in control spinal nerve ligated rats but was without effect in rats lacking medullary facilitatory cells

behavioural hypersensitivities to mechanical stimuli during the late stages of nerve injury. Moreover, these medullary neurones are essential for the state-dependent inhibitory actions of PGB in nerve-ligated rats. Thus during the early stages of injury (i.e. when the descending facilitaory neurones have little influence on the neuropathic phenotype), or following medullary MOR cell ablation, PGB is ineffective at inhibiting spinal neuronal responses, possibly due to quiescent activity at spinal $5HT_3$ receptors. This can however be overcome, and PGB's efficacy restored, by pharmacologically mimicking the descending drive at the spinal level with a $5HT_3$ receptor agonist (Bee and Dickenson, 2008). This is similar to the situation in nerve-ligated rats that had

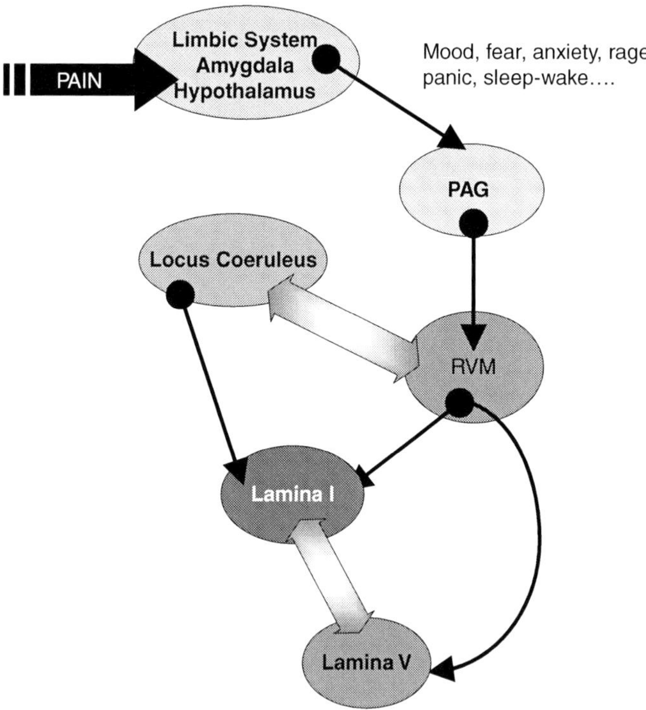

Fig. 14.4 A schematic outline of the triggering of descending controls by painful inputs into limbic areas of the brain, as well as by cognitive and affective processes. See text for full details

been intrathecally injected with saporin-Substance P as described above, since in this situation too the inhibitory actions of GBP could be shifted from ineffective to effective by spinal $5HT_3$ receptor stimulation (Suzuki et al., 2005). Lamina I/III projection neurones and RVM facilitatory neurones are integral to a spino-bulbo-spinal loop that reaches brain areas co-ordinating the sensory and affective components of pain, hence activity therein may influence painful outcome following nerve injury, and responsiveness to treatment with alpha-2 delta ligands.

The descending serotonergic pathway that terminates on spinal $5HT_3$ receptors is also enhanced in animal models of cancer-induced bone pain (CIBP) (Donovan-Rodriguez et al., 2006) where the pain state is thought to include inflammatory and neuropathic components, as well as factors that are unique to the CIBP phenotype (Urch et al., 2003). It is likely that the balance in inhibitory versus excitatory output from the brainstem varies in different pain states, and indeed, a look at inflammatory pain appears to bear out the idea of differential activity in different pathophysiological situations. Hence, in formalin-injected rats relative to naïve rats, there is both an ongoing α_2-adrenoceptor-mediated inhibition and an enhanced

$5HT_3$ receptor-mediated facilitation of spinal neurones. In mice lacking central serotonin, there is an increase in behavioural responses to this inflammatory agent and enhanced sensitivities to peripheral capsaicin, which suggests that usually there is an overall enhancement of 5HT inhibitory controls after chemical inflammation. Which receptors mediate the descending 5HT inhibitory pathway is unclear; $5HT_1$ receptors are a likely candidate, yet WAY-100635, a $5HT_{1A}$ receptor antagonist appears to have no effect on the formalin response. This indicates a lack of modulation by this specific receptor subtype, yet in mice lacking the $5HT_{1B}$ receptor there was an increase in the formalin response. This increase in sensitivity is in direct contrast to the *reduced* formalin sensitivity in littermate mice lacking the $5HT_{3A}$ receptor, a further example of 5HT's functional duality (Kayser et al., 2007).

In the Complete-Freunds Adjuvant (CFA) model of inflammation, there is evidence for a shift in the descending excitatory/inhibitory balance based on changes that link together *in vivo* AMPA receptor phosphorylation in RVM pain modulatory circuits, to enhanced descending modulations of nociception after inflammation (Guan et al., 2004). The other main glutamate receptor, the NMDA receptor, appears to enhance descending facilitations at early time points, but over time the dose-response curve of NMDA shifts to the left and descending inhibitions dominate to mask descending facilitations. This is similar to what is seen after chemical inflammation, and is diametrically opposite to what occurs in neuropathy, since studies have shown the temporal development of a dominant descending facilitatory drive that maintains neuropathic behaviours at late post-injury time points, which therefore implicates the descending facilitatory system in the maintenance of neuropathic pain, with peripheral mechanisms responsible for its genesis (Porreca et al., 2001; Burgess, 2002). One such mechanism is spontaneous afferent drive from the periphery; within 24 h of nerve injury there is an estimated four-to-six fold increase in ectopic firing from injured and adjacent nerves (Liu et al., 2000), which correlates with the early onset of behavioural hypersensitivities (Sun et al., 2005). However, this erroneous discharge tapers off over the course of a week (Han et al., 2000) and becomes insufficient to maintain neuropathic behaviours that can persist beyond 7 days, and possibly up to four months in animal models of neuropathy (Kim and Chung, 1992); thus, as ectopic activities diminish by post-operative day 5, descending facilitatory neurones acquire greater pathophysiological status. It should be noted however that deep dorsal horn neurones exhibit a high ongoing discharge more than 14 days after nerve injury (Chapman et al., 1998; Chu et al., 2004; Suzuki and Dickenson, 2006). This spontaneous activity, although significantly reduced with respect to activity at earlier time points, may still be important for maintaining the neuropathic phenotype (Sheen and Chung, 1993; Yoon et al., 1996).

The close interplay between peripheral and supraspinal mechanisms in neuropathic pain has been shown experimentally, since at later time points in nerve-injured animals, enhanced evoked transmitter release from peripheral nerves depends on descending facilitatory input from the RVM (Gardell et al., 2003).

This demonstrates the continued contribution of the peripheral nervous system to the pain state, and moreover that supraspinal systems act in concert with the peripheral nervous system to influence nociceptive transmission. The convergence of peripheral and central mechanisms likely explains why the application of an inflammatory agent increased On cell firing in the RVM, yet produced hyperalgesia in the treated limb only (Kincaid et al., 2006). A possible explanation for this restriction of effect which occurs despite the On cells' global and bilateral influence at multiple spinal segments (Fields et al., 1995) is that activation of these facilitatory neurones may be insufficient to produce hyperalgesia in the absence of peripheral drive (just as peripheral drive alone is insufficient to maintain hypersensitivities, as can be demonstrated by targeted ablation of medullary pain facilitating neurones, Porreca et al., 2001). Alternatively, descending inhibitory neurones may be recruited in parallel with descending facilitatory neurones to suppress excitability in circuits serving other (uninjured) areas of the body. This hypothesised heterotopic inhibition is akin to diffuse noxious inhibitory control whereby noxious stimuli activate inhibitory controls to sharpen the contrast between the stimulus zone and adjacent areas, having a net enhancing effect on the perceived intensity of the painful stimulus (Dickenson et al., 1981). Descending facilitations from the RVM can however induce hyperalgesia in both sides of the body after unilateral muscle insult (e.g. injecting acidic saline into the gastrocnemius muscle of the left hindlimb) (Tillu et al., 2007). This animal model is thought to replicate many features of human chronic widespread muscle pain (CWMP) (Gran, 2003), a condition characterised by alterations in descending modulatory controls, and in particular a loss of descending inhibitions (Kosek and Hansson, 1997). Hence, according to the theory above, this loss could explain the bilateral nature of the hyperalgesia. In the CWMP model, descending facilitations are thought to be critical for both the initiation and maintenance phases of pain, since local anaesthetic injections in the RVM at given time-points prevented the development of behavioural hypersensitivities and reversed established pain behaviours (Tillu et al., 2007). In addition to the quality of the nociceptive stimulus (i.e. inflammatory versus neuropathic), differences in the pattern of descending controls may relate to the location of the injury, since formalin injections in the muscle or in the skin induce neuronal activity in different areas of the midbrain (Keay and Bandler, 1993).

Finally, these spinal-supraspinal circuits that allow higher cognitive and emotional events to modulate spinal pain processing may also be relevant to visceral pain. Electrophysiological recordings from the abdominal muscles in anaesthetised rats during colonic distension have shown that descending serotonergic pathways that terminate on $5HT_3$ receptors have a key role in spinal processing of the noxious visceral input. Thus the pathways that impart a 'state-dependency' in neuropathy that enable $\alpha_2\delta$ ligands to alter pathophysiology, also appear active in visceral conditions.

The differential output and influences of descending controls from the brainstem could represent a homeostatic response of the CNS, and may relate to the pathological basis of neuropathy and the loss of afferent input which may drive compensatory increases in neuronal activity at peripheral, spinal and supraspinal levels. By contrast, peripheral sensitisation in inflammation favours central spinal excitability, which consequently triggers restitutive inhibitory mechanisms, both with respect to the enhanced inhibitory effects of spinal opioids, and enhanced descending inhibitory output from the brainstem.

14.12 Centrally Based Pains

Some of these non-localised changes in pain responsivity through descending influences could relate to fibromyalgia syndrome (FMS) a unique state characterised by diffuse, wide-spread pain in all quadrants of the body (Wolfe et al., 1990). In the absence of any peripheral pathology, it is thought that the pathogenesis of FMS is centrally based, and in particular that descending controls are altered without any concomitant drive from the periphery. Pain in this situation may be akin to endogenous depression where, unlike reactive depression, there is no explicit external cause for the mood changes. It is generally accepted that FMS is a clinical expression of a quasi rheumatological disorder in which the pain has a centrally-disruptive origin that is not necessarily driven by peripheral processes (Perrot et al., 2008). It is fitting therefore that effective drugs, including SNRIs (duloxetine and milnacipram), tramadol (an atypically acting opioid with additional monoamine actions) and more recently pregabalin, the $\alpha_2\delta$ ligand, all share the ability to interact with descending controls. In addition to improving pain scores, these drugs improve fatigue and sleep, which again implicates midbrain monoamine systems in the pathophysiology. FMS is one of a group of syndromes (which also includes irritable bowel, overactive bowel and chronic tension headaches) that suggests neurophysiological changes indicative of central sensitisation. As a result, peripheral nerves become sensitised, spinal cord neurones are rendered hyperexcitable, and ascending projections to higher centres can further trigger modulatory changes via tracts that descend through midbrain and brainstem areas. This trio of sensitising mechanisms act in parallel to effect 'bottom-up' and 'top-down' modulation. Given the nature of the links between the sensory and affective components of pain in midbrain and brainstem circuits, aberrant activity in the descending modulatory system may explain the hypothesised psychological and psychogenic basis of fibromyalgia (Rubin, 2005).

Abnormal supraspinal activity may additionally explain opioid-induced hyperalgesia and the related phenomenon of paradoxical pain. Many chronic pain states require increasing doses of opioids to maintain adequate pain relief, which, in addition to disease progression, is thought to be the consequence of opioid tolerance (defined as a decrease in efficacy after previous exposure to the same, or similar, drugs, Way et al., 1969). Moreover, many clinical reports have noted that opioids administered through different routes can unexpectedly produce hypersensitivities that often manifest in locationally and qualitatively distinct manners to the original pain complaint, particularly during rapid dose escalation (Blum et al., 2003) (Mercadante et al., 2003). Related observations have been validated pre-clinically in animal models (King et al., 2007), with early mechanistic interpretations citing cellular and peripheral adaptations. However, it is increasingly thought that opioid-induced hyperalgesia may be secondary to neuroplastic changes at supraspinal sites, with up-regulated descending facilitations from the RVM offsetting spinal opioid inhibitions (Gardell et al., 2006). Consequently, physical disruptions of the DLF pathway can block hyperaesthetic behaviours associated with sustained opioid delivery (Gardell et al., 2006). Sustained triggering of descending facilitatory influences from the RVM may act in concert with spinal mechanisms such as wind-up to enhance central sensitisation, giving rise to the diffuse pains associated with fibromyalgia and prolonged opioid use, which in each circumstance can be treated with antagonists acting at spinal NMDA or $5HT_3$ receptors (Lutfy et al., 1996; Farber et al., 2000; Graven-Nielsen et al., 2000).

14.13 Concluding Remarks

A cascade of consequences that may arise in the wake of tissue injury, nerve damage and malignant disease may set the precedent for chronic pain which persists longer than the temporal course of natural healing. These changes can occur peripherally (involving nociceptors, inflammatory mediators and ion channels for example), spinally (loss of inhibition and/or increased excitation induces facets of central sensitization) and supraspinally. Indeed with respect to the latter, the specialized role of midbrain and brainstem areas in pain processing has not only been demonstrated in animal studies but in human subjects too; human imaging studies have shown that supraspinal activations during painful stimulation of areas of secondary hyperalgesia are significantly localized in regions consistent with the nucleus cuneiformis and periaqueductal grey (Zambreanu et al., 2005), areas that collectively form the major source of input to the RVM. Furthermore, when the pattern of brain activity evoked by cold pain versus cold allodynia is compared in human volunteers, areas within the dorsolateral pons (that are consistent with the parabrachial area) are recruited in the processing of the latter, which verifies the role of the brainstem during central sensitization (Seifert and Maihofner, 2007). This is consistent with analyses of brainstem activity during dynamic mechanical allodynia, which highlighted neuronal changes and increased activity in the ipsilateral dorsal medulla/upper cervical cord in an area consistent with the location of the RVM (Mainero et al., 2007). Thus together with diffusion tractography techniques that confirm anatomical pathways mediating top-down control of nociceptive processing in humans (Hadjipavlou et al., 2006), these imaging studies provide a clear picture of the brainstem's representation of pain and central sensitization in humans, and concede the transfer of knowledge acquired in animal studies of chronic pain, to human correlates.

Such brainstem mechanisms include monoaminergic controls. Hence in the spinal dorsal horn, noradrenaline acting via α_2-adrenoceptors suppresses the release of excitatory transmitters from the central terminals of primary afferent fibres, and can also postsynaptically suppress the responses of spinal relay neurons. However, following inflammation or injury, the noradrenergic system is subject to various plastic changes that influence its antinociceptive efficacy. Likewise, 5HT acting at certain receptors in the spinal cord can mediate inhibition of nociception during acute pain, yet during states of chronic pain, the overriding and maladaptive influence of 5HT is excitation, which occurs at least partly through ionontropic $5HT_3$ receptors. Notwithstanding a 'magic bullet' approach of finding a single drug that acts at multiple sites to maximize pain control (for example a drug that synergistically targets both α_2-adrenoceptors and $5HT_3$ receptors), the multiplicity of mechanisms involved in chronic pain provides an empirical and rational basis for combination therapy – different drugs acting through different targets modulate more than one mechanism at more than one site. Finally, given the many underlying processes and the occasionally refractory nature of chronic pain, the concept of mechanisms-based treatments need to be appraised so that the relationship between signs and symptoms and treatment outcomes can lead pain control from the margins of medicine to the forefront of clinician's attention.

References

Arvidsson, U., Riedl, M., Chakrabarti, S., Lee, J. H., Nakano, A. H., Dado, R.J., Loh, H. H., Law, P. Y., Wessendorf, M. W. and Elde, R., 1995. Distribution and targeting of a mu-opioid receptor (MORI) in brain and spinal cord. J Neurosci. 15, 3328–3341.

Bee, L. A. and Dickenson, A. H., 2008. Descending facilitation from the brainstem determines behavioural and neuronal hypersensitivity following nerve injury and efficacy of pregabalin. Pain 140, 209–223.

Barbaro, N. M., Heinricher, M. M. and Fields, H. L., 1986. Putative pain modulating neurons in the rostral ventral medulla: reflex-related activity predicts effects of morphine. Brain Res. 366, 203–210.

Basbaum, A. I., 1981. Descending control of pain transmission: possible serotonergic–enkephalinergic interactions. Adv Exp Med Biol. 133, 177–189.

Basbaum, A. I. and Fields, H. L., 1978. Endogenous pain control mechanisms: review and hypothesis. Ann Neurol. 4, 451–462.

Benedetti, F., Amanzio, M., Baldi, S., Casadio, C. and Maggi, G., 1999. Inducing placebo respiratory depressant responses in humans via opioid receptors. Eur J Neurosci. 11, 625–631.

Blum, R. H., Novetsky, D., Shasha, D. and Fleishman, S., 2003. The multidisciplinary approach to bone metastases. Oncology (Williston Park). 17, 845–857; discussion 862–843, 867.

Braz, J. M. and Basbaum, A. I., 2008. Genetically expressed transneuronal tracer reveals direct and indirect serotonergic descending control circuits. J Comp Neurol. 507, 1990–2003.

Burgess, S. E., Gardell, L. R., Ossipov, M. H., Malan, T. P., Jr., Vanderah, T. W., Lai, J. and Porreca, F., 2002. Time-dependent facilitation from the rostral ventromedial medulla maintains, but does not initiate, neuropathic pain. J Neurosci. 22, 5129–5138.

Celerier, E., Laulin, J. P., Corcuff, J. B., Le Moal, M. and Simonnet, G., 2001. Progressive enhancement of delayed hyperalgesia induced by repeated heroin administration: a sensitization process. J Neurosci. 21, 4074–4080.

Chapman, V., Suzuki, R. and Dickenson, A. H., 1998. Electrophysiological characterization of spinal neuronal response properties in anaesthetized rats after ligation of spinal nerves L5–L6. J Physiol. 507 (Pt 3), 881–894.

Cheunsuang, O., Maxwell, D. and Morris, R., 2002. Spinal lamina I neurones that express neurokinin 1 receptors: II. Electrophysiological characteristics, responses to primary afferent stimulation and effects of a selective mu-opioid receptor agonist. Neuroscience. 111, 423–434.

Chu, K. L., Faltynek, C. R., Jarvis, M. F. and McGaraughty, S., 2004. Increased WDR spontaneous activity and receptive field size in rats following a neuropathic or inflammatory injury: implications for mechanical sensitivity. Neurosci Lett. 372, 123–126.

Cleary, D. R., Neubert, M. J. and Heinricher, M. M., 2008. Are opioid-sensitive neurons in the rostral ventromedial medulla inhibitory interneurons? Neuroscience. 151, 564–571.

Coutinho, S. V., Urban, M. O. and Gebhart, G. F., 1998. Role of glutamate receptors and nitric oxide in the rostral ventromedial medulla in visceral hyperalgesia. Pain. 78, 59–69.

Crisp, T., Stafinsky, J. L., Spanos, L. J., Uram, M., Perni, V. C. and Donepudi, H. B., 1991. Analgesic effects of serotonin and receptor-selective serotonin agonists in the rat spinal cord. Gen Pharmacol. 22, 247–251.

D'Mello, R. and Dickenson, A. H., 2008. Spinal cord mechanisms of pain. Br J Anaesth. 101, 8–16.

Dickenson, A. H., Fardin, V., Le Bars, D and Besson, J. M., 1979. Antinociceptive action following microinjection of methionine-enkephalin in the nucleus raphe magnus of the rat. Neurosci Lett. 15, 265–270.

Dickenson, A. H., Rivot, J. P., Chaouch, A., Besson, J. M. and Le Bars, D., 1981. Diffuse noxious inhibitory controls (DNIC) in the rat with or without pCPA pretreatment. Brain Res. 216, 313–321.

Donovan-Rodriguez, T., Urch, C. E. and Dickenson, A. H., 2006. Evidence of a role for descending serotonergic facilitation in a rat model of cancer-induced bone pain. Neurosci Lett. 393, 237–242.

Duan, J. and Sawynok, J., 1987. Enhancement of clonidine-induced analgesia by lesions induced with spinal and intracerebroventricular administration of 5,7-dihydroxytryptamine. Neuropharmacology. 26, 323–329.

Farber, L., Stratz, T., Bruckle, W., Spath, M., Pongratz, D., Lautenschlager, J., Kotter, I., Zoller, B., Peter, H. H., Neeck, G., Alten, R. and Muller, W., 2000. Efficacy and tolerability of tropisetron in primary fibromyalgia – a highly selective and competitive 5-HT3 receptor antagonist. German Fibromyalgia Study Group. Scand J Rheumatol Suppl. 113, 49–54.

Fields, H., 2004. State-dependent opioid control of pain. Nat Rev Neurosci. 5, 565–575.

Fields, H. L. and Basbaum, A. I., 1978. Brainstem control of spinal pain-transmission neurons. Annu Rev Physiol. 40, 217–248.

Fields, H. L., Bry, J., Hentall, I. and Zorman, G., 1983a. The activity of neurons in the rostral medulla of the rat during withdrawal from noxious heat. J Neurosci. 3, 2545–2552.

Fields, H. L., Heinricher, M. M. and Mason, P., 1991. Neurotransmitters in nociceptive modulatory circuits. Annu Rev Neurosci. 14, 219–245.

Fields, H. L., Malick, A. and Burstein, R., 1995. Dorsal horn projection targets of ON and OFF cells in the rostral ventromedial medulla. J Neurophysiol. 74, 1742–1759.

Fields, H. L., Vanegas, H., Hentall, I. D. and Zorman, G., 1983b. Evidence that disinhibition of brain stem neurones contributes to morphine analgesia. Nature. 306, 684–686.

Gao, K., Chen, D. O., Genzen, J. R. and Mason, P., 1998. Activation of serotonergic neurons in the raphe magnus is not necessary for morphine analgesia. J Neurosci 18, 1860–1868.

Gao, K. and Mason, P., 2001. Physiological and anatomic evidence for functional subclasses of serotonergic raphe magnus cells. J Comp Neurol. 439, 426–439.

Gardell, L. R., King, T., Ossipov, M. H., Rice, K. C., Lai, J., Vanderah, T. W. and Porreca, F., 2006. Opioid receptor-mediated hyperalgesia and antinociceptive tolerance induced by sustained opiate delivery. Neurosci Lett. 396, 44–49.

Gardell, L. R., Vanderah, T. W., Gardell, S. E., Wang, R., Ossipov, M. H., Lai, J. and Porreca, F., 2003. Enhanced evoked excitatory transmitter release in experimental neuropathy requires descending facilitation. J Neurosci. 23, 8370–8379.

Gauriau, C. and Bernard, J. F., 2004. A comparative reappraisal of projections from the superficial laminae of the dorsal horn in the rat: the forebrain. J Comp Neurol. 468, 24–56.

Gee, N. S., Brown, J. P., Dissanayake, V. U., Offord, J., Thurlow, R. and Woodruff, G. N., 1996. The novel anticonvulsant drug, gabapentin (Neurontin), binds to the alpha2delta subunit of a calcium channel. J Biol Chem. 271, 5768–5776.

Glaum, S. R. and Anderson, E. G., 1988. Identification of 5-HT3 binding sites in rat spinal cord synaptosomal membranes. Eur J Pharmacol. 156, 287–290.

Glaum, S. R., Proudfit, H. K. and Anderson, E. G., 1990. 5-HT3 receptors modulate spinal nociceptive reflexes. Brain Res. 510, 12–16.

Gran, J. T., 2003. The epidemiology of chronic generalized musculoskeletal pain. Best Pract Res Clin Rheumatol. 17, 547–561.

Graven-Nielsen, T., Aspegren Kendall, S., Henriksson, K. G., Bengtsson, M., Sorensen, J., Johnson, A., Gerdle, B. and Arendt-Nielsen, L., 2000. Ketamine reduces muscle pain, temporal summation, and referred pain in fibromyalgia patients. Pain. 85, 483–491.

Green, G. M., Lyons, L. and Dickenson, A. H., 1998. Alpha2-adrenoceptor antagonists enhance responses of dorsal horn neurones to formalin induced inflammation. Eur J Pharmacol. 347, 201–204.

Green, G. M., Scarth, J. and Dickenson, A., 2000. An excitatory role for 5-HT in spinal inflammatory nociceptive transmission; state-dependent actions via dorsal horn 5-HT(3) receptors in the anaesthetized rat. Pain. 89, 81–88.

Guan, Y., Guo, W., Robbins, M. T., Dubner, R. and Ren, K., 2004. Changes in AMPA receptor phosphorylation in the rostral ventromedial medulla after inflammatory hyperalgesia in rats. Neurosci Lett. 366, 201–205.

Hadjipavlou, G., Dunckley, P., Behrens, T. E. and Tracey, I., 2006. Determining anatomical connectivities between cortical and brainstem pain processing regions in humans: a diffusion tensor imaging study in healthy controls. Pain. 123, 169–178.

Han, H. C., Lee, D. H. and Chung, J. M., 2000. Characteristics of ectopic discharges in a rat neuropathic pain model. Pain. 84, 253–261.

Hansson, P. T. and Dickenson, A. H., 2005. Pharmacological treatment of peripheral neuropathic pain conditions based on shared commonalities despite multiple etiologies. Pain. 113, 251–254.

Heinricher, M. M., Morgan, M. M. and Fields, H. L., 1992. Direct and indirect actions of morphine on medullary neurons that modulate nociception. Neuroscience. 48, 533–543.

Heinricher, M. M., Morgan, M. M., Tortorici, V. and Fields, H. L., 1994. Disinhibition of off-cells and antinociception produced by an opioid action within the rostral ventromedial medulla. Neuroscience. 63, 279–288.

Holstege, G. and Kuypers, H. G., 1982. The anatomy of brain stem pathways to the spinal cord in cat. A labeled amino acid tracing study. Prog Brain Res. 57, 145–175.

Hosobuchi, Y., Adams, J. E. and Linchitz, R., 1977. Pain relief by electrical stimulation of the central gray matter in humans and its reversal by naloxone. Science. 197, 183–186.

Howe, J. R., Wang, J. Y. and Yaksh, T. L., 1983. Selective antagonism of the antinociceptive effect of intrathecally applied alpha adrenergic agonists by intrathecal prazosin and intrathecal yohimbine. J Pharmacol Exp Ther. 224, 552–558.

Hunt, S. P., 2000. Pain control: breaking the circuit. Trends Pharmacol Sci. 21, 284–287.

Hylden, J. L. and Wilcox, G. L., 1983. Intrathecal serotonin in mice: analgesia and inhibition of a spinal action of substance P. Life Sci. 33, 789–795.

Hwang, J. H. and Yaksh, T. L., 1997. Effect of subarachnoid gabapentin on tactile-evoked allodynia in a surgically induced neuropathic pain model in the rat. Reg Anesth. 22, 249–256.

Ikeda, H., Heinke, B., Ruscheweyh, R. and Sandkuhler, J., 2003. Synaptic plasticity in spinal lamina I projection neurons that mediate hyperalgesia. Science. 299, 1237–1240.

Johansen, J. P. and Fields, H. L., 2004. Glutamatergic activation of anterior cingulate cortex produces an aversive teaching signal. Nat Neurosci. 7, 398–403.

Jones, S. L. and Light, A. R., 1990. Electrical stimulation in the medullary nucleus raphe magnus inhibits noxious heat-evoked fos protein-like immunoreactivity in the rat lumbar spinal cord. Brain Res. 530, 335–338.

Kaneko, M., Mestre, C., Sanchez, E. H. and Hammond, D. L., 2000. Intrathecally administered gabapentin inhibits formalin-evoked nociception and the expression of Fos-like immunoreactivity in the spinal cord of the rat. J Pharmacol Exp Ther. 292, 743–751.

Karlsson, G. A., Preuss, C. V., Chaitoff, K. A., Maher, T. J. and Ally, A., 2006. Medullary monoamines and NMDA-receptor regulation of cardiovascular responses during peripheral nociceptive stimuli. Neurosci Res. 55, 316–326.

Kayser, V., Elfassi, I. E., Aubel, B., Melfort, M., Julius, D., Gingrich, J. A., Hamon, M. and Bourgoin, S., 2007. Mechanical, thermal and formalin-induced nociception is differentially altered in 5-HT1A–/–, 5-HT1B–/–, 5-HT2A–/–, 5-HT3A–/– and 5-HTT–/– knock-out male mice. Pain. 130, 235–248.

Keay, K. A. and Bandler, R., 1993. Deep and superficial noxious stimulation increases Fos-like immunoreactivity in different regions of the midbrain periaqueductal grey of the rat. Neurosci Lett. 154, 23–26.

Kim, S. H. and Chung, J. M., 1992. An experimental model for peripheral neuropathy produced by segmental spinal nerve ligation in the rat. Pain. 50, 355–363.

Kincaid, W., Neubert, M. J., Xu, M., Kim, C. J. and Heinricher, M. M., 2006. Role for medullary pain facilitating neurons in secondary thermal hyperalgesia. J Neurophysiol. 95, 33–41.

King, T., Vardanyan, A., Majuta, L., Melemedjian, O., Nagle, R., Cress, A. E., Vanderah, T. W., Lai, J. and Porreca, F., 2007. Morphine treatment accelerates sarcoma-induced bone pain, bone loss, and spontaneous fracture in a murine model of bone cancer. Pain.

Kosek, E. and Hansson, P., 1997. Modulatory influence on somatosensory perception from vibration and heterotopic noxious conditioning stimulation (HNCS) in fibromyalgia patients and healthy subjects. Pain. 70, 41–51.

Liu, X., Eschenfelder, S., Blenk, K. H., Janig, W. and Habler, H., 2000. Spontaneous activity of axotomized afferent neurons after L5 spinal nerve injury in rats. Pain. 84, 309–318.

Luo, Z. D., Chaplan, S. R., Higuera, E. S., Sorkin, L. S., Stauderman, K. A., Williams, M. E. and Yaksh, T. L., 2001. Upregulation of dorsal root ganglion (alpha)2(delta) calcium channel subunit and its correlation with allodynia in spinal nerve-injured rats. J Neurosci. 21, 1868–1875.

Lutfy, K., Shen, K. Z., Woodward, R. M. and Weber, E., 1996. Inhibition of morphine tolerance by NMDA receptor antagonists in the formalin test. Brain Res. 731, 171–181.

Mainero, C., Zhang, W. T., Kumar, A., Rosen, B. R. and Sorensen, A. G., 2007. Mapping the spinal and supraspinal pathways of dynamic mechanical allodynia in the human trigeminal system using cardiac-gated fMRI. Neuroimage. 35, 1201–1210.

Mao, J., Price, D. D. and Mayer, D. J., 1995. Mechanisms of hyperalgesia and morphine tolerance: a current view of their possible interactions. Pain. 62, 259–274.

Marinelli, S., Schnell, S. A., Hack, S. P., Christie, M. J., Wessendorf, M. W. and Vaughan, C. W., 2004. Serotonergic and Nonserotonergic Dorsal Raphe Neurons Are Pharmacologically and Electrophysiologically Heterogeneous. J Neurophysiol. 92, 3532–3537.

Marinelli, S., Vaughan, C. W., Schnell, S. A., Wessendorf, M. W. and Christie, M. J., 2002. Rostral ventromedial medulla neurons that project to the spinal cord express multiple opioid receptor phenotypes. J Neurosci. 22, 10847–10855.

Mason, P., 1997. Physiological identification of pontomedullary serotonergic neurons in the rat. J Neurophysiol. 77, 1087–1098.

Mason, P., Back, S. A. and Fields, H. L., 1992. A confocal laser microscopic study of enkephalin-immunoreactive appositions onto physiologically identified neurons in the rostral ventromedial medulla. J Neurosci. 12, 4023–4036.

Matos, F. F., Rollema, H., Brown, J. L. and Basbaum, A. I., 1992. Do opioids evoke the release of serotonin in the spinal cord? An in vivo microdialysis study of the regulation of extracellular serotonin in the rat. Pain. 48, 439–447.

Max, M. B., Lynch, S. A., Muir, J., Shoaf, S. E., Smoller, B. and Dubner, R., 1992. Effects of desipramine, amitriptyline, and fluoxetine on pain in diabetic neuropathy. N Engl J Med. 326, 1250–1256.

McCleane, G. J., Suzuki, R. and Dickenson, A. H., 2003. Does a single intravenous injection of the 5HT3 receptor antagonist ondansetron have an analgesic effect in neuropathic pain? A double-blinded, placebo-controlled cross-over study. Anesth Analg. 97, 1474–1478.

Meng, I. D. and Harasawa, I., 2007. Chronic morphine exposure increases the proportion of on-cells in the rostral ventromedial medulla in rats. Life Sci. 80, 1915–1920.

Mercadante, S., Ferrera, P., Villari, P. and Arcuri, E., 2003. Hyperalgesia: an emerging iatrogenic syndrome. J Pain Symptom Manage. 26, 769–775.

Merskey, H., 1994. Logic, truth and language in concepts of pain. Qual Life Res. 3 Suppl 1, S69–S76.

Meyer-Rosberg, K., Kvarnstrom, A., Kinnman, E., Gordh, T., Nordfors, L. O. and Kristofferson, A., 2001. Peripheral neuropathic pain – a multidimensional burden for patients. Eur J Pain. 5, 379–389.

Millan, M. J., 2002. Descending control of pain. Prog Neurobiol. 66, 355–474.

Monconduit, L., Desbois, C. and Villanueva, L., 2002. The integrative role of the rat medullary subnucleus reticularis dorsalis in nociception. Eur J Neurosci. 16, 937–944.

Montagne-Clavel, J., Oliveras, J. L., 1994. Are ventromedial medulla neuronal properties modified by chronic peripheral inflammation? A single-unit study in the awake, freely moving polyarthritic rat. Brain Res. 657, 92?104.

Neubert, M. J., Kincaid, W. and Heinricher, M. M., 2004. Nociceptive facilitating neurons in the rostral ventromedial medulla. Pain. 110, 158–165.

Nicholas, A. P., Pieribone, V. and Hokfelt, T., 1993. Distributions of mRNAs for alpha-2 adrenergic receptor subtypes in rat brain: an in situ hybridization study. J Comp Neurol. 328, 575–594.

Nichols, M. L., Allen, B. J., Rogers, S. D., Ghilardi, J. R., Honore, P., Luger, N. M., Finke, M. P., Li, J., Lappi, D. A., Simone, D. A. and Mantyh, P. W., 1999. Transmission of chronic nociception by spinal neurons expressing the substance P receptor. Science. 286, 1558–1561.

Nicholson, R., Dixon, A. K., Spanswick, D. and Lee, K., 2005. Noradrenergic receptor mRNA expression in adult rat superficial dorsal horn and dorsal root ganglion neurons. Neurosci Lett. 380, 316–321.

Oatway, M. A., Chen, Y. and Weaver, L. C., 2004. The 5-HT3 receptor facilitates at-level mechanical allodynia following spinal cord injury. Pain. 110, 259–268.

Olave, M. J. and Maxwell, D. J., 2003. Axon terminals possessing the alpha 2c-adrenergic receptor in the rat dorsal horn are predominantly excitatory. Brain Res. 965, 269–273.

Ossipov, M. H., Lai, J., King, T., Vanderah, T. W. and Porreca, F., 2005. Underlying mechanisms of pronociceptive consequences of prolonged morphine exposure. Biopolymers. 80, 319–324.

Perrot, S., Dickenson, A. H. and Bennett, R. M., 2008. Fibromyalgia: harmonizing science with clinical practice considerations. Pain Pract. 8, 177–189.

Petroff, O. A., Rothman, D. L., Behar, K. L., Lamoureux, D. and Mattson, R. H., 1996. The effect of gabapentin on brain gamma-aminobutyric acid in patients with epilepsy. Ann Neurol. 39, 95–99.

Petrovic, P., Kalso, E., Petersson, K. M. and Ingvar, M., 2002. Placebo and opioid analgesia – imaging a shared neuronal network. Science. 295, 1737–1740.

Porreca, F., Burgess, S. E., Gardell, L. R., Vanderah, T. W., Malan, T. P., Jr., Ossipov, M. H., Lappi, D. A. and Lai, J., 2001. Inhibition of neuropathic pain by selective ablation of brainstem medullary cells expressing the mu-opioid receptor. J Neurosci. 21, 5281–5288.

Proudfit, H. K., 1988. Pharmacologic evidence for the modulation of nociception by noradrenergic neurons. Prog Brain Res. 77, 357–370.

Proudfit, H. K., 1980. Reversible inactivation of raphe magnus neurons: effects on nociceptive threshold and morphine-induced analgesia. Brain Res 201, 459–464.

Rahman, W., Sikander, S., Suzuki, R., Hunt, S. P. and Dickenson, A. H., 2007. Superficial NK1 expressing spinal dorsal horn neurones modulate inhibitory neurotransmission mediated by spinal GABA(A) receptors. Neurosci Lett. 419, 278–283.

Rahman, W., Suzuki, R., Webber, M., Hunt, S. P. and Dickenson, A. H., 2006. Depletion of endogenous spinal 5-HT attenuates the behavioural hypersensitivity to mechanical and cooling stimuli induced by spinal nerve ligation. Pain. 123, 264–274.

Ren, K. and Dubner, R., 2002. Descending modulation in persistent pain: an update. Pain. 100, 1–6.

Ren, K. and Ruda, M. A., 1996. Descending modulation of Fos expression after persistent peripheral inflammation. Neuroreport. 7, 2186–2190.

Reynolds, D. V., 1969. Surgery in the rat during electrical analgesia induced by focal brain stimulation. Science. 164, 444–445.

Rivot, J. P., Pointis, D. and Besson, J. M., 1988. Morphine increases 5-HT metabolism in the nucleus raphe magnus: an in vivo study in freely moving rats using 5-hydroxyindole electrochemical detection. Brain Res 446, 333–342.

Rivot, J. P., Pointis, D. and Besson, J. M., 1989. A comparison of the effects of morphine on 5-HT metabolism in the periaqueductal gray, ventromedial medulla and medullary dorsal horn: in vivo electrochemical studies in freely moving rats. Brain Res 495, 140–144.

Rubin, J. J., 2005. Psychosomatic pain: new insights and management strategies. South Med J. 98, 1099–1110; quiz 1111–1092, 1138.

Rygh, L. J., Suzuki, R., Rahman, W., Wong, Y., Vonsy, J. L., Sandhu, H., Webber, M., Hunt, S. and Dickenson, A. H., 2006. Local and descending circuits regulate long-term potentiation and zif268 expression in spinal neurons. Eur J Neurosci. 24, 761–772.

Saarto, T. and Wiffen, P. J., 2005. Antidepressants for neuropathic pain. Cochrane Database Syst Rev, CD005454.

Sasaki, M., Obata, H., Kawahara, K., Saito, S. and Goto, F., 2006. Peripheral 5-HT2A receptor antagonism attenuates primary thermal hyperalgesia and secondary mechanical allodynia after thermal injury in rats. Pain. 122, 130–136.

Seifert, F. and Maihofner, C., 2007. Representation of cold allodynia in the human brain – a functional MRI study. Neuroimage. 35, 1168–1180.

Sheen, K. and Chung, J. M., 1993. Signs of neuropathic pain depend on signals from injured nerve fibers in a rat model. Brain Res. 610, 62–68.

Shi, T. J., Winzer-Serhan, U., Leslie, F. and Hokfelt, T., 1999. Distribution of alpha2-adrenoceptor mRNAs in the rat lumbar spinal cord in normal and axotomized rats. Neuroreport. 10, 2835–2839.

Shimoyama, M., Shimoyama, N. and Hori, Y., 2000. Gabapentin affects glutamatergic excitatory neurotransmission in the rat dorsal horn. Pain. 85, 405–414.

Sindrup, S. H. and Jensen, T. S., 2000. Pharmacologic treatment of pain in polyneuropathy. Neurology. 55, 915–920.

Sindrup, S. H., Otto, M., Finnerup, N. B. and Jensen, T. S., 2005. Antidepressants in the treatment of neuropathic pain. Basic Clin Pharmacol Toxicol. 96, 399–409.

Skagerberg, G., Bjorklund, A. and Lindvall, O., 1985. Further studies on the use of the fluorescent retrograde tracer True Blue in combination with monoamine histochemistry. J Neurosci Methods. 14, 25–40.

Sun, Q., Tu, H., Xing, G. G., Han, J. S. and Wan, Y., 2005. Ectopic discharges from injured nerve fibers are highly correlated with tactile allodynia only in early, but not late, stage in rats with spinal nerve ligation. Exp Neurol. 191, 128–136.

Sung, B., Na, H. S., Kim, Y. I., Yoon, Y. W., Han, H. C., Nahm, S. H. and Hong, S. K., 1998. Supraspinal involvement in the production of mechanical allodynia by spinal nerve injury in rats. Neurosci Lett. 246, 117–119.

Suzuki, R. and Dickenson, A. H., 2006. Differential pharmacological modulation of the spontaneous stimulus-independent activity in the rat spinal cord following peripheral nerve injury. Exp Neurol. 198, 72–80.

Suzuki, R., Morcuende, S., Webber, M., Hunt, S. P. and Dickenson, A. H., 2002. Superficial NK1-expressing neurons control spinal excitability through activation of descending pathways. Nat Neurosci. 5, 1319–1326.

Suzuki, R., Rahman, W., Hunt, S. P. and Dickenson, A. H., 2004. Descending facilitatory control of mechanically evoked responses is enhanced in deep dorsal horn neurones following peripheral nerve injury. Brain Res. 1019, 68–76.

Suzuki, R., Rahman, W., Rygh, L. J., Webber, M., Hunt, S. P. and Dickenson, A. H., 2005. Spinal-supraspinal serotonergic circuits regulating neuropathic pain and its treatment with gabapentin. Pain. 117, 292–303.

Svensson, C. I., Tran, T. K., Fitzsimmons, B., Yaksh, T. L. and Hua, X. Y., 2006. Descending serotonergic facilitation of spinal ERK activation and pain behavior. FEBS Lett. 580, 6629–6634.

Tanabe, M., Takasu, K., Kasuya, N., Shimizu, S., Honda, M. and Ono, H., 2005. Role of descending noradrenergic system and spinal alpha2-adrenergic receptors in the effects of gabapentin on thermal and mechanical nociception after partial nerve injury in the mouse. Br J Pharmacol. 144, 703–714.

Tillu, D. V., Gebhart, G. F. and Sluka, K. A., 2007. Descending facilitatory pathways from the RVM initiate and maintain bilateral hyperalgesia after muscle insult. Pain.

Todd, A. J., 2002. Anatomy of primary afferents and projection neurones in the rat spinal dorsal horn with particular emphasis on substance P and the neurokinin 1 receptor. Exp Physiol. 87, 245–249.

Todd, A. J., McGill, M. M. and Shehab, S. A., 2000. Neurokinin 1 receptor expression by neurons in laminae I, III and IV of the rat spinal dorsal horn that project to the brainstem. Eur J Neurosci. 12, 689–700.

Todd, A. J., Puskar, Z., Spike, R. C., Hughes, C., Watt, C. and Forrest, L., 2002. Projection neurons in lamina I of rat spinal cord with the neurokinin 1 receptor are selectively innervated by substance p-containing afferents and respond to noxious stimulation. J Neurosci. 22, 4103–4113.

Urban, M. O., Coutinho, S. V. and Gebhart, G. F., 1999. Biphasic modulation of visceral nociception by neurotensin in rat rostral ventromedial medulla. J Pharmacol Exp Ther. 290, 207–213.

Urch, C. E., Donovan-Rodriguez, T. and Dickenson, A. H., 2003. Alterations in dorsal horn neurones in a rat model of cancer-induced bone pain. Pain. 106, 347–356.

Vera-Portocarrero, L. P., Yie, J. X., Kowal, J., Ossipov, M. H., King, T. and Porreca, F., 2006. Descending facilitation from the rostral ventromedial medulla maintains visceral pain in rats with experimental pancreatitis. Gastroenterology. 130, 2155–2164.

Way, E. L., Loh, H. H. and Shen, F. H., 1969. Simultaneous quantitative assessment of morphine tolerance and physical dependence. J Pharmacol Exp Ther. 167, 1–8.

Wessendorf, M. W. and Anderson, E. G., 1983. Single unit studies of identified bulbospinal serotonergic units. Brain Res. 279, 93–103.

Wigdor, S. and Wilcox, G. L., 1987. Central and systemic morphine-induced antinociception in mice: contribution of descending serotonergic and noradrenergic pathways. J Pharmacol Exp Ther 242, 90–95.

Wolfe, F., Smythe, H. A., Yunus, M. B., Bennett, R. M., Bombardier, C., Goldenberg, D. L., Tugwell, P., Campbell, S. M., Abeles, M., Clark, P., et al., 1990. The American College of Rheumatology 1990 criteria for the classification of fibromyalgia. Report of the Multicenter Criteria Committee. Arthritis Rheum. 33, 160–172.

Yaksh, T. L., 1985. Pharmacology of spinal adrenergic systems which modulate spinal nociceptive processing. Pharmacol Biochem Behav. 22, 845–858.

Yaksh, T. L., DuChateau, J. C. and Rudy, T. A., 1976. Antagonism by methysergide and cinanserin of the antinociceptive action of morphine administered into the periaqueductal gray. Brain Res 104, 367–372.

Yaksh, T. L. and Wilson, P. R., 1979. Spinal serotonin terminal system mediates antinociception. J Pharmacol Exp Ther 208, 446–453.

Yoon, Y. W., Na, H. S. and Chung, J. M., 1996. Contributions of injured and intact afferents to neuropathic pain in an experimental rat model. Pain. 64, 27–36.

Zambreanu, L., Wise, R. G., Brooks, J. C., Iannetti, G. D. and Tracey, I., 2005. A role for the brainstem in central sensitisation in humans. Evidence from functional magnetic resonance imaging. Pain. 114, 397–407.

Zeitz, K. P., Guy, N., Malmberg, A. B., Dirajlal, S., Martin, W. J., Sun, L., Bonhaus, D. W., Stucky, C. L., Julius, D. and Basbaum, A. I., 2002. The 5-HT3 subtype of serotonin receptor contributes to nociceptive processing via a novel subset of myelinated and unmyelinated nociceptors. J Neurosci. 22, 1010–1019.

Zhang, L., Sykes, K. T., Buhler, A. V. and Hammond, D. L., 2006. Electrophysiological heterogeneity of spinally projecting serotonergic and nonserotonergic neurons in the rostral ventromedial medulla. J Neurophysiol 95, 1853–1863.

Zhao, Z. Q., Chiechio, S., Sun, Y. G., Zhang, K. H., Zhao, C. S., Scott, M., Johnson, R. L., Deneris, E. S., Renncr, K. J., Gereau, R. W. t. and Chen, Z. F., 2007. Mice lacking central serotonergic neurons show enhanced inflammatory pain and an impaired analgesic response to antidepressant drugs. J Neurosci. 27, 6045–6053.

Zhuo, M., Sengupta, J. N. and Gebhart, G. F., 2002. Biphasic modulation of spinal visceral nociceptive transmission from the rostroventral medial medulla in the rat. J Neurophysiol. 87, 2225–2236.

Chapter 15
Cannabinoid Receptor Mediated Analgesia: Novel Targets for Chronic Pain States

Victoria Chapman, David Kendall, and Devi Rani Sagar

Abstract Cannabinoid receptors are present at key sites involved in the relay and modulation of nociceptive responses. The analgesic effects of cannabinoid CB_1 receptor are well described. The wide-spread distribution of these receptors in the brain does, however, also explain the side-effects associated with CB_1 receptor agonists. The cannabinoid CB_2 receptor also produces analgesic effects in models of acute, inflammatory and neuropathic pain. The sites and mechanisms of CB_2 receptor mediated analgesia are described herein. In addition to directly targeting cannabinoid receptors, protection of endocannabinoids from metabolism also produces analgesic effects. Indeed, reports that noxious stimulation elevates levels of endocannabinoids in the spinal cord and brain provide further rationale for this approach. The effects of inhibition of fatty acid amide hydrolase on nociceptive responses in models of inflammatory and neuropathic pain are discussed.

Abbreviations

2-AG	2-arachidonoyl glycerol
AEA	N-arachidonoyl ethanolamine; anandamide
CB_1	cannabinoid 1 receptor
CB_2	cannabinoid 2 receptor
CNS	central nervous system
COX-2	cyclooxygenase type 2
DRG	dorsal root ganglion
FAAH	fatty acid amide hydrolase
i.p.	intraperitoneal
MAGL	monoacylglycerol lipase
NAAA	N-acylethanolamine hydrolysing acid amidase

V. Chapman (✉)
School of Biomedical Sciences, University of Nottingham, Nottingham
NG7 2UH, UK
e-mail: victoria.chapman@nottingham.ac.uk

M. Malcangio (ed.), *Synaptic Plasticity in Pain,*
DOI 10.1007/978-1-4419-0226-9_15, © Springer Science+Business Media, LLC 2009

NADA   *N*-arachidonoyl dopamine
NAE   *N*-acylethanolamines
NAPE   *N*-arachidonoyl-phosphatidyl-ethanolamine ethanolamine
OEA   *N*-oleoyl ethanolamine
p.o.   oral
PEA   *N*-palmitoyl ethanolamine
PAG   periaqueductal grey
PGE2-G   prostaglandin E2 glycerol ester
PLC   phospholipase c
PLD   phospholipase D

15.1 Introduction

The cannabinoid receptor system, which consists of cannabinoid receptors (CB_1 and CB_2 receptors), endogenous cannabinoid ligands and their synthesizing and metabolising enzymes, modulates a broad spectrum of physiological and patho-physiological responses (for reviews see Howlett et al., 2002; Mackie, 2006). As a consequence there are a number of different therapeutic targets for cannabinoid-based medicines, including pain, neuroprotection, obesity and psychiatric disorders. There are numerous anecdotal reports of the analgesic effects of cannabis, more recently cannabinoid-based medicines have been shown to mod-ulate nociception in patients with multiple sclerosis (Iskedjian et al., 2007; Conte et al., 2009) and rheumatoid arthritis (Blake et al., 2006). Cannabinoids have well described analgesic effects in animal models of acute and chronic pain (Pertwee, 2001; Hohmann, 2002; Iversen and Chapman, 2002; Rice et al., 2002; Walker and Huang, 2002; Jhaveri et al., 2007a, 2007b). Nevertheless, the potential therapeutic effects of cannabinoids may be offset by centrally mediated side-effects. Recent advances in the understanding of the role of components of the cannabinoid system have identified different targets for cannabinoid-mediated analgesia, which potentially have a reduced side-effect profile. This chapter will discuss the sites and mechanisms of action of cannabinoids on nociceptive processing and how selective targeting of some components of the cannabinoid receptor system may dissociate beneficial versus adverse effects, which will have implica-tions for the development of novel analgesics for the treatment of chronic pain states.

15.2 Multiple Sites of Action Mediate the Analgesic Effects of CB_1 Agonists

CB_1 receptors are associated with neuronal tissue (Herkenham et al., 1991; Tsou et al., 1998; Egertova and Elphick, 2000) with a moderate to high density in regions involved in pain transmission and modulation, such as dorsal root ganglia (DRG), spinal cord, thalamus, periaqueductal grey (PAG), amygdala and rostroventromedial medulla (Tsou et al., 1998). The analgesic effects produced by systemic administration of cannabinoid ligands, via the activation of CB_1 receptors, are well described and extensively reviewed (for reviews see Pertwee, 2001; Iversen and Chapman, 2002; Walker and Huang, 2002). CB_1 receptors are located at multiple key regions involved in the relay and modulation of nociceptive inputs and, therefore, can inhibit nociceptive processing via both peripheral and central sites of action. Activation of CB_1 receptors in peripheral cutaneous tissue (Richardson et al., 1998; Kelly et al., 2003; Nackley et al., 2003; Kelly and Donaldson, 2008) inhibits nociceptive responses in models of acute and persistent pain. Peripheral effects of CB_1 agonists may be enhanced in the presence of inflammation as CB_1 receptor expression is increased in DRG neurones in the presence of hindpaw inflammation (Amaya et al., 2006). Direct spinal administration of cannabinoids also attenuates nociceptive responses in naïve rats and models of inflammatory and neuropathic pain, via activation of CB_1 receptors (Welch and Stevens, 1992; Hohmann et al., 1998; Martin et al., 1999; Drew et al., 2000; Kelly and Chapman, 2001, 2003; Scott et al., 2004; Sagar et al., 2005; Rahn et al., 2007). At higher levels, discrete activation of CB_1 receptors in the PAG, amygdala, thalamus, hypothalamus and rostroventromedial medulla produces antinociceptive effects in models of acute/tonic pain (Lichtman et al., 1996; Meng et al., 1998; Welch et al., 1998; Martin et al., 1999; Finn et al., 2003). The CB_1 receptor has been shown to play an important role in modulating descending inhibitory controls (Finn et al., 2003; de Novellis et al., 2005) as well as contributing to stress-induced analgesia (Suplita et al., 2005) and aversive (Finn et al., 2004) responses. Although the brain distribution of CB_1 receptors is ideal for the modulation of pain, this can be offset by centrally mediated side-effects which compromise these therapeutic benefits. One way of overcoming the psychoactive side-effects is to use peripherally restricted CB_1 agonists. Work using mice with the gene deletion of the CB_1 receptor in peripheral nociceptors reported a substantial reduction in the analgesic effects of local and systemic, but not intrathecally, administered cannabinoids (Agarwal et al., 2007). Thus, it appears that CB_1 receptor agonists which do not cross the blood brain barrier, and thus selectively activate peripheral CB_1 receptors, may be a promising analgesic strategy.

15.3 Analgesic Potential for CB_2 Receptor Agonists

CB_2 receptors are primarily located, at high densities on immune cells. Although early studies (Munro et al., 1993) and more recent studies (Elphick et al., 2008) report an absence of CB_2 mRNA in the central nervous system, CB_2 mRNA has been reported in the spinal cord of control rats (Beltramo et al., 2006). By contrast, CB_2 protein was not detectable in the spinal cord of control rats (Wotherspoon et al., 2005). Recent studies have reported widespread expression of CB_2 receptor protein in the brain of control rats, based on data from immunocytochemical studies (Van Sickle et al., 2005; Gong et al., 2006). Since a recent functional imaging study reported no contribution of CB_2 receptors to the effects of a non-selective cannabinoid ligand on brain activation in control rats (Chin et al., 2008), the functional relevance of basal CB_2 receptor expression in the brain is unclear. Indeed, basal expression of CB_2 receptors in the brain may be at a very low level or, if present, restricted to small populations of neurons (Van Sickle et al., 2005).

Behavioural studies have shown that systemic administration of CB_2 agonists attenuates nociceptive responses in models of acute and chronic pain (Malan et al., 2001; Elmes et al., 2005; Valenzano et al., 2005; reviewed by Jhaveri et al., 2006; Ibrahim et al., 2006). Studies in our group have shown that there is no evidence that spinal (Sagar et al., 2005) or supra-spinal (Jhaveri et al., 2008a) CB_2 receptors modulate nociceptive responses in naïve rats, consistent with a low-level basal expression of CB_2 receptors in the CNS of naïve rats (see above). Importantly, systemically administered CB_2 receptor agonists are devoid of CNS-mediated side effects (Malan et al., 2003), further suggesting limited CNS mediated effects of these compounds in naïve rats. Overall, effects of CB_2 agonists in models of acute and inflammatory pain appear to be mediated by peripheral sites of action. CB_2 receptors are present in the skin (Stander et al., 2005) and local hindpaw administration of CB_2 agonists attenuates nociceptive responses in naïve rats (Malan et al., 2001; Elmes et al., 2004) and in models of inflammatory pain (Elmes et al., 2004; Gutierrez et al., 2007). It is unclear whether a direct effect of CB_2 agonists on peripheral nerve activity contributes to the observed inhibitory effects. Another better characterised mechanism underlying the inhibitory effects of CB_2 agonists involves the release of endorphins from keratinocytes which act via μ opioid receptors to produce analgesia (Ibrahim et al., 2005).

Pathological states, such as brain inflammation (Benito et al., 2008), are associated with an increased expression of CB_2 receptors in the brain. Similarly, there is evidence for the increased/novel expression of CB_2 mRNA and/or protein in the spinal cord in models of neuropathic pain (Zhang et al., 2003; Wotherspoon et al., 2005; Beltramo et al., 2006; Walczak et al., 2006, Fig. 15.1). Work in our group (Sagar et al., 2005) and by others (Romero-Sandoval et al., 2008; Yamamoto et al., 2008) has provided evidence for a novel functional role of CB_2 receptors in the spinal cord of neuropathic rats. Indeed, we have shown that spinal administration of the CB_2 receptor agonist JWH133 attenuates evoked responses of spinal neurones in neuropathic rats, but not sham-operated rats (Fig. 15.1). These effects of JWH133 were blocked by the CB_2 receptor antagonist SR144528, but not a CB_1 receptor antagonist. The lack of effect of JWH133 in sham-operated rats is in stark contrast to the robust inhibitory effects of spinally administered CB_1 receptor agonists in control and sham-operated rats (Drew et al., 2000), and corroborates the lack of evidence for the presence of CB_2 receptor protein in the spinal cord of sham-operated rats (Wotherspoon et al.,

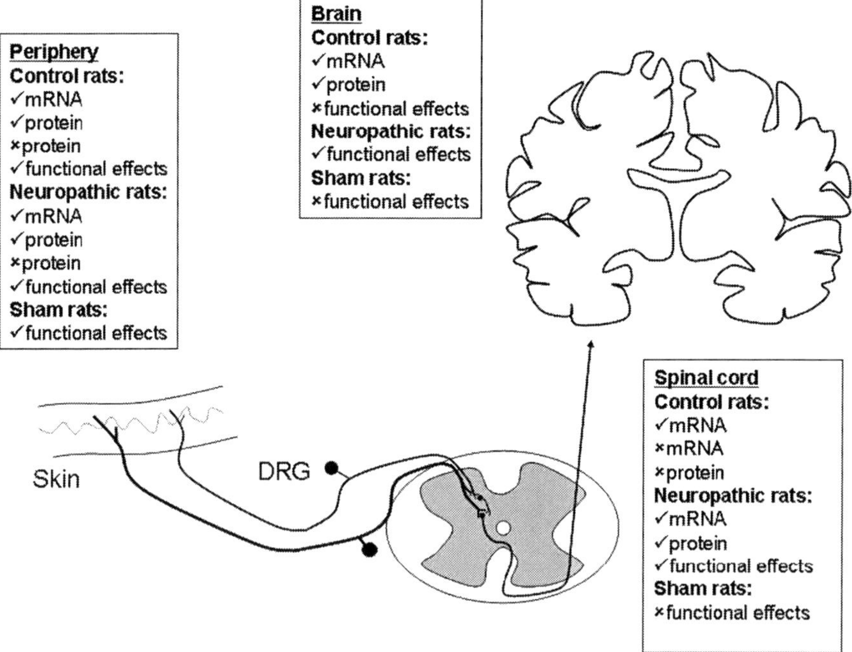

Fig. 15.1 Summary of evidence for the presence ($\checkmark$) or absence (✗) of CB2 mRNA, CB2 receptor protein or functional effects from activation of CB_2 receptors in control, neuropathic or sham-operated rats in either the periphery, spinal cord or brain, based on outcomes of the following studies (Ashton et al., 2006; Beltramo et al., 2006; Chin et al., 2008; Elmes et al., 2004; Gong et al., 2006; Jhaveri et al., 2008a; Ross et al., 2001; Sagar et al., 2005; Suarez et al., 2008; Van Sickle et al., 2005; Wotherspoon et al., 2005; Zhang et al., 2003)

2005). Inhibitory effects of spinally administered CB_2 agonists in neuropathic mice were absent in CB_2 knockout mice (Yamamoto et al., 2008), further supporting the concept that there is a novel functional role of CB_2 receptors in the spinal cord in models of neuropathic pain (Fig. 15.1). These changes do not appear to be limited to the spinal cord as we have shown that intra-thalamic administration of the CB_2 agonist JWH133 attenuates thalamic neuronal responses to peripheral mechanical stimulation in neuropathic rats, but not sham-operated rats (Jhaveri et al., 2008a).

15.4 Endocannabinoids

Currently, five endogenous cannabinoid receptor ligands or endocannabinoids have been discovered, of which anandamide (AEA) was the first to be identified (Devane et al., 1992). Since then, 2-arachidonoyl glycerol (2-AG; Mechoulam et al., 1995), noladin ether (Hanus et al., 2001), virodhamine (Porter et al., 2002) and *N*-arachidonoyl dopamine (NADA; Huang et al., 2002) have been identified. In addition, the related ethanolamines, *N*-oleoyl ethanolamine (OEA) and *N*-palmitoyl ethanolamine (PEA) have biological effects which are not mediated by CB_1 or CB_2 receptors.

Endocannabinoids are thought to be synthesised *de novo* in a calcium dependent manner. AEA, OEA and PEA can be synthesised from *N*-arachidonol phosphatidyl ethanolamine (NAPE) by the NAPE-phospholipase D (PLD) pathway. In addition, phospholipase-C (PLC)-PTPN22 (Liu et al., 2008) and αβ hydrolase (αβH4)-GDE1 (Simon and Cravatt, 2008) are also able to generate NAEs. The role of these multiple pathways is unclear, but may contribute to the differential synthesis of NAEs as NAPE-PLD mainly generates saturated *N*-acylethanolamines (NAEs) such as PEA (Leung et al., 2006). Potential differences in the tissue distribution of the synthetic enzymes require further investigation and may also have biological implications. AEA and PEA biosynthesis in the CNS is suggested to be predominantly via the αβH4-GDE1 pathway (Simon and Cravatt, 2008). In macrophages, PLC mediated cleavage of NAPE to phosphoanandamide prior to PTPN22-mediated dephosphorylation to NAE has been described (Liu et al., 2006).

The actions of the endocannabinoids are rapidly terminated via enzymatic hydrolysis. AEA and other NAEs are mainly hydrolysed by fatty acid amide hydrolase (FAAH; Deutsch and Chin, 1993; Cravatt et al., 1996) whilst 2-AG is mainly metabolised by monoacylglycerol lipase (MAGL; Dinh et al., 2002). Although, FAAH and MAGL are the main enzymes for metabolism of AEA and 2-AG, enzymes such as cyclooxygenase type 2 (COX-2; for review see; Fowler, 2007) also metabolise AEA and 2-AG. In addition, *N*-acylethanolamine hydrolysing acid amidase (NAAA) can also metabolise AEA and PEA (Tsuboi et al., 2007).

15.5 Endocannabinoids and Pain Processing

Exogenous administration of endocannabinoids inhibits nociceptive processing via the activation of cannabinoid receptors. For example, AEA is anti-nociceptive in behavioural models of acute and chronic pain (for review see Pertwee, 2001). We have demonstrated anti-nociceptive effects of spinally (Harris et al., 2000) and peripherally (Sokal et al., 2003) administered AEA in the carrageenan model of inflammatory pain. Similarly, 2-AG reduces pain behaviour in the tail-flick test (Mechoulam et al., 1995) and the formalin test (Guindon et al., 2007).

AEA and 2-AG are present in key regions involved in the detection, relay and integration of nociceptive inputs. One of the roles of the endocannabinoids appears to be the tonic inhibition of nociceptive responses which contributes to the setting of nociceptive thresholds. We have shown that spinal administration of a selective CB_1 receptor antagonist increases evoked-firing of dorsal horn neurones (Chapman, 1999). Studies using global CB_1 knockout mice have reported unaltered spontaneous nociceptive thresholds (Ledent et al., 1999) and that the development of neuropathic pain behaviour is also unaltered (Castane et al., 2006). By contrast, mice with deletion of CB_1 receptors in peripheral nociceptors exhibited exaggerated pain thresholds and responses to noxious stimuli such as capsaicin and formalin (Agarwal et al., 2007). Thus the role of CB_1 receptors in modulating nociceptive thresholds and the development of chronic pain responses remains unclear but might be the result of compensatory mechanisms in global CB_1 knockout mice masking any potential tonic inhibitory role.

Recent studies have demonstrated that levels of endocannabinoids are altered in models of persistent pain. We have demonstrated a significant reduction in levels of AEA and PEA in the hindpaw of rats with a carrageenan-induced hindpaw inflammation (Jhaveri et al., 2008b). Similarly, levels of AEA, 2-AG and PEA were decreased in the hindpaw following intraplantar injection of formalin (Maione et al., 2007). The basis for the decreased levels of endocannabinoids in the area of inflammation is unclear, but may contribute to the hyperalgesia associated with this condition. Note, however, that a lack of significant alteration in levels of AEA, 2-AG and PEA in the hindpaw of formalin-treated rats has also been reported (Beaulieu et al., 2000). Although there are inconsistencies in the literature, it is clear that hindpaw inflammation produced by injection of carrageenan or formalin is not associated with increased levels of endocannabinoids in the hindpaw as may be expected on the basis of the activity-dependent nature of their synthesis. This is, however, the case when considering the effects of noxious stimulation on levels of endocannabinoids in the brain. Formalin-evoked hindpaw inflammation increased levels of endocannabinoids in the periaqueductal grey implicating a role for endocannabinoids in descending control of pain processing (Walker et al., 1999).

Models of neuropathic pain are associated with aberrant primary afferent firing, spinal hyperexcitability and increased levels of endocannabinoids in the spinal cord (Petrosino et al., 2007) and dorsal root ganglia (Mitrirattanakul et al., 2006). Interestingly, there does not appear to be a global increase in NAEs *per se* under these conditions as we observed increased levels of AEA alongside decreased levels of PEA in the spinal cord of neuropathic rats (unpublished observations). The presence of activated microglia in the spinal cord in models of neuropathic pain contributes to aberrant pain responses. Microglia are able to contribute to the synthesis and catabolism of endocannabinoids; thus, changes in the cellular composition of the spinal cord are likely to contribute to the complexity of changes in spinal levels of NAEs in neuropathic pain states.

15.6 Facilitating Endocannabinoid-Mediated Analgesia

Although the endocannabinoids produce analgesia, these effects are short-lived due to their rapid catabolism. In order to prolong these effects, research has investigated the effects of inhibiting the breakdown of endocannabinoids (see

Jhaveri et al., 2007a). One of the benefits of this approach is that regions with elevated levels of endocannabinoids, for example as a result of noxious stimulation, are targeted as opposed to the global effects of receptor agonists.

15.6.1 Targeting Fatty Acid Amide Hydrolase

Genetic deletion of FAAH results in a 15 fold elevated levels of AEA, compared to wild-type mice, and an hypoalgesia phenotype in models of acute and inflammatory pain (Cravatt et al., 2001; Lichtman et al., 2004b), highlighting the important contribution of FAAH to the catabolism of the endocannabinoids. Similarly, pharmacological inhibition of FAAH, with compounds such as URB597 and OL135, produces antinociceptive effects in models of acute and inflammatory pain (Kathuria et al., 2003; Lichtman et al., 2004a; Fegley et al., 2005; Chang et al., 2006; Jayamanne et al., 2006; Russo et al., 2007). Work in our group has demonstrated that local hindpaw inhibition of FAAH with URB597 reduces carrageenan-induced hyperalgesia and increases levels of AEA and 2-AG in hindpaw skin of these rats (Jhaveri et al., 2008b).

Although models of neuropathic pain are associated with elevated levels of endocannabinoids in the spinal cord and the novel functional expression of CB_2 receptors in the spinal cord (see earlier), inhibition of FAAH has less consistent effects on neuropathic pain behaviour compared to inflammatory pain states. A systemic dose of URB597 (0.3 mg/kg, i.p.), which is effective in inflammatory pain states, did not alter mechanical allodynia in a model of peripheral neuropathic pain (Jayamanne et al., 2006). A far higher dose of the reversible FAAH inhibitor OL135 (ED_{50} 9 mg/kg i.p.) reduced mechanical allodynia in neuropathic rodents (Chang et al., 2006). Repeated dosing with URB597 (10 mg/kg, for 4 days p.o.) significantly reduced thermal and mechanical hyperalgesia in neuropathic mice, whereas the effects of a single oral dose were limited (Russo et al., 2007). In our hands, local hindpaw inhibition of FAAH with URB597 (25 µg in 50 µl) reduced mechanically-evoked responses of spinal cord dorsal horn neurones in sham-operated rats, did not alter responses in neuropathic rats (Jhaveri et al., 2006). Indeed, effects of URB597 in neuropathic rats were only observed following administration of a four-fold higher dose (100 µg in 50 µl, intraplantar) of URB597 (Jhaveri et al., 2006). In the same study, spinal administration of URB597 (10–50 µg in 50 µl) was equi-effective at reducing mechanically-evoked responses of dorsal horn neurones in neuropathic and sham-operated rats. Collectively, there is evidence for changes in the synthesis/metabolism of endocannabinoids in models of neuropathic pain.

15.6.2 Targeting Monoacylglycerol Lipase

The predominant route of catabolism of 2-AG is via MAGL, which has a localised distribution in the brain (Dinh et al., 2002). There is, however, some *in vitro* and *in vivo* evidence suggesting the involvement of FAAH in the metabolism of 2-AG (see Jhaveri et al., 2007a; Vandevoorde, 2008). There are far fewer studies of the effects of inhibitors for MAGL on nociceptive processing, compared to inhibitors of FAAH, mainly due to the lack of selective pharmacological agents. URB602 was described as a selective, but low potency inhibitor of MAGL, which did not alter the catabolism of AEA (Suplita et al., 2005). Despite this study, URB602 has subsequently been reported to inhibit both MAGL and FAAH *in vitro* (Vandevoorde et al., 2006). Thus, whether elevating levels of 2-AG produces anti-nociceptive effects remains to be determined once selective inhibitors of MAGL are available.

Overall, understanding of the role of FAAH in the catabolism of AEA and related compounds is by far the most advanced compared to the other potential routes of degradation. There is evidence that endocannabinoids such as AEA can be catabolised by cyclooxygenase 2 (COX2) to form prostamides (Yu et al., 1997; for review see Burstein et al., 2000). Although the biological importance of this pathway under control conditions remains unclear, it is evident that exogenausly administered AEA is catabolised to prostamides in FAAH knockout mice (Weber et al., 2004). This route of catabolism may become increasingly important under conditions associated with increased COX2 expression such as inflammation. COX2 can also break-down 2-AG to form prostaglandin E2 glycerol ester (PGE2-G) which, although rapidly metabolised is pro-nociceptive inducing mechanical allodynia and thermal hyperalgesia (Hu et al., 2008).

15.7 Concluding Remarks

The location of inhibitory cannabinoid receptors at key sites involved in the relay and modulation of nociceptive responses and their ability to attenuate acute and chronic nociceptive processing makes them an ideal target for the development of novel analgesics. Although the wide-spread distribution of CB_1 receptors in the brain does result in adverse side-effects, the targeting of peripherally located CB_1 receptors with peripherally restricted drugs may produce analgesia in absence of side-effects. CB_2 receptors increasingly appear to be an important target for the development of novel analgesics, in particular for chronic inflammatory and neuropathic pain states. Tissue specific changes in levels of the endocannabinoids following persistent noxious stimulation can also present another novel target for the treatment of pain through the use of specific inhibitors of catabolic enzymes which maximise levels of endocannabinoids and the associated analgesic effects.

Acknowledgments We would like to thank the Wellcome Trust, Medical Research Council and GlaxoSmithKline for financial support towards the original research discussed in this review.

References

Agarwal, N., Pacher, P., Tegeder, I., Amaya, F., Constantin, C. E., Brenner, G. J., Rubino, T., Michalski, C. W., Marsicano, G., Monory, K., Mackie, K., Marian, C., Batkai, S., Parolaro, D., Fischer, M. J., Reeh, P., Kunos, G., Kress, M., Lutz, B., Woolf, C. J., Kuner, R., 2007. Cannabinoids mediate analgesia largely via peripheral type 1 cannabinoid receptors in nociceptors. Nat Neurosci 10, 870–879.

Amaya, F., Shimosato, G., Kawasaki, Y., Hashimoto, S., Tanaka, Y., Ji, R. R., Tanaka, M., 2006. Induction of CB1 cannabinoid receptor by inflammation in primary afferent neurons facilitates antihyperalgesic effect of peripheral CB1 agonist. Pain 124, 175–183.

Beaulieu, P., Bisogno, T., Punwar, S., Farquhar-Smith, W. P., Ambrosino, G., Di Marzo, V., Rice, A. S., 2000. Role of the endogenous cannabinoid system in the formalin test of persistent pain in the rat. Eur J Pharmacol 396, 85–92.

Beltramo, M., Bernardini, N., Bertorelli, R., Campanella, M., Nicolussi, E., Fredduzzi, S., Reggiani, A., 2006. CB2 receptor-mediated antihyperalgesia: possible direct involvement of neural mechanisms. Eur J Neurosci 23, 1530–1538.

Benito, C., Tolon, R. M., Pazos, M. R., Nunez, E., Castillo, A. I., Romero, J., 2008. Cannabinoid CB2 receptors in human brain inflammation. Br J Pharmacol 153, 277–285.

Blake, D. R., Robson, P., Ho, M., Jubb, R. W., McCabe, C. S., 2006. Preliminary assessment of the efficacy, tolerability and safety of a cannabis-based medicine (Sativex) in the treatment of pain caused by rheumatoid arthritis. Rheumatology (Oxford) 45, 50–52.

Burstein, S. H., Rossetti, R. G., Yagen, B., Zurier, R. B., 2000. Oxidative metabolism of anandamide. Prostaglandins Other Lipid Mediat 61, 29–41.

Castane, A., Celerier, E., Martin, M., Ledent, C., Parmentier, M., Maldonado, R., Valverde, O., 2006. Development and expression of neuropathic pain in CB1 knockout mice. Neuropharmacology 50, 111–122.

Chang, L., Luo, L., Palmer, J. A., Sutton, S., Wilson, S. J., Barbier, A. J., Breitenbucher, J. G., Chaplan, S. R., Webb, M., 2006. Inhibition of fatty acid amide hydrolase produces analgesia by multiple mechanisms. Br J Pharmacol 148, 102–113.

Chapman, V., 1999. The cannabinoid CB1 receptor antagonist, SR141716A, selectively facilitates nociceptive responses of dorsal horn neurones in the rat. Br J Pharmacol 127, 1765–1767.

Chin, C. L., Tovcimak, A. E., Hradil, V. P., Seifert, T. R., Hollingsworth, P. R., Chandran, P., Zhu, C. Z., Gauvin, D., Pai, M., Wetter, J., Hsieh, G. C., Honore, P., Frost, J. M., Dart, M. J., Meyer, M. D., Yao, B. B., Cox, B. F., Fox, G. B., 2008. Differential effects of cannabinoid receptor agonists on regional brain activity using pharmacological MRI. Br J Pharmacol 153, 367–379.

Conte A., Bettolo, C. M., Onesti, E., Frasca, V., Iacovelli, E., Gilio, F., Giacomelli, E., Gabriele, M., Aragona, M., Tomassini, V., Pantano, P., Pozzilli, C., Inghilleri, M., 2009. Cannabinoid-induced effects on the nociceptive system: a neurophysiological study in patients with secondary progressive multiple sclerosis. Eur J Pain 13(5):472–7.

Cravatt, B. F., Demarest, K., Patricelli, M. P., Bracey, M. H., Giang, D. K., Martin, B. R., Lichtman, A. H., 2001. Supersensitivity to anandamide and enhanced endogenous cannabinoid signaling in mice lacking fatty acid amide hydrolase. Proc Natl Acad Sci USA 98, 9371–9376.

Cravatt, B. F., Giang, D. K., Mayfield, S. P., Boger, D. L., Lerner, R. A., Gilula, N. B., 1996. Molecular characterization of an enzyme that degrades neuromodulatory fatty-acid amides. Nature 384, 83–87.

de Novellis, V., Mariani, L., Palazzo, E., Vita, D., Marabese, I., Scafuro, M., Rossi, F., Maione, S., 2005. Periaqueductal grey CB1 cannabinoid and metabotropic glutamate subtype 5 receptors modulate changes in rostral ventromedial medulla neuronal activities induced by subcutaneous formalin in the rat. Neuroscience 134, 269–281.

Deutsch, D. G., Chin, S. A., 1993. Enzymatic synthesis and degradation of anandamide, a cannabinoid receptor agonist. Biochem Pharmacol 46, 791–796.

Devane, W. A., Hanus, L., Breuer, A., Pertwee, R. G., Stevenson, L. A., Griffin, G., Gibson, D., Mandelbaum, A., Etinger, A., Mechoulam, R., 1992. Isolation and structure of a brain constituent that binds to the cannabinoid receptor. Science 258, 1946–1949.

Dinh, T. P., Carpenter, D., Leslie, F. M., Freund, T. F., Katona, I., Sensi, S. L., Kathuria, S., Piomelli, D., 2002. Brain monoglyceride lipase participating in endocannabinoid inactivation. Proc Natl Acad Sci USA 99, 10819–10824.

Drew, L. J., Harris, J., Millns, P. J., Kendall, D. A., Chapman, V., 2000. Activation of spinal cannabinoid 1 receptors inhibits C-fibre driven hyperexcitable neuronal responses and increases [35S]GTPgammaS binding in the dorsal horn of the spinal cord of noninflamed and inflamed rats. Eur J Neurosci 12, 2079–2086.

Egertova, M., Elphick, M. R., 2000. Localisation of cannabinoid receptors in the rat brain using antibodies to the intracellular C-terminal tail of CB. J Comp Neurol 422, 159–171.

Elmes, S. J., Jhaveri, M. D., Smart, D., Kendall, D. A., Chapman, V., 2004. Cannabinoid CB2 receptor activation inhibits mechanically evoked responses of wide dynamic range dorsal horn neurons in naive rats and in rat models of inflammatory and neuropathic pain. Eur J Neurosci 20, 2311–2320.

Elmes, S. J., Winyard, L. A., Medhurst, S. J., Clayton, N. M., Wilson, A. W., Kendall, D. A., Chapman, V., 2005. Activation of CB1 and CB2 receptors attenuates the induction and maintenance of inflammatory pain in the rat. Pain 118, 327–335.

Elphick, M. R., Guasti, L., Jhaveri, M. D., Kendall, D. A., Chapman, V., Michael, G. J., 2008. Analysis of CB2 receptor expression in the nervous system. International Cannabinoid Research Society, Aviemore, Scotland, p. P113.

Fegley, D., Gaetani, S., Duranti, A., Tontini, A., Mor, M., Tarzia, G., Piomelli, D., 2005. Characterization of the fatty acid amide hydrolase inhibitor cyclohexyl carbamic acid 3'-carbamoyl-biphenyl-3-yl ester (URB597): effects on anandamide and oleoylethanolamide deactivation. J Pharmacol Exp Ther 313, 352–358.

Finn, D. P., Beckett, S. R., Richardson, D., Kendall, D. A., Marsden, C. A., Chapman, V., 2004. Evidence for differential modulation of conditioned aversion and fear-conditioned analgesia by CB1 receptors. Eur J Neurosci 20, 848–852.

Finn, D. P., Jhaveri, M. D., Beckett, S. R., Roe, C. H., Kendall, D. A., Marsden, C. A., Chapman, V., 2003. Effects of direct periaqueductal grey administration of a cannabinoid receptor agonist on nociceptive and aversive responses in rats. Neuropharmacology 45, 594–604.

Fowler, C. J., 2007. The contribution of cyclooxygenase-2 to endocannabinoid metabolism and action. Br J Pharmacol 152, 594–601.

Gong, J. P., Onaivi, E. S., Ishiguro, H., Liu, Q. R., Tagliaferro, P. A., Brusco, A., Uhl, G. R., 2006. Cannabinoid CB2 receptors: immunohistochemical localization in rat brain. Brain Res 1071, 10–23.

Guindon, J., Desroches, J., Beaulieu, P., 2007. The antinociceptive effects of intraplantar injections of 2-arachidonoyl glycerol are mediated by cannabinoid CB2 receptors. Br J Pharmacol 150, 693–701.

Gutierrez, T., Farthing, J. N., Zvonok, A. M., Makriyannis, A., Hohmann, A. G., 2007. Activation of peripheral cannabinoid CB1 and CB2 receptors suppresses the maintenance of inflammatory nociception: a comparative analysis. Br J Pharmacol 150, 153–163.

Hanus, L., Abu-Lafi, S., Fride, E., Breuer, A., Vogel, Z., Shalev, D. E., Kustanovich, I., Mechoulam, R., 2001. 2-Arachidonyl glyceryl ether, an endogenous agonist of the cannabinoid CB1 receptor. Proc Natl Acad Sci USA 98, 3662–3665.

Harris, J., Drew, L. J., Chapman, V., 2000. Spinal anandamide inhibits nociceptive transmission via cannabinoid receptor activation in vivo. Neuroreport 11, 2817–2819.

Herkenham, M., Groen, B. G., Lynn, A. B., De Costa, B. R., Richfield, E. K., 1991. Neuronal localization of cannabinoid receptors and second messengers in mutant mouse cerebellum. Brain Res 552, 301–310.

Hohmann, A. G., 2002. Spinal and peripheral mechanisms of cannabinoid antinociception: behavioral, neurophysiological and neuroanatomical perspectives. Chem Phys Lipids 121, 173–190.

Hohmann, A. G., Tsou, K., Walker, J. M., 1998. Cannabinoid modulation of wide dynamic range neurons in the lumbar dorsal horn of the rat by spinally administered WIN55, 212-2. Neurosci Lett 257, 119–122.

Howlett, A. C., Barth, F., Bonner, T. I., Cabral, G., Casellas, P., Devane, W. A., Felder, C. C., Herkenham, M., Mackie, K., Martin, B. R., Mechoulam, R., Pertwee, R. G., 2002. International Union of Pharmacology. XXVII. Classification of cannabinoid receptors. Pharmacol Rev 54, 161–202.

Hu, S. S., Bradshaw, H. B., Chen, J. S., Tan, B., Walker, J. M., 2008. Prostaglandin E2 glycerol ester, an endogenous COX-2 metabolite of 2-arachidonoylglycerol, induces hyperalgesia and modulates NFkappaB activity. Br J Pharmacol 153, 1538–1549.

Huang, S. M., Bisogno, T., Trevisani, M., Al-Hayani, A., De Petrocellis, L., Fezza, F., Tognetto, M., Petros, T. J., Krey, J. F., Chu, C. J., Miller, J. D., Davies, S. N., Geppetti, P., Walker, J. M., Di Marzo, V., 2002. An endogenous capsaicin-like substance with high potency at recombinant and native vanilloid VR1 receptors. Proc Natl Acad Sci USA 99, 8400–8405.

Ibrahim, M. M., Porreca, F., Lai, J., Albrecht, P. J., Rice, F. L., Khodorova, A., Davar, G., Makriyannis, A., Vanderah, T. W., Mata, H. P., Malan, T. P., Jr., 2005. CB2 cannabinoid receptor activation produces antinociception by stimulating peripheral release of endogenous opioids. Proc Natl Acad Sci USA 102, 3093–3098.

Ibrahim, M. M., Rude, M. L., Stagg, N. J., Mata, H. P., Lai, J., Vanderah, T. W., Porreca, F., Buckley, N. E., Makriyannis, A., Malan, T. P., Jr., 2006. CB2 cannabinoid receptor mediation of antinociception. Pain 122, 36–42.

Iskedjian, M., Bereza, B., Gordon, A., Piwko, C., Einarson, T. R., 2007. Meta-analysis of cannabis based treatments for neuropathic and multiple sclerosis-related pain. Curr Med Res Opin 23, 17–24.

Iversen, L., Chapman, V., 2002. Cannabinoids: a real prospect for pain relief? Curr Opin Pharmacol 2, 50–55.

Jayamanne, A., Greenwood, R., Mitchell, V. A., Aslan, S., Piomelli, D., Vaughan, C. W., 2006. Actions of the FAAH inhibitor URB597 in neuropathic and inflammatory chronic pain models. Br J Pharmacol 147, 281–288.

Jhaveri, M. D., Elmes, S. J., Richardson, D., Barrett, D. A., Kendall, D. A., Mason, R., Chapman, V., 2008a. Evidence for a novel functional role of cannabinoid CB receptors in the thalamus of neuropathic rats. Eur J Neurosci 27, 1722–1730.

Jhaveri, M. D., Richardson, D., Chapman, V., 2007a. Endocannabinoid metabolism and uptake: novel targets for neuropathic and inflammatory pain. Br J Pharmacol 152, 624–632.

Jhaveri, M. D., Richardson, D., Kendall, D. A., Barrett, D. A., Chapman, V., 2006. Analgesic effects of fatty acid amide hydrolase inhibition in a rat model of neuropathic pain. J Neurosci 26, 13318–13327.

Jhaveri, M. D., Richardson, D., Robinson, I., Garle, M. J., Patel, A., Sun, Y., Sagar, D. R., Bennett, A. J., Alexander, S. P., Kendall, D. A., Barrett, D. A., Chapman, V., 2008b. Inhibition of fatty acid amide hydrolase and cyclooxygenase-2 increases levels of endocannabinoid related molecules and produces analgesia via peroxisome proliferator-activated receptor-alpha in a model of inflammatory pain. Neuropharmacology 55, 85–93.

Jhaveri, M. D., Sagar, D. R., Elmes, S. J., Kendall, D. A., Chapman, V., 2007b. Cannabinoid CB(2) receptor-mediated anti-nociception in models of acute and chronic pain. Mol Neurobiol 36, 26–35.

Kathuria, S., Gaetani, S., Fegley, D., Valino, F., Duranti, A., Tontini, A., Mor, M., Tarzia, G., La Rana, G., Calignano, A., Giustino, A., Tattoli, M., Palmery, M., Cuomo, V., Piomelli, D., 2003. Modulation of anxiety through blockade of anandamide hydrolysis. Nat Med 9, 76–81.

Kelly, S., Chapman, V., 2001. Selective cannabinoid CB1 receptor activation inhibits spinal nociceptive transmission in vivo. J Neurophysiol 86, 3061–3064.

Kelly, S., Chapman, V., 2003. Cannabinoid CB(1) receptor inhibition of mechanically evoked responses of spinal neurones in control rats, but not in rats with hindpaw inflammation. Eur J Pharmacol 474, 209–216.

Kelly, S., Donaldson, L. F., 2008. Peripheral cannabinoid CB1 receptors inhibit evoked responses of nociceptive neurones in vivo. Eur J Pharmacol 586, 160–163.

Kelly, S., Jhaveri, M. D., Sagar, D. R., Kendall, D. A., Chapman, V., 2003. Activation of peripheral cannabinoid CB1 receptors inhibits mechanically evoked responses of spinal neurons in noninflamed rats and rats with hindpaw inflammation. Eur J Neurosci 18, 2239–2243.

Ledent, C., Valverde, O., Cossu, G., Petitet, F., Aubert, J. F., Beslot, F., Bohme, G. A., Imperato, A., Pedrazzini, T., Roques, B. P., Vassart, G., Fratta, W., Parmentier, M., 1999. Unresponsiveness to cannabinoids and reduced addictive effects of opiates in CB1 receptor knockout mice. Science 283, 401–404.

Leung, D., Saghatelian, A., Simon, G. M., Cravatt, B. F., 2006. Inactivation of N-acyl phosphatidylethanolamine phospholipase D reveals multiple mechanisms for the biosynthesis of endocannabinoids. Biochemistry 45, 4720–4726.

Lichtman, A. H., Cook, S. A., Martin, B. R., 1996. Investigation of brain sites mediating cannabinoid-induced antinociception in rats: evidence supporting periaqueductal gray involvement. J Pharmacol Exp Ther 276, 585–593.

Lichtman, A. H., Leung, D., Shelton, C. C., Saghatelian, A., Hardouin, C., Boger, D. L., Cravatt, B. F., 2004a. Reversible inhibitors of fatty acid amide hydrolase that promote analgesia: evidence for an unprecedented combination of potency and selectivity. J Pharmacol Exp Ther 311, 441–448.

Lichtman, A. H., Shelton, C. C., Advani, T., Cravatt, B. F., 2004b. Mice lacking fatty acid amide hydrolase exhibit a cannabinoid receptor-mediated phenotypic hypoalgesia. Pain 109, 319–327.

Liu, J., Wang, L., Harvey-White, J., Huang, B. X., Kim, H. Y., Luquet, S., Palmiter, R. D., Krystal, G., Rai, R., Mahadevan, A., Razdan, R. K., Kunos, G., 2008. Multiple pathways involved in the biosynthesis of anandamide. Neuropharmacology 54, 1–7.

Liu, J., Wang, L., Harvey-White, J., Osei-Hyiaman, D., Razdan, R., Gong, Q., Chan, A. C., Zhou, Z., Huang, B. X., Kim, H. Y., Kunos, G., 2006. A biosynthetic pathway for anandamide. Proc Natl Acad Sci USA 103, 13345–13350.

Mackie, K., 2006. Cannabinoid receptors as therapeutic targets. Annu Rev Pharmacol Toxicol 46, 101–122.

Maione, S., De Petrocellis, L., de Novellis, V., Moriello, A. S., Petrosino, S., Palazzo, E., Rossi, F. S., Woodward, D. F., Di Marzo, V., 2007. Analgesic actions of N-arachidonoyl-serotonin, a fatty acid amide hydrolase inhibitor with antagonistic activity at vanilloid TRPV1 receptors. Br J Pharmacol 150, 766–781.

Malan, T. P., Jr., Ibrahim, M. M., Deng, H., Liu, Q., Mata, H. P., Vanderah, T., Porreca, F., Makriyannis, A., 2001. CB2 cannabinoid receptor-mediated peripheral antinociception. Pain 93, 239–245.

Malan, T. P., Jr., Ibrahim, M. M., Lai, J., Vanderah, T. W., Makriyannis, A., Porreca, F., 2003. CB2 cannabinoid receptor agonists: pain relief without psychoactive effects? Curr Opin Pharmacol 3, 62–67.

Martin, W. J., Coffin, P. O., Attias, E., Balinsky, M., Tsou, K., Walker, J. M., 1999. Anatomical basis for cannabinoid-induced antinociception as revealed by intracerebral microinjections. Brain Res 822, 237–242.

Mechoulam, R., Ben-Shabat, S., Hanus, L., Ligumsky, M., Kaminski, N. E., Schatz, A. R., Gopher, A., Almog, S., Martin, B. R., Compton, D. R., et al., 1995. Identification of an endogenous 2-monoglyceride, present in canine gut, that binds to cannabinoid receptors. Biochem Pharmacol 50, 83–90.

Meng, I. D., Manning, B. H., Martin, W. J., Fields, H. L., 1998. An analgesia circuit activated by cannabinoids. Nature 395, 381–383.

Mitrirattanakul, S., Ramakul, N., Guerrero, A. V., Matsuka, Y., Ono, T., Iwase, H., Mackie, K., Faull, K. F., Spigelman, I., 2006. Site-specific increases in peripheral cannabinoid receptors and their endogenous ligands in a model of neuropathic pain. Pain 126, 102–114.

Munro, S., Thomas, K. L., Abu-Shaar, M., 1993. Molecular characterization of a peripheral receptor for cannabinoids. Nature 365, 61–65.

Nackley, A. G., Suplita, R. L., 2nd, Hohmann, A. G., 2003. A peripheral cannabinoid mechanism suppresses spinal fos protein expression and pain behavior in a rat model of inflammation. Neuroscience 117, 659–670.

Pertwee, R. G., 2001. Cannabinoid receptors and pain. Prog Neurobiol 63, 569–611.

Petrosino, S., Palazzo, E., de Novellis, V., Bisogno, T., Rossi, F., Maione, S., Di Marzo, V., 2007. Changes in spinal and supraspinal endocannabinoid levels in neuropathic rats. Neuropharmacology 52, 415–422.

Porter, A. C., Sauer, J. M., Knierman, M. D., Becker, G. W., Berna, M. J., Bao, J., Nomikos, G. G., Carter, P., Bymaster, F. P., Leese, A. B., Felder, C. C., 2002. Characterization of a novel endocannabinoid, virodhamine, with antagonist activity at the CB1 receptor. J Pharmacol Exp Ther 301, 1020–1024.

Rahn, E. J., Makriyannis, A., Hohmann, A. G., 2007. Activation of cannabinoid CB1 and CB2 receptors suppresses neuropathic nociception evoked by the chemotherapeutic agent vincristine in rats. Br J Pharmacol 152, 765–777.

Rice, A. S., Farquhar-Smith, W. P., Nagy, I., 2002. Endocannabinoids and pain: spinal and peripheral analgesia in inflammation and neuropathy. Prostaglandins Leukot Essent Fatty Acids 66, 243–256.

Richardson, J. D., Kilo, S., Hargreaves, K. M., 1998. Cannabinoids reduce hyperalgesia and inflammation via interaction with peripheral CB1 receptors. Pain 75, 111–119.

Romero-Sandoval, A., Nutile-McMenemy, N., DeLeo, J. A., 2008. Spinal microglial and perivascular cell cannabinoid receptor type 2 activation reduces behavioral hypersensitivity without tolerance after peripheral nerve injury. Anesthesiology 108, 722–734.

Russo, R., Loverme, J., La Rana, G., Compton, T., Parrot, J., Duranti, A., Tontini, A., Mor, M., Tarzia, G., Calignano, A., Piomelli, D., 2007. The fatty-acid amide hydrolase inhibitor URB597 (cyclohexyl carbamic acid 3'-carbamoyl-biphenyl-3-yl ester) reduces neuropathic pain after oral administration in mice. J Pharmacol Exp Ther 322, 236–242.

Sagar, D. R., Kelly, S., Millns, P. J., O'Shaughnessey C, T., Kendall, D. A., Chapman, V., 2005. Inhibitory effects of CB1 and CB2 receptor agonists on responses of DRG neurons and dorsal horn neurons in neuropathic rats. Eur J Neurosci 22, 371–379.

Scott, D. A., Wright, C. E., Angus, J. A., 2004. Evidence that CB-1 and CB-2 cannabinoid receptors mediate antinociception in neuropathic pain in the rat. Pain 109, 124–131.

Simon, G. M., Cravatt, B. F., 2008. Anandamide biosynthesis catalyzed by the phosphodiesterase GDE1 and detection of glycerophospho-N-acyl ethanolamine precursors in mouse brain. J Biol Chem 283, 9341–9349.

Sokal, D. M., Elmes, S. J., Kendall, D. A., Chapman, V., 2003. Intraplantar injection of anandamide inhibits mechanically-evoked responses of spinal neurones via activation of CB2 receptors in anaesthetised rats. Neuropharmacology 45, 404–411.

Stander, S., Schmelz, M., Metze, D., Luger, T., Rukwied, R., 2005. Distribution of cannabinoid receptor 1 (CB1) and 2 (CB2) on sensory nerve fibers and adnexal structures in human skin. J Dermatol Sci 38, 177–188.

Suplita, R. L., 2nd, Farthing, J. N., Gutierrez, T., Hohmann, A. G., 2005. Inhibition of fatty-acid amide hydrolase enhances cannabinoid stress-induced analgesia: sites of action in the dorsolateral periaqueductal gray and rostral ventromedial medulla. Neuropharmacology 49, 1201–1209.

Tsou, K., Brown, S., Sanudo-Pena, M. C., Mackie, K., Walker, J. M., 1998. Immunohisto-chemical distribution of cannabinoid CB1 receptors in the rat central nervous system. Neuroscience 83, 393–411.

Tsuboi, K., Takezaki, N., Ueda, N., 2007. The *N*-acylethanolamine-hydrolyzing acid amidase (NAAA). Chem Biodivers 4, 1914–1925.

Valenzano, K. J., Tafesse, L., Lee, G., Harrison, J. E., Boulet, J. M., Gottshall, S. L., Mark, L., Pearson, M. S., Miller, W., Shan, S., Rabadi, L., Rotshteyn, Y., Chaffer, S. M., Turchin, P. I., Elsemore, D. A., Toth, M., Koetzner, L., Whiteside, G. T., 2005. Pharmacological and pharmacokinetic characterization of the cannabinoid receptor 2 agonist, GW405833, utilizing rodent models of acute and chronic pain, anxiety, ataxia and catalepsy. Neuro-pharmacology 48, 658–672.

Van Sickle, M. D., Duncan, M., Kingsley, P. J., Mouihate, A., Urbani, P., Mackie, K., Stella, N., Makriyannis, A., Piomelli, D., Davison, J. S., Marnett, L. J., Di Marzo, V., Pittman, Q. J., Patel, K. D., Sharkey, K. A., 2005. Identification and functional characterization of brainstem cannabinoid CB2 receptors. Science 310, 329–332.

Vandevoorde, S., 2008. Overview of the chemical families of fatty acid amide hydrolase and monoacylglycerol lipase inhibitors. Curr Top Med Chem 8, 247–267.

Vandevoorde, S., Jonsson, K. O., Labar, G., Persson, E., Lambert, D. M., Fowler, C. J., 2006. Lack of selectivity of URB602 for 2-oleoylglycerol compared to anandamide hydrolysis in vitro. Br J Pharmacol 150, 186–191.

Walczak, J. S., Pichette, V., Leblond, F., Desbiens, K., Beaulieu, P., 2006. Characterization of chronic constriction of the saphenous nerve, a model of neuropathic pain in mice showing rapid molecular and electrophysiological changes. J Neurosci Res 83, 1310–1322.

Walker, J. M., Huang, S. M., 2002. Cannabinoid analgesia. Pharmacol Ther 95, 127–135.

Walker, J. M., Huang, S. M., Strangman, N. M., Tsou, K., Sanudo-Pena, M. C., 1999. Pain modulation by release of the endogenous cannabinoid anandamide. Proc Natl Acad Sci USA 96, 12198–12203.

Weber, A., Ni, J., Ling, K. H., Acheampong, A., Tang-Liu, D. D., Burk, R., Cravatt, B. F., Woodward, D., 2004. Formation of prostamides from anandamide in FAAH knockout mice analyzed by HPLC with tandem mass spectrometry. J Lipid Res 45, 757–763.

Welch, S. P., Huffman, J. W., Lowe, J., 1998. Differential blockade of the antinociceptive effects of centrally administered cannabinoids by SR141716A. J Pharmacol Exp Ther 286, 1301–1308.

Welch, S. P., Stevens, D. L., 1992. Antinociceptive activity of intrathecally administered cannabinoids alone, and in combination with morphine, in mice. J Pharmacol Exp Ther 262, 10–18.

Wotherspoon, G., Fox, A., McIntyre, P., Colley, S., Bevan, S., Winter, J., 2005. Peripheral nerve injury induces cannabinoid receptor 2 protein expression in rat sensory neurons. Neuroscience 135, 235–245.

Yamamoto, W., Mikami, T., Iwamura, H., 2008. Involvement of central cannabinoid CB2 receptor in reducing mechanical allodynia in a mouse model of neuropathic pain. Eur J Pharmacol 583, 56–61.

Yu, M., Ives, D., Ramesha, C. S., 1997. Synthesis of prostaglandin E2 ethanolamide from anandamide by cyclooxygenase-2. J Biol Chem 272, 21181–21186.

Zhang, J., Hoffert, C., Vu, H. K., Groblewski, T., Ahmad, S., O'Donnell, D., 2003. Induction of CB2 receptor expression in the rat spinal cord of neuropathic but not inflammatory chronic pain models. Eur J Neurosci 17, 2750–2754.

Chapter 16
Spinal Dynorphin and Neuropathic Pain

Josephine Lai, Ruizhong Wang, and Frank Porreca

Abstract The endogenous neuropeptides dynorphins are proteolytic products of prodynorphin which are characterized by their high affinity for opioid receptors. Dynorphin A is widely distributed in the CNS and in the spinal cord is found predominantly in neurons of laminae I/II and V. Intrathecal dynorphin A displays predominantly non-opioid activities which can be reversed by NMDA antagonists as well as by bradykinin B2 receptor antagonists. The levels of spinal dynorphin expression can be easily perturbed; elevated levels of dynorphin A in the spinal cord are essential for the expression of chronic pain. Descending modulatory pain pathways from the rostral ventromedial medulla contribute to dynorphin up-regulation and the maintenance of neuropathic pain. Recovery from neuropathic pain may depend not only on recovery from the peripheral injury but also on reversing the injury-induced adaptive changes to the central nervous system such as dynorphin up-regulation.

Abbreviations

cAMP	cyclic adenosine monophosphate
CGRP	calcitonin gene-related peptide
DALBK	des-Arg^{9}Leu-bradykinin
DLF	dorsolateral funiculus
DREAM	downstream regulatory element antagonistic modulator
DRG	dorsal root ganglion
i.c.v.	intracerebroventricularly
i.t.	intrathecally
NMDA	N-methyl-d-aspartate
NPY	neuropeptide Y

F. Porreca (✉)
Department of Pharmacology, University of Arizona Health Sciences Center, Tucson, AZ 85724, USA
e-mail: frankp@emailarizona.edu

M. Malcangio (ed.), *Synaptic Plasticity in Pain*,
DOI 10.1007/978-1-4419-0226-9_16, © Springer Science+Business Media, LLC 2009

RVM   rostral ventral medulla
SNL   spinal nerve ligation

16.1 Introduction

Endogenous opioid neuropeptides have been classified into three families, the enkephalins, the endorphins and the dynorphins. These are proteolytic products of proenkephalin, proopiomelanocortin, and prodynorphin, respectively (Rossier, 1982). The peptides are characterized by their high affinity for opioid mu, delta and kappa receptors as well as the ability of the opioid receptor antagonist, naloxone, to block or reverse agonist actions of these substances. The physiological actions of enkephalins and endorphins are the result of inhibitory effects, predominately on neuronal cells, to elicit analgesia, to alter motor and secretory functions of the gastrointestinal tract, as well as other functions including respiration, cardiovascular activity, hormonal balance, temperature and responses to stress (see Marvizon, Chapter 7). The physiological role of the dynorphins, on the other hand, remains somewhat more obscure.

The dynorphins are all C terminal extensions of leu-enkephalin (Hughes et al., 1975). Dynorphin A(1–17) is one of the major proteolytic fragments of prodynorphin (Civelli et al., 1985) and is widely distributed in the CNS (Civelli et al., 1985; Watson et al., 1982). In the spinal cord, dynorphin A immunoreactivity is found in the dorsal horn primarily in laminae I/II and V (Cruz and Basbaum, 1985; Miller and Seybold, 1987) in short-axon neurons (Botticelli et al., 1981) that are fusiform, pyramidal or flattened (Lima et al., 1993). These neurons may receive direct input from calcitonin gene related peptide (CGRP)-containing primary afferent (Carlton and Hayes, 1989; Takahashi et al., 1990; Takahashi et al., 1988). In addition, dynorphin is also present in the autonomic regions of the lumbar and sacral spinal cord that may be involved in nociception and the autonomic regulation of pelvic viscera (Sasek and Elde, 1986; Sasek et al., 1984). Dynorphin immunoreactivity is also found in spinoreticular tract neurons, located predominantly near the central canal (Nahin, 1987), that project to the medial thalamus (Nahin, 1988).

16.2 Structure–Activity Relationship of Dynorphin A

Dynorphin A was so named when it was first identified because it was far more potent than leu-enkephalin in inhibiting smooth muscle contractility in the guinea pig ileum preparation (Goldstein et al., 1979) primarily through the kappa opioid receptor (Chavkin et al., 1982; Huidobro-Toro et al., 1981). However, in brain tissues the actions of dynorphin A are complex, suggesting its interactions with multiple receptors. The latter has been confirmed with radioligand binding analyses demonstrating that dynorphin A interacts with all three opioid receptor types in brain membranes (Garzon et al., 1984) and more recently in cells that heterologously express opioid receptors (Zhang et al., 1998). The high affinity of dynorphin A for the opioid receptors in binding analysis is conferred by the N terminal tyrosine residue because the *des*-tyrosyl fragments of dynorphin A (e.g., dynorphin A(2–13)) have very low affinity for opioid receptors (Lai et al., 2001). In vivo, dynorphin A is converted rapidly to the *des*-tyrosyl fragment by aminopeptidase activity, and is considered to be the major inactivation mechanism of its opioid actions upon release (Young et al., 1987).

Dynorphin A is distinct from other prodynorphin-derived peptides, namely dynorphin B and neo-endorphin, by its neuronal excitatory actions and excitotoxicity that are not mediated by opioid receptors (Tan-No et al., 2002). Unlike the enkephalins, pharmacological administration of dynorphin A produces little or no antinociception (Piercey et al., 1982; Walker et al., 1982b, 1980). This unexpected property of an endogenous opioid was thought to be due to the rapid breakdown of dynorphin A upon release (Chavkin et al., 1982), but subsequent analyses showed that dynorphin A and its proteolytic products had other, non-opioid mediated effects that may confound its actions at the opioid receptors. In fact, some early observations of antinociception by intrathecal dynorphin may have been confused with dynorphin-induced motor dysfunction as discussed below.

16.3 The Opioid and Non-Opioid Activities of Dynorphin A

The inhibitory effect of dynorphin A on smooth muscle contractility is characteristic of opioid agonists, and is blocked by naloxone. Naloxone is also effective in blocking the ability of dynorphin A to modulate the spontaneous or evoked activity of individual neurons in many regions of the CNS (Vidal et al., 1984). Dynorphin A given intracerebroventricularly (*i.c.v.*) (Walker et al., 1982a) or intrathecally (*i.t.*) (Faden and Jacobs, 1984; Stevens and Yaksh, 1986) produces little or no antinociception. Relatively low doses of dynorphin A (e.g., < 10 nmol of dynorphine A(1–17)) induce pro-nociceptive behaviors including biting, licking and scratching (Tan-No et al., 2002), or transient hypersensitivity to innocuous touch and noxious heat stimuli to the hind paw (Lai et al., 2006). At non-paralytic doses (e.g., 10–30 nmol of dynorphin A(1–17)), dynorphin A elicits long-lasting tactile hypersensitivity and thermal hyperalgesia after a period of hind limb flaccidity (Laughlin et al., 1997; Vanderah et al., 1996). At higher doses (e.g., > 30 nmol of dynorphin A (1–17)), intrathecal dynorphin A produces paralysis which is indicative of severe motor effects (Faden and Jacobs, 1984; Stevens and Yaksh, 1986). At these doses, dynorphin A is neurotoxic, depleting sensory neurons, motor neurons, and interneurons in the spinal cord (Long et al., 1988), and potentiating excitatory neurotransmitter release (Faden, 1992; Skilling et al., 1992). The pronociceptive and excitotoxic effects of dynorphin A are not reversed by naloxone, and can be elicited by the *des*-tyrosyl fragments of dynorphin, indicating that they are non-opioid effects.

16.4 Putative Non-Opioid Targets of Dynorphin A

Blockade of the NMDA receptor could reverse many non-opioid effects of dynorphin; thus the pronociceptive and neuroexcitatory actions of dynorphin A have been proposed to be a direct agonist action of dynorphin A at the NMDA receptor (for recent review, see Lai et al., 2008). Although the evidence suggests that NMDA receptor may ultimately mediate the neuroexcitatory effects of dynorphin A, it is not sufficient to differentiate between direct *versus* indirect actions of dynorphin A at the NMDA receptor. The evidence that implicates a direct interaction of dynorphin A at the NMDA receptor came from radioligand binding studies. The inhibition constants of dynorphin against NMDA receptor ligands (e.g., [^{3}H]MK-801 or [^{3}H]CGP39,653) (Dumont and Lemaire, 1994; Massardier and Hunt, 1989; Shukla et al., 1992) suggest moderate to low affinity. Direct binding using [^{125}I]dynorphin A(2–17) suggest the existence of a low capacity, high affinity binding site (~10 nM) whose capacity is enhanced by a number of NMDA receptor selective antagonists including AP5, chlorokynurenic acid and ifenprodil (Tang et al., 1999). Functional assays using recombinant, heteromeric NMDA receptors expressed in Xenopus oocytes showed that dynorphin inhibited NMDA-induced currents (Brauneis et al., 1996) except under conditions of low extracellular concentrations of glycine (<0.1 μM), where dynorphin potentiated the NMDA-induced currents probably by mimicking the action of glycine at the NMDA receptor (Zhang et al., 1997). A direct interaction of dynorphin at these isolated NMDA receptors on the oocytes, therefore, appears to be primarily inhibitory. Thus, the excitatory effects of dynorphin seen in vivo may not be a consequence of this interaction.

An earlier study showed that another class of G protein coupled receptor, the melanocortin receptors, may also be a putative site of non-opioid action of dynorphin. Dynorphin A(1–13) and its des-tyrosyl fragment, dynorphin A(2–13), antagonized the melanocortin receptor-mediated cAMP formation in transfected HEK cells (Quillan and Sadee, 1997). The inhibitory effect of dynorphin was competitive, with dissociation constants ranging from 40 to 150 nM. It was proposed that some of the paradoxical actions of dynorphin, such as motor effects, neurotoxicity and inflammation may be due to its antagonism of the neuroprotective role of the melanocortins. Another study suggested that high concentrations of dynorphin may inhibit NPY YY receptor (Miura et al., 1994), however, the authors could not detect a direct interaction of dynorphin at the YY receptor. The validity of these putative sites of action of dynorphin remains to be established. Both models propose that the neuroexcitatory effects of dynorphin result from a blockade of an inhibitory transmitter rather than a direct excitatory effect on neurons.

We recently reported that dynorphin A and its des-tyrosyl fragment, dynorphin A(2–13), stimulate calcium influx via L-type and P/Q type voltage sensitive calcium channels through interaction with bradykinin receptors in a dorsal root ganglion X neuroblastoma hybrid cell line, F-11 (Lai et al., 2006). This novel, non-opioid agonist action of dynorphin was not predicted by any structural similarity either between dynorphin and bradykinin, or between the opioid receptor and the bradykinin receptor. Dynorphin A displaced the binding of [^{3}H]bradykinin and [^{3}H]kallidin in brain tissues as well as cell lines that express either endogenous or heterologous bradykinin B1 and B2 receptors with apparent moderate affinity (~ 1 µM). Despite the moderate affinity, the interaction between dynorphin and bradykinin receptors appears to have physiological relevance because intrathecal injection of a B2 selective bradykinin receptor antagonist, HOE 140, blocks the hyperalgesia induced by intrathecal injection of dynorphin A(2–13), suggesting that spinal bradykinin receptors ultimately mediate the pronociceptive actions of spinal dynorphin. In addition, intrathecal dynorphin does not induce hyperalgesia in transgenic mice that lack B2 receptors, further supporting bradykinin receptors as a site of action of spinal dynorphin. Our observation that dynorphin A may activate bradykinin receptors identifies an unexpected putative direct neuroexcitatory target of dynorphin.

16.5 The Pathophysiological Relevance of Agonist Actions of Dynorphin A at Bradykinin Receptors

The level of spinal dynorphin expression appears to be easily perturbed (Cho and Basbaum, 1988) (for review, see Smith and Lee, 1988). It is also likely that changes in the expression of endogenous dynorphin A contribute to abnormal pain states because many experimental models of pathological pain all show a significant regional elevation of dynorphin A levels in the spinal cord. These include inflammatory pain (Nahin et al., 1989; Noguchi et al., 1991; Ruda et al., 1988), neuropathic pain (Kajander et al., 1990; Malan et al., 2000; Wagner et al., 1993), bone cancer pain (Peters et al., 2004), chronic pancreatitis (Vera-Portocarrero et al., 2006), abnormal pain (hyperalgesia) following sustained exposure to morphine (Vanderah et al., 2000) or nicotine (Lough et al., 2007), spinal cord trauma (Abraham et al., 2001; Faden et al., 1985; Tachibana et al., 1998), and arthritis (Weihe et al., 1989).

Elevated levels of dynorphin A are critical for the expression of chronic pain because approaches that inhibit dynorphin A activity consistently normalize enhanced sensory responses (i.e., diminish hyperalgesia). For example, spinal administration of an anti-dynorphin A antiserum reduces neurological impairment after nerve injury (Faden, 1990), blocks the increased sensitivity to noxious thermal and innocuous mechanical stimuli and attenuates trauma induced changes in the spinal cord (Winkler et al., 2002), but does not alter normal sensory thresholds in non-injured rats (Malan et al., 2000; Wagner and Deleo, 1996). Intrathecal administration of an anti-dynorphin antiserum also blocks the inflammatory hyperalgesia induced by Complete Freund's Adjuvant (Luo et al., 2008). Transgenic mice that carry a null mutation in the prodynorphin gene do not exhibit persistent pain states after peripheral nerve injury when compared with their wild type littermates (Wang et al., 2001). These findings support the hypothesis that the upregulation of spinal dynorphin is pronociceptive and important in the maintenance of experimental neuropathic pain, and suggest that approaches that prevent or reverse the excitatory actions of dynorphin A in the spinal cord are likely to have potential therapeutic relevance.

The agonist activity of spinal dynorphin A at bradykinin receptors is a possible mechanism for the pronociceptive actions of dynorphin A based on the following considerations. First, prodynorphin and bradykinin receptors are both localized to the superficial laminae of the dorsal horn of the spinal cord making potential interactions between the two anatomically feasible. Second, bradykinin receptors are localized to small diameter DRG neurons that express substance P or CGRP; the receptors' distribution is relevant to dynorphin A's enhancing effect on neuropeptide release from the central terminals of the primary afferent in spinal cord tissues (Gardell et al., 2002). Third, activation of bradykinin receptors activates sensory neurons. Fourth, the calcium channels that are modulated by dynorphin A via the bradykinin receptors are associated with neuronal excitability and transmitter release.

Indeed, transient hyperalgesia induced by the intrathecal administration of the non-opioid fragment, dynorphin A(2–13), was blocked by intrathecal HOE140 (Lai et al., 2006) suggesting that spinal B2 receptors are essential for the pronociceptive actions of spinal dynorphin. In rats with experimental neuropathic pain, both HOE140 and a B1 receptor selective antagonist, des-Arg[9]Leu-bradykinin (DALBK), acutely reversed allodynia and hyperalgesia when given intrathecally, seen about a week after nerve injury. At the earlier time points (first 4–6 days), however, the antagonists had no significant effect. This delayed effect of bradykinin receptor antagonists correlated with the time course of the upregulation of spinal dynorphin but not with changes in expression of bradykinin receptors. An important consideration is that there is no evidence for de novo synthesis of bradykinin precursors in the spinal cord either in control rats or nerve injured rats. The data support the postulation that activation of spinal bradykinin receptors promotes pain, and they are likely activated by dynorphin when the latter is present in elevated concentrations. Elevated concentrations of spinal dynorphin may also enhance its opioid activity at the kappa opioid receptors as proposed by Xu et al. (2004), showing that transgenic mice that lack kappa opioid receptors exhibited enhanced hyperalgesia upon injury when compared to that of injured wild-type mice. The enhanced opioid activity, however, is apparently insufficient to counter the peptide's non-opioid, pronociceptive actions.

Despite the in vitro and in vivo evidence, some important questions remain regarding the putative binding site for dynorphin at the bradykinin receptors and its therapeutic potential. First and foremost is the apparent moderate affinity of dynorphin for the bradykinin receptor based on indirect competitive binding analysis, which raises the issue about the specificity of this interaction in vivo, and whether a low affinity agonist has physiological relevance. In principle, having a moderate affinity for bradykinin receptor is consistent with the pronociceptive action of dynorphin only upon its elevated production, and presumably enhanced release, in the spinal cord upon an injurious insult, but dynorphin's action at spinal bradykinin receptors would not occur under normal, physiological conditions. Thus, dynorphin may be considered a "pathogenic" endogenous ligand for bradykinin receptors (Altier and Zamponi, 2006). This concept, however, raises a second question of whether the extent of upregulation of spinal dynorphin in vivo is sufficient to directly activate bradykinin receptors. The therapeutic potential of a pharmacological intervention of dynorphin at spinal bradykinin receptor in chronic pain states can be tested by novel design of bradykinin receptor antagonists based on the prototypic structures of dynorphin and known bradykinin antagonists.

In a related study, a transgenic mouse generated by a null mutation of the transcription repressor protein, DREAM (downstream regulatory element antagonistic modulator) exhibited much less pronounced neuropathic pain behavior when compared with similarly injured wild-type mice (Cheng et al., 2002). On the basis that DREAM was previously shown to be a repressor of prodynorphin, and that these knock out mice had elevated level of spinal dynorphin A, the authors concluded that the latter enhanced kappa opioid receptor activation to suppress neuropathic pain behavior in these mice. Thus, in this transgenic animal model, elevated spinal dynorphin A appears to be antihyperalgesic and not pronociceptive. This conclusion may be misleading because the action of DREAM is not exclusive to regulating the transcription of prodynorphin, and thus the behavioral outcome of injury in these mice cannot be inferred to prodynorphin expression or its opioid activity alone. A recent analysis using human synovial fibroblast-like cells and peripheral blood mononuclear cells found that the expression of prodynorphin in these cells was not influenced by the level of DREAM expression (Reisch et al., 2008).

16.6 Descending Pain Modulatory Pathway Is Essential for Spinal Dynorphin A Upregulation

As elevated level of spinal dynorphin A is a pre-requisite for its pronociceptive, non-opioid actions under a number of pathological conditions, mechanisms that enhance the expression of prodynorphin and release of dynorphin A likely promote the pathological actions of spinal dynorphin. Our current understanding of the regulatory mechanisms of dynorphin A expression is limited. As mentioned above, despite the identification of DREAM as a *trans* regulatory element, its role in the transcriptional regulation of prodynorphin, especially under conditions of injury, remains to be established. Another study suggests that inhibitors of the ubiquitin/proteasome system block nerve injury induced pain and upregulation of spinal dynorphin and may attenuate its release from spinal cord neurons (Ossipov et al., 2007). However, because the function of the ubiquitin/proteasome system has broad impact, how this system regulates spinal dynorphin expression and release has to be substantiated.

The time course of the observed changes in the expression of dynorphin and its distribution in the dorsal horn of the spinal cord after unilateral L5/L6 spinal nerve ligation (SNL) injury suggests that increased synthesis is the primary cause of elevated spinal dynorphin. Increased transcription of prodynorphin is evident within 48 h after injury (Lai et al., 2006), followed by an upregulation of prodynorphin immunoreactivity in both superficial (lamina I/II) and deep (lamina V) laminae of the ipsilateral dorsal horn of lumbar spinal cord, evident by about day 3 and reaches maximal levels by about day 10 (Fig. 16.1). The latter coincides with the maximal upregulation of dynorphin A(1–17) content in the lumbar spinal cord, and remains elevated for many weeks (Malan et al., 2000).

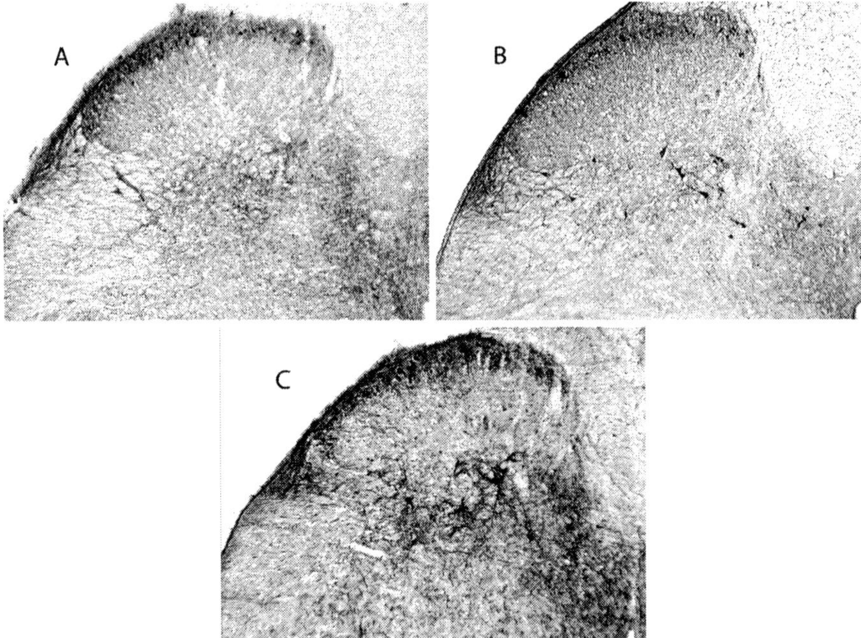

Fig. 16.1 Immunoreactivity of spinal prodynorphin in the doral horn of the lumbar 5 segment of the spinal cord from naïve rat (**A**), rat 3 days after SNL (**B**), and rat 10 days after SNL (**C**). The dorsal horn sections shown in **B** and **C** are ipsilateral to the spinal nerve injury. In naïve tissue, prodynorphin labeling is largely confined to the superficial lamina (I/II), and is found in both cell soma and fibers. On day 3 after SNL, a significant number of immunoreactive pyramidal cell bodies are seen in lamina V. Day 10 after SNL shows peak level of immunoreactivity; the increase is found in both laminae I/II and lamina V and is associated with both cell bodies and fibers when compared with samples from naïve rats (and sham operated rats, data not shown)

While elevated prodynorphin transcripts in the ipsilateral dorsal horn after SNL suggests a local production of spinal dynorphin, we have consistently found that the bulbospinal descending projections from the rostral ventromedial medulla (RVM) via the dorsolateral funiculus (DLF) are necessary for this upregulation. Bilateral lesion of the DLF prevents the upregulation of spinal dynorphin despite subsequent SNL injury (Burgess et al., 2002) or sustained morphine exposure (Vanderah et al., 2001). In the SNL injured rats, the DLF lesion did not prevent the initiation of neuropathic pain, which is established by day 2 after injury, however, the hypersensitivity to both noxious and innocuous stimuli gradually diminish such that by about day 8, the sensory thresholds are not different from that seen in sham-operated control rats despite the permanent injury to the sciatic nerve (Burgess et al., 2002). This reversal of hypersensitivity follows a time course that is very similar to the upregulation of spinal dynorphin in injured animals that did not receive prior DLF lesion. The findings substantiate the evidence that enhanced output of spinal dynorphin is a critical mediator of persistent pain states, and that the descending pain modulatory pathway from the RVM is essential for the upregulation of spinal dynorphin and the maintenance (but not the initiation) of neuropathic pain. These findings also imply that under pathological conditions, the descending pathway is pain facilitatory, which is contrary to the role of this pathway under physiological conditions (for recent review, see Mason, 2005). Consistent with this concept is the observation that microinjection of lidocaine into the RVM acutely blocks SNL induced neuropathic pain (Burgess et al., 2002) as well as hyperalgesia induced by sustained morphine (Vanderah et al., 2001) and pancreatitis (Vera-Portocarrero et al., 2006).

The RVM neurons that mediate this descending pain facilitation are characterized by their responsiveness to mu opioids. Lesion of these neurons by RVM administration of the toxin conjugate, dermorphin-saporin, which specifically targets mu opioid receptor expressing cells, has the same effects on spinal dynorphin levels and neuropathic pain as DLF lesion as described above (Burgess et al., 2002). When dermorphin-saporin is given into the RVM of SNL injured rats with established allodynia and hyperalgesia, the treatment reverses established neuropathic pain states over time, and by day 14 after dermorphin-saporin injection, the sensory thresholds of these animals are not statistically different from that of sham injured rats which had received either vehicle or dermorphin-saporin in the RVM (Porreca et al., 2001).

Although neuropathic pain is a consequence of injuries to the peripheral nerves, central adaptations which have been loosely termed "central sensitization" constitute essential mechanisms for the manifestation of chronic pain (Suzuki and Dickenson, 2005) (see Randic, Chapter 10; Drdla and Sandkuhler, Chapter 9; Thompson, Chapter 11) . The spinal cord has been a major focus as the site of action of these adaptive changes (D'Mello and Dickenson, 2008; Ji et al., 2003), and which include alterations of expression of spinal dynorphin and its putative action at bradykinin receptors as presented in this review. Furthermore, we have discovered that these critical changes in spinal dynorphin are regulated by descending projections from the RVM, whose function is essential in the maintenance of neuropathic pain. The expression of dynorphin therefore effectively serves as an indicator for further characterizing the anatomical, neurophysiological and neurochemical processes by which descending inputs from the RVM regulate spinal cord function upon nerve injury. In addition, RVM neurons that act to facilitate neuropathic pain are potential therapeutic targets (see Bee and Dickenson, Chapter 19); characterization of the phenotype of these neurons may provide innovative strategies. By the same token, phenotypic characterization of RVM neurons that promote descending pain inhibitory inputs is also critical.

16.7 Concluding Remarks

As the paradox of neuropathic pain is that the pain persists despite the apparent recovery from the injurious insult to the peripheral nerve, these emerging insights further strengthen the argument that recovery from neuropathic pain may depend not only on recovery from the peripheral injury but also on minimizing the onset of adaptive changes to the central nervous system, or effective intervention that promotes a reversal of these changes.

Acknowledgments The work presented in this review has been supported by grants from the National Institute on Drug Abuse (F.P.) and the National Institute of Dental and Craniofacial Research (J.L.).

References

Abraham KE, McGinty JF, Brewer KL. The role of kainic acid/AMPA and metabotropic glutamate receptors in the regulation of opioid mRNA expression and the onset of pain-related behavior following excitotoxic spinal cord injury. Neuroscience, 2001; 104: 863–74.

Altier C, Zamponi GW. Opioid, cheating on its receptors, exacerbates pain. Nat Neurosci, 2006; 9: 1465–7.

Botticelli LJ, Cox BM, Goldstein A. Immunoreactive dynorphin in mammalian spinal cord and dorsal root ganglia. Proc Natl Acad Sci USA, 1981; 78: 7783–6.

Brauneis U, Oz M, Peoples RW, Weight FF, Zhang L. Differential sensitivity of recombinant N-methyl-d-aspartate receptor subunits to inhibition by dynorphin. J Pharmacol Exp Ther, 1996; 279: 1063–8.

Burgess SE, Gardell LR, Ossipov MH, Malan TP, Jr., Vanderah TW, Lai J, Porreca F. Time-dependent descending facilitation from the rostral ventromedial medulla maintains, but does not initiate, neuropathic pain. J Neurosci, 2002; 22: 5129–36.

Carlton SM, Hayes ES. Dynorphin A(1–8) immunoreactive cell bodies, dendrites and terminals are postsynaptic to calcitonin gene-related peptide primary afferent terminals in the monkey dorsal horn. Brain Res, 1989; 504: 124–8.

Chavkin C, James IF, Goldstein A. Dynorphin is a specific endogenous ligand of the kappa opioid receptor. Science, 1982; 215: 413–5.

Cheng HY, Pitcher GM, Laviolette SR, Whishaw IQ, Tong KI, Kockeritz LK, Wada T, Joza NA, Crackower M, Goncalves J, Sarosi I, Woodgett JR, Oliveira-dos-Santos AJ, Ikura M, van der Kooy D, Salter MW, Penninger JM. DREAM is a critical transcriptional repressor for pain modulation. Cell, 2002; 108: 31–43.

Cho HJ, Basbaum AI. Increased staining of immunoreactive dynorphin cell bodies in the deafferented spinal cord of the rat. Neurosci Lett, 1988; 84: 125–30.

Civelli O, Douglass J, Goldstein A, Herbert E. Sequence and expression of the rat prodynorphin gene. Proc Natl Acad Sci USA, 1985; 82: 4291–5.

Cruz L, Basbaum AI. Multiple opioid peptides and the modulation of pain: immunohistochemical analysis of dynorphin and enkephalin in the trigeminal nucleus caudalis and spinal cord of the cat. J Comp Neurol, 1985; 240: 331–48.

D'Mello R, Dickenson AH. Spinal cord mechanisms of pain. Br J Anaesth, 2008; 101: 8–16.

Dumont M, Lemaire S. Dynorphin potentiation of [3H]CGP-39653 binding to rat brain membranes. Eur J Pharmacol, 1994; 271: 241–4.

Faden AI. Dynorphin increases extracellular levels of excitatory amino acids in the brain through a non-opioid mechanism. J Neurosci, 1992; 12: 425–9.

Faden AI. Opioid and nonopioid mechanisms may contribute to dynorphin's pathophysiological actions in spinal cord injury. Ann Neurol, 1990; 27: 67–74.

Faden AI, Jacobs TP. Dynorphin-related peptides cause motor dysfunction in the rat through a non-opiate action. Br J Pharmacol, 1984; 81: 271–6.

Faden AI, Molineaux CJ, Rosenberger JG, Jacobs TP, Cox BM. Increased dynorphin immunoreactivity in spinal cord after traumatic injury. Regul Pept, 1985; 11: 35–41.

Gardell LR, Burgess SE, Dogrul A, Ossipov MH, Malan TP, Lai J, Porreca F. Pronociceptive effects of spinal dynorphin promote cannabinoid-induced pain and antinociceptive tolerance. Pain, 2002; 98: 79–88.

Garzon J, Sanchez-Blazquez P, Gerhart J, Loh HH, Lee NM. Dynorphin-1–13: interaction with other opiate ligand bindings in vitro. Brain Res, 1984; 302: 392–6.

Goldstein A, Tachibana S, Lowney LI, Hunkapiller M, Hood L. Dynorphin-(1–13), an extraordinarily potent opioid peptide. Proc Natl Acad Sci USA, 1979; 76: 6666–70.

Hughes J, Smith TW, Kosterlitz HW, Fothergill LA, Morgan BA, Morris HR. Identification of two related pentapeptides from the brain with potent opiate agonist activity. Nature, 1975; 258: 577–80.

Huidobro-Toro JP, Yoshimura K, Lee NM, Loh HH, Way EL. Dynorphin interaction at the kappa-opiate site. Eur J Pharmacol, 1981; 72: 265–6.

Ji RR, Kohno T, Moore KA, Woolf CJ. Central sensitization and LTP: do pain and memory share similar mechanisms? Trends Neurosci, 2003; 26: 696–705.

Kajander KC, Sahara Y, Iadarola MJ, Bennett GJ. Dynorphin increases in the dorsal spinal cord in rats with a painful peripheral neuropathy. Peptides, 1990; 11: 719–28.

Lai J, Luo MC, Chen Q, Ma S, Gardell LR, Ossipov MH, Porreca F. Dynorphin A activates bradykinin receptors to maintain neuropathic pain. Nat Neurosci, 2006; 9: 1534–40.

Lai J, Luo MC, Chen Q, Porreca F. Pronociceptive actions of dynorphin via bradykinin receptors. Neurosci Lett, 2008; 437: 175–9.

Lai J, Ossipov MH, Vanderah TW, Malan TP, Jr., Porreca F. Neuropathic pain: the paradox of dynorphin. Mol Interv, 2001; 1: 160–7.

Laughlin TM, Vanderah TW, Lashbrook J, Nichols ML, Ossipov M, Porreca F, Wilcox GL. Spinally administered dynorphin A produces long-lasting allodynia: involvement of NMDA but not opioid receptors. Pain, 1997; 72: 253–60.

Lima D, Avelino A, Coimbra A. Morphological characterization of marginal (lamina I) neurons immunoreactive for substance P, enkephalin, dynorphin and gamma-aminobutyric acid in the rat spinal cord. J Chem Neuroanat, 1993; 6: 43–52.

Long JB, Petras JM, Mobley WC, Holaday JW. Neurological dysfunction after intrathecal injection of dynorphin A (1–13) in the rat. II. Nonopioid mechanisms mediate loss of motor, sensory and autonomic function. J Pharmacol Exp Ther, 1988; 246: 1167–74.

Lough C, Young T, Parker R, Wittenauer S, Vincler M. Increased spinal dynorphin contributes to chronic nicotine-induced mechanical hypersensitivity in the rat. Neurosci Lett, 2007; 422: 54–8.

Luo MC, Chen Q, Porreca F, Lai J. Dynorphin maintains inflammatory hyperalgesia by the activation of spinal bradykinin receptors. Journal of Pain, 2008; 9(12): 1096–1105.

Malan TP, Ossipov MH, Gardell LR, Ibrahim M, Bian D, Lai J, Porreca F. Extraterritorial neuropathic pain correlates with multisegmental elevation of spinal dynorphin in nerve-injured rats. Pain, 2000; 86: 185–94.

Mason P. Ventromedial medulla: pain modulation and beyond. J Comp Neurol, 2005; 493: 2–8.

Massardier D, Hunt PF. A direct non-opiate interaction of dynorphin-(1–13) with the *N*-methyl-d-aspartate (NMDA) receptor. Eur J Pharmacol, 1989; 170: 125–6.

Miller KE, Seybold VS. Comparison of met-enkephalin-, dynorphin A-, and neurotensin-immunoreactive neurons in the cat and rat spinal cords: I. Lumbar cord. J Comp Neurol, 1987; 255: 293–304.

Miura M, Inui A, Sano K, Ueno N, Teranishi A, Hirosue Y, Nakajima M, Okita M, Togami J, Koshiya K, et al. Dynorphin binds to neuropeptide Y and peptide YY receptors in human neuroblastoma cell lines. Am J Physiol, 1994; 267: E702–9.

Nahin RL. Immunocytochemical identification of long ascending peptidergic neurons contributing to the spinoreticular tract in the rat. Neuroscience, 1987; 23: 859–69.

Nahin RL. Immunocytochemical identification of long ascending, peptidergic lumbar spinal neurons terminating in either the medial or lateral thalamus in the rat. Brain Res, 1988; 443: 345–9.

Nahin RL, Hylden JL, Iadarola MJ, Dubner R. Peripheral inflammation is associated with increased dynorphin immunoreactivity in both projection and local circuit neurons in the superficial dorsal horn of the rat lumbar spinal cord. Neurosci Lett, 1989; 96: 247–52.

Noguchi K, Kowalski K, Traub R, Solodkin A, Iadarola MJ, Ruda MA. Dynorphin expression and Fos-like immunoreactivity following inflammation induced hyperalgesia are colocalized in spinal cord neurons. Brain Res, 1991; 10: 227–33.

Ossipov MH, Bazov I, Gardell LR, Kowal J, Yakovleva T, Usynin I, Ekstrom TJ, Porreca F, Bakalkin G. Control of chronic pain by the ubiquitin proteasome system in the spinal cord. J Neurosci, 2007; 27: 8226–37.

Peters CM, Lindsay TH, Pomonis JD, Luger NM, Ghilardi JR, Sevcik MA, Mantyh PW. Endothelin and the tumorigenic component of bone cancer pain. Neuroscience, 2004; 126: 1043–52.

Piercey MF, Varner K, Schroeder LA. Analgesic activity of intraspinally administered dynorphin and ethylketocyclazocine. Eur J Pharmacol, 1982; 80: 283–4.

Porreca F, Burgess SE, Gardell LR, Vanderah TW, Malan TP, Jr., Ossipov MH, Lappi DA, Lai J. Inhibition of neuropathic pain by selective ablation of brainstem medullary cells expressing the mu-opioid receptor. J Neurosci, 2001; 21: 5281–8.

Quillan JM, Sadee W. Dynorphin peptides: antagonists of melanocortin receptors. Pharm Res, 1997; 14: 713–9.

Reisch N, Engler A, Aeschlimann A, Simmen BR, Michel BA, Gay RE, Gay S, Sprott H. DREAM is reduced in synovial fibroblasts of patients with chronic arthritic pain: is it a suitable target for peripheral pain management? Arthritis Res Ther, 2008; 10: R60.

Rossier J. Opioid peptides have found their roots. Nature, 1982; 298: 221–2.

Ruda MA, Iadarola MJ, Cohen LV, Young WS, 3rd. In situ hybridization histochemistry and immunocytochemistry reveal an increase in spinal dynorphin biosynthesis in a rat model of peripheral inflammation and hyperalgesia. Proc Natl Acad Sci USA, 1988; 85: 622–6.

Sasek CA, Elde RP. Coexistence of enkephalin and dynorphin immunoreactivities in neurons in the dorsal gray commissure of the sixth lumbar and first sacral spinal cord segments in rat. Brain Res, 1986; 381: 8–14.

Sasek CA, Seybold VS, Elde RP. The immunohistochemical localization of nine peptides in the sacral parasympathetic nucleus and the dorsal gray commissure in rat spinal cord. Neuroscience, 1984; 12: 855–73.

Shukla VK, Bansinath M, Dumont M, Lemaire S. Selective involvement of kappa opioid and phencyclidine receptors in the analgesic and motor effects of dynorphin-A-(1–13)-Tyr-Leu-Phe-Asn-Gly-Pro. Brain Res, 1992; 591: 176–80.

Skilling SR, Sun X, Kurtz HJ, Larson AA. Selective potentiation of NMDA-induced activity and release of excitatory amino acids by dynorphin: possible roles in paralysis and neurotoxicity. Brain Res, 1992; 575: 272–8.

Smith AP, Lee NM. Pharmacology of dynorphin. Annu Rev Pharmacol Toxicol, 1988; 28: 123–40.

Stevens CW, Yaksh TL. Dynorphin A and related peptides administered intrathecally in the rat: a search for putative kappa opiate receptor activity. J Pharmacol Exp Ther, 1986; 238: 833–8.

Suzuki R, Dickenson A. Spinal and supraspinal contributions to central sensitization in peripheral neuropathy. Neurosignals, 2005; 14: 175–81.

Tachibana T, Miki K, Fukuoka T, Arakawa A, Taniguchi M, Maruo S, Noguchi K. Dynorphin mRNA expression in dorsal horn neurons after traumatic spinal cord injury: temporal and spatial analysis using in situ hybridization. J Neurotrauma, 1998; 15: 485–94.

Takahashi O, Shiosaka S, Traub RJ, Ruda MA. Ultrastructural demonstration of synaptic connections between calcitonin gene-related peptide immunoreactive axons and dynorphin A(1–8) immunoreactive dorsal horn neurons in a rat model of peripheral inflammation and hyperalgesia. Peptides, 1990; 11: 1233–7.

Takahashi O, Traub RJ, Ruda MA. Demonstration of calcitonin gene-related peptide immunoreactive axons contacting dynorphin A(1–8) immunoreactive spinal neurons in a rat model of peripheral inflammation and hyperalgesia. Brain Res, 1988; 475: 168–72.

Tan-No K, Esashi A, Nakagawasai O, Niijima F, Tadano T, Sakurada C, Sakurada T, Bakalkin G, Terenius L, Kisara K. Intrathecally administered big dynorphin, a prodynorphin-derived peptide, produces nociceptive behavior through an N-methyl-d-aspartate receptor mechanism. Brain Res, 2002; 952: 7–14.

Tang Q, Gandhoke R, Burritt A, Hruby VJ, Porreca F, Lai J. High-affinity interaction of (des-Tyrosyl) dynorphin A(2–17) with NMDA receptors. J Pharmacol Exp Ther, 1999; 291: 760–5.

Vanderah TW, Gardell LR, Burgess SE, Ibrahim M, Dogrul A, Zhong CM, Zhang ET, Malan TP, Jr., Ossipov MH, Lai J, Porreca F. Dynorphin promotes abnormal pain and spinal opioid antinociceptive tolerance. J Neurosci, 2000; 20. 7074–9.

Vanderah TW, Laughlin T, Lashbrook JM, Nichols ML, Wilcox GL, Ossipov MH, Malan TP, Jr., Porreca F. Single intrathecal injections of dynorphin A or des-Tyr-dynorphins produce long-lasting allodynia in rats: blockade by MK-801 but not naloxone. Pain, 1996; 68: 275–81.

Vanderah TW, Suenaga NM, Ossipov MH, Malan TP, Jr., Lai J, Porreca F. Tonic descending facilitation from the rostral ventromedial medulla mediates opioid-induced abnormal pain and antinociceptive tolerance. J Neurosci, 2001; 21: 279–86.

Vera-Portocarrero LP, Xie JY, Kowal J, Ossipov MH, King T, Porreca F. Descending facilitation from the rostral ventromedial medulla maintains visceral pain in rats with experimental pancreatitis. Gastroenterology, 2006; 130: 2155–64.

Vidal C, Maier R, Zieglgansberger W. Effects of dynorphin A (1–17), dynorphin A (1–13) and D-ala2-D-leu5-enkephalin on the excitability of pyramidal cells in CA1 and CA2 of the rat hippocampus in vitro. Neuropeptides, 1984; 5: 237–40.

Wagner R, Deleo JA. Pre-emptive dynorphin and N-methyl-d-aspartate glutamate receptor antagonism alters spinal immunocytochemistry but not allodynia following complete peripheral nerve injury. Neuroscience, 1996; 72: 527–34.

Wagner R, DeLeo JA, Coombs DW, Willenbring S, Fromm C. Spinal dynorphin immunoreactivity increases bilaterally in a neuropathic pain model. Brain Res, 1993; 629: 323–6.

Walker JM, Katz RJ, Akil H. Behavioral effects of dynorphin 1–13 in the mouse and rat: initial observations. Peptides, 1980; 1: 341–5.

Walker JM, Moises HC, Coy DH, Baldrighi G, Akil H. Nonopiate effects of dynorphin and des-Tyr-dynorphin. Science, 1982a; 218: 1136–8.

Walker JM, Moises HC, Coy DH, Young EA, Watson SJ, Akil H. Dynorphin (1–17): lack of analgesia but evidence for non-opiate electrophysiological and motor effects. Life Sci, 1982b; 31: 1821–4.

Wang Z, Gardell LR, Ossipov MH, Vanderah TW, Brennan MB, Hochgeschwender U, Hruby VJ, Malan TP, Jr., Lai J, Porreca F. Pronociceptive actions of dynorphin maintain chronic neuropathic pain. J Neurosci, 2001; 21: 1779–86.

Watson SJ, Khachaturian H, Akil H, Coy DH, Goldstein A. Comparison of the distribution of dynorphin systems and enkephalin systems in brain. Science, 1982; 218: 1134–6.

Weihe E, Millan MJ, Hollt V, Nohr D, Herz A. Induction of the gene encoding pro-dynorphin by experimentally induced arthritis enhances staining for dynorphin in the spinal cord of rats. Neuroscience, 1989; 31: 77–95.

Winkler T, Sharma HS, Gordh T, Badgaiyan RD, Stalberg E, Westman J. Topical application of dynorphin A (1–17) antiserum attenuates trauma induced alterations in spinal cord evoked potentials, microvascular permeability disturbances, edema formation and cell injury: an experimental study in the rat using electrophysiological and morphological approaches. Amino Acids, 2002; 23: 273–81.

Xu M, Petraschka M, McLaughlin JP, Westenbroek RE, Caron MG, Lefkowitz RJ, Czyzyk TA, Pintar JE, Terman GW, Chavkin C. Neuropathic pain activates the endogenous kappa opioid system in mouse spinal cord and induces opioid receptor tolerance. J Neurosci, 2004; 24: 4576–84.

Young EA, Walker JM, Houghten R, Akil H. The degradation of dynorphin A in brain tissue in vivo and in vitro. Peptides, 1987; 8: 701–7.

Zhang L, Peoples RW, Oz M, Harvey-White J, Weight FF, Brauneis U. Potentiation of NMDA receptor-mediated responses by dynorphin at low extracellular glycine concentrations. J Neurophysiol, 1997; 78: 582–90.

Zhang S, Tong Y, Tian M, Dehaven RN, Cortesburgos L, Mansson E, Simonin F, Kieffer B, Yu L. Dynorphin A as a potential endogenous ligand for four members of the opioid receptor gene family. J Pharmacol Exp Ther, 1998; 286: 136–41.

Chapter 17
Microglia, Cytokines and Pain

E.D. Milligan, Ryan G. Soderquist, and Melissa J. Mahoney

Abstract Chronic pain is a significant national health problem which afflicts more than 25% of adults in the United States, alone, and is the most common reason individuals seek medical care. Chronic and recurrent pain, which persists or recurs for more than 3 months, is itself a disease condition. Historically, our understanding of the creation and maintenance of neuropathic pathological pain has focused on neuronal mechanisms in the pain pathway. However, research conducted during the past ~15 years has indicated that many of the neuronal and biochemical changes in the dorsal spinal cord are in part, initiated by and consequences of immune and glial cell signaling. Thus, conditions that activate and/or maintain activation of primary sensory neurons and dorsal spinal cord pain transmission neurons also involve surrounding glial activation. Well-characterized proinflammatory cytokines, derived from glia are critically involved in pathological pain. The most studied cytokines in pathological pain conditions are tumor necrosis factor-alpha (TNF-α), interleukin-1beta (IL-1β) and IL-6. The anti-inflammatory cytokine, interleukin-10 (IL-10), is one of the most powerful counter-regulatory controls over proinflammatory function. Novel and promising viral and non-viral gene therapeutic approaches that employ the actions of anti-inflammatory cytokines such as interleukin-4 and IL-10 are being developed as novel therapeutics to treat chronic neuropathic pain conditions.

Abbreviations

ATP	adenosine 5′-triphosphate
BDNF	brain-derived neurotrophic factor
CCL2	a chemokine of the 'CC' class. Also named monocyte chemo-attractant protein; MCP-1

E.D. Milligan (✉)
Department of Neurosciences, Health Sciences Center, University of New Mexico, Albuquerque, NM 87131-0001, USA
e-mail: emilligan@salud.unm.edu

M. Malcangio (ed.), *Synaptic Plasticity in Pain*,
DOI 10.1007/978-1-4419-0226-9_17, © Springer Science+Business Media, LLC 2009

CCI	chronic constriction injury
CNS	central nervous system
CSF	cerebrospinal fluid
CX3CL1	a chemokine of the CX3C class. Also named fractalkine
DRG	dorsal root ganglia
GABA	gamma-aminobutyric acid
GLAST	glutamate (Glu)-aspartate (Asp) transporter; Glu-Asp transporter
GLT-1	glutamate (Glu)-transporter-1; Glu-transporter-1
HSP	heat-shock proteins
HSV	herpes simplex virus
IL-10	interleukin-10
IL-1β	interleukin 1beta
IL6	interlukin-6
IFN-γ	interferon-gamma
JAK	Janus Kinases
MyD88	myeloid differentiation 88
MAP3K	mitogen-activated protein kinase kinase kinase
MMPs	matrix metalloproteases
NF-kB	nuclear factor-kappaB
P2X	ATP-gated cation channels of the P2 purinergic receptor family
pDNA	plasmid DNA
PI3K	phosphoinositide-3 kinase
PLGA	poly(lactic-co-glycolic) acid copolymer
SOCS	suppressors of cytokine signaling
STAT	signal transducers and activators of transcription
sTNFR	TNF soluble receptor (p55)
TGF-®	transforming growth factor-beta
TIR	toll/interleukin-1 receptor
TLR	toll-like receptors
TNF-α	tumor necrosis factor-alpha

17.1 Introduction

17.1.1 Physiological Pain Processing

Pain is a sensory system that serves protective and adaptive functions. During healthy conditions, pain signals from the body occur by well-known multi-synaptic pathways initiated in the peripheral nervous system and processed at multiple sites within the central nervous system (CNS) (Woolf and Ma, 2007). Peripheral nerve terminals within the body (e.g. glabrous skin) have afferent projecting axons, whose sensory cell body is located in the dorsal root ganglia (DRG), that synapse onto nociceptive-specific neurons in lamina I, II and V of the superficial dorsal horn spinal cord. Pain projection neurons in the CNS that first receive incoming signals from the body, referred to as second order neurons within the superficial dorsal horn of the spinal cord, ascend and synapse onto brainstem relay nuclei as well as discrete brain nuclei. All the components within the pain pathway work in concert to serve a protective, adaptive response function to the host organism.

17.1.2 Pathological Pain Processing: Neuropathic Pain

Tissue damage (e.g. in the foot) activates specialized, pain-responsive peripheral nerve endings. After tissue damage, a local inflammatory response ensues that is maintained by a variety of factors from surrounding cells as well as from the nerve endings, themselves. These mediators are capable of *sensitizing* pain responsive nerve endings (functional nociceptors), and continued inflammation during disease states leads to chronic excitation, further sensitizing nerve endings (McMahon, 2005). In addition, centrally projecting nociceptor terminals to the spinal cord dorsal horn release of a variety of factors that dramatically alter pain spinal or brain neuron activity upon chronic excitation. (Woolf and Mannion, 1999; Hucho and Levine, 2007). These chemical (e.g. increased glutamate and Substance P) and physiological changes within the dorsal spinal cord are a key aspect of enhanced neuronal excitability in response to incoming nociceptive information. The ultimate outcome is exaggerated responses of pain neurons that lead to hyperalgesia (i.e. exaggerated responses to pain stimuli) and allodynia (non-pain stimuli that become painful). Although a number of conditions can lead to chronic pathological pain (e.g. osteoarthritis), understanding chronic neuropathic pain has gained recent considerable attention. Chronic neuropathic pain results from disease or trauma of the nervous system. Often, pain expands from the site of insult, and other parts of the originally unaffected pain pathway, such as the dorsal horn and higher cortical areas, will contribute to pain signaling (Zimmermann, 2001).

17.2 Glial Role in Neuropathic Pain

Until recently, most research focused on the role of neurons in neuropathic pain without considering glia (Watkins et al., 2006). However, greater attention to the role of both astrocytes and microglia (glia) in the creation and maintenance of neuropathic pain has recently been reviewed (DeLeo et al., 2007; Scholz and Woolf, 2007). A number of neuropathic conditions (e.g. diabetic neuropathy, peripheral nerve trauma and inflammation, spinal cord injury) may involve spinal cord glial activation as a common underlying mechanism for producing neuropathic pain. Both astrocytes and microglia are capable of directly altering synaptic communication in the pain pathway because glia encapsulate synapses, respond to neuromodulators such as nitric oxide, and express receptors for, and respond to a variety of pain-related neurotransmitters such as glutamate and substance P (Haydon, 2001; Halassa et al., 2007; Pocock and Kettenmann, 2007). Glia further modulate neuronal function by regulating extracellular excitatory neurotransmitters levels via the glutamate excitatory amino acid transporters, GLT-1 and GLAST, and by clearing cellular debris upon damage (Kreutzberg, 1996; Haydon, 2001; Faulkner et al., 2004). Although both cell types perform additional functions that can modulate neuronal communication in pain signaling, microglia in particular, known as the macrophage of the CNS, resemble many functional aspects of the innate immune system (e.g. immediate responders to tissue damage and/or pathogen invasion) (Nguyen et al., 2002) that can ultimately influence pain transmission. Indeed, many factors released from activated microglia are strongly associated with enhanced pain states (Guo and Schluesener, 2007).

17.3 Cellular Signaling of TLR Activated Glia

One way which microglia become activated is by a family of phylogenetically conserved pattern recognition receptors such as toll-like receptors (TLR) that these glia express (Aravalli et al., 2007), which were originally identified on peripheral innate immune cells, like macrophage and dendritic cells (Medzhitov and Janeway, 2000). These receptors recognize molecular patterns, commonly referred to as pathogen-associated motifs, which are components of pathogens. Microglial cell activation includes regulation by TLRs in the CNS during inflammation and neuronal injury (Aderem and Ulevitch, 2000). Once TLRs recognize pathogen-associated molecular patterns, TLR activation induces the initiation of signaling pathways that (1) arise from intracytoplasmic Toll/ interleukin-1 receptor (TIR) domains that interact with specific adaptor molecules (e.g. MyD88) which are used by the TLR complex to ultimately regulate transcription factor activation (O'Neill and Bowie, 2007), (2) diverge at the intracellular signaling level of mitogen-activated protein kinase kinase kinase (MAP3Ks) that sets into motion several transcription factors including nuclear factor-kappaB (NF-kB) (Banerjee and Gerondakis, 2007), and (3) upon NF-kB activation, controls the expression of a variety of inflammatory cytokine genes including the pro-inflammatory cytokines, interleukin-1beta (IL-1β), tumor necrosis factor-alpha (TNF-α) and interleukin-6 (IL-6) (Kawai and Akira, 2007). Several pathways activated by TLRs that result in NF-kB activation include inhibitory kappaB (IkB) proteins, which must be phosphorylated and degraded for NF-kB signaling cascades to occur (Kawai and Akira, 2007) for the expression of interferon-gamma (IFN-γ), known as a proinflammatory cytokine, and IL-1β, TNF-α and IL-6 from microglia (Olson and Miller, 2004). Despite the large number of signaling cascades that are induced upon TLR glial activation, these events are distilled down to a handful of intracellular mechanisms that finely tune discrete cellular responses (Natarajan et al., 2006). Such intracellular pathways include not only NF-kB and MAP3Ks, but also Janus Kinases (JAK), signal transducers and activators of transcription (STAT), and phosphoinositide-3 kinase (PI3K) signaling cascades upon cytokine signaling and TLR stimulation (Baetz et al., 2004). The response to cytokine and TLR-induced activation of these cascades is an upregulation of the intracellular proteins, suppressors of cytokine signaling (SOCS), as well as other associated proteins, that exert negative regulation of the cytokine and TLR signaling JAK/STAT pathway (for review, see Yoshimura et al., 2004). These reports collectively support that if microglial activation becomes dysregulated, several opportunities exist for proinflammatory signaling pathways to exert downstream effects on nearby astrocytes and neuronal function in the pain pathway.

A number of intracellular signaling molecules that are strongly implicated in mediating neuropathic pain are increased in activated microglia that include phosphorylated p38-MAPK and matrix metalloproteases (MMPs) (Ji and Suter, 2007; Kawasaki et al., 2008a). Indeed, spinal inhibition of TNF-α and IL-1β activity blocks neuropathic pain and p38-MAPK activation (Svensson et al., 2003; Sung et al., 2005; Svensson et al., 2005). Further, p-38MAPK controls IL-1β and TNF-α signaling through transcriptional regulation of NF-kB in cultured activated microglia (Wilms et al., 2003). More recently, distinct roles of MMP2 and MMP9 have been described, whereby MMP9 in activated microglia was found to be critical for the development of neuropathic pain, while MMP2 in astrocytes played a critical role during ongoing neuropathic pain, both via cleavage of IL-1β (Kawasaki et al., 2008a). Thus, although several intracellular signaling pathways occur after microglial activation that lead to common signaling molecules, their functional output can exert complicated and diverse effects during peripheral neuropathic pain conditions.

While TLR activation is best known to occur upon pathogen invasion, TLRs on microglia are also stimulated by non-pathogen endogenous factors (Guo and Schluesener, 2007). This is an important consideration because TLR signaling has been strongly implicated in neuropathic pain states (Tanga et al., 2005) where endogenous factors during neuronal-glial communication must occur that activate TLRs, but are poorly understood. However, several candidate factors exist. One example of such includes the cell stress proteins, heat-shock proteins (HSP) (Roelofs et al., 2006; van Noort, 2008). Indeed, without CNS pathogen stimulation, the role for TLR4 was identified in initiating spinal nerve trauma-induced neuropathic pain because mice lacking the TLR4 receptor did not develop neuropathic pain nor activation of spinal innate immune cells or glial activation (Tanga et al., 2005). Further, blocking spinal cord TLR4 production in neuropathic rats also blocked pathological pain and microglial/macrophage activation (Tanga et al., 2005), supporting that TLR4 signaling in the spinal cord is important for mediating neuropathic pain. More recently, spinal TLR4 antagonism during sciatic nerve damage that leads to neuropathic pain in rats potently alleviated pathological pain (Hutchinson et al., 2007). This study demonstrated that, like neuropathic pain, opioids activate glia via TLR4 activation, which was suggested to play a critical role in opioid-induced tolerance (Hutchinson et al., 2007). Collectively, these findings not only impact future treatment strategies for clinically utilized drugs like opioids, but also novel strategies that act on TLR activation, as endogenous signals appear to play a significant role in neuropathic pain.

Aside from a TLR-specific role, microglia rapidly become activated in response to a variety of endogenous signals (Murphy, 1993; Kreutzberg, 1996; Husemann et al., 2002; Olson and Miller, 2004) that may be responsible for ongoing neuropathic pain signaling. Once microglia are activated, a number of microglial derived pain-relevant substances can produce further stimulation of nearby astrocytes, microglia and neurons. For example, microglial activation leads to the release of nitric oxide, leukotriens, arachidonic acid, and prosta-glandins, which can excite nearby astrocytes and neurons (Kreutzberg, Bezzi, Marchand). Microglia respond to these substances as well as the transmitters, glutamate and adenosine $5'$-triphosphate (ATP). Microglia express glutamate metabotropic group I-III receptors (Biber et al., 1999; Taylor et al., 2002, 2003), and respond to glutamate by releasing TNF-α (Taylor et al., 2005). In turn, TNF-α excites nearby astrocytes, neurons as well as microglia (Ohtori et al., 2004; Stellwagen and Malenka, 2006). Upon IL-1β, TNF-α and IL-6 produc-tion and release, these cytokines can further stimulate microglial-derived IL-1β, TNF-α and IL-6 release (Hanisch, 2002), as well as induce neuro-excitability because neurons also express receptors and respond to IL-1β, TNF-α and IL-6 (Haydon, 2001; Kawasaki et al., 2008b). Measuring the increased production of the proinflammatory cytokines, IL-1β, TNF-α and IL-6 from glia has been a reliable method to assess glial activation (Benveniste, 1997; Hanisch, 2002). Studies blocking the action of IL-1β, TNF-α and IL-6 elucidated a critical role that proinflammatory cytokines indeed play to produce enhanced pain states in various animal models (Watkins et al., 2001). Further, IL-1β, TNF-α or IL-6 mediate spinal cord dorsal horn changes such as enhanced neuronal excitability with consequent neuropathic pain (Kawasaki et al., 2008b). Thus, growing documentation supports that activation of microglia can coordinate inflamma-tory signaling ultimately producing ongoing neuropathic pain changes.

Prior discussions have been primarily speculative in terms of what is under-stood about neuronally-derived signals that trigger glial activation after per-ipheral nerve injury that leads to neuropathic pain. A number of candidate signals exist that are released from primary afferent terminals or intrinsic spinal neurons capable of communicating to surrounding microglia and astrocytes in pain-relevant areas of the spinal cord. Of the possibilities, recent data strongly support a role for the extracellular transmitter purine, ATP, that upon binding to cationic purinoceptors (P2X), mediates neuropathic pain.

17.4 Purinoreceptors: Glial Signals in Neuropathic Pain

Further, microglia express ionotropic and metabotropic purinoceptors (Tsuda, 2003; Bianco et al., 2005) and are activated in response to ATP (Suzuki et al., 2004). Coull and colleagues demonstrated that one effect of ATP is the induced release of brain-derived neurotrophic factor (BDNF) from microglia, which in turn inverts GABA inhibition to GABA excitation in spinal lamina I neurons (Coull et al., 2005). Other reports have shown that the ionotropic P2X4 receptor, expressed solely on microglia, up-regulate in response to peripheral neuropathy, and blocking this microglial P2X4 receptor blocks neuropathic pain (Tsuda, 2003). Activation of this microglial P2X4 receptor not only releases BDNF as noted above, but also induces the release of proinflammatory cytokines from microglia (Le Feuvre et al., 2002). Thus, activated microglia respond to and release transmitters and other substances that act in an autocrine/paracrine fashion, and stimulate astrocytes as well as neurons (Castonguay et al., 2001).

17.5 A Unique Role for Innate Immune System Cells

Responses of immune stimulation begin with the rapid production of a wide variety of cytokines and chemokines, which are chemo-attractant cytokines (Asensio and Campbell, 1999best known as having a role in the maturation and trafficking of leukocytes during inflammatory conditions (Rossi and Zlotnik, 2000. Indeed, a number of studies have now demonstrated significant trafficking of innate immune cells into the DRG (DeLeo et al., 2007; Dubovy et al., 2007; Morin et al., 2007) and dorsal spinal cord (Zhang et al., 2007) during neuropathic conditions. It is important to note that infiltrating lymphocytes (T lymphocytes) into the DRG and spinal cord have also been observed during neuropathy (Sweitzer et al., 2002; Cao and DeLeo, 2008). The chemokine, CCL2 (also named monocyte chemo-attractant protein, MCP-1), is expressed in rat spinal cord neurons (White et al., 2005, 2007) as well as on innate immune cells like glia, macrophages and dendtritic cells (Ransohoff et al., 2007; White et al., 2007). Signaling from CCL2 upregulated in dorsal root ganglia and released from their central nerve terminals in the dorsal horn spinal cord (Dansereau et al., 2008; Jung et al., 2008) was recently shown to be important for spinal dorsal horn innate immune cell influx and microglial activation after peripheral nerve damage with corresponding neuropathic pain (Zhang and De Koninck, 2006; Zhang et al., 2007). In addition to CCL2, the chemokine, fractalkine expressed by neurons (CX_3CL1), is enzymatically cleaved and binds to its unique receptor (CX_3CR1) identified mostly on microglia in pain relevant lamina of the spinal cord during neuropathic pain conditions (Lindia et al., 2005; Clark et al., 2007; White et al., 2007). It was further demonstrated that spinal proinflammatory cytokines, IL-1β and IL-6 act to mediate fractalkine-induced neuropathic pain signaling (Milligan et al., 2004, 2005c). While innate immune cells have been shown to infiltrate DRG and dorsal spinal cord, a large number of resident macrophage and dendritic cells occur within the CSF-filled subarachnoid matrix of the meninges. Upon innate immune stimulation, these innate immune cells are capable of producing IL-1β, TNF-α and IL-6 as well as nitric oxide (Wieseler-Frank et al., 2007), which strongly supports that activated meningeal cells in addition to nearby spinal cord microglia can significantly contribute to pathological pain.

17.6 Anti-Inflammatory Cytokines to Treat Neuropathic Pain

Since the early 1990s, a number of reports have demonstrated that activated spinal cord glia and associated proinflammatory cytokines are critical players in diverse pain states (Watkins et al., 2006 for a comprehensive review). In terms of clinical applications, although, short-term disruption of proinflammatory cytokine action is sufficient to identify their role in pain enhancement, prolonged disruption is required if clinical pain control is the goal. Changes in proinflammatory *and anti-inflammatory cytokine* levels have been reported in several chronic, widespread pain conditions in people. For example, people with fibromyalgia show serum cytokine profiles that are significantly different from those without chronic pain (Uceyler et al., 2006). Further, chronic widespread pain in people is associated with lower gene and protein expression levels of anti-inflammatory cytokines such as interleukin-10 (IL-10) and interleukin-4 (IL-4). And, people with complex regional pain syndrome showed increased IL-1β and IL-6 in CSF (Alexander et al., 2005), suggesting that negative feedback suppression of proinflammatory cytokine activity with anti-inflammatory cytokines, such as interleukin-2 (IL-2), IL-4, transforming growth factor-beta (TGF-β) (Janeway et al., 2005), and IL-10 (Moore et al., 2001) could be a promising clinical approach. Indeed, several anti-inflammatory cytokines such as IL-4 and IL-10 are currently being explored as possible as drug therapies for clinical pain control.

Given sustained suppression of proinflammatory cytokine action could be beneficial in people with chronic neuropathic pain conditions refractory to currently available pain drugs, exploring chronic over-expression of anti-inflammatory cytokines or antagonist to cytokine actions via gene delivery has been a recent approach by some investigators. For example, IL-4 or TNF soluble receptor (p55 sTNFR) was delivered by viral delivery of a replication-defective genomic herpes simplex virus (HSV) encoding IL-4 or p55 sTNFR (Mata et al., 2008). These studies characterized IL-4 or p55 sTNFR gene expression in the dorsal root ganglion after subcutaneous injection into the plantar surface of the hind paw and demonstrated that this procedure produced attenuated mechanical allodynia and fully reversed thermal hyperalgesia in a rat model of neuropathic pain. In addition, IL-4 treatment greatly reduced touch-induced expression of c-Fos-like immunoreactivity in the dorsal horns, a histological marker of neuropathic pain. In addition, IL-1β, microglial p38 activation and prostaglandin expression were blocked in IL-4-treated neuropathic rats (Hao et al., 2006).

17.7 Interleukin-10 Trasngene Delivery to Control Pathological Pain

In addition to IL-4, the potential therapeutic benefit for controlling pathological pain states (enhanced responses to pain stimuli as well as to non-painful stimuli like light touch) by gene transfer to achieve chronic IL-10 transgene expression in the spinal cord is currently being pursued. IL-10 is a good candidate to control glial products that act to enhance pain transmission because the function of IL-10 is well-characterized to terminate pro-inflammatory processes. For example, IL-10 suppresses the production and function of IL-1β, TNF-α and IL-6 cytokines at multiple levels, by preventing p38 MAP kinase activation, NFkB activation, translocation and DNA binding. In addition, IL-10 destabilizes IL-1β and TNF-α mRNA by decreasing its half-life, and counter regulates the actions of IL-1β and TNF-α by increasing IL1 receptor antagonist and TNF decoy receptors and decreases membrane-bound inflammatory cytokine receptors. Thus, a comprehensive suppression of at least IL-1β and TNF-α production and signaling can be achieved through the actions of IL-10 (Moore et al., 2001 for review). IL-10 binds to receptors expressed by glia (Ledeboer et al., 2003) and innate immune cells such as macrophage and dendritic cells (Moore et al., 2001) found in the subarachnoid matrix (Wu et al., 2005) and meninges.

Current evidence indicates that spinal cord neurons do not express IL-10 receptors (Ledeboer et al., 2003), thus disruption of neuronal activity may be less likely by the presence of chronic IL-10. Spinal (intrathecal) delivery of recombinant IL-10 protein blocked the onset of mechanical allodynia induced by either dynorphin (Laughlin et al., 2000) or peri-sciatic phospholipase A2 (Chacur et al., 2004). Allodynia in both of these models were mediated by spinal IL-1β. Spinal cord excitotoxic injury, which is associated with increases in spinal cord glial activation and proinflammatory cytokines produces pathological pain behaviors that are blocked by IL-10 (Bethea et al., 1999; Brewer et al., 1999; Plunkett et al., 2001; Yu C-G et al., 2003; Abraham et al., 2004). Thus, in diverse pain models, IL-10 produced strong pain control effects possibly by its anti-inflammatory mechanism.

In support of these findings, spinal injection of the IL-10 gene for prolonged control of neuropathic pain was initially explored using viral vectors (Milligan et al., 2005a,b). Although spinal cord chronic expression of the IL-10 transgene was delivered into the subarachnoid matrix (i.e. intrathecal) with either adenovirus or adeno-associated viral vectors, pathological pain from spinal inflammation produced by intrathecal gp120, or neuropathic pain produced by sciatic inflammatory neuropathy or chronic constriction injury was robustly prevented and/or reversed (Milligan et al., 2005a,b). These single intrathecal viral vector injections are known to infect cells in the subarachnoid matrix within the spinal meninges without expression in the spinal parenchyma (Iadarola et al., 1997; Mannes et al., 1998; Milligan et al., 2005b). However, the pain control produced by IL-10 gene delivery using adenovirus or adeno-associated viral vectors was transient, possibly due to recognition by the immune system (Jooss and Chirmule, 2003; Liu and Muruve, 2003). Thus, neither viral approach produced sufficiently sustained pain reversal to be clinically relevant.

While non-viral transgene delivery methods are well-known to produce low levels of therapeutic transgene expression, with levels substantially inferior to viral vector methods (Kaplitt and During, 2006), non-viral vector (e.g. plasmid DNA; pDNA) delivery into the spinal subarachnoid compartment (intrathecal) has significant advantages because (1) it avoids invasive surgery required for injections into spinal tissue, (2) it is easily performed across years as needed to maintain pain control, (3) intrathecal injections can lead to retrograde transport to other pain processing anatomical areas such as the DRG (Ledeboer et al., 2007), (4) non-viral gene delivery methods readily transfect cells of the pia and arachnoid (meninges surrounding CSF), both on the spinal and dural surfaces, and (5) it leads to a constant release of IL-10 into the CSF, with IL10 diffusion into the spinal cord creating pain control (Milligan et al., 2006a,b; Ledeboer et al., 2007). All of these actions can be achieved by the use of non-viral vectors while avoiding the potential dangers that are present with the use of viral vectors. For a thorough perspective on DNA-based non-viral gene transfer, see van Gaal and colleagues (2006).

In studies to date, subarachnoid (intrathecal) delivery of plasmid DNA encoding the IL-10 gene (pDNA-IL-10) produced prolonged reversal of neuropathic pain induced by chronic constriction injury (Milligan et al., 2006a; Sloane et al., 2006; Ledeboer et al., 2007; Sloane et al., 2008). Importantly, while intrathecal gene therapy to treat neuropathic pain is not unique (Wu et al., 2001a,2001b; Eaton et al., 2002; Yao et al., 2002a,2002b, 2003; Hao et al., 2006), non-viral IL-10 gene delivery shows excellent promise. This subarachnoid non-viral DNA-based approach is unique because two subarachnoid injections were required to achieve long-lasting neuropathic pain control effects of IL-10 gene therapy (Milligan et al., 2006a). This gene therapy protocol has established a period of sensitization between 2 sequential, subarachnoid injections of non-viral plasmid DNA, where the 2nd injection must encode the IL-10 gene. The initial subarachnoid injection of plasmid DNA vector acts to 'sensitize' the subarachnoid site that enables the second, necessary plasmid DNA encoding IL-10 injection to produce enduring neuropathic pain control. The sensitization period is discrete because the 2nd subarachnoid plasmid DNA injection does not lead to long-duration pain control when injected more than 72 h after the first injection (Milligan et al., 2006a). Both CCI-induced mechanical allodynia (Milligan et al., 2006a; Ledeboer et al., 2007) and thermal hyperalgesia (Milligan et al., 2007) have been reversed in rats upon applying this unique non-viral gene transfer method. Further, well-established paclitaxel-induced allodynia was also attenuated with this pDNA-IL-10 treatment for greater than 35 days (Milligan et al., 2006a; Ledeboer et al., 2007). Thus, pathological pain states such as inflammatory, traumatic and chemotherapy-induced neuropathies, may be controlled by subarachnoid injection of the gene encoding the anti-inflammatory cytokine, IL-10. The action of IL-10 was presumably due to blocking the actions of at least IL-1β, as the increased protein expression of IL-1β from CSF of IL-10-treated rats was suppressed (Ledeboer et al., 2007).

17.8 Innate Immune Cells in the Subarachnoid Matrix May Facilitate Transgene Delivery

Based on the cell types that occur in the spinal subarachnoid matrix such as macrophage and dendritic cells, plasmid DNA delivered to the subarachnoid compartment may be an ideal site for non-viral mediated gene transfer because macrophage and dendritic cells in the meningeal subarachnoid compartment form an extensive and functional network (McMenamin et al., 2003), and express receptors that have been strongly implicated in the cellular process of phagocytosis. The process of phagocytosis is distinct in that particulate material is ingested by active membrane expansion around the material, which results in the formation of large intracellular phagosomes. In addition to TLRs, scavenger receptors like CD163 have been characterized augment the process of phagocytosis in microglia, astrocytes, macrophage and dendritic cells (Buechler et al., 2000; Sulahian et al., 2000; Husemann et al., 2002; Faulkner et al., 2004; Sarrias et al., 2004). Upon intracellular phagosome formation, intracellular TLR9 activation, which is well characterized to bind DNA, can produce profound and long-lasting cellular activation changes (Latz et al., 2004; Wagner, 2004). Further, constitutive levels of a wide range of TLR and chemokine receptors (discussed above) are expressed on microglia and astrocytes (Hanisch, 2002; Olson and Miller, 2004; White et al., 2005) that can become activated in response to released factors from nearby innate immune and glial cells and lead to the production of a panel of cytokines and chemokines (Ghosh et al., 2006).

Activated microglia strongly up-regulate IL-10 production upon TLR stimulation (Olson and Miller, 2004). Further, TLR 9 stimulation by DNA vectors (Krieg, 2002) on dendritic cells and macrophage has been shown to induce endogenous IL-10 production (Yi et al., 2002). And, IL-10 itself can enhance the capacity for phagocytosis of immune cells (Lingnau et al., 2007). Although currently speculative, this process may be one way in which plasmid DNA may act to sensitize innate immune and glial cells in the subarachnoid matrix that then leads to enhanced uptake upon cell contact, gene expression and long-duration pain control after a second plasmid DNA vector injection.

17.9 A Copolymer of Lactic and Glycolic Acid, Poly(Lactic-co-Glycolic) (PLGA) for Targeted Spinal Cord Transgene IL-10 Delivery

While a number of non-viral gene delivery vehicles are being developed to improve efficiency for gene transfer (Amiji, 2005; Huang et al., 2006), synthetic, biocompatible polymers are promising as they are safe, more flexible in manipulating their chemistry, and easier to manufacture (Pack et al., 2005). Although pain reversal by subarachnoid free plasmid DNA encoding IL10 has been very successful in several clinically relevant animals of pathological pain, several potential limitations exist. These include the dose required to achieve therapeutic benefit (125 µg total in rat), and the fact that two injections are required. Poly(lactic-co-glycolic acid) (PLGA), approved by the US Food and Drug Administration, has an established history of successful clinical applications for slow release of large molecules, peptides and proteins (Hedley, 2003). PLGA exploits the body's innate immune system for transgene uptake via the process of phagocytosis, a major advantage of using such polymers for therapeutic transgene delivery. These particles are created to microencapsulate plasmid DNA encoding the IL-10 gene. Upon subarachnoid injection in neuropathic rats, the PLGA microparticles begin to break down to components of the citric acid cycle, releasing their plasmid DNA-IL-10 contents into the surrounding cellular environment. Indeed, these microparticles were identified, using confocal microscopy, in close association with cellular nuclei in the subarachnoid meninges 2 weeks after the second plasmid DNA-IL-10 injection (Milligan et al., 2006b). This is particularly attractive for spinal pDNA-IL10 gene therapy in future clinical application because, in the same report, pain reversal was observed at doses ~100-fold lower than free plasmid DNA (Milligan et al., 2006b).

Further, continued development of PLGA microparticles, formulated with pDNA encoding IL-10, has led to very promising outcomes to control chronic neuropathic pain. After a *single* injection of pDNA encoding IL-10 and microencapsulated in PLGA microparticles, this newer PLGA formulation is capable of targeted spinal cord delivery of up to 8 µg pDNA-IL-10 that leads to robust pain relief in chronic constriction injury-induced neuropathic rats for at least 30 days in an ongoing study (*unpublished observations*). In the spinal cord meninges, this formulation of PLGA microparticles (red) interact with cells that express the classic monocyte/macrophage marker, MHC II (Fig. 17.1A). In adjacent tissue sections, microglial cells, identified by staining for microglial Cd11b expression using classic OX-42 antibody (green) IHC procedures, illustrate red-stained microparticles clearly interact with microglial cells (Fig. 17.1B). These data are highly promising as we progress toward a clinical application. PLGA microparticles are well-documented to become phagocytosed (Hedley, 2003). One question that stands out is whether PLGA microparticles that encapsulate pDNA-IL-10 can also exploit cellular processes like

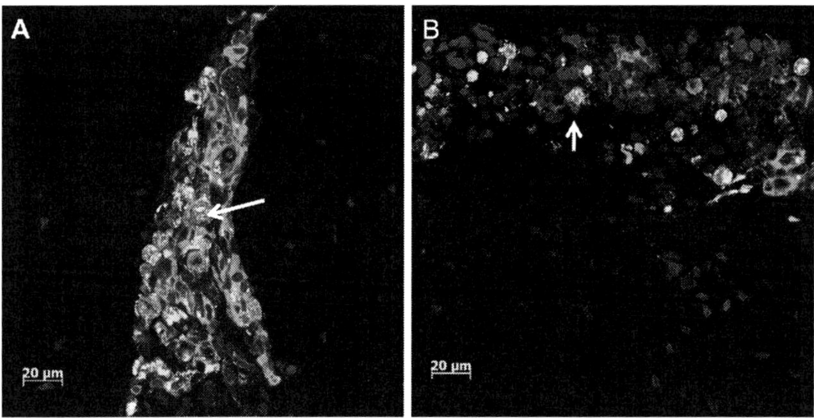

Fig. 17.1 Biosynthetic PLGA microparticles associate with innate immune cells and microglia in the spinal subarachnoid matrix. For verification that the microparticles could associate with phagocytic immune-cells in the subarachnoid matrix, after a single intrathecal injection of poly-L-glycolide-D-lactide (PLGA), a co-polymer (50:50 MW 75,000) used to microencapsulate a plasmid DNA vector encoding the anti-inflammatory gene, interleukin-10 (IL-10) was conducted. At 24 h after subarachnoid (intrathecal) delivery of fluorescently labeled (rhodamine) microparticles encapsulating pDNA-IL-10, animals were deeply anesthetized, and transcardial saline followed by 4% paraformaldehyde perfusion procedures were conducted. Isolated spinal cords were post-fixed and 30 μm cryosections were collected. Spinal cord cross-sections and labeled microparticles (*red*; rhodamine), cellular nuclei (*blue*; DAPI), and either MHC II (*green*; OX-6) of microglial (*green*; OX-42) were examined and imaged using confocal microscopy (40X). (**A**) Microparticles (*red*) appeared to embed themselves (white arrow) in meningeal tissue surrounding spinal cord. In localized areas containing high microparticle concentrations, cells expressing major histocompatibility complex class II (MHC II), as identified by OX-6, the antibody that recognizes MHC II, were observed to be closely associated with microparticles. MHC II is a classic marker for phagocytic macrophage and dendritic cells. (**B**) Microparticles (*red*) also appear to be closely associated with microglia (white arrow), as identified by OX-42 staining (*green*), the antibody directed against complement receptor 3/cluster of differentiation 11b (CR3/CD11b) in the meninges surrounding the spinal cord

phagocytosis to enhance IL-10 transgene expression with subsequent long-duration pain relief upon a single subarachnoid injection. Simply manipulating particle size has resulted in enhanced phagocytosis (Hedley, 2003). Of the many possibilities, long-term pDNA exposure coupled with enhanced phagocytosis is the most straight forward method for further improving transgene delivery targeted to the spinal cord subarachnoid matrix.

17.10 Concluding Remarks

Chronic pain is a significant national health problem that afflicts more than 25% of adults in the United States (Centers for Disease Control and Prevention, 2006) and is not only a major hallmark of many chronic clinical conditions but is the most common reason individuals seek medical care (Centers for Disease Control and Prevention, 2006). Historically, our understanding of the creation and maintenance of chronic neuropathic pain, which persists or recurs for more than 3 months and is itself a disease condition (Loeser, 2006), has focused on neuronal mechanisms in the pain pathway. However, research conducted during the past decade revealed that many of the neuronal and biochemical changes in the dorsal spinal cord are in part, initiated by and consequences of immune and glial cell signaling (DeLeo et al., 2007). Thus, conditions that activate and/or maintain activation of primary sensory neurons and dorsal spinal cord pain transmission neurons also involve surrounding glial activation (DeLeo et al., 2007). In response to a number of cellular activation pathways that are initiated during neuropathic pain states, glia are activated and release the proinflammatory cytokines, IL-1β, TNF-α, and IL-6. These cytokines are the most characterized and ultimately participate in underlying pathological pain signaling that includes chronic neuropathic pain. The anti-inflammatory cytokine, IL-10, is one of the most powerful counter-regulatory controls over proinflammatory function. Novel and promising gene therapeutic approaches that employ the actions of anti-inflammatory cytokines such as interleukin-4 and IL-10 are being developed as novel therapeutics to treat chronic neuropathic pain conditions. Targeted therapeutic gene delivery using biodegradable and biocompatible synthetic polymers is currently being developed for the improvement of transgene expression in the spinal cord and meninges. The goal in these and future studies is to develop therapeutic gene delivery methods for future clinical utility to treat chronic neuropathic pain.

References

Abraham, KE, D McMillen and KL Brewer (2004). "The effects of endogenous interleukin-10 on gray matter damage and pain behaviors following excitotoxic spinal cord injury in the mouse." *Neuroscience* 124: 945–922.

Aderem, A and RJ Ulevitch (2000). "Toll-like receptors in the induction of the innate immune response." *Nature* 406: 782–787.

Alexander, GM, MA Rijn, JJ van-Hilten, et al. (2005). "Changes in cerebrospinal fluid levels of pro-inflammatory cytokines in CRPS." *Pain* 116: 213–219.

Amiji, MM (2005). *Polymeric Gene Delivery: Principles and Applications*. Boca Raton, CRC Press.

Aravalli, RN, PK Peterson and JR Lokensgard (2007). "Toll-like receptors in defense and damage of the central nerovus system." *J Neuroimmune Pharmacol* 2: 297–312.

Asensio, VC and IL Campbell (1999). "Chemokines in the CNS: plurifunctional mediators in diverse states." *Trends Neurosci* 22: 504–512.

Baetz, A, M Frey, K Heeg, et al. (2004). "Suppressor of cytokine signaling (SOCS) proteins indirectly regulate toll-like receptor signaling in innate immune cells." *J Biol Chem* 279(52): 54708–54715.

Banerjee, A and S Gerondakis (2007). "Coordinating TLR-activated signaling pathways in cells of the immune system." *Immunol Cell Biol* 85(6): 420–424.

Benveniste, EN (1997). "Cytokine expression in the nervous system." *Immunology of the Nervous System*. RW Keane and WF Hickey (eds.). New York, Oxford University Press: 419–459.

Bethea, JR, H Nagashima, MC Acosta, et al. (1999). "Systemically administered interleukin-10 reduces tumor necrosis factor-alpha production and significantly improves functional recovery following traumatic spinal cord injury in rats." *Neurotrauma* 16: 851–863.

Bianco, F, M Fumagalli, E Pravettoni, et al. (2005). "Pathophysiological roles of extracellular nucleotides in glial cells: differential expression of purinergic receptors in resting and activated microglia." *Brain Res Rev* 48: 144–156.

Biber, K, DJ Laurie, A Berthele, et al. (1999). "Expression and signalling of group I metabotropic glutamate receptors in astrocytes and microglia." *J Neurochem* 72: 1671–1680.

Brewer, KL, JR Bethea and RP Yezierski (1999). "Neuroprotective effects of interleukin-10 following spinal cord injury." *Exp Neurol* 159: 484–493.

Buechler, C, M Ritter, E Orso, et al. (2000). "Regulation of scavenger receptor CD163 expression in human monocytes and macrophages by pro- and antiinflammatory stimuli." *J Leukoc Biol* 67(1): 97–103.

Cao, L and JA DeLeo (2008). "CNS-infiltrating CD4+ T lymphocytes contribute to murine spinal nerve transection-induced neuropathic pain." *Eur J Immunol* 38: 448–458.

Castonguay, A, S Levesque and R Robitaille (2001). "Glial cells as active partners in synaptic functions." *Glial Cell Function.* B Castellano-Lopez and M Nieto-Sampedro (eds.). Amsterdam, Elsevier Science B.V. 132: 227–240.

Centers for Disease Control and Prevention (2006). *Health, United States, 2006, with Special Feature on Pain.* Centers for Disease Control and Prevention's (CDC) National Center for Health Statistics: 559.

Chacur, M, ED Milligan, EM Sloane, et al. (2004). "Snake venom phospholipase A2s (Asp49 and Lys49) induce mechanical allodynia upon peri-sciatic administration: involvement of spinal cord glia, proinflammatory cytokines and nitric oxide." *Pain* 108: 180–191.

Clark, AK, PK Yip, J Grist, et al. (2007). "Inhibition of spinal microglial cathepsin S for the reversal of neuropathic pain." *Proc Natl Acad Sci USA* 104(25): 10655–10660.

Coull, JA, S Beggs, D Boudreau, et al. (2005). "BDNF from microglia causes the shift in neuronal anion gradient underlying neuropathic pain." *Nature* 438: 923–925.

Dansereau, MA, RD Gosselin, M Pohl, et al. (2008). "Spinal CCL2 pronociceptive action is no longer effective in CCR2 receptor antagonist-treated rats." *J Neurochem* 106(2): 757–769.

DeLeo, JA, LS Sorkin and LR Watkins, Eds. (2007). *Immune and Glial Regulation of Pain.* Seattle, IASP Press.

Dubovy, P, L Tuckova, R Jancalek, et al. (2007). "Increased invasion of ED-1 positive macrophages in both ipsi- and contralateral dorsal root ganglia following unilateral nerve injuries." *Neurosci Lett* 427(2): 88–93.

Eaton, MJ, B Blits, MJ Ruitenberg, et al. (2002). "Amerlioration of chronic neuropathic pain after partial nerve injury by adeno-assocaited viral (AAV) vector-mediated over-expression of BDNF in the rat spinal cord." *Gene Ther* 9: 1387–1395.

Faulkner, JR, JE Herrmann, MJ Woo, et al. (2004). "Reactive astrocytes protect tissue and preserve function after spinal cord injury." *J Neurosci* 24(9): 2143–2155.

Ghosh, TK, DJ Mickelson, J Fink, et al. (2006). "Toll-like receptor (TLR) 2–9 agonists-induced cytokines and chemokines: I. Comparison with T cell receptor-induced responses." *Cell Immunol* 243(1): 48–57.

Guo, L-H and HJ Schluesener (2007). "The innate immunity of the central nervous system in chronic pain: The role of Toll-like receptors." *Cell Mol Life Sci* 64: 1128–1136.

Halassa MM, T Fellin and PG Haydon (2007). "The tripartite synapse: roles for gliotransmission in health and disease." *Trends Mol Med* 13(2): 54–63.

Hanisch, U-K (2002). "Microglia as a source and target of cytokines." *Glia* 40: 140–155.

Hao, S, M Mata, JC Glorioso, et al. (2006). "HSV-mediated expression of interleukin-4 in dorsal root ganglion neurons reduces neuropathic pain." *Mol Pain* 2(6).

Haydon, PG (2001). "GLIA: Listening and talking to the synapse." *Nat Rev Neurosci* 2: 185–193.

Hedley, ML (2003). "Formulations containing poly-lactide-co-glycolide and plasmid DNA expression vectors." *Exper Opin Biol Ther* 3(6): 903–910.

Huang, D, F-D Shi, S Jung, et al. (2006). "The neural chemokine CX3CL1/fractalkine selectively recruits NK cells that modify experimental autoimmune encephalomyelitis within the central nervous system." *FASEB J* 20: 896–905.

Hucho, T and JD Levine (2007). "Signaling pathways in sensitization: toward a nociceptor cell biology." *Neuron* 55(3): 365–376.

Husemann, J, JD Loike, R Anankov, et al. (2002). "Scavenger receptors in neurobiology and neuropathology: their role on microglia and other cells of the nervous system." *Glia* 40(2): 195–205.

Hutchinson, MR, ST Bland, KW Johnson, et al. (2007). "Opioid-induced glial activation: Mechanisms of activation and implications for opioid analgesia, dependence and reward." *Sci World J* 7(S2): 98–111.

Iadarola, MJ, S Lee and AJ Mannes (1997). Gene transfer approaches to pain control. *Molecular Neurobiology of Pain*. D Borsook (ed.). Seattle, IASP Press: 337–360.

Janeway, CA, P Travers, M Walport, et al. (2005). *Immunobiology: The Immune System in Health and Disease*. New York, NY, Garland Science Publishing.

Ji, RR and MR Suter (2007). "p38 MAPK, microglial signaling, and neuropathic pain." *Mol Pain* 3: 33.

Jooss, K and N Chirmule (2003). "Immunity to adenovirus and adeno-associated viral vectors: implication for gene therapy." *Gene Ther* 10: 995–963.

Jung, H, PT Toth, FA White, et al. (2008). "Monocyte chemoattractant protein-1 functions as a neuromodulator in dorsal root ganglia neurons." *J Neurochem* 104(1): 254–263.

Kaplitt, MG and MJ During (2006). *Gene Therapy in the Central Nervous System: From Bench to Bedside*. San Diego, CA, Elsevier, Inc.

Kawai, T and S Akira (2007). "Signaling to NF-kappaB by Toll-like receptors." *Trends Mol Med* 13(11): 460–469.

Kawasaki, Y, ZZ Xu, X Wang, et al. (2008a). "Distinct roles of matrix metalloproteases in the early- and late-phase development of neuropathic pain." *Nat Med* 14(3): 331–336.

Kawasaki, Y, L Zhang, J-K Cheng, et al. (2008b). "Cytokine mechanisms of central sensitization: Distinct and overlapping role of interleukin-1b, interleukin-6, and tumor necrosis factor-a in regulating syntapic and neuronal activity in the superficial spinal cord." *J Neurosci* 28(20): 5189–5194.

Kreutzberg, GW (1996). "Microglia: a sensor for pathological events in the CNS." *Trends Neurosci* 19: 312–318.

Krieg, AM (2002). "CpG motifs in bacterial DNA and thier immune effects." *Annu Rev Immunol* 20: 709–760.

Latz, E, A Schoenemeyer, A Visintin, et al. (2004). "TLR9 signals after translocating from the ER to CpG DNA in the lysosome." *Nat Immunol* 5: 190–198.

Laughlin, TM, JR Bethea, RP Yezierski, et al. (2000). "Cytokine involvement in dynorphin-induced allodynia." *Pain* 84: 159–167.

Le Feuvre, R, D Brough and N Rothwell (2002). "Extracellular ATP and P2X7 receptors in neurodegeneration." *Eur J Pharmacol* 447: 261–269.

Ledeboer, AM, BM Jekich, EM Sloane, et al. (2007). "Intrathecal interleukin-10 gene therapy attenuates paclitaxel-induced mechanical allodynia and proinflammatory cytokine expression in dorsal root ganglia in rats." *Brain Behav Immun* 21(5): 686–698.

Ledeboer, A, A Wierinckx, JGJM Bol, et al. (2003). "Regional and temporal expression patterns of interleukin-10, interleukin-10 receptor and adhesion molecules in the rat spinal cord during chronic relapsing EAE." *J Neuroimmunol* 136: 94–103.

Lindia, JA, E McGowan, N Jochnowitz, et al. (2005). "Induction of CX3CL1 expression in astocytes and CX3CR1 in microglia in the spinal cord of a rat model of neuropathic pain." *J Pain* 6: 434–438.

Lingnau, M, C Hoflich, HD Volk, et al. (2007). "Interleukin-10 enhances the CD14-dependent phagocytosis of bacteria and apoptotic cells by human monocytes." *Hum Immunol* 68(9): 730–738.

Liu, Q and DA Muruve (2003). "Molecular basis of the inflammatory response to adenovirus vectors." *Gene Ther* 10: 935–940.

Loeser, JD (2006). "Pain as a disease." *Pain*. F Cervero and TS Jensen (eds.). Amsterdam, Elsevier B.V. 81: 11–20.

Mannes, AJ, RM Caudle and BC O'Connell, Iadarola, MJ (1998). "Adenoviral gene transfer to spinal cord neurons: intrathecal vs. intraparenchymal administration." *Brain Res* 793: 1–6.

Mata, M, S Hao and DJ Fink (2008). "Gene therapy directed at the neuroimmune component of chronic pain with particular attention to the role of TNF alpha." *Neurosci Lett* 437(3): 209–213.

McMahon, SB, WBJ Cafferty and F Marchand (2005). "Immune and glial cell factors as pain mediators and modulators." *Exp Neurol* 192: 444–462.

McMenamin, PG, RJ Wealthall, M Deverall, et al. (2003). "Macrophages and dendritic cells in the rat meninges and choroid plexus: three-dimensional localisation by environmental scanning electron microscopy and confocal microscopy." *Cell Tissue Res* 313: 259–269.

Medzhitov, R and C Janeway, Jr. (2000). "The Toll receptor family and microbial recognition. " *Trends Microbiol* 8(10): 452–456.

Milligan, ED, SJ Langer, EM Sloane, et al. (2005a). "Controlling pathological pain by adenovirally driven spinal production of the anti-inflammatory cytokine, Interleukin-10." *Eur J Neurosci* 21: 2136–2148.

Milligan, ED, A Ledeboer, EM Sloane, et al. (2007). "Glially driven enhancement of pain and its control by anti-inflammatory cytokines." *Immune and Glial Regulation of Pain*. JA DeLeo, LS Sorkin and LR Watkins (eds.). Seattle, IASP Press: 319–337.

Milligan, ED, EM Sloane, SJ Langer, et al. (2005b). "Controlling neuropathic pain by adeno-associated virus driven production of the anti-inflammatory cytokine, interleukin-10." *Mol Pain* 1: 9–22.

Milligan, ED, EM Sloane, SJ Langer, et al. (2006a). "Repeated intrathecal injections of plasmid DNA encoding interleukin-10 produce prolonged reversal of neuropathic pain." *Pain* 126: 294–308.

Milligan, ED, RG Soderquist, SM Malone, et al. (2006b). "Intrathecal polymer-based interleukin-10* gene delivery for neuropathic pain." *Neuron Glia Biol* 2: 293–308.

Milligan, ED, V Zapata, M Chacur, et al. (2004). "Evidence that exogenous and endogenous fractalkine can induce spinal nociceptive facilitation in rats." *Eur J Neurosci* 20: 2294–2302.

Milligan, ED, V Zapata, D Schoeniger, et al. (2005c). "An initial investigation of spinal mechanisms underlying pain enhancement induced by fractalkine, a neuronally released chemokine." *Eur J Neurosci* 22: 2775–2782.

Moore, KW, R de Waal Malefyt, RL Coffman, et al. (2001). "Interleukin-10 and the interleukin-10 receptor." *Annu Rev Immunol* 19: 683–765.

Morin, N, SA Owolabi, MW Harty, et al. (2007). "Neutrophils invade lumbar dorsal root ganglia after chronic constriction injury of the sciatic nerve." *J Neuroimmunol* 184(1–2): 164–171.

Murphy, S, Ed. (1993). *Astrocytes: Pharmacology and Function*. San Diego, Academic Press.

Natarajan, M, K-M Lin, RC Hsueh, et al. (2006). "A global analysis of cross-talk in a mammalian cellular signalling network." *Nat Cell Biol* 8(6): 571–580.

Nguyen, MD, JP Julien and S Rivest (2002). "Innate immunity: the missing link in neuro-protection and neurodegeneration." *Nat Rev Neurosci* 3: 216–227.

Ohtori, S, K Takahashi, H Moriya, et al. (2004). "TNF-alpha and TNF-alpha receptor type 1 upregulation in glia and neurons after peripheral nerve injury: studies in murine DRG and spinal cord." *Spine* 29: 1082–1088.

Olson, JK and SD Miller (2004). "Microglia initiate central nervous system innate and adaptive immune responses through multiple TLRs." *J Immunol* 173: 3916–3924.

O'Neill, LA and AG Bowie (2007). "The family of five: TIR-domain-containing adaptors in Toll-like receptor signalling." *Nat Rev Immunol* 7(5): 353–364.

Pack, DW, AS Hoffman, S Pun, et al. (2005). "Design and development of polymers for gene delivey." *Nat Rev Drug Dis* 4: 581–593.

Plunkett, JA, C-G Yu, JM Easton, et al. (2001). "Effects of interleukin-10 (IL-10) on pain behavior and gene expression following excitotoxic spinal cord injury in the rat." *Exp Neurol* 168: 144–154.

Pocock, JM and H Kettenmann (2007). "Neurotransmitter receptors on microglia." *Trends Neurosci* 30(10): 527–535.

Ransohoff, RM, L Liu and AE Cardona (2007). "Chemokines and chemokine receptors: multipurpose players in neuroinflammation." *Int Rev Neurobiol* 82: 187–204.

Roelofs, MF, WC Boelens, LA Joosten, et al. (2006). "Identification of small heat shock protein B8 (HSP22) as a novel TLR4 ligand and potential involvement in the pathogenesis of rheumatoid arthritis." *J Immunol* 176(11): 7021–7027.

Rossi, D and A Zlotnik (2000). "The biology of chemokines and their receptors." *Annu Rev Immunol* 18: 217–242.

Sarrias, MR, J Gronlund, O Padilla, et al. (2004). "The Scavenger Receptor Cysteine-Rich (SRCR) domain: an ancient and highly conserved protein module of the innate immune system." *Crit Rev Immunol* 24(1): 1–37.

Scholz, J and CJ Woolf (2007). "The neuropathic pain triad: neurons, immune cells, and glia." *Nat Neurosci* 10(11): 1361–1368.

Sloane, EM, SJ Langer, BM Jekich, et al. (2009). "Immunological priming potentiates non-viral anti-inflammatory gene therapy treatment of neuropathic pain." *Gene Therapy*, submitted.

Sloane, EM, SJ Langer, ED Milligan, et al. (2006). *A novel anti-inflammatory cytokine based non-viral gene therapy: Controlling neuropathic pain and beyond.* Immunology 2006, Boston, MA, The American Association of Immunologists.

Stellwagen, D and RC Malenka (2006). "Synaptic scaling mediated by glial TNF-alpha." *Nature* 440: 1054–1059.

Sulahian, TH, P Hogger, AE Wahner, et al. (2000). "Human monocytes express CD163, which is upregulated by IL-10 and identical to p155." *Cytokine* 12(9): 1312–1321.

Sung, CS, ZH Wen, WK Chang, et al. (2005). "Inhibition of p38 mitogen-activated protein kinase attenuates interleukin-1beta-induced thermal hyperalgesia and inducible nitric oxide synthase expression in the spinal cord." *J Neurochem* 94(3): 742–752.

Suzuki, T, I Hide, K Ido, et al. (2004). "Production and release of neuroprotective tumor necrosis factor by P2X7 receptor activated microglia." *J Neurosci* 24: 1–7.

Svensson, CI, B Fitzsimmons, S Azizi, et al. (2005). "Spinal p38beta isoform mediates tissue injury-induced hyperalgesia and spinal sensitization." *J Neurochem* 92: 1508–1520.

Svensson CI, M Marsala, A Westerlund, et al. (2003). "Activation of p38 mitogen-activated protein kinase in spinal microglia is a critical link in inflammation-induced spinal pain processing." *J Neurochem* 86(6): 1534–1544.

Sweitzer, SM, WF Hickey, MD Rutkowski, et al. (2002). "Focal peripheral nerve injury induces leukocyte trafficking into the central nervous system: potenital relationship to neuropathic pain." *Pain* 100: 163–170.

Tanga, FY, N Nutile-McMenemy and JA DeLeo (2005). "The CNS role of Toll-like receptor 4 in innate neuroimmunity and painful neuropathy." *PNAS* 102: 16.

Taylor, DL, LT Diemel, ML Cuzner, et al. (2002). "Activation of group II glutamate receptors underlies microglial reactivity and neurotoxicity following stimulation with peptides upregulated Alzheimer's disease." *J Neurochem* 82: 1179–1191.

Taylor, DL, LT Diemel and JM Pocock (2003). "Activation of microglial group III metabotropic glutamate receptors protects neurons against microglial neurotoxicity." *J Neurosci* 23: 2150–2160.

Taylor, DL, F Jones, ES Kubata, et al. (2005). "Stimulation of microglial metabotropic glutamate receptor mGlu2 triggers tumor necrosis factor alpha-induced neurotoxicity in concert with microglial-derived Fas ligand." *J Neurosci* 25: 2952–2964.

Tsuda, M (2003). "P2X4 receptors induced in spinal microglia gate tactile allodynia after nerve injury." *Nature* 424: 778–783.

Uceyler, N, R Valenza, M Stock, et al. (2006). "Reduced levels of antiinflammatory cytokines in patients with chronic widespread pain." *Arthrit Rheum* 54: 2656–2664.

van Gaal, EVB, WE Hennink, DJA Crommelin, et al. (2006). "Plasmid engineering for controlled and sustained gene expression for non-viral gene therapy." *Pharm Res* 23(6): DOI: 10.1007/s11095-006-0164-2.

van Noort, JM (2008). "Stress proteins in CNS inflammation." *J Pathol* 214(2): 267–275.

Wagner, H (2004). "The immunobiology of the TLR9 subfamily." *Trends Immunol* 25(7): 381–386.

Watkins, LR, ED Milligan and SF Maier (2001). "Spinal cord glia: new players in pain." *Pain* 93: 201–205.

Watkins, LR, J Wieseler-Frank, ED Milligan, et al. (2006). "Contribution of glia to pain processing in health and disease." *Handbook of Clinical Neurology*. F Cervero and TS Jensen (eds.). Amsterdam, Elsevier. 81: 309–323.

White, FA, SK Bhangoo and RD Miller (2005). "Chemokines: Integrators of pain and inflammation." *Nature Rev* 4: 834–844.

White, FA, H Jung and RJ Miller (2007). "Chemokines and the pathophysiology of neuropathic pain." *Proc Natl Acad Sci USA* 104(51): 20151–20158.

Wieseler-Frank, J, BM Jekich, JH Mahoney, et al. (2007). "A novel immune-to-CNS communication pathway: cells of the meninges surrounding the spinal cord CSF space produce proinflammatory cytokines in response to an inflammatory stimulus." *Brain Behav Immun* 21(5): 711–718.

Wilms, H, P Rosenstiel, J Sievers, et al. (2003). "Activation of microglia by human neuromelanin is NF-kappaB dependent and involves p38 mitogen-activated protein kinase: implications for Parkinson's disease." *FASEB J* 17(3): 500–502.

Woolf, CJ and Q Ma (2007). "Nociceptors – noxious stimulus detectors." *Neuron* 55(3): 353–364.

Woolf, CJ and RJ Mannion (1999). "Neuropathic pain: aetiology, symptoms, mechanisms, and management." *Lancet* 353(9168): 1959–1964.

Wu, CL, MG Garry, RA Zollo, et al. (2001a). "Gene therapy for the management of pain: Part I: Methods and strategies." *Anesthes* 94: 1119–1132.

Wu CL,, MG Garry, RA Zollo, et al. (2001b). "Gene therapy for the management of pain. Part II: Molecular targets." *Anesthes* 95: 216–240.

Wu, Z, J Zhang and H Nakanishi (2005). "Leptomeningeal cells activate microglia and astrocytes to induce IL-10 production by releasing pro-inflammatory cytokines during systemic inflammation." *J Neuroimmunol* 167(1–2): 90–98.

Yao, MZ, JF Gu, JH Wang, et al. (2002a). "Interleukin-2 gene therapy of chronic neuropathic pain." *Neuroscience* 112(2): 409–416.

Yao, MZ, JF Gu, HJ Wang, et al. (2003). "Adenovirus-mediated interleukin-2 gene therapy of nociception." *Gene Ther* 10: 1392–1399.

Yao, MZ, JH Wang, JF Gu, et al. (2002b). "Interleukin-2 gene has superior antinociceptive effects when delivered intrathecally." *Clin Neurosci Neuropathol* 13(6): 791–794.

Yi, A-K, J-G Yoon, S-J Yeo, et al. (2002). "Role of mitogen-activated protein kinases in CpG DNA-mediated IL-10 and IL-12 production: Central role of extracellular signal-regulated kinase in the negative feedback loop of the CpG DNA-mediated Th1 response." *J Immunol* 168: 4711–4720.

Yoshimura, A, HM Ohishi, D Aki, et al. (2004). "Regulation of TLR signaling and inflammation by SOCS family proteins." *J Leukoc Biol* 75(3): 422–427.

Yu, C-G, CA Fairbanks, GL Wilcox, et al. (2003). "Effects of agmatine, interleukin-10 and cyclosporin on spontaneous pain behavior following excitotoxic spinal cord injury in rats." *J Pain* 4: 129–140.

Zhang, J and Y De Koninck (2006). "Spatial and temporal relationship between monocyte chemoattractant protein-1 expression and spinal glial activation following peripheral nerve injury." *J Neurochem* 97(3): 772–783.

Zhang, J, XQ Shi, S Echeverry, et al. (2007). "Expression of CCR2 in both resident and bone marrow-derived microglia plays a critical role in neuropathic pain." *J Neurosci* 27(45): 12396–12406.

Zimmermann, M (2001). "Pathobiology of neuropathic pain." *Eur J Pharmacol* 429: 23–37.

Chapter 18
The Role of Astrocytes in the Modulation of Pain

Vivianne L. Tawfik and Joyce A. DeLeo

Abstract Astrocytes are central nervous system (CNS) glia that were originally considered mere support cells; functioning to support the metabolic and energy needs of the neuron. A large body of work has now demonstrated that astrocytes are not only fundamental to a variety of CNS processes but actually act as orchestrators of synaptic communication. Newly discovered roles for astrocytes include the exquisite control of synaptic glutamate concentrations through glutamate transporter-mediated uptake and the release of gliotransmitters such as D-serine, ATP and glutamate. These gliotransmitters may alter neuronal excitability directly and/or by the secretion of factors that alter the synaptic microenvironment to guide axonal growth. In the context of chronic pain, spinal cord synaptic plasticity, known as central sensitization, is thought to contribute to the persistence of allodynia and hyperalgesia months to years after injury. Astrocytes have been implicated in the maintenance of this aberrant transmission. In this chapter we will discuss the role of astrocytes in modulating synaptic plasticity and demonstrate that a view of the synapse expanded to include the astrocyte is crucial to our understanding of chronic pain.

Abbreviations

ATP	adenosine triphosphate
CCI	chronic constriction injury
Cx	connexins
EphA4R	ephrin-A4 receptor
GABA	gamma amino butyric acid
GFAP	glial fibrillary acid protein
IL-1	interleukin-1

J.A. DeLeo (✉)
Department of Pharmacology and Toxicology, HB 7650, Dartmouth Medical School, Hanover, NH 03755, USA
e-mail: joyce.a.deleo@dartmouth.edu

M. Malcangio (ed.), *Synaptic Plasticity in Pain*,
DOI 10.1007/978-1-4419-0226-9_18, © Springer Science+Business Media, LLC 2009

MSO   methionine sulfoximine
PDE IV   phosphodiesterase type IV
UNC-6   netrin
NO   nitric oxide
RAGE   receptor for advanced glycation end products
TLR   toll like receptors

18.1 Introduction

18.1.1 Astrocytes and Pain

Neuropathic pain, characterized by hyperalgesia (a heightened response to noxious stimuli) and allodynia (pain perceived following non-noxious stimuli) affects millions of people with conditions as diverse as post-herpetic neuralgia and diabetic neuropathy (Dworkin et al. 2003). Over the years there has been much research into the mechanisms of chronic pain with a strong focus on the development of central sensitization following nerve injury. Such a theory suggests that a host of mediators including (but not limited to) adenosine triphosphate (ATP), gamma-aminobutyric acid (GABA), glutamate, nitric oxide (NO), prostaglandins and substance P produce alterations in the central nervous system (CNS) milieu that lead to permanent changes in synaptic processing. Our laboratory, among others, has proposed that central neuroimmune activation is a player in this complicated nociceptive signaling cascade. Specifically, we have postulated that spinal glial cell reactivity with subsequent cytokine and chemokine production may induce the expression of final common pain mediators such as glutamate and NO- implicating non-neuronal cells in spinal cord plasticity leading to chronic neuropathic pain (DeLeo and Yezierski 2001).

Classically, pain development and perception has been considered a purely neuron-mediated phenomenon. However, glial cells (microglia, astrocytes and oligodendrocytes), which constitute over 70% of the total cell population in the CNS, have recently been examined as active contributors to neuromodulatory, neurotrophic and neuroimmune events in the brain and spinal cord (DeLeo et al. 2006; Fellin et al. 2006; Pellerin 2005). The fundamental importance of glial cells to CNS development was clearly demonstrated by studies in which *Drosophila* were engineered without glia. These mutant flies died during the embryonic stage as a result of major defects in neuronal survival and differentiation (Hosoya et al. 1995; Jones et al. 1995). Phylogenetically, an increased astrocyte-to-neuron ratio is observed in parallel with increased brain complexity related to species hierarchy (Nedergaard et al. 2003). This implies a greater requirement for network integration in higher mammals that is likely reliant on the functional domains imparted by astrocytic processes.

Microglia are the first non-neuronal cell to respond to a CNS perturbation such as nerve injury or chronic opioid administration (Raghavendra et al. 2002; Tanga et al. 2004). Astrocytes, the focus of this chapter, along with oligodendrocytes and ependymal cells, constitute the macroglia of the CNS. Astrocytes are responsible for a plethora of tasks including their well-known role in the formation of the blood-brain barrier, which limits the entry of circulating elements into the nervous system. In fact, the selective ablation of reactive astrocytes in the process of CNS restoration leads to failure of blood-brain barrier repair, an enhanced infiltration of leukocytes and subsequent excitotoxic neuronal death (Bush et al. 1999). Such extreme consequences resulting from astrocytic loss underscore their crucial role in homeostasis and regeneration after injury. However, as discussed below, proper regulation of such functions is key to maintaining synaptic integrity.

Astrocytes derive from the neuroectoderm and express a series of "marker antigens" during development such as vimentin and nestin (Eliasson et al. 1999). Once they reach their adult phenotype, other markers become expressed including S100β and glial fibrillary acidic protein (GFAP). The latter is an intermediate filament commonly considered to be astrocyte-specific, although it may also be found on reactive choroid plexus epithelium cells (Reichenbach and Wolburg 2005). GFAP functions as a structural protein and has been shown to increase in hypertrophic, reactive astrocytes (Eng et al. 2000). While enhancement of GFAP remains the mainstay for demonstrating astrocytic reactivity, it is important to note that estimates suggest that only 15% of the total cell volume is labeled with this marker (Bushong et al. 2002). In addition, there exist populations of astrocytes that are not $GFAP^+$; approximately 40% of cerebral cortex astrocytes are $GFAP^-$ and $S100\beta^+$, however the physiological relevance of these distinct groups is not well understood (Kimelberg 2004). S100β is an EF-hand domain Ca^{2+}-binding protein that is produced primarily by CNS glia and is often used as an alternate marker for astrocytic reactivity in pathologic states (Van Eldik and Wainwright 2003). Nanomolar concentrations of S100β released from astrocytes stimulate neurite outgrowth and promote neuronal survival through binding the neuronal receptor for advanced glycation end products (RAGE) (Donato 2003). In contrast, higher levels of S100β have been associated with neuroinflammation and neuronal dysfunction, making it an interesting target for further studies (Mrak and Griffin 2001).

A significant feature of astrocytes is their proclivity to function as part of a syncytium and utilize intercellular communication ports, i.e. gap junctions formed of two hemichannel connexons, to transfer information in the form of ATP and Ca^{2+} (Cotrina et al. 1998). Functionally, astrocytic gap junctions also facilitate the dilution of substances, such as glutamate, taken up from the synapse. These communication ports are formed of hexameric protein subunits called connexins (Cx), the most important of which is Cx43. In fact, knock out of Cx43 impaired the spread of Ca^{2+} waves in a glial network and decreased dye coupling between neighboring astrocytes (Naus et al. 1997). In disease states, this ease of communication may lead to the transfer of toxic substances to distant sites and to the extension of neuronal injury (Kielian and Esen 2004). Interestingly, the release of pro-inflammatory cytokines such as IL-1β leads to the downregulation of Cx43 mRNA and protein and causes a marked reduction in intercellular communication (John et al. 1999). In addition, recent work using *S. aureus*, a common bacterium causing brain abscesses, suggests that a decrease of specific connexins leads to a consequent inhibition in gap junction communication which may have a protective function (Esen et al. 2007).

18.2 Astrocytes and Synaptic Plasticity

It is clear that astrocytes are well poised in the CNS to modulate synaptic communication. For example, it is estimated that 56% of rat cortical synapses are ensheathed by astrocyte domains (Chao et al. 2002). A single astrocyte can make contact with more than a hundred thousand synapses suggesting an integral role of these complex cells in synchronization of neuronal activity (Bushong et al. 2002). Interestingly, neuronal glutamate release has been shown to activate AMPA receptors on Bergmann glia in the cerebellum leading to ensheathment of synapses with astrocytes processes (Gallo and Chittajallu 2001; Iino et al. 2001). Additionally, inhibition of glial AMPA receptor activation leads to a decrease in complexity of synapses and differential innervation of Purkinje cells. Oliet et al. (2001) demonstrated that the absence of GLT-1 mediated astrocytic glutamate uptake reduced EPSC amplitude in supraoptic neurons by enhancing activation of presynaptic mGluRIII. Taken together, these results highlight the importance of astrocytes in the regulation of synaptic efficacy and neuronal integrity.

In neuropathic pain specifically, spinal cord plasticity is thought to underlie the maintenance of aberrant neuronal signaling that leads to enhanced nociception in the absence of a continued stimulus known as central sensitization (Ji et al. 2003) (see Drdla and Sandkehler Chapter 9 and Randic Chapter 10). Astrocytes contribute to this synaptic plasticity in a variety of ways including: (1) Exquisite control of synaptic glutamate concentrations through glutamate transporter-mediated uptake; (2) Release of gliotransmitters such as D-serine, ATP and glutamate that may alter neuronal excitability directly (Inoue et al. 2003; Parpura et al. 1994; Snyder and Kim 2000); (3) Secretion of factors that alter the synaptic microenvironment to guide axonal growth; and (4) Through cell–cell communication with neurons via specific receptor-ligand interactions.

In order for efficient excitatory signaling to take place in the nervous system, the levels of synaptic glutamate need to be exquisitely controlled. Astrocytes are thought to perform the majority of this function by clearing synaptic glutamate via specific sodium-dependent glutamate transporters (Danbolt 2001). Glutamate transporter-1 (GLT-1/EAAT2) is the essential CNS transporter, responsible for over 90% of uptake and primarily localized on astrocytes (Tanaka et al. 1997). The glutamate-aspartate transporter (GLAST/EAAT1) is also located on astrocytes while the excitatory amino acid carrier (EAAC/EAAT3) is found in neurons (Kanai and Hediger 1992; Pines et al. 1992; Storck et al. 1992).

Several studies in animal models have linked allodynia and hyperalgesia to excess levels of excitatory amino acids in the cerebral spinal fluid (Malmberg and Yaksh 1995; Sluka and Westlund 1992). It then follows that second order neurons would be exposed to increased levels of glutamate in the synapse after nerve injury which would lead to aberrant firing and strengthening of additional synapses. Injury to the central or peripheral nervous system increases GFAP immunoreactivity in astrocytes and in vitro studies have demonstrated that this phenotype includes reduced levels of the transporter GLT-1 (Schlag et al. 1998). Therefore, the potential for glutamate transporter downregulation in neuropathic pain states exists.

Using the chronic constriction injury (CCI) model of neuropathic pain, Sung et al. (2003) demonstrated that GLT-1, GLAST and EAAC1 were all decreased by day 7 post-nerve injury. In addition, after spinal nerve ligation, glutamate uptake in the lumbar spinal cord is attenuated (Binns et al. 2005). Interestingly, pharmacological inhibition of glutamate transporters has been shown to produce spontaneous nociceptive behaviors that could be attenuated by NMDA antagonists (Liaw et al. 2005). In our own studies, we have found that propentofylline, a glial-modulating agent, is capable of reversing nerve transection-induced reductions in glutamate transporters, GLT-1 and GLAST (Tawfik et al. 2008). Using primary astrocyte cultures, we showed that propentofylline shifted astrocytes from a high GFAP, low GLT-1 expressing state to a quiescent, high GLT-1 phenotype that demonstrated enhanced glutamate clearance (Tawfik et al. 2006 and Fig. 18.1). In addition, we found that propentofylline was able to decrease LPS-induced chemokine release from these astrocytes, providing further evidence for a mechanism related to neuroimmune suppression. This body of work highlights a role for deficient glutamate transport in synaptic plasticity and the importance of returning astrocytes to a quiescent phenotype, competent for maintenance of excitatory amino acid homeostasis.

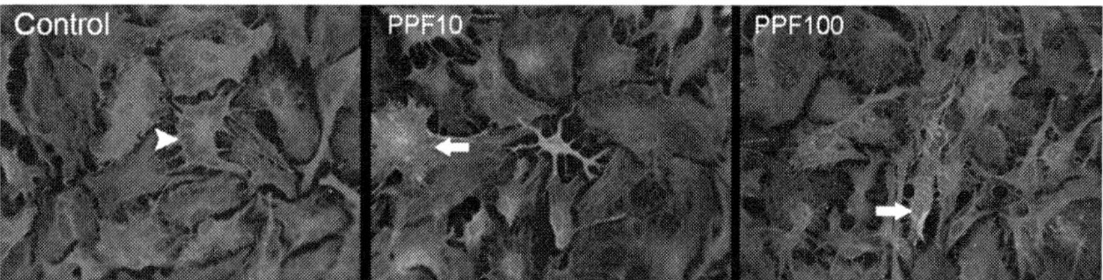

Fig. 18.1 Seven-day treatment with propentofylline enhances GLT-1 expression in cultured astrocytes. *Arrows* identify astrocytes which are low GFAP, high GLT-1 expressing. *Arrowheads* identify astrocytes which are high GFAP, low GLT-1 expressing. PPF10, 100: Propentyfylline-treated 1 or 100 µM for 7 days in vitro

Once glutamate is taken up by astrocytes it is converted to glutamine by glutamine synthetase (GS) and subsequently released to neurons for the regeneration of glutamate stores (Hertz and Zielke 2004). Inhibition of GS with methionine sulfoximine (MSO) reduces astrocyte swelling and increases in GFAP induced by hyperammonemia (Tanigami et al. 2005). Furthermore, Chiang et al. (2007) recently determined a role for the astrocytic glutamate-glutamine shuttle in central sensitization. Using a model of inflammatory pain consisting of application of mustard oil to the rat tooth pulp it was demonstrated that intrathecal superfusion of MSO led to the reversal of mustard oil-induced increases in mechanoreceptive field size and decreases in the mechanical activation threshold. In addition, it was demonstrated that inhibition of glutamine uptake by neurons was sufficient to prevent the development of central sensitization in rat medullary dorsal horn neurons (Chiang et al. 2008).

The discovery that astrocytes are active participants in synaptic communication via the release of gliotransmitters has changed the way these "support cells" are viewed (Bezzi and Volterra 2001; Jourdain et al. 2007). Astrocytes play a more extensive and integral role in the regulation of glutamate than previously suspected (Araque et al. 1999) by responding to neuronal activity by exhibiting calcium oscillations (Hassinger et al. 1995; Parpura et al. 1994). These calcium waves have been shown to lead to astrocytic glutamate release (Bezzi et al. 1998) that appears to modulate synaptic strength in the hippocampus (Jourdain et al. 2007) by increasing the probability of neurotransmitter release and evoking long-term potentiation at single hippocampal synapses in a mGluR-mediated fashion (Perea and Araque 2007). In addition, several studies have now established that astrocytes contain vesicular glutamate stores (Montana et al. 2006) which can be triggered to undergo exocytosis by mechanical stimulation (Araque et al. 2000), to activate ionotropic glutamate receptors (Pasti et al. 2001) and to activate metabotropic glutamate receptors (Bezzi et al. 2004). Finally, exciting new work in the ferret visual cortex has shown that astrocytes exhibit robust calcium responses to visual stimuli that are tuned for orientation and spatial frequency more sharply than neurons (Schummers et al. 2008). These astrocytic responses have a key role in coupling neuronal organization and coordinating the temporal dynamics of neuronal signaling. Taken together, these findings suggest a direct modulatory function of astrocytes in synaptic transmission. However, the role of this astrocytic glutamate release in chronic pain-related central sensitization remains to be determined.

Astrocytes also release other substances that may affect neuronal excitation, such as D-serine which may sensitize neurons in pain states by decreasing the threshold for NMDA receptor activation (Guo et al. 2006). Another potential gliotransmitter involved in synaptic plasticity is the Ca^{2+}- binding protein S100β. Nishiyama et al. (2002) demonstrated that S100β-null mice exhibit enhanced long-term potentiation which was reversible with perfusion of recombinant S100β, signifying negative regulation by this protein. In the context of pain transmission, our laboratory established that S100β is elevated in spinal cord dorsal horn astrocytes in inflammatory and neuropathic pain models and that S100β over expression sensitizes mice to tactile stimuli (Raghavendra et al. 2004; Tanga et al. 2006). Given that S100β expression is enhanced in several brain pathologic states (Griffin et al. 1989; Hardemark et al. 1989; Hinkle et al. 1997), it may act as a sensor to prevent neuronal excitotoxicity.

A large body of work outlining a potential role for glial-derived cytokines in the generation of persistent pain also exists (DeLeo et al. 2004; Wieseler-Frank et al. 2005). Recent work from Ikeda et al. (2007) has proposed a potential role of these pro-inflammatory molecules in long-term potentiation. Specifically, they demonstrate using optical imaging of the spinal cord dorsal horn that enhanced neuronal excitability after application of a P2X receptor agonist, αβ-methylene-ATP, is dependent on astrocytic activation and release of IL-6 and TNF-α. The contribution of a plethora of gliotransmitters to the development of central sensitization continues to be explored.

The formation of the 100 trillion synapses within the CNS is a complex process that involves the simultaneous coordination of axonal growth and target recognition known as synaptic specificity (Sanes and Lichtman 2001). The involvement of astrocytes in synapse formation has been elucidated using simple models of neural circuit formation such as *C. elegans*. Colon-Ramos et al. (2007) demonstrated that connectivity between two interneurons in *C. elegans* was dependent on the secreted chemotropic factor, UNC-6 (netrin) from astrocytes. UNC-6 signaling through neuronal UNC-40 (DCC) induced distinct cellular responses in the interneurons, directing axonal outgrowth in the postsynaptic neuron and presynaptic assembly of terminals in the presynaptic neuron in proximity to the glial end feet.

The Ephrin family of ligands and receptor tyrosine kinases has recently been implicated in astrocyte-mediated synaptic remodeling and development of mature excitatory synapses (Takasu et al. 2002). Nestor et al. (2007) demonstrated that hippocampal astrocytes in culture express ephrin-A3 and the EphA receptor 4 (EphA4R) and activation of this receptor resulted in astrocytic process outgrowth and morphological changes. In addition, astrocytic ephrins have been shown to modulate neuronal synapse formation by repulsive mechanisms; activation of EphA4R on dendritic spines by astrocytic ephrin-A3 causes them to retract (Murai et al. 2003). Finally, ephrin-A and its receptors have been shown to be increased in activated astrocytes after spinal cord injury (Irizarry-Ramirez et al. 2005), suggesting a mechanism for alterations in neuronal connectivity leading to central sensitization.

18.3 Pain, Central Sensitization and the Role of Astrocyte Modulatory Strategies

It is evident that astrocytes are well poised in the CNS to modulate synaptic communication and as such they perform a multitude of tasks that both directly and indirectly affect neuronal firing. Astrocytes also have the ability to adapt to their synaptic milieu; a quality which can be problematic as it may cause them to lose their principal role of orchestrator in the tetrapartite synapse (DeLeo et al. 2006). In the context of neuropathic pain, astrocytes exhibit increased expression of GFAP in the spinal cord following several types of nerve injury that result in behavioral hypersensitivity including chronic constriction injury (Garrison et al. 1991) and L5 nerve ligation and transection (Colburn et al. 1999 and Fig. 18.2). It has been postulated that this increased GFAP and hypertrophy of astrocytes represents aberrant activation and results in the loss of homeostatic mechanisms along with the expression of algesic mediators such as cytokines and chemokines (Abbadie et al. 2003; DeLeo et al. 1997) which may sensitize neurons (see above).

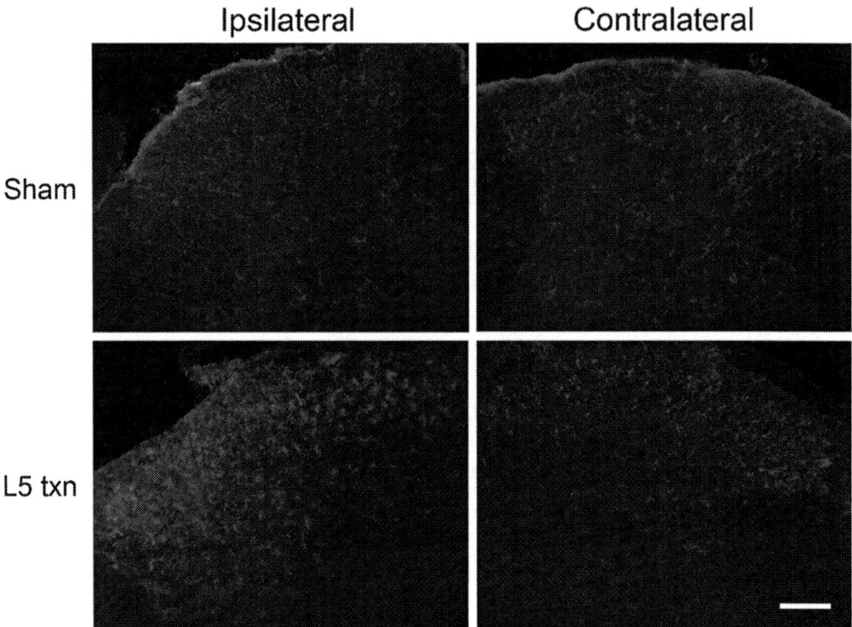

Fig. 18.2 L5 spinal nerve transection in mice induces increased GFAP reactivity in astrocytes in the spinal cord dorsal horn. GFAP immunoreactivity in the dorsal horn was notably enhanced ipsilateral to the nerve injury on day 12 (L5 txd) compared to sham surgery. On treatment day 12, mice were transcardially perfused and crytostat sections were stained with mouse anti-GFAP (1:200) followed by an Alexa-Fluor-405 linked secondary (1:250). Images were taken at a magnification of 20x. Scale bar, 100 μm

In spite of this large body of work supporting an essential role for astrocytes in central sensitization, there are a limited number of pharmacologic agents targeting astrocytes specifically (Romero-Sandoval et al. 2008). Fluorocitrate, an inhibitor of glial aconitase, was first used to demonstrate that astrocytic

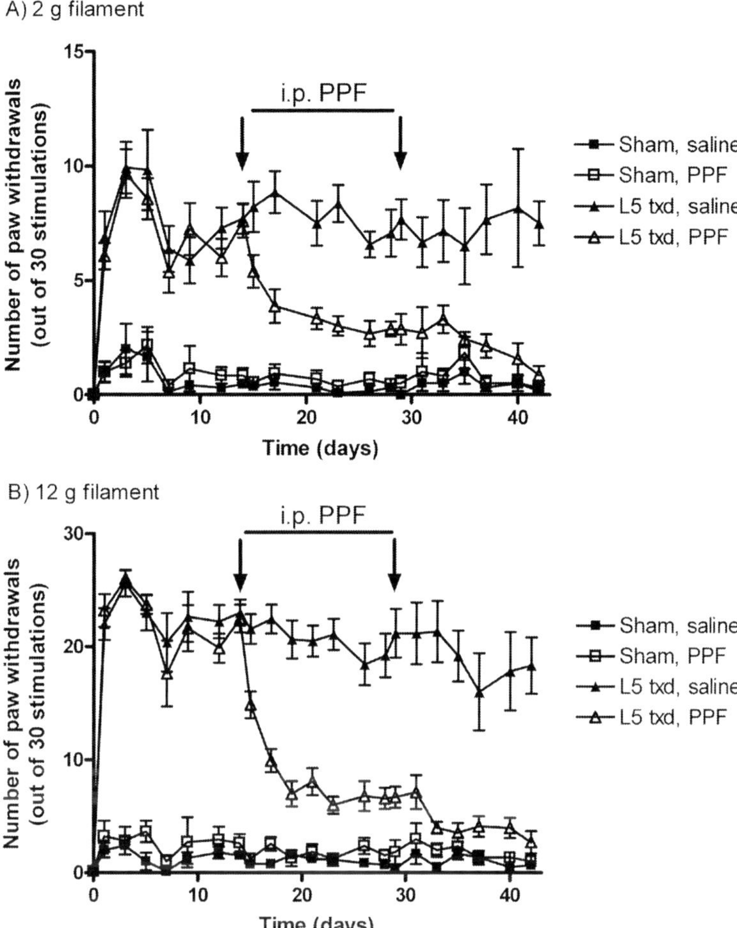

Fig. 18.3 Propentofylline reverses existing allodynia to 2 and 12 g von Frey filaments in an L5 spinal nerve transection model of neuropathic pain. Rats received L5 spinal nerve transection (L5 txd) or sham surgery on day 0 and the development of mechanical allodynia to a 2 g (**A**) and 12 g (**B**) von Frey filament was assessed. On days 14–27 post-surgery, animals received either 10 mg/kg propentofylline (PPF) or saline by intraperitoneal injection and behavioral responses were tested 15 h later. Mechanical allodynia is reported as the average number of paw withdrawals out of $30 \pm$ S.E.M. (n $= 6$–8/treatment). Administration of propentofylline to L5 spinal nerve transected rats (L5 txd, PPF) resulted in a significant decrease in mechanical allodynia compared with L5 spinal nerve transected, saline (L5 txd, sal) controls ($P< 0.001$ for days 15–42, "L5 txd, sal" vs "L5 txd, PPF") revealing a significant effect of propentofylline on mechanical allodynia during treatment and post-drug/washout phases. Day 0 represents pre-injury baseline responses

activation was essential to hyperalgesia after intraplantar zymosan (Meller et al. 1994), however, its use has been limited by toxicity (Fonnum et al. 1997). As discussed above, methionine sulfoximine (MSO) has more recently been used to inhibit astrocytic glutamine synthetase directly thus preventing formation of metabolic substrates key to neuronal activity (Chiang et al. 2007). While these data are encouraging, the use of MSO as a therapeutic is controversial since it can cause behavioral changes and seizures (Sellinger et al. 1968; Tanigami et al. 2005).

Another potential astrocytic modulator is propentofylline, an atypical methylxaNthine derivative suggested to act via inhibition of adenosine transport and the cyclic-adenosine-5',3'-monophophate (cAMP)-specific phosphodiesterase type IV (PDE IV) (Meskini et al. 1994; Nagata et al. 1985; Parkinson and Fredholm 1991). Propentofylline was first shown to attenuate astrocytic activation in a rodent model of ischemia (DeLeo et al. 1987). As a result of this action, it has been used extensively to study the role of glial reactivity in behavioral sensitization after nerve injury. In an initial study, it was shown that pre-emptive intrathecal or systemic treatment with propentofylline decreased mechanical allodynia after L5 spinal nerve transection (Sweitzer et al. 2001). In concert with this anti-allodynic effect, the expression of microglial and astrocytic activation markers (OX-42 and GFAP) was reduced. Further work established that propentofylline suppressed glial activation-induced cytokine release and had an anti-hyperalgesic effect after peripheral nerve injury (Raghavendra et al. 2003). In relation to a therapeutically relevant action, we recently administered propentofylline 14 days after L5 spinal nerve transection to determine if it was capable of reversing existing, chronic pain (Tawfik et al. 2007). We discovered that daily, systemic propentofylline reversed mechanical allodynia for the duration of dosing and furthermore, that this anti-allodynic effect was maintained for at least 14 days after the final dose (Fig. 18.3, adapted from Brain, Behavior, and Immunity 2007 with permission from Elsevier), suggestive of a disease-modifying effect.

18.4 Concluding Remarks

The active role played by astrocytes in the modulation of synaptic plasticity is just beginning to be uncovered. However, it is already apparent that these cells are crucial in orchestrating CNS processes and modulating neurotransmission (Fig. 18.4). The potential already exists to target astrocytes to inhibit or reverse central sensitization in neuropathic pain but as the specific contribution of these glial cells becomes clearer, focused therapies may be developed. A view of synaptic plasticity expanded to include the "gliapse" (Ren and Dubner 2008) will certainly alter our current approaches to the treatment of pain and likely lead to pivotal advances that will improve patient care.

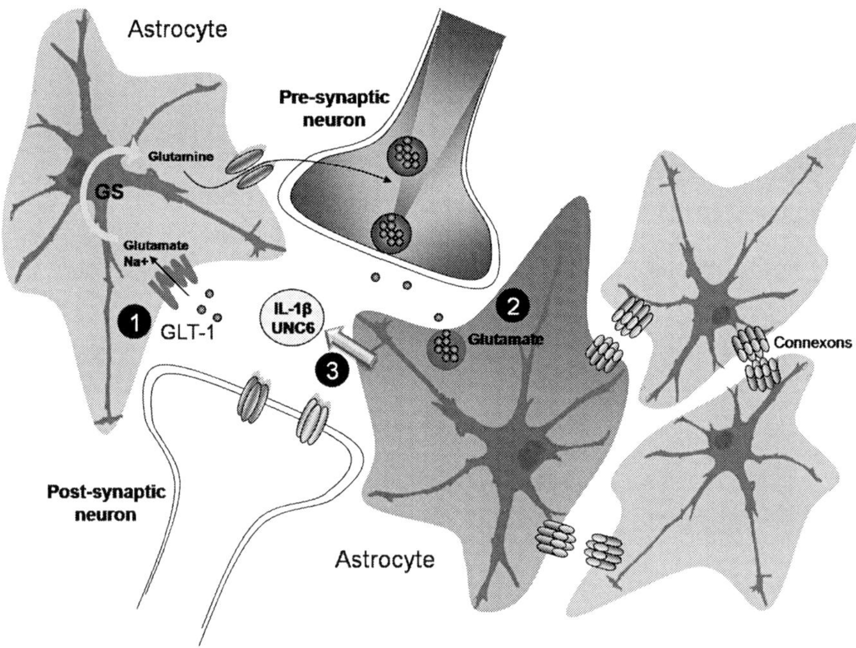

Fig. 18.4 The Role of Astrocytes in the Modulation of Central Sensitization and Pain. Astrocyte processes ensheath neuronal synapses making them well poised to orchestrate synaptic plasticity in a multitude of ways including: 1 Exquisite control of synaptic glutamate concentrations through glutamate transporter-mediated uptake which may be disrupted in neuropathic pain leading to increased synaptic glutamate; 2 Release of gliotransmitters such as D-serine, ATP and glutamate that may alter neuronal excitability directly; and 3 Secretion of factors that alter the synaptic microenvironment and may direct neuronal synapse formation. In addition, astrocytes form a syncytial network allowing for the integration of signals spatially and temporally through the transfer of calcium and other substances via connexon hemi-channels; in pain states this communication may be compromised. Given this plethora of modulatory roles, astrocytes are an attractive and novel target for the control of chronic pain

Acknowledgments The authors would like to acknowledge the support from NIH-NIDA (5 RO1 DA11276 (DeLeo)) and Elan, Inc. Pharmaceuticals (Tawfik).

References

Abbadie C, Lindia JA, Cumiskey AM, Peterson LB, Mudgett JS, Bayne EK, DeMartino JA, MacIntyre DE, Forrest MJ (2003) Impaired neuropathic pain responses in mice lacking the chemokine receptor CCR2. Proc Natl Acad Sci USA 100: 7947–52.
Araque A, Li N, Doyle RT, Haydon PG (2000) SNARE protein-dependent glutamate release from astrocytes. J Neurosci 20: 666–73.
Araque A, Sanzgiri RP, Parpura V, Haydon PG (1999) Astrocyte-induced modulation of synaptic transmission. Can J Physiol Pharmacol 77: 699–706.

Bezzi P, Carmignoto G, Pasti L, Vesce S, Rossi D, Rizzini BL, Pozzan T, Volterra A (1998) Prostaglandins stimulate calcium-dependent glutamate release in astrocytes. Nature 391: 281–5.

Bezzi P, Gundersen V, Galbete JL, Seifert G, Steinhauser C, Pilati E, Volterra A (2004) Astrocytes contain a vesicular compartment that is competent for regulated exocytosis of glutamate. Nat Neurosci 7: 613–20.

Bezzi P, Volterra A (2001) A neuron-glia signalling network in the active brain. Curr Opin Neurobiol 11: 387–94.

Binns BC, Huang Y, Goettl VM, Hackshaw KV, Stephens RL, Jr. (2005) Glutamate uptake is attenuated in spinal deep dorsal and ventral horn in the rat spinal nerve ligation model. Brain Res 1041: 38–47.

Bush TG, Puvanachandra N, Horner CH, Polito A, Ostenfeld T, Svendsen CN, Mucke L, Johnson MH, Sofroniew MV (1999) Leukocyte infiltration, neuronal degeneration, and neurite outgrowth after ablation of scar-forming, reactive astrocytes in adult transgenic mice. Neuron 23: 297–308.

Bushong EA, Martone ME, Jones YZ, Ellisman MH (2002) Protoplasmic astrocytes in CA1 stratum radiatum occupy separate anatomical domains. J Neurosci 22: 183–92.

Chao TI, Rickmann M, Wolff JR (2002) The synapse-astrocyte boundary: an anatomical basis for an integrative role of glial in synaptic transmission. In: Volterra A, Magistretti P, Haydon P (eds) The Tripartite Synapse: glia in synaptic transmission. Oxford University Press, Oxford, pp 3–23.

Chiang CY, Li Z, Dostrovsky JO, Hu JW, Sessle BJ (2008) Glutamine uptake contributes to central sensitization in the medullary dorsal horn. Neuroreport 19: 1151–4.

Chiang CY, Wang J, Xie YF, Zhang S, Hu JW, Dostrovsky JO, Sessle BJ (2007) Astroglial glutamate-glutamine shuttle is involved in central sensitization of nociceptive neurons in rat medullary dorsal horn. J Neurosci 27: 9068–76.

Colburn RW, Rickman AJ, DeLeo JA (1999) The effect of site and type of nerve injury on spinal glial activation and neuropathic pain behavior. Exp Neurol 157: 289–304.

Colon-Ramos DA, Margeta MA, Shen K (2007) Glia promote local synaptogenesis through UNC-6 (netrin) signaling in C. elegans. Science 318: 103–6.

Cotrina ML, Lin JH, Alves-Rodrigues A, Liu S, Li J, Azmi-Ghadimi H, Kang J, Naus CC, Nedergaard M (1998) Connexins regulate calcium signaling by controlling ATP release. Proc Natl Acad Sci USA 95: 15735–40.

Danbolt NC (2001) Glutamate uptake. Prog Neurobiol 65: 1–105.

DeLeo J, Toth L, Schubert P, Rudolphi K, Kreutzberg GW (1987) Ischemia-induced neuronal cell death, calcium accumulation, and glial response in the hippocampus of the Mongolian gerbil and protection by propentofylline (HWA 285). J Cereb Blood Flow Metab 7: 745–51.

DeLeo JA, Colburn RW, Rickman AJ (1997) Cytokine and growth factor immunohistochemical spinal profiles in two animal models of mononeuropathy. Brain Res 759: 50–7.

DeLeo JA, Tanga FY, Tawfik VL (2004) Neuroimmune activation and neuroinflammation in chronic pain and opioid tolerance/hyperalgesia. Neuroscientist 10: 40–52.

DeLeo JA, Tawfik VL, LaCroix-Fralish ML (2006) The tetrapartite synapse: path to CNS sensitization and chronic pain. Pain 122: 17–21.

DeLeo JA, Yezierski RP (2001) The role of neuroinflammation and neuroimmune activation in persistent pain. Pain 90: 1–6.

Donato R (2003) Intracellular and extracellular roles of S100 proteins. Microsc Res Tech 60: 540–51.

Dworkin RH, Backonja M, Rowbotham MC, Allen RR, Argoff CR, Bennett GJ, Bushnell MC, Farrar JT, Galer BS, Haythornthwaite JA, Hewitt DJ, Loeser JD, Max MB, Saltarelli M, Schmader KE, Stein C, Thompson D, Turk DC, Wallace MS, Watkins LR, Weinstein SM (2003) Advances in neuropathic pain: diagnosis, mechanisms, and treatment recommendations. Arch Neurol 60: 1524–34.

Eliasson C, Sahlgren C, Berthold CH, Stakeberg J, Celis JE, Betsholtz C, Eriksson JE, Pekny M (1999) Intermediate filament protein partnership in astrocytes. J Biol Chem 274: 23996–4006.

Eng LF, Ghirnikar RS, Lee YL (2000) Glial fibrillary acidic protein: GFAP-thirty-one years (1969-2000). Neurochem Res 25: 1439–51.

Esen N, Shuffield D, Syed MM, Kielian T (2007) Modulation of connexin expression and gap junction communication in astrocytes by the gram-positive bacterium S. aureus. Glia 55: 104–17.

Fellin T, Pascual O, Haydon PG (2006) Astrocytes coordinate synaptic networks: balanced excitation and inhibition. Physiology (Bethesda) 21: 208–15.

Fonnum F, Johnsen A, Hassel B (1997) Use of fluorocitrate and fluoroacetate in the study of brain metabolism. Glia 21: 106–13.

Gallo V, Chittajallu R (2001) Neuroscience. Unwrapping glial cells from the synapse: what lies inside? Science 292: 872–3.

Garrison CJ, Dougherty PM, Kajander KC, Carlton SM (1991) Staining of glial fibrillary acidic protein (GFAP) in lumbar spinal cord increases following a sciatic nerve constriction injury. Brain Res 565: 1–7.

Griffin WS, Stanley LC, Ling C, White L, MacLeod V, Perrot LJ, White CL, 3rd, Araoz C (1989) Brain interleukin 1 and S-100 immunoreactivity are elevated in Down syndrome and Alzheimer disease. Proc Natl Acad Sci USA 86: 7611–5.

Guo JD, Wang H, Zhang YQ, Zhao ZQ (2006) Distinct effects of D-serine on spinal nociceptive responses in normal and carrageenan-injected rats. Biochem Biophys Res Commun 343: 401–6.

Hardemark HG, Ericsson N, Kotwica Z, Rundstrom G, Mendel-Hartvig I, Olsson Y, Pahlman S, Persson L (1989) S-100 protein and neuron-specific enolase in CSF after experimental traumatic or focal ischemic brain damage. J Neurosurg 71: 727–31.

Hassinger TD, Atkinson PB, Strecker GJ, Whalen LR, Dudek FE, Kossel AH, Kater SB (1995) Evidence for glutamate-mediated activation of hippocampal neurons by glial calcium waves. J Neurobiol 28: 159–70.

Hertz L, Zielke HR (2004) Astrocytic control of glutamatergic activity: astrocytes as stars of the show. Trends Neurosci 27: 735–43.

Hinkle DA, Baldwin SA, Scheff SW, Wise PM (1997) GFAP and S100beta expression in the cortex and hippocampus in response to mild cortical contusion. J Neurotrauma 14: 729–38.

Hosoya T, Takizawa K, Nitta K, Hotta Y (1995) glial cells missing: a binary switch between neuronal and glial determination in Drosophila. Cell 82: 1025–36.

Iino M, Goto K, Kakegawa W, Okado H, Sudo M, Ishiuchi S, Miwa A, Takayasu Y, Saito I, Tsuzuki K, Ozawa S (2001) Glia-synapse interaction through Ca2+-permeable AMPA receptors in Bergmann glia. Science 292: 926–9.

Ikeda H, Tsuda M, Inoue K, Murase K (2007) Long-term potentiation of neuronal excitation by neuron-glia interactions in the rat spinal dorsal horn. Eur J Neurosci 25: 1297–306.

Inoue K, Koizumi S, Tsuda M, Shigemoto-Mogami Y (2003) Signaling of ATP receptors in glia-neuron interaction and pain. Life Sci 74: 189–97.

Irizarry-Ramirez M, Willson CA, Cruz-Orengo L, Figueroa J, Velazquez I, Jones H, Foster RD, Whittemore SR, Miranda JD (2005) Upregulation of EphA3 receptor after spinal cord injury. J Neurotrauma 22: 929–35.

Ji RR, Kohno T, Moore KA, Woolf CJ (2003) Central sensitization and LTP: do pain and memory share similar mechanisms? Trends Neurosci 26: 696–705.

John GR, Scemes E, Suadicani SO, Liu JS, Charles PC, Lee SC, Spray DC, Brosnan CF (1999) IL-1beta differentially regulates calcium wave propagation between primary human fetal astrocytes via pathways involving P2 receptors and gap junction channels. Proc Natl Acad Sci USA 96: 11613–8.

Jones BW, Fetter RD, Tear G, Goodman CS (1995) glial cells missing: a genetic switch that controls glial versus neuronal fate. Cell 82: 1013–23.

Jourdain P, Bergersen LH, Bhaukaurally K, Bezzi P, Santello M, Domercq M, Matute C, Tonello F, Gundersen V, Volterra A (2007) Glutamate exocytosis from astrocytes controls synaptic strength. Nat Neurosci 10: 331–9.

Kanai Y, Hediger MA (1992) Primary structure and functional characterization of a high-affinity glutamate transporter. Nature 360: 467–71.

Kielian T, Esen N (2004) Effects of neuroinflammation on glia-glia gap junctional intercellular communication: a perspective. Neurochem Int 45: 429–36.

Kimelberg HK (2004) The problem of astrocyte identity. Neurochem Int 45: 191–202.

Liaw WJ, Stephens RL, Jr., Binns BC, Chu Y, Sepkuty JP, Johns RA, Rothstein JD, Tao YX (2005) Spinal glutamate uptake is critical for maintaining normal sensory transmission in rat spinal cord. Pain 115: 60–70.

Malmberg AB, Yaksh TL (1995) Cyclooxygenase inhibition and the spinal release of prostaglandin E2 and amino acids evoked by paw formalin injection: a microdialysis study in unanesthetized rats. J Neurosci 15: 2768–76.

Meller ST, Dykstra C, Grzybycki D, Murphy S, Gebhart GF (1994) The possible role of glia in nociceptive processing and hyperalgesia in the spinal cord of the rat. Neuropharmacology 33: 1471–8.

Meskini N, Nemoz G, Okyayuz-Baklouti I, Lagarde M, Prigent AF (1994) Phosphodiesterase inhibitory profile of some related xanthine derivatives pharmacologically active on the peripheral microcirculation. Biochem Pharmacol 47: 781–8.

Montana V, Malarkey EB, Verderio C, Matteoli M, Parpura V (2006) Vesicular transmitter release from astrocytes. Glia 54: 700–15.

Mrak RE, Griffin WS (2001) The role of activated astrocytes and of the neurotrophic cytokine S100B in the pathogenesis of Alzheimer's disease. Neurobiol Aging 22: 915–22.

Murai KK, Nguyen LN, Irie F, Yamaguchi Y, Pasquale EB (2003) Control of hippocampal dendritic spine morphology through ephrin-A3/EphA4 signaling. Nat Neurosci 6: 153–60.

Nagata K, Ogawa T, Omosu M, Fujimoto K, Hayashi S (1985) In vitro and in vivo inhibitory effects of propentofylline on cyclic AMP phosphodiesterase activity. Arzneimittelforschung 35: 1034–6.

Naus CC, Bechberger JF, Zhang Y, Venance L, Yamasaki H, Juneja SC, Kidder GM, Giaume C (1997) Altered gap junctional communication, intercellular signaling, and growth in cultured astrocytes deficient in connexin43. J Neurosci Res 49: 528–40.

Nedergaard M, Ransom B, Goldman SA (2003) New roles for astrocytes: redefining the functional architecture of the brain. Trends Neurosci 26: 523–30.

Nestor MW, Mok LP, Tulapurkar ME, Thompson SM (2007) Plasticity of neuron-glial interactions mediated by astrocytic EphARs. J Neurosci 27: 12817–28.

Nishiyama H, Knopfel T, Endo S, Itohara S (2002) Glial protein S100B modulates long-term neuronal synaptic plasticity. Proc Natl Acad Sci USA 99: 4037–42.

Oliet SH, Piet R, Poulain DA (2001) Control of glutamate clearance and synaptic efficacy by glial coverage of neurons. Science 292: 923–6.

Parkinson FE, Fredholm BB (1991) Effects of propentofylline on adenosine A1 and A2 receptors and nitrobenzylthioinosine-sensitive nucleoside transporters: quantitative autoradiographic analysis. Eur J Pharmacol 202: 361–6.

Parpura V, Basarsky TA, Liu F, Jeftinija K, Jeftinija S, Haydon PG (1994) Glutamate-mediated astrocyte-neuron signalling. Nature 369: 744–7.

Pasti L, Zonta M, Pozzan T, Vicini S, Carmignoto G (2001) Cytosolic calcium oscillations in astrocytes may regulate exocytotic release of glutamate. J Neurosci 21: 477–84.

Pellerin L (2005) How astrocytes feed hungry neurons. Mol Neurobiol 32: 59–72.

Perea G, Araque A (2007) Astrocytes potentiate transmitter release at single hippocampal synapses. Science 317: 1083–6.

Pines G, Danbolt NC, Bjoras M, Zhang Y, Bendahan A, Eide L, Koepsell H, Storm-Mathisen J, Seeberg E, Kanner BI (1992) Cloning and expression of a rat brain L-glutamate transporter. Nature 360: 464–7.

Raghavendra V, Rutkowski MD, DeLeo JA (2002) The role of spinal neuroimmune activation in morphine tolerance/hyperalgesia in neuropathic and sham-operated rats. J Neurosci 22: 9980–9.

Raghavendra V, Tanga F, Rutkowski MD, DeLeo JA (2003) Anti-hyperalgesic and morphine-sparing actions of propentofylline following peripheral nerve injury in rats: mechanistic implications of spinal glia and proinflammatory cytokines. Pain 104: 655–64.

Raghavendra V, Tanga FY, DeLeo JA (2004) Complete Freunds adjuvant-induced peripheral inflammation evokes glial activation and proinflammatory cytokine expression in the CNS. Eur J Neurosci 20: 467–73.

Reichenbach A, Wolburg H (2005) Astrocytes and ependymal glia. In: Kettenmann H, Ransom BR (eds) Neuroglia. Oxford University Press, Oxford, pp 19–35.

Ren K, Dubner R (2008) Neuron-glia crosstalk gets serious: role in pain hypersensitivity. Curr Opin Anes 21: 570–579.

Romero-Sandoval EA, Horvath RJ, DeLeo JA (2008) Neuroimmune interactions and pain: focus on glial-modulating targets. Curr Opin Investig Drugs 9: 726–34.

Sanes JR, Lichtman JW (2001) Induction, assembly, maturation and maintenance of a postsynaptic apparatus. Nat Rev Neurosci 2: 791–805.

Schlag BD, Vondrasek JR, Munir M, Kalandadze A, Zelenaia OA, Rothstein JD, Robinson MB (1998) Regulation of the glial Na + -dependent glutamate transporters by cyclic AMP analogs and neurons. Mol Pharmacol 53: 355–69.

Schummers J, Yu H, Sur M (2008) Tuned responses of astrocytes and their influence on hemodynamic signals in the visual cortex. Science 320: 1638–43.

Sellinger OZ, Azcurra JM, Ohlsson WG (1968) Methionine sulfoximine seizures. 8. The dissociation of the convulsant and glutamine synthetase inhibitory effects. J Pharmacol Exp Ther 164: 212–22.

Sluka KA, Westlund KN (1992) An experimental arthritis in rats: dorsal horn aspartate and glutamate increases. Neurosci Lett 145: 141–4.

Snyder SH, Kim PM (2000) D-amino acids as putative neurotransmitters: focus on D-serine. Neurochem Res 25: 553–60.

Storck T, Schulte S, Hofmann K, Stoffel W (1992) Structure, expression, and functional analysis of a Na(+)-dependent glutamate/aspartate transporter from rat brain. Proc Natl Acad Sci USA 89: 10955–9.

Sung B, Lim G, Mao J (2003) Altered expression and uptake activity of spinal glutamate transporters after nerve injury contribute to the pathogenesis of neuropathic pain in rats. J Neurosci 23: 2899–910.

Sweitzer SM, Schubert P, DeLeo JA (2001) Propentofylline, a glial modulating agent, exhibits antiallodynic properties in a rat model of neuropathic pain. J Pharmacol Exp Ther 297: 1210–7.

Takasu MA, Dalva MB, Zigmond RE, Greenberg ME (2002) Modulation of NMDA receptor-dependent calcium influx and gene expression through EphB receptors. Science 295: 491–5.

Tanaka K, Watase K, Manabe T, Yamada K, Watanabe M, Takahashi K, Iwama H, Nishikawa T, Ichihara N, Kikuchi T, Okuyama S, Kawashima N, Hori S, Takimoto M, Wada K (1997) Epilepsy and exacerbation of brain injury in mice lacking the glutamate transporter GLT-1. Science 276: 1699–702.

Tanga FY, Raghavendra V, DeLeo JA (2004) Quantitative real-time RT-PCR assessment of spinal microglial and astrocytic activation markers in a rat model of neuropathic pain. Neurochem Int 45: 397–407.

Tanga FY, Raghavendra V, Nutile-McMenemy N, Marks A, Deleo JA (2006) Role of astrocytic S100beta in behavioral hypersensitivity in rodent models of neuropathic pain. Neuroscience 140: 1003–10.

Tanigami H, Rebel A, Martin LJ, Chen TY, Brusilow SW, Traystman RJ, Koehler RC (2005) Effect of glutamine synthetase inhibition on astrocyte swelling and altered astroglial protein expression during hyperammonemia in rats. Neuroscience 131: 437–49.

Tawfik VL, Lacroix-Fralish ML, Bercury KK, Nutile-McMenemy N, Harris BT, Deleo JA (2006) Induction of astrocyte differentiation by propentofylline increases glutamate transporter expression in vitro: heterogeneity of the quiescent phenotype. Glia 54: 193–203.

Tawfik VL, Nutile-McMenemy N, Lacroix-Fralish ML, Deleo JA (2007) Efficacy of propentofylline, a glial modulating agent, on existing mechanical allodynia following peripheral nerve injury. Brain Behav Immun 21: 238–46.

Tawfik VL, Regan MR, Haenggeli C, Lacroix-Fralish ML, Nutile-McMenemy N, Perez N, Rothstein JD, DeLeo JA (2008) Propentofylline-induced astrocyte modulation leads to alterations in glial glutamate promoter activation following spinal nerve transection. Neuroscience 152: 1086–92.

Van Eldik LJ, Wainwright MS (2003) The Janus face of glial-derived S100B: beneficial and detrimental functions in the brain. Restor Neurol Neurosci 21: 97–108.

Wieseler-Frank J, Maier SF, Watkins LR (2005) Central proinflammatory cytokines and pain enhancement. Neurosignals 14: 166–74.

Chapter 19
Spinal Cord Phospholipase A_2 and Prostanoids in Pain Processing

Camilla I. Svensson

Abstract Prostanoids are lipid mediators formed by the action of phospholipase A2 (PLA_2), cyclooxygenase (COX) and prostanoid synthases, which converts arachidonic acid into various prostanoids. Advances in our understanding of how prostanoids are synthesized, regulated and functions have led to a new appreciation of their actions in health and disease. The prostanoids serve pivotal functions in pain signaling, not only at the site of inflammation but also at the level of the spinal cord. The roles of spinal prostanoids are thought to be particular important in inflammation-induced pain, but there are indications that this system may also play out in postoperative pain and in the early phase of nerve-injury induced pain. Agents that inhibit COX have long been used for pain relief but it is possible that inhibition of other steps in the PLA_2-COX-prostanoid-receptor axis would provide a more potent or a less side effect-associated pain relief. As there are several PLA_2 isoforms and prostanoid receptors expressed in the spinal cord, the purpose of this chapter is to describe current knowledge of the different spinal PLA_2's involvement in pain signaling, as well as the action of spinal prostanoids in this process.

Abbreviations

AACOCF3	arachidonyl trifluoromethylketone
AX048	ethyl 4-[(2-oxohexadecanoyl)amino] butanoate
BEL	bromoenol lactone
$cPLA_2$	calcium dependent cytosolic phospholipase A_2
CSF	cerebrospinal fluid
COX	cyclooxygenase
EP	PGE_2 receptor
DP	prostaglandin D receptor

C.I. Svensson (✉)
Department of Physiology and Pharmacology, Karolinska Institutet, Stockholm, Sweden
e-mail: camilla.svensson@ki.se

M. Malcangio (ed.), *Synaptic Plasticity in Pain*,
DOI 10.1007/978-1-4419-0226-9_19, © Springer Science+Business Media, LLC 2009

DRG	dorsal root ganglia
EET	epoxyeicosatrienoic acid
EP	prostaglandin E receptor
FP	prostaglandin F receptor
HETE	hydroxyeicosatetraenoic acid
IP	prostaglandin I receptor
IL-1	interleukin-1
iPLA$_2$	calcium independent cytosolic phospholipase A$_2$
LO	lipoxygenase
LPA	lysophosphatidic acid
MAFP	methyl arachidonyl fluorophosphonate
MAPK	mitogen activated protein kinases
NK-1	neurokinin-1
NMDA	N-methyl-D-aspartate
NSAID	non-steroial anti-inflammatory drugs
PLA$_2$	phospholipase A$_2$
PG	prostaglandin
PPAR	peroxisoem proliferators activated receptor
sPLA$_2$	secretory phospholipase A$_2$
TNF	tumor necrosis factor alpha
TP	tromboxane receptor
TXB2	tromboxane B2

19.1 Introduction

The prostanoids (prostaglandins, prostacyclins and thromboxanes) is a family of biologically active lipids that coordinate a multitude of physiologic and pathologic processes in response to specific stimuli. Prostanoid synthesis starts with PLA$_2$-mediated liberation of arachidonic acid from cellular membranes, which then COX converts to prostanoids. Soon after they were identified, prostanoids were shown to influence inflammation and immune responses, and the local increase of prostanoids following tissue injury and inflammation was linked to sensitization of primary afferents and a subsequent augmentation of the pain perception. More recently it has been revealed that prostanoids modulate nociception not only at the site of inflammation, but also at the level of the spinal cord. The roles of spinal prostanoids are thought to be particular important in inflammation-induced pain, but there are indications that this system may also play out in postoperative pain (Prochazkova et al., 2006; Kroin et al., 2008) and in the early phase of nerve-injury induced pain (Hefferan et al., 2003b,2003a; Zhu and Eisenach, 2003; O'Rielly and Loomis, 2007).

Agents that inhibit COX have long been used for pain relief. Derivatives of salicylic acid, structurally similar to aspirin, found in willow plants, have been in medical use since ancient times. By 1899, the company Bayer marketed

Aspirin for pain relief, however, it was first in the early 1970's, when sir John Vane and colleagues discovered that aspirin blocks the activity of COX-1 and COX-2, and thus prostaglandin formation, that the mechanism of action was established (Vane, 1971). Since then the body of evidence supporting an important role for COX and prostanoids in pain modulation has been constantly growing. During the 1990's it was found that spinal COX and the formation of spinal prostanoids are part of the regulatory system in nociceptive processing, and this moved the field into the central nervous system. The purpose of this chapter is to describe the current knowledge of *spinal* PLA$_2$'s involvement in pain signaling and the action of prostanoids in this process. The central enzyme in this scheme is cyclooxygenase, however, as there are many excellent reviews describing the role of COX in spinal sensitization, this enzyme will only be briefly discussed here.

19.2 Spinal Actions of the PLA$_2$/COX/Prostanoids Pathway in Hyperalgesia

The formation of prostanoids depends upon the initial activation of PLA$_2$ isozymes that cleave arachidonic acid from membrane phospholipids, and the subsequent action of COX which convert arachidonic acid into PGH$_2$. Cyclooxygenases are membrane-bound enzymes possessing dual properties, which provide both cyclooxygenase and hydroperoxidase activities and thereby the transformation of arachidonic acid to the prostaglandin endoperoxide intermediates PGG$_2$ and PGH$_2$. Prostaglandin H$_2$ is converted to the final active products, prostaglandins and thromboxanes, by individual prostaglandin and thromboxane synthases (Hamberg and Samuelsson, 1967; Blackwell et al., 1975; Ogino et al., 1977; Smith et al., 2000) (Fig. 19.1). The prostanoids exert most of their actions via G-protein coupled receptors and these have different agonist selectivity, tissue distribution and downstream signal transduction pathways (see below).

It is well established that prostaglandins drive the sensitization produced by tissue damage and inflammation. Local intraplantar injection of prostaglandins reproduces the major signs of inflammation, including augmented pain, and the hypersensitivity is observed as an increased neuronal firing rate in response to peripheral stimulus (Ferreira et al., 1978; Meves, 2006). In the spinal cord, prostanoids affects neuronal excitability and synaptic transmission. The prostaglandins act on prostaglandin receptors to facilitate the release of glutamate and other neurotransmitters from the central terminals of primary afferents (Nicol et al., 1992; Hingtgen et al., 1995; Malmberg et al., 1995; Minami et al., 1999), depolarize deep dorsal horn neurons (Baba et al., 2001) and block inhibitory glycinergic neurotransmission (Ahmadi et al., 2002; Harvey et al., 2004; Reinold et al., 2005). Behavioral experiments have demonstrated that prostaglandins injected spinally (intrathecally) gives rise to nociceptive

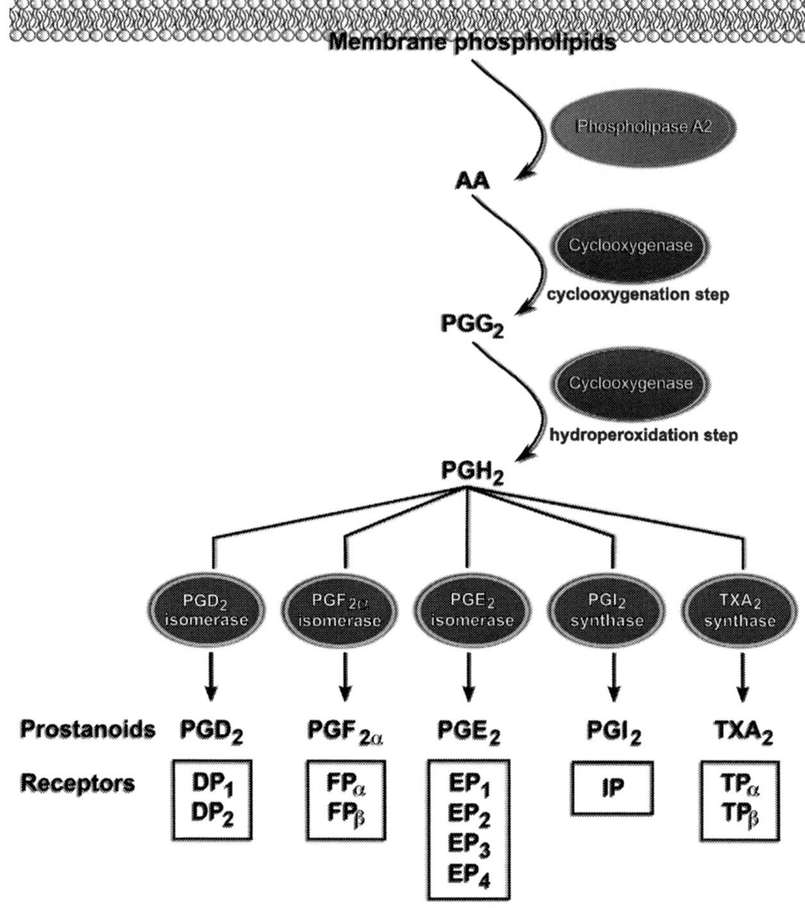

Fig. 19.1 Prostanoid biosynthetic pathway. Arachidonic acid is released from the *sn*2-position in membrane phospholipids by phospholipase A_2. Arachidonic acid is then converted in a two-step reaction, first to prostaglandin G2 (PGG2), then to PGH2, by the action of cyclooxygenase (COX), which have both COX and hydroperoxidase activity. PGH_2 is converted by tissue-specific enzymes to prostanoids. The prostanoids activate specific cell-membrane receptors of the superfamily of G-protein–coupled receptors. IP indicates prostacyclin receptor, TP, thromboxane receptor, DP, PGD_2 receptor, EP, PGE_2 receptor, and FP, $PGF_{2\alpha}$ receptor

behavior and intrathecal injection of COX-inhibitors, which attenuates prostaglandin formation, have antihyperalgesic effect in a number of experimental models of pain (for review, see Svensson and Yaksh, 2002). From a drug development stand point, blocking the flux through the PLA_2/COX/prostanoid cascade, at any level, might be considered a strategy for pain relief, with a global blockade of arachidonic acid products on one end and interference with a single receptor on the other end of the spectrum.

19.3 Spinal Phospholipase A$_2$

Prostanoids are not stored within cells, but are synthesized as required. The arachidonic acid concentration in cell cytosol is also low under normal conditions, and thus must be released from phospholipids before it can be used for prostanoid synthesis. Formation of arachidonic acid and the expression level of COX-2 are generally considered to be the rate-limiting step in prostanoid synthesis. In the spinal cord, both COX-1 and COX-2 are constitutively expressed and as PLA$_2$s regulates the formation of arachidonic acid, its activity is potentially very important in the regulation of spinal prostanoid generation and pain processing. The PLA$_2$ super family of enzymes is defined by their ability to cleave fatty acid from the sn-2 position of phospholipids, producing free fatty acids and lysophospholipids. This family can be divided into at least 19 groups based on amino acid structure and other characteristics. Focusing on biological properties the enzymes are divided into three major subgroups: calcium-dependent cytosolic PLA$_2$ (cPLA$_2$), calcium-independent cytosolic PLA$_2$ (iPLA$_2$), and secretory PLA$_2$ (sPLA$_2$) (Dennis, 1997). Below is an overview of these three PLA$_2$ groups and reported data related to the a role of these enzymes in spinal pain processing.

19.4 Calcium-Dependent Cytosolic PLA$_2$

To date four cPLA$_2$ paralogs has been identified in mammals (cPLA$_2$ groups IVA, IVB, IVC and IVD, also referred to as cPLA$_2\alpha$, cPLA$_2\beta$, cPLA$_2\gamma$ and cPLA$_2\delta$). Among them, cPLA$_2$ group IVA is the only PLA$_2$ enzyme that shows significant selectivity toward phospholipids containing arachidonic acid in the sn-2 position. cPLA$_2$ group IVD is not expressed in the central nervous system (Farooqui et al., 2006) and as there is no information is available on cPLA$_2$ groups IVB and IVC in pain modulation, the text in the following sections refers to cPLA$_2$ group IVA.

In neuronal cells, cPLA$_2$ is activated in response to extracellular stimuli such as cytokines (TNF and IL-1), growth factors, neurotransmitters (including substance P, glutamate (NMDA), serotonin, dopamine and ATP) and endotoxins (Sun et al., 2004). Specific subcellular location of cPLA$_2$ is important in regulation of arachidonic release and eicosanoid production. While cPLA$_2$ does not use Ca^{2+} for catalysis, an increase in intracellular Ca^{2+} concentration (sub μM range) is needed for translocation of cPLA$_2$ from the cytosol to the perinuclear region and for membrane binding (Glover et al., 1995; Schievella et al., 1995; Hirabayashi et al., 1999; Evans et al., 2001). The translocation of cPLA$_2$ brings the enzyme in proximity to the substrate, phospholipids, as well as to the downstream enzymes that act on freed fatty acids. Cyclooxygenase and lipoxygenase (LO), enzymes in the arachidonic metabolic cascade, are thus localized to the perinuclear region, COX constitutively and LO upon cell stimulation

(Hirabayashi and Shimizu, 2000; Radmark et al., 2007). Accordingly, in the central nervous system, where both $cPLA_2$ and COX are constitutively expressed, prostanoids can rapidly be synthesized in response to specific extracellular stimuli.

19.5 Expression of $cPLA_2$ in the Spinal Cord

$cPLA_2$ gene and protein expression is readily detected in rodent spinal cord (Kishimoto et al., 1999; Ong et al., 1999; Samad et al., 2001; Lucas et al., 2005; Ayoub et al., 2008; Kim et al., 2008). In naïve animals the expression is restricted to neurons in the gray matter and oligodendrocytes in the white matter (Liu et al., 2006; Kim et al., 2008) (Fig. 19.2). It should be noted that an increase in $cPLA_2$ protein expression has been reported following spinal cord injury (Liu et al., 2006), global forebrain ischemia (Clemens et al., 1996) and kainate-induced neuronal injury in the brain (Ong et al., 1999). This increase has been observed both in neuronal cells as well as microglia and astrocytes, suggesting that in certain conditions $cPLA_2$ is subjected to transcriptional control, as an additional level of activity regulation, and that the enzyme can be expressed de novo in glia (Clark et al., 1995). However, de novo expression of spinal $cPLA_2$ has thus far not been reported in models of inflammatory pain (Samad et al., 2001; Lucas et al., 2005) and the expression of $cPLA_2$ appears restricted to neurons and oligodendrocytes (Liu et al., 2006; Kim et al., 2008).

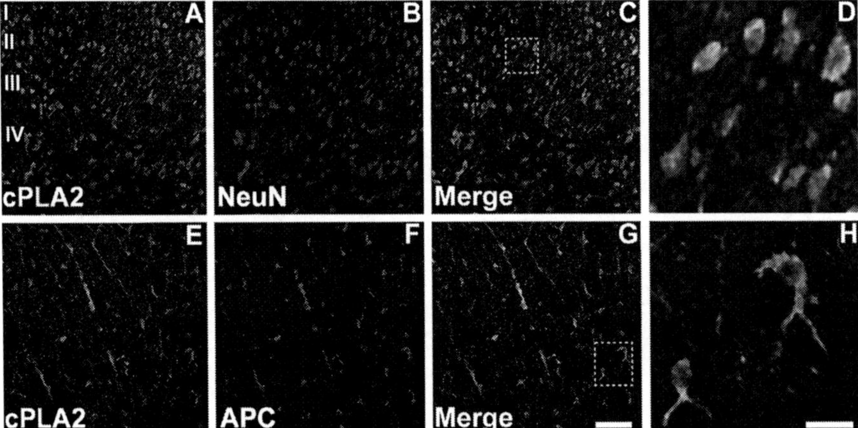

Fig. 19.2 Distribution and cellular localization of $cPLA_2\alpha$ in the spinal dorsal horn of naïve rat. $cPLA_2\alpha$ immunoreactivity is distributed throughout dorsal horn laminae I–IV (A, C). Double immunofluorescence confocal micrographs show that $cPLA_2\alpha$ (A) and the neuronal marker NeuN (B) co-localize in the majority of neurons in the spinal cord (C, D). $cPLA_2\alpha$ immunostaining also seen in white matter (E, F), where it co-localized with the oligodendrocyte marker APC (G, H). Scale bar represents 20 μm for figures A–C and E–G and 60 μm for D and H

19.6   Is Inhibition of Spinal cPLA$_2$ an Effective Target for Pain Relief?

Due to the lack of highly specific cPLA$_2$ inhibitors it has been a challenge to study the role of cPLA$_2$ in spinal sensitization. However, inhibitors such as arachidonyl trifluoromethylketone (AACOCF3), methyl arachidonyl fluoro-phosphonate (MAFP) and ethyl 4-[(2-oxohexadecanoyl)amino] butanoate (AX048), which inhibits both cPLA$_2$ and iPLA$_2$ group VI, collectively suggest that PLA$_2$'s participate in the regulation of spinal hyperexcitability. When administrated intrathecally, AACOCF3, MAFP and AX048 attenuate hypersensitivity induced by tissue injury and inflammation (formalin induced flinching and carrageenan induced thermal hyperalgesia) (Lucas et al., 2005; Yaksh et al., 2006), intrathecal injection of NMDA (Lucas et al., 2005) and constriction induced nerve injury (Sung et al., 2003). Importantly, bromoenol lactone (BEL), which selectively inhibits iPLA$_2$ group VI (see below), does not have anti-nociceptive effect (Lucas et al., 2005). Thus, taken together the pharmacological data points to cPLA$_2$ being the dominating PLA$_2$ in this process. Also brain cPLA$_2$ has been indicated in hypersensitivity as intracerebroventricular administration of AACOCF3 blocks allodynia after facial injection of carrageenan in mice (Yeo et al., 2004). In addition, intrathecal injection of AACOCF3 and MAFP, but not BEL, attenuate spinal PGE$_2$ release in experimental models of pain (Lucas et al., 2005) further linking cPLA$_2$ activity to spinal prostanoids formation and pain. One caveat with the inhibitors listed above, is that they also act on several other systems, for example, MAFP inhibits fatty acid amide hydrolase and activates cannabinoid receptor 1 (Deutsch et al., 1997; Fernando and Pertwee, 1997). Hence, it is difficult to rule out the possibility that the antinociceptive effect of these inhibitors could be achieved through cPLA$_2$ independent mechanisms. However, more direct evidence for the involvement of cPLA$_2$ in spinal pain processing has been generated by the use of specific cPLA$_2$-targeted antisense oligonucleotides. Knock-down of spinal cPLA$_2$ gene and protein expression, which had no effect on sPLA$_2$ or iPLA$_2$ expression, attenuated formalin-induced hyperalgesia (Kim et al., 2008).

19.7   Phosphorylation – An Additional Level of cPLA$_2$ Activity Regulation

Noteworthy, cPLA$_2$ activity is not only regulated through cytosol-perinuclear translocation and de novo protein expression, but also through phosphorylation (Clark et al., 1995; Hirabayashi et al., 2004). Depending on the cell type and stimulus, cPLA$_2$ is phosphorylated by the mitogen activated protein kinases (MAPK) ERK and p38 (Lin et al., 1993; Kramer et al., 1996), Ca^{2+}/calmodulin-dependent protein kinase II (Muthalif et al., 2001) and by the p38 activated kinase MNK1 (Hefner et al., 2000). The mechanism by which phosphorylation of cPLA$_2$ increases its enzymatic activity is not fully understood but it has been suggested that phosphorylation improves the affinity of cPLA$_2$ for perinuclear membranes by slowing down cPLA$_2$-membrane dissociation (Das et al., 2003). At present there are very few reports regarding the phosphorylation status of cPLA$_2$ in experimental models of pain. Of note, an increase in cPLA$_2$ phosphorylation at Serine505 (phosphorylation site for p42/44 and/or p38 MAPK) in DRGs has been observed in a model of nerve injury, and this activation was linked to ATP and P2X3 receptors (Tsuda et al., 2007). In addition, inhibition of spinal p38 MAPK attenuates PGE$_2$ release evoked by intrathecal injection of substance P, NMDA and dynorhin, as well as injection of formalin to the paw (Svensson et al., 2003a,2003b, 2005b). Importantly, the MAPK phosphorylation site is only present in cPLA$_2$ group IVA and is not conserved in the other cPLA$_2$ paralogs. Thus to the extent p38 MAPK inhibition can be directly linked to a reduction in cPLA$_2$ activity, cPLA$_2$ group IVA is most likely the candidate.

19.8 Calcium-Independent PLA$_2$

The iPLA$_2$ (group VI) includes iPLA$_2$ groups IVA and IVB, with five splice variants for group VIA iPLA$_2$. iPLA$_2$ shows no strict *sn*-2 fatty acid specificity of the substrate phospholipids and is fully active in the absence of Ca^{2+}. iPLA$_2$ gene and protein expression is detected in human and rodent spinal cord (Larsson Forsell et al., 1999; Lucas et al., 2005), however, its function has not been elucidated. iPLA$_2$ overexpression in HEK293 cells leads to an increase in non-selective fatty acid release, which is reversed by BEL (Murakami et al., 1998; Atsumi et al., 2000) and under certain conditions iPLA$_2$ liberates arachidonic acid and lysophospholipids thought to regulate glucose-induced insulin secretion and cellular proliferation (Akiba and Sato, 2004). However, the arachidonic acid that is spontaneously released by iPLA$_2$ is for the most part poorly linked with downstream COX activity and prostanoid formation (Winstead et al., 2000; Akiba et al., 2002). Intrathecal injection of BEL does not attenuate tissue injury and inflammation-induced pain and does not prevent spinal PGE$_2$ release evoked by intrathecal NMDA (Lucas et al., 2005). In general, the main function of iPLA$_2$ is considered to be phospholipid remodeling including the incorporation of arachidonic acid and other fatty acids into membrane phospholipids (Balsinde et al., 1998; Winstead et al., 2000).

19.9 Secretory PLA$_2$

Ten mammalian sPLA$_2$'s have been identified: Groups IB, IIA, IIC, IID, IIE, IIF, III, V, X, XII (Gelb et al., 2000; Ho et al., 2001). sPLA$_2$s hydrolyze the ester bond at the *sn*-2 position of glycerophospholipids with no strict fatty acid selectivity (Schaloske and Dennis, 2006) and the enzymatic reaction requires µM concentrations of Ca^{2+}. With a molecular weight of 14–18 kDa, this group of phospholipases is smaller than the cPLA$_2$'s (85–114 kDa) and iPLA$_2$'s (85–90 kDa). In addition, sPLA$_2$'s are released from cells and can be translocated from the extracellular milieu into the cytosol (Matsuzawa et al., 1996), giving them further distinct characteristics. Upon secretion, sPLA$_2$s may act in a paracrine as well as in an autocrine manner to release fatty acids (Schaloske and Dennis, 2006). The physiological function of most of sPLA$_2$ isoforms remains to be determined. Out of the ten, sPLA$_2$ group IIA has been the most extensively studied because of its involvement in inflammatory processes in peripheral systems (Murakami et al., 1997). While there is only a handful reports on the role of spinal sPLA$_2$s, the role of this class of phospholipases have been somewhat more studied in the brain. Interestingly, in the brain, sPLA$_2$ group IIA mRNA is expressed in astrocytes and neurons, is induced in response to proinflammatory cytokines (TNF, IL-1) and interferon gamma (Oka and Arita, 1991; Li et al., 1999; Xu et al., 2003) and can be released into the extracellular space following neuronal depolarization (Matsuzawa et al., 1996). External application of human synovial sPLA$_2$ to PC12 cells and primary cultures of hippocampal neurons has been shown to cause neurotransmitter release (Wei et al., 2003).

19.9.1 Secretory PLA$_2$ in the Spinal Cord

In the spinal cord, both protein and gene expression of sPLA$_2$ group IIA and sPLA$_2$ group V (Svensson et al., 2005a; Kim et al., 2008) has been documented, though the cellular distribution is yet to be determined. Inhibition of spinal sPLA$_2$, using the sPLA$_2$ specific inhibitor LY311727 (Schevitz et al., 1995), prevents hypersensitivity induced by inflammation/tissue injury (carrageenan-induced thermal hyperalgesia, formalin-induced flinching), and intrathecal injection of substance P (Svensson et al., 2005a) pointing to a role of this group of phospholipases in spinal pain processing. Of note, LY311727 does not distinguish between the different sPLA$_2$ isoforms, hence, no conclusions can be drawn with regards to the roles of the different sPLA$_2$ isoforms in pain processing using this inhibitor.

19.9.2 Potential Non-Arachidonic Acid-Mediated Nociceptive Actions of sPLA$_2$

In addition to contributing to the release of arachidonic acid, sPLA$_2$ may participate in spinal sensitization through hydrolysis of other fatty acids. Recent work has shown that lysophosphatidic acid (LPA) and the LPA1 receptor in DRGs and spinal cord play an important role in hyperalgesia (Inoue et al., 2004). Group IIA sPLA$_2$ hydrolyzes phosphatidic acid, which promotes the formation of LPA (Snitko et al., 1997). On another note, using human and snake $sPLA_2$s as ligands, two mammalian cell surface $sPLA_2$ receptors have been identified. Several lines of evidence indicate that $sPLA_2$s may exert physiological roles through acting in a cytokine-like fashion independent of their catalytic function (Valentin and Lambeau, 2000). Groups IIA, IIB and X $sPLA_2$, bind and activate the M-type receptor, which has been associated with cell proliferation, cell migration and hormone release (Lambeau and Lazdunski, 1999). While the presence of the M-type receptor has been demonstrated in the brain (Copic et al., 1999), it still remains to determine if the M-type receptor is present in the spinal cord, and if its biological function can be linked to spinal facilitation of nociception. Intraplantar and peri-sciatic injection of both enzymatically active and enzymatically inactive $sPLA_2$s isolated from *Bothrops asper* snake venom lead to identical behavior and activation of spinal glia, suggesting that at least the in the periphery, $sPLA_2$s do not require enzymatic release of fatty acids for its nociceptive action (Chacur et al., 2004a,2004b).

While most studies point to a potential pro-nociceptive contribution of $sPLA_2$, contrasting findings has been reported from work utilizing a $sPLA_2$ group IIA knockout mouse. In this particular mouse strain the *$sPLA_2$-IIA* gene is intrinsically disrupted due to frame shift mutation, giving rise to a "natural" knockout. Assessment of carrageenan-induced hypersensitivity in these mice displayed an increased sensitivity (Kennedy et al., 1995). This rather unexpected finding complicates the assessment of $sPLA_2$'s role in pain. However, these data may have to be interpreted with caution as in other experiments, these animals responded normally to inflammatory stimuli (Kennedy et al., 1995; MacPhee et al., 1995) which is contradictory to a large pool of data demonstrating a proinflammatory role of $sPLA_2$ group IIA. It is possible that the gene deficiency is compensated for by induction or increased activity of other $sPLA_2$ isoforms.

There is limited availability of specific pharmacological tools for studies of the role of the different PLA_2 isoforms. Taken together, available data points to involvement of $cPLA_2$ group IVA and $sPLA_2$ in the regulation of spinal pain processing. Blocking the activity of PLA_2s prevents the formation of not only prostanoids, but also other arachidonic acid derived products and may therefore be associated with more severe side effects than those noted for COX inhibitors. Hence it may be argued that going down-stream of COX and identify the prostanoid or prostanoid receptor that is critical for spinal hyperexcitability is a "safer" strategy.

19.10 Prostanoids and Prostanoid Receptors in Spinal Pain Signaling

19.10.1 Prostanoids in Cerebrospinal Fluid

A variety of studies examining spinal cord slices in vitro have shown that PGE_2 concentrations are increased by local application of NK-1, vanilloid (TRPV1), and glutamate receptor agonists (Dirig and Yaksh, 1999) and that this release is blocked by COX-2 inhibition (Dirig and Yaksh, 1999). A close correlation exists between manipulations that induce hyperalgesia and those that evoke the spinal release of prostanoids. Guay and colleagues (Guay et al., 2004) have assayed PGE$_2$, PGF$_{2\alpha}$, PGD$_2$, 6-keto-PGF$_{1\alpha}$ (PGI$_2$ metabolite) and TXB$_2$ in cerebrospinal fluid (CSF) collected after injection of carrageenan to the paw. They found that PGE$_2$ is increased during the first 12 hours in this model. An increase, though much smaller, in PGD$_2$, 6-keto-PGF$_{1\alpha}$ and TXB$_2$ concentration was also detected. In agreement with this finding, our recent study, in which CSF sampled from the same model was assayed for approximately 120 different lipid mediators, also showed that PGE$_2$ is the most prominent spinal prostanoid after peripheral inflammation (M. Buczynski and CI. Svensson, unpublished observations). In studies where PGE$_2$ has been the single analyte in CSF collected by in vivo spinal microdialysis and direct cerebrospinal fluid sampling, an increase in PGE$_2$ has been shown following (a) acute activation of small afferents (intraplantar formalin, heat) (Coderre et al., 1990; Malmberg et al., 1995; Shi et al., 2006); (b) persistent inflammation (carrageenan in knee joint, intraplantar Freund's adjuvant, intraplantar zymosan) (Yang et al., 1996b; Smith et al., 1998; Ebersberger et al., 1999; Guhring et al., 2000; Samad et al., 2001; Bianchi et al., 2007); (c) surgery (thoracic muscle incision) (Kroin et al., 2006); (d) intrathecal injection of substance P, NMDA, dynorphin, kainate or cytokines (Yang et al., 1996a; Svensson et al., 2003a,2003b; Lucas et al., 2005; Svensson et al., 2005b,a; Shi et al., 2006); or (e) systemic cytokines (Samad et al., 2001). The functional relevance of this release is emphasized by the observations that the spinal PGE$_2$ release blocked by IT and systemic doses of COX inhibitors that reverse the respective hyperalgesia (for review, see Svensson and Yaksh, 2002). Also in humans have increased PGE$_2$ concentrations been detected in CSF and associated with pain. PGE$_2$ is increased in CSF in cancer patients (Inada et al., 2007) and during and after lower extremity (Reuben et al., 2006) and hip replacement (Buvanendran et al., 2006) surgery. When examined, administration of COX inhibitors reduced both the pain and the concentration of PGE$_2$ in the CSF.

While it appears clear that PGE$_2$ plays an important role in spinal sensitization, the contribution of the other prostanoids is still debated (see also Schaible and Ebersberger, Chapter 12). Intrathecal injection of PGE$_2$ uniformly produces pronociceptive effects (Uda et al., 1990; Minami et al., 1994; Reinold et al., 2005), however, opposing results, i.e. pronociceptive, no effect or antinociceptive effect, have been reported for PGF$_{2\alpha}$, PGD$_2$ and PGI$_2$ (Uda et al., 1990; Minami et al., 1992, 1996; Doi et al., 2002; Turnbach et al., 2002; Reinold et al., 2005; Telleria-Diaz et al., 2008). There is no information available that points to nociceptive actions of spinal thromboxanes. Noteworthy, in a recent study it by Telleria-Diaz et al. (2008) it was demonstrated that PGD$_2$ has potentially two opposite actions in the spinal cord. Depending on the state of the spinal cord, spinally applied PGD$_2$ was shown to act as a pronociceptive mediator leading to hyperalgesia (normal condition), while under inflammatory conditions in which spinal cord neurons are hyperexcitable, PGD_2 acted in an antinociceptive fashion. Changes in the effect of spinal PGE_2 also occur during inflammation, with the inhibitory EP3 receptor coming into play when there is a prevailing inflammation (Bar et al., 2004). Though the regulation and mechanisms for these types of switches have not been established, it illustrates the complexity of prostanoids and may explain in part the difficulties associated with the dissection of the exact contribution of the different prostanoids to spinal sensitization.

19.10.2 Prostanoid Synthases

Inhibition of prostanoid synthases may be considered as an alternative strategy for pain relief as it falls between a global reduction of prostanoid formation through inhibition of PLA_2 or COX and blockade of a single receptor. However, because of the lack of selective prostanoid synthase inhibitors the role of these enzymes in spinal nociceptive processing is largely unexplored, and has only been addressed in one study by the use of antisense oligonucleotides. In this study, knock-down of spinal cytosolic PGE synthase attenuated formalin-induced flinching and the early face of zymosan-induced hyperalgesia (Hofacker et al., 2005) pointing to possibility to interfere with prostanoid formation at the level of prostanoid synthases. Mice deficient in PGE synthases and PGI synthases have been described, but published data on their nociceptive phenotypes is sparse. Only mice deficient in microsomal PGE-1 synthase has been studied, and then only in the writing test. These mice showed a reduction in acetic acid-evoked response (Trebino et al., 2003; Kamei et al., 2004).

19.10.3 Prostanoid Receptors

Prostanoids exert their actions by binding to prostanoid receptors. There are at least 9 known prostanoid receptors, as well as several additional splice variants. Four of the receptor subtypes bind PGE_2 (EP_1–EP_4), two bind PGD_2 (DP_1 and DP_2), and PGF_2, PGI_2, and TXA_2 binds to FP, IP, and TP, respectively (Funk, 2001). The receptors couple via G-proteins to enzymes such as adenylate cyclase and phospholipase C (PLC). IP, DP_1, EP_2, and EP_4 signal through G_s-proteins to activate adenylate cyclase leading to an increase in cAMP and subsequent activation of protein kinase A or exchange protein activated by cAMP (Epac). EP_1, FP, and TP form a second group that signals through G_q-proteins and which leads to an increase in intracellular calcium (Hingtgen et al., 1995; Negishi et al., 1995a; Narumiya et al., 1999; Wang et al., 2007). The EP_3 receptor is regarded as an "inhibitory" receptor that couples to G_i and decreases cAMP formation (Negishi et al., 1995b). In situ hybridization and immunohistochemical studies have localized EP1, EP2, EP3, EP4 (Oida et al., 1995; Kawamura et al., 1997; Beiche et al., 1998; Donaldson et al., 2001), IP (Matsumura et al., 1995; Oida et al., 1995) and DP receptors (Wright et al., 1999) in the superficial layers of spinal cord, and therefore it is of considerable interest to explore their role in spinal hyper excitability. Unfortunately, like with the PLA$_2$ and prostanoid synthase inhibitors, the tools are limited and truly selective agonist and antagonists for the various prostanoid receptors are rare. One of the more specific prostanoid receptor antagonist, ONO-8711, blocks the EP1 receptor and intrathecal injection of this agent attenuates hypersensitivity in models of postoperative and inflammation (carrageenan-induced) pain (Nakayama et al., 2002; Omote et al., 2002). Many studies exploring the role of individual prostanoid receptors in pain signaling have relied on genetically modified mice. So far, these mice are exclusively non-conditional knockouts, and as the receptor then is absent in all tissues, and through development, it is difficult to draw any conclusions with regards to the spinal role of the targeted prostanoid receptor.

To make it more complex, postanoids may also acts as ligands for other receptor systems. In in vitro studies, 15-deoxy-PGJ$_2$, (a dehydration meta-bolite of PGD$_2$) and 6-keto-PGF$_{1\alpha}$ analogues have been shown to bind nuclear peroxisome proliferators activated receptor (PPAR)γ and PPARδ, respectively (for review, see Funk, 2001). Though such binding has not yet been demonstrated in the spinal cord, PPARγ gene and protein expression is induced in response to peripheral nerve injury and intrathecal injection of 15-deoxy-PGJ$_2$, attenuates nerve-injury induced allodynia (Churi et al., 2008).

19.11   Implications for Spinal PLA$_2$ and Prostanoids as Drug Targets

The aim with therapy targeted at prostanoid production is to reduce the hyperexcitatory accumulation of prostanoids without affecting normal prosta-noid-dependent homeostatic functions. There are several potential ways of achieving this: reducing prostanoid synthesis by inhibition of COX, PLA$_2$ or prostanoid synthase activity or by blocking the action of prostanoids on their receptors. COX inhibitors have long been used for pain relief, however, the search for alternative prostanoid-targeted therapeutics have continued with the goal to identify agents that either (i) have a more potent or effective profile, or (ii) are better tolerated as analgesics. The later is based on the gastrointestinal side effects associated with the classical NSAIDs and the increased risk of cardiovascular complications associated with the COX-2 selective inhibitors. Targeting a single prostanoid synthase or prostanoid receptor could potentially reduce unwanted effects. This chapter has focused on spinal prostanoids, and perhaps agents with high specificity, and with the ability to cross the blood-brain barrier, would show a reduced side effect profile. However, based on the available preclinical data it is still difficult to pin point which prostanoid or prostanoid receptor to target. A better understanding of the nature of these systems is needed, in particular as their roles in regulation of hypersensitivity may shift dependent on the condition. One intriguing approach may be to explore the prostanoids and prostanoid receptors with anti-nociceptive properties.

Though not the focus of this review, it is important to note that inhibition of PLA_2s not only prevent formation of prostanoids, but also generation of other lipid mediators derived from arachidonic acid. Lipoxygenase isozymes generate pro-inflammatory leukotrienes and anti-inflammatory lipoxins and cytochrome P450 hydroxylases form hydroxyeicosatetraenoic acids (HETEs) and epoxygenases generate epoxyeicosatrienoic acids (EETs) (for review, see Phillis et al., 2006). Several of these products have been implicated to play a role in spinal sensitization. In addition, arachidonic acid itself can directly modulate neuronal function by various mechanisms, such as altering membrane fluidity and polarization state, activating protein kinase C, and regulating gene transcription (Sun et al., 2004; Farooqui et al., 2006). Hence, inhibition spinal PLA_2 may provide a powerful way of dampening production of lipid mediators, and may prove useful if persistent pain when driven by an excess of spinal arachidonic acid and arachidonic acid derived products. The question becomes if such approach is feasible considering the potential anti-nociceptive, anti-inflammatory nature of some of these factors as well as the contribution of PLA_2 derived lipid mediators to normal functions. It remains a challenging task to determine if specific PLA_2 isoforms can be linked to pain pathophysiology.

19.12 Concluding Remarks

In conclusion, the role of the different PLA_2's and prostanoids in spinal pain processing cannot be appropriately addressed until potent, specific, and clinically useful PLA_2 and prostanoid synthase inhibitors and/or prostanoid receptor antagonists are available. Hence, the development of small molecule inhibitors, together with molecular biological tools such as antisense oligonucleotides and siRNA and alterations in signal transduction processes in knockout mice are critical, as they may provide the information needed to identify the feasible targets. There is little doubt that PLA_2's and prostanoids play important roles in the regulation of spinal pain processing, however, we are still in the dark when it comes to the details of how, which and where to interfere with these factors.

Acknowledgments The author's work was supported by NIH grant R21 DA021654, the Arthritis Foundation and the Swedish Research Council. Figure 19.2 was kindly provided by David Kim.

References

Ahmadi S, Lippross S, Neuhuber WL, Zeilhofer HU (2002) PGE(2) selectively blocks inhibitory glycinergic neurotransmission onto rat superficial dorsal horn neurons. Nat Neurosci 5:34–40.

Akiba S, Ohno S, Chiba M, Kume K, Hayama M, Sato T (2002) Protein kinase Calpha-dependent increase in Ca2+-independent phospholipase A2 in membranes and arachidonic acid liberation in zymosan-stimulated macrophage-like P388D1 cells. Biochem Pharmacol 63:1969–1977.

Akiba S, Sato T (2004) Cellular function of calcium-independent phospholipase A2. Biol Pharm Bull 27:1174–1178.

Atsumi G, Murakami M, Kojima K, Hadano A, Tajima M, Kudo I (2000) Distinct roles of two intracellular phospholipase A2s in fatty acid release in the cell death pathway. Proteolytic fragment of type IVA cytosolic phospholipase A2alpha inhibits stimulus-induced arachidonate release, whereas that of type VI Ca2+-independent phospholipase A2 augments spontaneous fatty acid release. J Biol Chem 275:18248–18258.

Ayoub SS, Yazid S, Flower RJ (2008) Increased susceptibility of annexin-A1 null mice to nociceptive pain is indicative of a spinal antinociceptive action of annexin-A1. Br J Pharmacol 154:1135–1142.

Baba H, Kohno T, Moore KA, Woolf CJ (2001) Direct activation of rat spinal dorsal horn neurons by prostaglandin E2. J Neurosci 21:1750–1756.

Balsinde J, Balboa MA, Dennis EA (1998) Functional coupling between secretory phospholipase A2 and cyclooxygenase-2 and its regulation by cytosolic group IV phospholipase A2. Proc Natl Acad Sci USA 95:7951–7956.

Bar KJ, Natura G, Telleria-Diaz A, Teschner P, Vogel R, Vasquez E, Schaible HG, Ebersberger A (2004) Changes in the effect of spinal prostaglandin E2 during inflammation: prostaglandin E (EP1–EP4) receptors in spinal nociceptive processing of input from the normal or inflamed knee joint. J Neurosci 24:642–651.

Beiche F, Klein T, Nusing R, Neuhuber W, Goppelt-Struebe M (1998) Localization of cyclooxygenase-2 and prostaglandin E2 receptor EP3 in the rat lumbar spinal cord. J Neuroimmunol 89:26–34.

Bianchi M, Martucci C, Ferrario P, Franchi S, Sacerdote P (2007) Increased tumor necrosis factor-alpha and prostaglandin E2 concentrations in the cerebrospinal fluid of rats with inflammatory hyperalgesia: the effects of analgesic drugs. Anesth Analg 104:949–954.

Blackwell GJ, Flower RJ, Vane JR (1975) Some characteristics of the prostaglandin synthesizing system in rabbit kidney microsomes. Biochim Biophys Acta 398:178–190.

Buvanendran A, Kroin JS, Berger RA, Hallab NJ, Saha C, Negrescu C, Moric M, Caicedo MS, Tuman KJ (2006) Upregulation of prostaglandin E2 and interleukins in the central nervous system and peripheral tissue during and after surgery in humans. Anesthesiology 104:403–410.

Chacur M, Gutierrez JM, Milligan ED, Wieseler-Frank J, Britto LR, Maier SF, Watkins LR, Cury Y (2004a) Snake venom components enhance pain upon subcutaneous injection: an initial examination of spinal cord mediators. Pain 111:65–76.

Chacur M, Milligan ED, Sloan EM, Wieseler-Frank J, Barrientos RM, Martin D, Poole S, Lomonte B, Gutierrez JM, Maier SF, Cury Y, Watkins LR (2004b) Snake venom phospholipase A2s (Asp49 and Lys49) induce mechanical allodynia upon peri-sciatic administration: involvement of spinal cord glia, proinflammatory cytokines and nitric oxide. Pain 108:180–191.

Churi SB, Abdel-Aleem OS, Tumber KK, Scuderi-Porter H, Taylor BK (2008) Intrathecal rosiglitazone acts at peroxisome proliferator-activated receptor-gamma to rapidly inhibit neuropathic pain in rats. J Pain 9:639–649.

Clark JD, Schievella AR, Nalefski EA, Lin LL (1995) Cytosolic phospholipase A2. J Lipid Mediat Cell Signal 12:83–117.

Clemens JA, Stephenson DT, Smalstig EB, Roberts EF, Johnstone EM, Sharp JD, Little SP, Kramer RM (1996) Reactive glia express cytosolic phospholipase A2 after transient global forebrain ischemia in the rat. Stroke 27:527–535.

Coderre TJ, Gonzales R, Goldyne ME, West J, Levine JD (1990) Noxious stimulus-induced increase in spinal prostaglandin E2 is noradrenergic terminal-dependent. Neurosci Lett 115:253–258.

Copic A, Vucemilo N, Gubensek F, Krizaj I (1999) Identification and purification of a novel receptor for secretory phospholipase A(2) in porcine cerebral cortex. J Biol Chem 274:26315–26320.

Das S, Rafter JD, Kim KP, Gygi SP, Cho W (2003) Mechanism of group IVA cytosolic phospholipase A(2) activation by phosphorylation. J Biol Chem 278:41431–41442.

Dennis EA (1997) The growing phospholipase A2 superfamily of signal transduction enzymes. Trends Biochem Sci 22:1–2.

Deutsch DG, Omeir R, Arreaza G, Salehani D, Prestwich GD, Huang Z, Howlett A (1997) Methyl arachidonyl fluorophosphonate: a potent irreversible inhibitor of anandamide amidase. Biochem Pharmacol 53:255–260.

Dirig DM, Yaksh TL (1999) In vitro prostanoid release from spinal cord following peripheral inflammation: effects of substance P, NMDA and capsaicin. Br J Pharmacol 126:1333–1340.

Doi Y, Minami T, Nishizawa M, Mabuchi T, Mori H, Ito S (2002) Central nociceptive role of prostacyclin (IP) receptor induced by peripheral inflammation. Neuroreport 13:93–96.

Donaldson LF, Humphrey PS, Oldfield S, Giblett S, Grubb BD (2001) Expression and regulation of prostaglandin E receptor subtype mRNAs in rat sensory ganglia and spinal cord in response to peripheral inflammation. Prostaglandins Other Lipid Mediat 63:109–122.

Ebersberger A, Grubb BD, Willingale HL, Gardiner NJ, Nebe J, Schaible HG (1999) The intraspinal release of prostaglandin E2 in a model of acute arthritis is accompanied by an up-regulation of cyclo-oxygenase-2 in the spinal cord. Neuroscience 93:775–781.

Evans JH, Spencer DM, Zweifach A, Leslie CC (2001) Intracellular calcium signals regulating cytosolic phospholipase A2 translocation to internal membranes. J Biol Chem 276:30150–30160.

Farooqui AA, Ong WY, Horrocks LA (2006) Inhibitors of brain phospholipase A2 activity: their neuropharmacological effects and therapeutic importance for the treatment of neurologic disorders. Pharmacol Rev 58:591–620.

Fernando SR, Pertwee RG (1997) Evidence that methyl arachidonyl fluorophosphonate is an irreversible cannabinoid receptor antagonist. Br J Pharmacol 121:1716–1720.

Ferreira SH, Nakamura M, de Abreu Castro MS (1978) The hyperalgesic effects of prostacyclin and prostaglandin E2. Prostaglandins 16:31–37.

Funk CD (2001) Prostaglandins and leukotrienes: advances in eicosanoid biology. Science 294:1871–1875.

Gelb MH, Valentin E, Ghomashchi F, Lazdunski M, Lambeau G (2000) Cloning and recombinant expression of a structurally novel human secreted phospholipase A2. J Biol Chem 275:39823–39826.

Glover S, de Carvalho MS, Bayburt T, Jonas M, Chi E, Leslie CC, Gelb MH (1995) Translocation of the 85-kDa phospholipase A2 from cytosol to the nuclear envelope in rat basophilic leukemia cells stimulated with calcium ionophore or IgE/antigen. J Biol Chem 270:15359–15367.

Guay J, Bateman K, Gordon R, Mancini J, Riendeau D (2004) Carrageenan-induced paw edema in rat elicits a predominant prostaglandin E2 (PGE2) response in the central nervous system associated with the induction of microsomal PGE2 synthase-1. J Biol Chem 279:24866–24872.

Guhring H, Gorig M, Ates M, Coste O, Zeilhofer HU, Pahl A, Rehse K, Brune K (2000) Suppressed injury-induced rise in spinal prostaglandin E2 production and reduced early thermal hyperalgesia in iNOS-deficient mice. J Neurosci 20:6714–6720.

Hamberg M, Samuelsson B (1967) On the mechanism of the biosynthesis of prostaglandins E-1 and F-1-alpha. J Biol Chem 242:5336–5343.

Harvey RJ, Depner UB, Wassle H, Ahmadi S, Heindl C, Reinold H, Smart TG, Harvey K, Schutz B, Abo-Salem OM, Zimmer A, Poisbeau P, Welzl H, Wolfer DP, Betz H, Zeilhofer HU, Muller U (2004) GlyR alpha3: an essential target for spinal PGE2-mediated inflammatory pain sensitization. Science 304:884–887.

Hefferan MP, O'Rielly DD, Loomis CW (2003a) Inhibition of spinal prostaglandin synthesis early after L5/L6 nerve ligation prevents the development of prostaglandin-dependent and prostaglandin-independent allodynia in the rat. Anesthesiology 99:1180–1188.

Hefferan MP, Carter P, Haley M, Loomis CW (2003b) Spinal nerve injury activates prostaglandin synthesis in the spinal cord that contributes to early maintenance of tactile allodynia. Pain 101:139–147.

Hefner Y, Borsch-Haubold AG, Murakami M, Wilde JI, Pasquet S, Schieltz D, Ghomashchi F, Yates JR, 3rd, Armstrong CG, Paterson A, Cohen P, Fukunaga R, Hunter T, Kudo I, Watson SP, Gelb MH (2000) Serine 727 phosphorylation and activation of cytosolic phospholipase A2 by MNK1-related protein kinases. J Biol Chem 275:37542–37551.

Hingtgen CM, Waite KJ, Vasko MR (1995) Prostaglandins facilitate peptide release from rat sensory neurons by activating the adenosine 3',5'-cyclic monophosphate transduction cascade. J Neurosci 15:5411–5419.

Hirabayashi T, Kume K, Hirose K, Yokomizo T, Iino M, Itoh H, Shimizu T (1999) Critical duration of intracellular Ca2+ response required for continuous translocation and activation of cytosolic phospholipase A2. J Biol Chem 274:5163–5169.

Hirabayashi T, Shimizu T (2000) Localization and regulation of cytosolic phospholipase A(2). Biochim Biophys Acta 1488:124–138.

Hirabayashi T, Murayama T, Shimizu T (2004) Regulatory mechanism and physiological role of cytosolic phospholipase A2. Biol Pharm Bull 27:1168–1173.

Ho IC, Arm JP, Bingham CO, 3rd, Choi A, Austen KF, Glimcher LH (2001) A novel group of phospholipase A2s preferentially expressed in type 2 helper T cells. J Biol Chem 276:18321–18326.

Hofacker A, Coste O, Nguyen HV, Marian C, Scholich K, Geisslinger G (2005) Down-regulation of cytosolic prostaglandin E2 synthase results in decreased nociceptive behavior in rats. J Neurosci 25:9005–9009.

Inada T, Kushida A, Sakamoto S, Taguchi H, Shingu K (2007) Intrathecal betamethasone pain relief in cancer patients with vertebral metastasis: a pilot study. Acta Anaesthesiol Scand 51:490–494.

Inoue M, Rashid MH, Fujita R, Contos JJ, Chun J, Ueda H (2004) Initiation of neuropathic pain requires lysophosphatidic acid receptor signaling. Nat Med 10:712–718.

Kamei D, Yamakawa K, Takegoshi Y, Mikami-Nakanishi M, Nakatani Y, Oh-Ishi S, Yasui H, Azuma Y, Hirasawa N, Ohuchi K, Kawaguchi H, Ishikawa Y, Ishii T, Uematsu S, Akira S, Murakami M, Kudo I (2004) Reduced pain hypersensitivity and inflammation in mice lacking microsomal prostaglandin e synthase-1. J Biol Chem 279:33684–33695.

Kawamura T, Yamauchi T, Koyama M, Maruyama T, Akira T, Nakamura N (1997) Expression of prostaglandin EP2 receptor mRNA in the rat spinal cord. Life Sci 61:2111–2116.

Kennedy BP, Payette P, Mudgett J, Vadas P, Pruzanski W, Kwan M, Tang C, Rancourt DE, Cromlish WA (1995) A natural disruption of the secretory group II phospholipase A2 gene in inbred mouse strains. J Biol Chem 270:22378–22385.

Kim DH, Fitzsimmons B, Hefferan MP, Svensson CI, Wancewicz E, Monia BP, Hung G, Butler M, Marsala M, Hua XY, Yaksh TL (2008) Inhibition of spinal cytosolic phospholipase A(2) expression by an antisense oligonucleotide attenuates tissue injury-induced hyperalgesia. Neuroscience 154:1077–1087.

Kishimoto K, Matsumura K, Kataoka Y, Morii H, Watanabe Y (1999) Localization of cytosolic phospholipase A2 messenger RNA mainly in neurons in the rat brain. Neuroscience 92:1061–1077.

Kramer RM, Roberts EF, Um SL, Borsch-Haubold AG, Watson SP, Fisher MJ, Jakubowski JA (1996) p38 mitogen-activated protein kinase phosphorylates cytosolic phospholipase A2 (cPLA2) in thrombin-stimulated platelets. Evidence that proline-directed phosphorylation is not required for mobilization of arachidonic acid by cPLA2. J Biol Chem 271:27723–27729.

Kroin JS, Buvanendran A, Watts DE, Saha C, Tuman KJ (2006) Upregulation of cerebrospinal fluid and peripheral prostaglandin E2 in a rat postoperative pain model. Anesth Analg 103:334–343, table of contents.

Kroin JS, Takatori M, Li J, Chen EY, Buvanendran A, Tuman KJ (2008) Upregulation of dorsal horn microglial cyclooxygenase-1 and neuronal cyclooxygenase-2 after thoracic deep muscle incisions in the rat. Anesth Analg 106:1288–1295, table of contents.

Lambeau G, Lazdunski M (1999) Receptors for a growing family of secreted phospholipases A2. Trends Pharmacol Sci 20:162–170.

Larsson Forsell PK, Kennedy BP, Claesson HE (1999) The human calcium-independent phospholipase A2 gene multiple enzymes with distinct properties from a single gene. Eur J Biochem 262:575–585.

Li W, Xia J, Sun GY (1999) Cytokine induction of iNOS and sPLA2 in immortalized astrocytes (DITNC): response to genistein and pyrrolidine dithiocarbamate. J Interferon Cytokine Res 19:121–127.

Lin LL, Wartmann M, Lin AY, Knopf JL, Seth A, Davis RJ (1993) cPLA2 is phosphorylated and activated by MAP kinase. Cell 72:269–278.

Liu NK, Zhang YP, Titsworth WL, Jiang X, Han S, Lu PH, Shields CB, Xu XM (2006) A novel role of phospholipase A2 in mediating spinal cord secondary injury. Ann Neurol 59:606–619.

Lucas KK, Svensson CI, Hua XY, Yaksh TL, Dennis EA (2005) Spinal phospholipase A2 in inflammatory hyperalgesia: role of group IVA cPLA2. Br J Pharmacol 144:940–952.

MacPhee M, Chepenik KP, Liddell RA, Nelson KK, Siracusa LD, Buchberg AM (1995) The secretory phospholipase A2 gene is a candidate for the Mom1 locus, a major modifier of ApcMin-induced intestinal neoplasia. Cell 81:957–966.

Malmberg AB, Hamberger A, Hedner T (1995) Effects of prostaglandin E2 and capsaicin on behavior and cerebrospinal fluid amino acid concentrations of unanesthetized rats: a microdialysis study. J Neurochem 65:2185–2193.

Matsumura K, Watanabe Y, Onoe H, Watanabe Y (1995) Prostacyclin receptor in the brain and central terminals of the primary sensory neurons: an autoradiographic study using a stable prostacyclin analogue [3H]iloprost. Neuroscience 65:493–503.

Matsuzawa A, Murakami M, Atsumi G, Imai K, Prados P, Inoue K, Kudo I (1996) Release of secretory phospholipase A2 from rat neuronal cells and its possible function in the regulation of catecholamine secretion. Biochem J 318 (Pt 2):701–709.

Meves H (2006) The action of prostaglandins on ion channels. Curr Neuropharmacol 4:41–57.

Minami T, Uda R, Horiguchi S, Ito S, Hyodo M, Hayaishi O (1992) Allodynia evoked by intrathecal administration of prostaglandin F2 alpha to conscious mice. Pain 50:223–229.

Minami T, Uda R, Horiguchi S, Ito S, Hyodo M, Hayaishi O (1994) Allodynia evoked by intrathecal administration of prostaglandin E2 to conscious mice. Pain 57:217–223.

Minami T, Okuda-Ashitaka E, Mori H, Ito S, Hayaishi O (1996) Prostaglandin D2 inhibits prostaglandin E2-induced allodynia in conscious mice. J Pharmacol Exp Ther 278:1146–1152.

Minami T, Okuda-Ashitaka E, Hori Y, Sakuma S, Sugimoto T, Sakimura K, Mishina M, Ito S (1999) Involvement of primary afferent C-fibres in touch-evoked pain (allodynia) induced by prostaglandin E2. Eur J Neurosci 11:1849–1856.

Murakami M, Nakatani Y, Atsumi G, Inoue K, Kudo I (1997) Regulatory functions of phospholipase A2. Crit Rev Immunol 17:225–283.

Murakami M, Shimbara S, Kambe T, Kuwata H, Winstead MV, Tischfield JA, Kudo I (1998) The functions of five distinct mammalian phospholipase A2S in regulating arachidonic acid release. Type IIa and type V secretory phospholipase A2S are functionally redundant and act in concert with cytosolic phospholipase A2. J Biol Chem 273:14411–14423.

Muthalif MM, Hefner Y, Canaan S, Harper J, Zhou H, Parmentier JH, Aebersold R, Gelb MH, Malik KU (2001) Functional interaction of calcium-/calmodulin-dependent protein kinase II and cytosolic phospholipase A(2). J Biol Chem 276:39653–39660.

Nakayama Y, Omote K, Namiki A (2002) Role of prostaglandin receptor EP1 in the spinal dorsal horn in carrageenan-induced inflammatory pain. Anesthesiology 97:1254–1262.

Narumiya S, Sugimoto Y, Ushikubi F (1999) Prostanoid receptors: structures, properties, and functions. Physiol Rev 79:1193–1226.

Negishi M, Sugimoto Y, Ichikawa A (1995a) Molecular mechanisms of diverse actions of prostanoid receptors. Biochim Biophys Acta 1259:109–119.

Negishi M, Irie A, Sugimoto Y, Namba T, Ichikawa A (1995b) Selective coupling of prostaglandin E receptor EP3D to Gi and Gs through interaction of alpha-carboxylic acid of agonist and arginine residue of seventh transmembrane domain. J Biol Chem 270:16122–16127.

Nicol GD, Klingberg DK, Vasko MR (1992) Prostaglandin E2 increases calcium conductance and stimulates release of substance P in avian sensory neurons. J Neurosci 12: 1917–1927.

Ogino N, Miyamoto T, Yamamoto S, Hayaishi O (1977) Prostaglandin endoperoxide E isomerase from bovine vesicular gland microsomes, a glutathione-requiring enzyme. J Biol Chem 252:890–895.

Oida H, Namba T, Sugimoto Y, Ushikubi F, Ohishi H, Ichikawa A, Narumiya S (1995) In situ hybridization studies of prostacyclin receptor mRNA expression in various mouse organs. Br J Pharmacol 116:2828–2837.

Oka S, Arita H (1991) Inflammatory factors stimulate expression of group II phospholipase A2 in rat cultured astrocytes. Two distinct pathways of the gene expression. J Biol Chem 266:9956–9960.

Omote K, Yamamoto H, Kawamata T, Nakayama Y, Namiki A (2002) The effects of intrathecal administration of an antagonist for prostaglandin E receptor subtype EP(1) on mechanical and thermal hyperalgesia in a rat model of postoperative pain. Anesth Analg 95:1708–1712, table of contents.

Ong WY, Horrocks LA, Farooqui AA (1999) Immunocytochemical localization of cPLA2 in rat and monkey spinal cord. J Mol Neurosci 12:123–130.

O'Rielly DD, Loomis CW (2007) Spinal prostaglandins facilitate exaggerated A- and C-fiber-mediated reflex responses and are critical to the development of allodynia early after L5–L6 spinal nerve ligation. Anesthesiology 106:795–805.

Phillis JW, Horrocks LA, Farooqui AA (2006) Cyclooxygenases, lipoxygenases, and epoxygenases in CNS: their role and involvement in neurological disorders. Brain Res Rev 52:201–243.

Prochazkova M, Dolezal T, Sliva J, Krsiak M (2006) Different patterns of spinal cyclooxygenase-1 and cyclooxygenase-2 mRNA expression in inflammatory and postoperative pain. Basic Clin Pharmacol Toxicol 99:173–177.

Radmark O, Werz O, Steinhilber D, Samuelsson B (2007) 5-Lipoxygenase: regulation of expression and enzyme activity. Trends Biochem Sci 32:332–341.

Reinold H, Ahmadi S, Depner UB, Layh B, Heindl C, Hamza M, Pahl A, Brune K, Narumiya S, Muller U, Zeilhofer HU (2005) Spinal inflammatory hyperalgesia is mediated by prostaglandin E receptors of the EP2 subtype. J Clin Invest 115:673–679.

Reuben SS, Buvanendran A, Kroin JS, Steinberg RB (2006) Postoperative modulation of central nervous system prostaglandin E2 by cyclooxygenase inhibitors after vascular surgery. Anesthesiology 104:411–416.

Samad TA, Moore KA, Sapirstein A, Billet S, Allchorne A, Poole S, Bonventre JV, Woolf CJ (2001) Interleukin-1beta-mediated induction of Cox-2 in the CNS contributes to inflammatory pain hypersensitivity. Nature 410:471–475.

Schaloske RH, Dennis EA (2006) The phospholipase A2 superfamily and its group numbering system. Biochim Biophys Acta 1761:1246–1259.

Schevitz RW, Bach NJ, Carlson DG, Chirgadze NY, Clawson DK, Dillard RD, Draheim SE, Hartley LW, Jones ND, Mihelich ED, et al. (1995) Structure-based design of the first potent and selective inhibitor of human non-pancreatic secretory phospholipase A2. Nat Struct Biol 2:458–465.

Schievella AR, Regier MK, Smith WL, Lin LL (1995) Calcium-mediated translocation of cytosolic phospholipase A2 to the nuclear envelope and endoplasmic reticulum. J Biol Chem 270:30749–30754.

Shi L, Smolders I, Umbrain V, Lauwers MH, Sarre S, Michotte Y, Zizi M, Camu F (2006) Peripheral inflammation modifies the effect of intrathecal IL-1beta on spinal PGE2 production mainly through cyclooxygenase-2 activity. A spinal microdialysis study in freely moving rats. Pain 120:307–314.

Smith CJ, Zhang Y, Koboldt CM, Muhammad J, Zweifel BS, Shaffer A, Talley JJ, Masferrer JL, Seibert K, Isakson PC (1998) Pharmacological analysis of cyclooxygenase-1 in inflammation. Proc Natl Acad Sci USA 95:13313–13318.

Smith WL, DeWitt DL, Garavito RM (2000) Cyclooxygenases: structural, cellular, and molecular biology. Annu Rev Biochem 69:145–182.

Snitko Y, Yoon ET, Cho W (1997) High specificity of human secretory class II phospholipase A2 for phosphatidic acid. Biochem J 321 (Pt 3):737–741.

Sun GY, Xu J, Jensen MD, Simonyi A (2004) Phospholipase A2 in the central nervous system: implications for neurodegenerative diseases. J Lipid Res 45:205–213.

Sung B, Lim G, Mao J (2003) Altered expression and uptake activity of spinal glutamate transporters after nerve injury contribute to the pathogenesis of neuropathic pain in rats. J Neurosci 23:2899–2910.

Svensson CI, Yaksh TL (2002) The spinal phospholipase-cyclooxygenase-prostanoid cascade in nociceptive processing. Annu Rev Pharmacol Toxicol 42:553–583.

Svensson CI, Hua XY, Protter AA, Powell HC, Yaksh TL (2003a) Spinal p38 MAP kinase is necessary for NMDA-induced spinal PGE(2) release and thermal hyperalgesia. Neuroreport 14:1153–1157.

Svensson CI, Marsala M, Westerlund A, Calcutt NA, Campana WM, Freshwater JD, Catalano R, Feng Y, Protter AA, Scott B, Yaksh TL (2003b) Activation of p38 mitogen-activated protein kinase in spinal microglia is a critical link in inflammation-induced spinal pain processing. J Neurochem 86:1534–1544.

Svensson CI, Lucas KK, Hua XY, Powell HC, Dennis EA, Yaksh TL (2005a) Spinal phospholipase A2 in inflammatory hyperalgesia: role of the small, secretory phospholipase A2. Neuroscience 133:543–553.

Svensson CI, Hua XY, Powell HC, Lai J, Porreca F, Yaksh TL (2005b) Prostaglandin E2 release evoked by intrathecal dynorphin is dependent on spinal p38 mitogen activated protein kinase. Neuropeptides 39:485–494.

Telleria-Diaz A, Ebersberger A, Vasquez E, Schache F, Kahlenbach J, Schaible HG (2008) Different effects of spinally applied prostaglandin D(2) on responses of dorsal horn neurons with knee input in normal rats and in rats with acute knee inflammation. Neuroscience.

Trebino CE, Stock JL, Gibbons CP, Naiman BM, Wachtmann TS, Umland JP, Pandher K, Lapointe JM, Saha S, Roach ML, Carter D, Thomas NA, Durtschi BA, McNeish JD, Hambor JE, Jakobsson PJ, Carty TJ, Perez JR, Audoly LP (2003) Impaired inflammatory and pain responses in mice lacking an inducible prostaglandin E synthase. Proc Natl Acad Sci USA 100:9044–9049.

Tsuda M, Hasegawa S, Inoue K (2007) P2X receptors-mediated cytosolic phospholipase A2 activation in primary afferent sensory neurons contributes to neuropathic pain. J Neurochem 103:1408–1416.

Turnbach ME, Spraggins DS, Randich A (2002) Spinal administration of prostaglandin E(2) or prostaglandin F(2alpha) primarily produces mechanical hyperalgesia that is mediated by nociceptive specific spinal dorsal horn neurons. Pain 97:33–45.

Uda R, Horiguchi S, Ito S, Hyodo M, Hayaishi O (1990) Nociceptive effects induced by intrathecal administration of prostaglandin D2, E2, or F2 alpha to conscious mice. Brain Res 510:26–32.

Valentin E, Lambeau G (2000) Increasing molecular diversity of secreted phospholipases A(2) and their receptors and binding proteins. Biochim Biophys Acta 1488:59–70.

Vane JR (1971) Inhibition of prostaglandin synthesis as a mechanism of action for aspirin-like drugs. Nat New Biol 231:232–235.

Wang C, Gu Y, Li GW, Huang LY (2007) A critical role of the cAMP sensor Epac in switching protein kinase signalling in prostaglandin E2-induced potentiation of P2X3 receptor currents in inflamed rats. J Physiol 584:191–203.

Wei S, Ong WY, Thwin MM, Fong CW, Farooqui AA, Gopalakrishnakone P, Hong W (2003) Group IIA secretory phospholipase A2 stimulates exocytosis and neurotransmitter release in pheochromocytoma-12 cells and cultured rat hippocampal neurons. Neuroscience 121:891–898.

Winstead MV, Balsinde J, Dennis EA (2000) Calcium-independent phospholipase A(2): structure and function. Biochim Biophys Acta 1488:28–39.

Wright DH, Nantel F, Metters KM, Ford-Hutchinson AW (1999) A novel biological role for prostaglandin D2 is suggested by distribution studies of the rat DP prostanoid receptor. Eur J Pharmacol 377:101–115.

Xu J, Chalimoniuk M, Shu Y, Simonyi A, Sun AY, Gonzalez FA, Weisman GA, Wood WG, Sun GY (2003) Prostaglandin E2 production in astrocytes: regulation by cytokines, extracellular ATP, and oxidative agents. Prostaglandins Leukot Essent Fatty Acids 69:437–448.

Yaksh TL, Kokotos G, Svensson CI, Stephens D, Kokotos CG, Fitzsimmons B, Hadjipavlou-Litina D, Hua XY, Dennis EA (2006) Systemic and intrathecal effects of a novel series of phospholipase A2 inhibitors on hyperalgesia and spinal prostaglandin E2 release. J Pharmacol Exp Ther 316:466–475.

Yang LC, Marsala M, Yaksh TL (1996a) Effect of spinal kainic acid receptor activation on spinal amino acid and prostaglandin E2 release in rat. Neuroscience 75:453–461.

Yang LC, Marsala M, Yaksh TL (1996b) Characterization of time course of spinal amino acids, citrulline and PGE2 release after carrageenan/kaolin-induced knee joint inflammation: a chronic microdialysis study. Pain 67:345–354.

Yeo JF, Ong WY, Ling SF, Farooqui AA (2004) Intracerebroventricular injection of phospholipases A2 inhibitors modulates allodynia after facial carrageenan injection in mice. Pain 112:148–155.

Zhu X, Eisenach JC (2003) Cyclooxygenase-1 in the spinal cord is altered after peripheral nerve injury. Anesthesiology 99:1175–1179.

Chapter 20
MAP Kinase and Cell Signaling in DRG Neurons and Spinal Microglia in Neuropathic Pain

Ru-Rong Ji

Abstract Nerve injury is known to produce neuropathic pain by inducing changes not only in neurons such as primary sensory neurons in the dorsal root ganglion (DRG), but also in non-neuronal cells such as microglia in the spinal cord. Increasing evidence suggests that mitogen-activated protein kinases (MAPKs) play important roles in neuropathic pain sensitization by regulating intracellular signaling in both DRG neurons and spinal cord microglia. Intrathecal injection of MAPK inhibitors for the extracellular signal-regulated kinase (ERK), p38, or c-Jun N-terminal kinase (JNK) pathway targets the MAPK pathways at both DRG and spinal cord levels and has been shown to attenuate neuropathic pain in different animal models. In particular, activation of p38 in DRG neurons by nerve growth factor and cytokines contributes to thermal hypersensitivity by increasing the expression and activity of sodium channels (e.g., Nav1.7/Nav1.8) and TRP channels (e.g., TRPV1 and TRPA1). Activation of p38 in spinal microglia by chemokines, cytokines, ATP, and proteases also contributes to neuropathic pain symptoms such as mechanical allodynia. Thus, activation of MAPK pathways in both neurons and glia and in both the peripheral and central nervous system is important for neuropathic pain sensitization, and blocking these pathways at multiple sites may lead to effective therapies for neuropathic pain.

Abbreviations

AMPA	α-amino-2,3-dihydro-5-methyl-3-oxo-4-isoxazolepropanic acid
BDNF	brain-derived neurotrophic factor
CatS	cysteine protease cathepsin S
DRG	dorsal root ganglion
GABA	gamma aminobutyric acid

R.-R. Ji (✉)
Department of Anesthesiology, Pain Research Center, Brigham and Women's
Hospital and Harvard Medical School, Boston, MA 02115, USA
e-mail: rrji@zeus.bwh.harvard.edu

M. Malcangio (ed.), *Synaptic Plasticity in Pain*,
DOI 10.1007/978-1-4419-0226-9_20, © Springer Science+Business Media, LLC 2009

ERK   extracelllular signal-regulated kinase
bFGF   basic fibroblast growth factor
FKN   fractalkine
IL-1   Interleukin-1
JNK   c-Jun-N-terminal kinase
MAPK   mitogen-activated protein kinases
MCP-1   monocyte chemoattractant protein-1
NMDA N-methyl-D-aspartate
MMP-9   matrix metalloproteinase-9
NGF   nerve growth factor
NT-3   neurotrophin 3
PGE2   prostaglandin E2
PTN   pain transmission neurons STZ, streptozotocin
TNF   tumor necrosis factor
TRP   transient receptor potential

20.1 Introduction

20.1.1 Peripheral and Central Sensitization After Nerve Injury

Nerve injury which often results in neuropathic pain is caused by diabetic neuropathy, viral infection, chemotherapy, and major surgeries such as amputation and thoracotomy (Woolf and Mannion, 1999; Ji and Strichartz, 2004; and Kehlet et al., 2006). In patients, neuropathic pain is characterized by spontaneous pain, such as shooting, lancinating or burning pain, and thermal hypersensitivity such as heat hyperalgesia and cold allodynia. In particular, neuropathic pain manifests as mechanical allodynia (painful responses to normally innocuous tactile stimuli) so that patients feel very painful during movement. It is generally believed that neuropathic pain results from neural plasticity, leading to a sensitization both in the peripheral nervous system (peripheral sensitization) and central nervous system (central sensitization). After nerve injury, Schwann cells, immune cells, and damaged neurons will release multiple inflammatory mediators such as proinflammatory cytokines tumor necrosis factor-alpha (TNF-α) and interlukin-1 beta (IL-1β) (Ji and Strichartz, 2004). These mediators will stimulate the axons and cell bodies of primary sensory neurons to generate ectopic or spontaneous activity that can drive neuropathic pain (Devor, 1991). Peripheral nerve injury also induces marked phenotypic changes in primary sensory neurons in the dorsal root ganglion (DRG) and many of these changes may promote neuropathic pain (Hokfelt et al., 1994; Costigan et al., 2002; Xiao et al., 2002). Spontaneous activity from primary afferents will also induce central sensitization, which is responsible for the persistence of neuropathic pain and spread of this pain beyond injury site, such as contralateral pain. It is well established that increase in excitatory synaptic transmission (e.g., AMPA and NMDA currents) in spinal cord dorsal horn neurons contributes to central sensitization (see Randic, Chapter 10). However, recent studies have also revealed an essential role of disinhibition (loss of inhibitory synaptic transmission such as GABA transmission) in central sensitization and neuropathic pain (Moore et al., 2002; Coull et al., 2003). Central sensitization is not only driven by primary afferent input but also enhanced by descending facilitation from the brain stem and cortex (Porreca et al., 2002) (see Bee and Dickenson, Chapter 19).

20.1.2 Neuronal-Glial Interaction and Central Sensitization

Recent progress in pain research has indicated that central sensitization is further enhanced and maintained by glial cells in the spinal cord (DeLeo and Yezierski, 2001; Watkins et al., 2001; Tsuda et al., 2005; Suter et al., 2007) (see Abbadie and Sullivan, Chapter 15; Milligan et al., Chapter 16; Tawfik and DeLeo, Chapter 17; Clark and Malcangio, Chapter 23). For a long time, glial cells were only regarded as supporting cells. A growing body of evidence indicates that nerve injury produces substantial changes in spinal cord glial cells (such as microglia and astrocytes). For example, peripheral nerve injury produces profound upregulation of the microglia glial markers CD11b (OX-42) and Iba-1 and the astrocyte marker GFAP. The activated glial cells in the spinal cord play an important role in neuropathic pain facilitation by communicating with pain transmission neurons (PTN) in the dorsal horn. In particular, these cells release several pronociceptive mediators such as TNF-α and IL-1β, prostaglandin E_2 (PGE_2), nitric oxide, and growth factors such as brain-derived growth factor (BDNF). These mediators induce central sensitization by increasing excitation (Kawasaki et al., 2008) and decreasing inhibition in dorsal horn neurons (Coull et al., 2005; Kawasaki et al., 2008) (see Beggs, Chapter 22). Several glia modifying drugs such as fluorocitrate, propentofylline, and minocycline can alter pain sensitivity (Watkins et al., 1997; Sweitzer et al., 2001; Raghavendra et al., 2003). In particular, the role of microglia in neuropathic pain has been intensively studied, because nerve injury induces substantial morphological changes and proliferation of microglia and upregulation of the ATP receptors (e.g., P2X4 and P2Y12) and the chemokine receptors (e.g., CX3CR1) specifically in spinal microglia (Tsuda et al., 2003; Suter et al., 2007; Zhuang et al., 2007; Kobayashi et al., 2008).

20.1.3 MAP Kinases and Peripheral and Central Sensitization

The mitogen-activated protein kinase (MAPK) family consists of three major members: extracellular signal-regulated kinase (ERK), p38, and c-Jun

N-terminal kinase (JNK), representing three separate signaling pathways, although ERK5 is regarded as a new family member. There are also different isoforms, such as ERK1/2, p38α/β/γ/δ, and JNK1/2/3. MAPKs play critical roles in intracellular signaling both in neurons and non-neuronal cells. These kinases are activated by phosphorylation and transduce a broad range of extracellular stimuli into diverse intracellular responses by transcriptional and non-transcriptional regulation such as translational and post-translational regulation (Ji et al., 2007). Accumulating evidence demonstrates that all three MAPK pathways contribute to neuropathic pain sensitization (Ji et al., 2007). However, the activation patterns of these MAPKs after nerve injury are quite different in the DRG and spinal cord. In the DRG, ERK is activated in non-neuronal cells (satellite cells, Zhuang et al., 2005) and JNK is activated in injured neurons (Obata et al., 2004; Zhuang et al., 2006), whereas p38 is activated in both injured and non-injured neurons (Jin et al., 2003; Obata et al., 2004). In the spinal cord, p38 is activated in microglia (Jin et al., 2003; Tsuda et al., 2004), whereas JNK is activated in astrocytes (Zhuang et al., 2006). Further, there is sequential activation of ERK from neurons, to microglia, and finally to astrocytes (Zhuang et al., 2005). To limit the scale of this chapter, I will focus on p38 MAPK and cell signaling in DRG neurons and spinal cord microglia after nerve injury.

20.2 p38 MAP Kinase and Cell Signaling in DRG Neurons in Neuropathic Pain

After peripheral nerve injury, not only the injured neurons but also the adjacent intact neurons play important role in the development of neuropathic pain (Campbell and Meyer, 2006). Ligation of the L5 spinal nerve provides an ideal model to study the distinct roles of injured neurons (L5-DRG neurons) and intact neurons (L4-DRG neurons) in neuropathic pain. Spontaneous discharge, which is thought to underlie the genesis of neuropathic pain, is found in both injured neurons and intact neurons. Interestingly, spontaneous discharge is mainly observed in large A-fiber neurons in the injured L5 DRG but found in small C-fiber neurons in the intact L4 DRG (Devor et al., 1992; Wu et al., 2001; Ma et al., 2003; Campbell and Meyer, 2006). A comparison between the spinal nerve ligation (SNL) neuropathic pain model and complete Freund's adjuvant (CFA) inflammatory pain model has revealed that the intact L4 DRG neurons in the SNL model share the same features as L4/L5 DRG neurons in the CFA model. First, they both exhibit spontaneous discharge in C-fiber neurons (Wu et al., 2001; Djouhri et al., 2006). Second, they both require nerve growth factor (NGF) as a retrograde signaling molecule to induce the expression of several pain mediators such as BDNF and transient reverse potential subtype V1 (TRPV1) and A1 (TRPA1) (Fukuoka et al., 2001; Ji et al., 2002; Obata et al., 2004, 2005).

20.2.1 p38 Activation in Intact DRG Neurons After Nerve Injury

The active form of p38, p-p38 in normally expressed in about 10% DRG neurons that have small sizes (NF-200-negative, Ji et al., 2002; Fig. 20.1a). SNL at L5 induces a robust increase in p-p38 levels in intact neurons of the L4 DRG (Obata et al., 2004). Like the CFA model, this p38 activation (p-p38 increase) occurs in small C-fiber neurons and requires NGF (Ji et al., 2002; Obata et al., 2004). p38 activation in intact C-fiber neurons leads increased

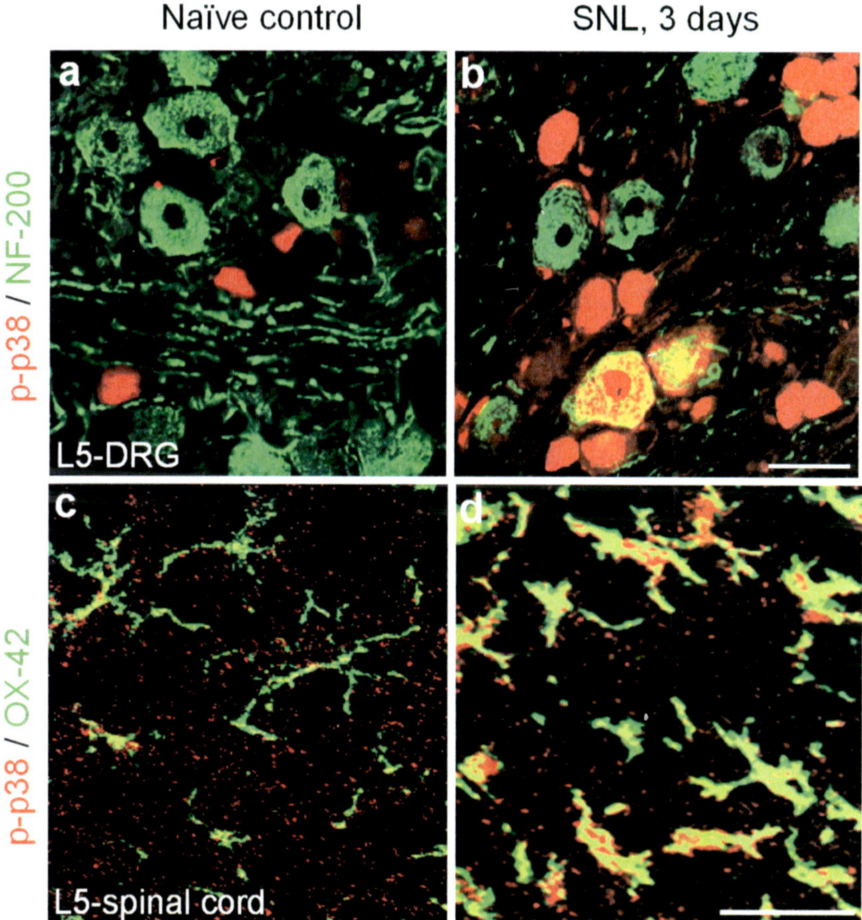

Fig. 20.1 Ligation of L5 spinal nerve (SNL) induces p38 activation in both DRG neurons and spinal cord microglial cells. (**a, b**) Double immunostaining of phosphorylated p38 (p-p38, *red*) with the myelinated A-fiber marker NF-200 (*green*) in the L5 DRG from a naïve control rat (**a**) and an injured rat with SNL at 3 days (**b**). (**c, d**) Double immunostaining of p-p38 (*red*) and the microglial marker OX-42 (*green*) in the spinal cord dorsal horn from a naïve control rat (**c**) and an injured rat with SNL at 3 days (**d**). Scales, 50 μm

expression of TRPV1, TRPA1, and BDNF (Ji et al., 2002; Obata et al., 2004, 2005). Moreover, intrathecal injection of p38 inhibitor can attenuate CFA- or SNL-induced heat and cold hypersensitivity that is mediated by TRPV1 and TRPA1, respectively (Ji et al., 2002; Obata et al., 2005). Consistently, intrathecal infusion of neurotrophin-3 (NT-3) reduces TRPV1 expression and heat hyperalgesia in the chronic constriction injury model by suppressing TrkA signaling and p38 activation (Wilson-Gerwing et al., 2005). In addition to L5 SNL, L5 ventral root transection also induces neuropathic pain symptoms and p38 activation in intact L5 DRG neurons, and this activation is prevented by thalidomide, an inhibitor of TNF-α synthesis inhibitor (Xu et al., 2007).

p38 activation can increase the sensitivity and excitability of DRG neurons via transcriptional and non-transcriptional regulation, and the latter includes post-transcriptional regulation (increasing the stability of mRNAs), translational regulation (increasing protein but not mRNA levels), and post-translational regulation (increasing the activity of target proteins within minutes) (Ji and Woolf, 2001). NGF can persistently increase TRPV1 protein levels via translational regulation (Ji et al., 2002). TNF-α may also transiently increase TRPV1 protein levels via the same mechanism (Constantin et al., 2008).

In particular, after SNL TNF-α induces immediate spontaneous discharge in both injured and intact DRG neurons (Schafers et al., 2003a). This fast post-translational regulation is likely to be mediated by p38 potentiation of the tetrodotoxin-resistant sodium channels (TTX-R Na +) $Na_V1.8$ (Jin and Gereau, 2006). Thus, acute application of TNF-α rapidly enhances TTX-R Na + currents in isolated DRG neurons, and this potentiation is blocked by a p38 inhibitor (Jin and Gereau, 2006). Importantly, p38 can directly phosphorylate $Na_V1.8$ at two serine residues on the L1 loop (S551 and S556), which causes an increase in $Na_v1.8$ current density without changing the gating properties of the channel (Hudmon et al., 2008). Another example of possible post-translational regulation is p38-mediated release of the neuropeptide substance P in cultured DRG neurons (Tang et al., 2007).

20.2.2 p38 Activation in Injured DRG Neurons After Nerve Injury

After nerve injury, p38 is also activated in injured L5 DRG neurons (Fig. 20.1b, Jin et al., 2003; Obata et al., 2004). While p38 is primarily activated in small size neurons in the intact L4 DRG, p38 is activated in both NF-200-negative (small size) and NF-200-positive (medium to large size) neurons in the L5-DRG (Fig. 20.1a,b, Jin et al., 2003; Obata et al., 2004). What is causing p38 activation in injured neurons? NGF is unlikely to cause p38 activation in these neurons, because injured neurons are deprived of NGF. However, cellular stress after axonal injury may cause p38 activation in injured neurons (reviewed in Ji and Woolf, 2001). Further, TNF-α appears to be responsible for an initial activation of p38 in injured neurons (Schafers et al., 2003b). Notably, basic fibroblast growth factor (bFGF) is persistently upregulated in injured DRG neurons (Ji et al., 1995) and responsible for persistent p38 activation in these neurons via an autocrine or paracrine mechanism (Yamanaka et al., 2007). It is of great interest that TRPV1 is downregulated in small size neurons but upregulated in medium to large size neurons in the L5 DRG after SNL (Ma et al., 2005). Thus p38 activation may also induce TRPV1 expression in injured myelinated DRG neurons. It is intriguing to propose that TRPV1 upregulation in these injured neurons may participate in the genesis of spontaneous activity.

Streptozotocin (STZ) is an agent extensively used to induce diabetes and diabetic peripheral neuropathy. Interestingly, STZ has a direct effect on DRG neurons by producing cellular stress and p38 activation. It modulates the expression of TRPV1 via p38 (Pabbidi et al., 2008). In a rodent model of diabetic neuropathy, increased expression of the sodium channel $Na_V1.7$ in DRG neurons is also blocked a p38 inhibitor (Chattopadhyay et al., 2008). Moreover, incubation of primary DRG neurons in vitro with glucose (45 mM for 18 hours) demonstrated increased p38 activation and $Na_V1.7$ expression (Chattopadhyay et al., 2008). A single mutation in this sodium channel can produce hyper- or hypoexcitability in different types of neurons (Rush et al., 2006).

Microarray studies have shown that nerve injury changes the expression of several hundred genes in the injured DRG neurons (Costigan et al., 2002; Xiao et al., 2002). It remains to be investigated how many of the upregulated genes are controlled by p38.

20.3 p38 MAP Kinase and Cell Signaling in Spinal Microglia in Neuropathic Pain

20.3.1 p38 Activation in Spinal Cord Microglia and Neuropathic Pain

Nerve injury activates p38 not only in the DRG but also in the spinal cord. Western blot analysis shows an increase in p-p38 levels in the spinal cord dorsal horn after SNL (Jin et al., 2003). Further, this increase in p38 phosphorylation is accompanied by an increase in p38 activity, as p-38 leads to elevated phosphorylation of its substrate, the transcription factor ATF-2 (Ji and Suter, 2007). Immunofluorescence shows that p38 is particularly activated in spinal cord microglial cells that labeled with CD11b (recognized by OX-42 antibody, Fig. 20.1c,d). In contrast, neurons (NeuN+) or astrocytes (GFAP+) do not express significant amount of p-p38 after nerve injury (Jin et al., 2003; Tsuda et al., 2004; Wen et al., 2007). Apart from nerve injury, p38 is also activated in spinal microglia after spinal cord injury (Hains and Waxman, 2006). It is worthwhile to point out that p-p38 antibody can recognize all the isoforms of p38 (α, β, γ and δ). However, α and β are the major isoforms in the mature nervous system (Kumar et al., 2003). It appears that p38 β may be important for p38 activation in spinal cord microglia (Svensson et al., 2005a). Importantly, intrathecal infusion of p38 inhibitor SB203580 or FR167653 has been shown to prevent and reverse neuropathic pain symptoms in different animal models (Jin et al., 2003; Obata et al., 2004; Schaefers et al., 2003b; Tsuda et al., 2004; Hains and Waxman, 2006; Wen et al., 2007), supporting a role of microglial p38 in neuropathic pain sensitization. In particular, minocycline, an inhibitor of microglia, inhibits spinal p38 activation in a neuropathic pain condition (Hains and Waxman, 2006).

20.3.2 p38 and Spinal Cord Microglial Signaling After Nerve Injury

The first question is how signal is conveyed, after peripheral nerve injury, to spinal cord microglia, leading to p38 activation in these cells. As discussed above, nerve injury-induced spontaneous activity in DRG neurons is crucial for the genesis of neuropathic pain (Devor et al., 1992). Spontaneous activity is also important for p38 activation in spinal microglia. Long-term blockade of nerve activity in the sciatic nerve for 3 days by bupivacaine microspheres prevents but does not reverse spinal cord p38 activation in the spared nerve injury model, suggesting that spontaneous activity is required for the induction but not maintenance of p38 activation in spinal microglia (Wen et al., 2007). Spontaneous activity can release several pain mediators such as ATP, cytokines, chemokines, and proteases from primary afferents, which can activate p38 in spinal microglia (Fig. 20.2). For example, the ATP receptor P2X4 and P2Y12 are induced in spinal microglia after nerve injury. Both type of ATP receptors are required for microglial p38 activation and neuropathic pain development (Tsuda et al., 2003; Kobayashi et al., 2008; Trang et al., 2009). The proinflamamtory cytokines TNF-α and IL-1β are also important for p38 activation in the spinal cord (Sung et al., 2005; Svensson et al., 2005b).

In particular, chemokines such as monocyte chemoattractant protein-1 (MCP-1) and fractalkine (CX3CL1) play an important role in regulating neural-glial interaction in the spinal cord. Nerve injury induces MCP-1 upregulation in injured DRG neurons, which causes microglia activation in the spinal cord (White et al., 2007; Zhang et al., 2007). Further, spinal p38 activation and neuropathic pain after nerve injury is reduced in mice that lack MCP-1 receptor CCR-2 (Abbadie et al., 2003). Nerve injury also upregulates the fractalkine receptor CX3CR1 in spinal microglia; and blocking CX3CR1 inhibits nerve injury-induced p38 activation in spinal microglia and neuropathic pain (Zhuang et al., 2007). Conversely, spinal infusion of fractalkine induces p38 activation in spinal microglia and mechanical allodynia. Thus, the fractalkine/CX3CR1/p38 pathway appears to be critical for the development of neuropathic pain (Zhuang et al., 2007).

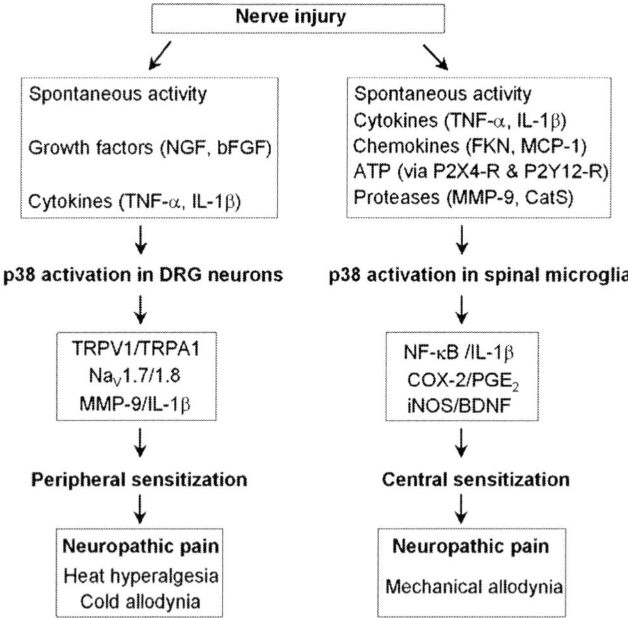

Fig. 20.2 Schematic of p38-mediated signal transduction in DRG neurons and spinal microglia after nerve injury. Abbreviations: BDNF, brain-derived neurotrophic factor; bFGF, basic fibroblast growth factor; CatS, cysteine protease cathepsin S; FKN, fractalkine; MCP-1, monocyte chemoattractant protein-1; MMP-9, matrix metalloproteinase-9; R, receptor

Recent studies have also demonstrated an important role of proteases in p38 activation and neuropathic pain. For example, the cysteine protease cathepsin S is induced in spinal microglia after nerve injury and promotes neuropathic pain by inducing cleavage of fractalkine from the membrane of DRG neurons, leading to p38 activation in spinal microglia (Clark et al., 2007) (see Clark and Malcangio, Chapter 23). Very recently, we have shown that the matrix metalloprotease-9 (MMP-9) is rapidly upregulated in injured DRG neurons after SNL. MMP-9 appears to induce neuropathic pain by active cleavage of IL-1β and activating p38 in spinal microglia (Kawasaki et al., 2008).

The second question is how p38 activation in spinal microglia results in pain hypersensitivity. p38 activation is known to activate the transcription factor NF-κB in cultured microglia, leading to the expression of the IL-1β, IL-6, and COX-2 (Fig. 20.2). For example, intrathecal injection of the p38 inhibitor SB203580 reduces nerve injury-induced upregulation of IL-1β (Ji and Suter, 2007). A recently study shows that p38 activation by ATP and P2X4 receptor in microglia further leads to the synthesis and release of BDNF (Trang et al., 2009). IL-1β and BDNF contribute importantly to the generation of central sensitization. Application of IL-1β to spinal neurons not only enhances AMPA and NMDA receptor-mediated excitatory synaptic transmission but also suppresses GABA and glycine receptor-mediated inhibitory synaptic transmission (Kawasaki et al., 2008). BDNF was shown to suppress inhibitory GABA transmission in dorsal horn neuron (Coull et al., 2005). p38 activation in spinal cord by intrathecal IL-1β is also required for the synthesis of inducible nitric oxide synthase (iNOS, Sung et al., 2005). p38 activation in microglia may also regulate the synthesis of inflammatory mediators by posttranslational regulation. For example, p38 may activate phospholipase A2 to produce PGE_2 or cause a rapid release of IL-1β (Clark et al., 2006; Ji and Suter, 2007).

20.4 Concluding Remarks

In summary, p38 is activated in both DRG neurons and spinal cord microglia in different neuropathic pain conditions and is critical for intracellular signaling in both neuronal and non-neuronal cells. The phosphorylated p38 (p-p38) may serve as a useful marker to reflect an activation status of these cells after nerve injury. Most studies have administrated p38 inhibitors via intrathecal route, which can affect both DRG cells and spinal cord cells (Ji et al., 2007). Therefore, the effects of intrathecal p38 inhibitors on neuropathic pain should be contributed by p38 activation in both DRG neurons and spinal microglia. However, p38 activation in DRG neurons and microglia may contribute to different features of neuropathic pain: p38 activation in C-fiber DRG neurons may be important for the development of heat and cold hypersensitivity, whereas p38 activation in spinal microglia may be more important for the development of mechanical allodynia (Fig. 20.2). Taken together, p38 inhibitors acting both peripherally and centrally will be more effective to alleviate neuropathic pain.

Acknowledgments The work was supported in part by NIH grants NS40698, DE17794, and TW7180.

References

Abbadie, C., Lindia, J.A., Cumiskey, A.M., Peterson, L.B., Mudgett, J.S., Bayne, E.K., DeMartino, J.A., MacIntyre, D.E., and Forrest, M.J. (2003). Impaired neuropathic pain responses in mice lacking the chemokine receptor CCR2. Proc. Natl. Acad. Sci. U.S.A. *100*, 7947–7952.

Campbell, J.N. and Meyer, R.A. (2006). Mechanisms of neuropathic pain. Neuron *52*, 77–92.

Chattopadhyay, M., Mata, M., and Fink, D.J. (2008). Continuous delta-opioid receptor activation reduces neuronal voltage-gated sodium channel (NaV1.7) levels through activation of protein kinase C in painful diabetic neuropathy. J. Neurosci. *28*, 6652–6658.

Clark, A.K., D'Aquisto, F., Gentry, C., Marchand, F., McMahon, S.B., and Malcangio, M. (2006). Rapid co-release of interleukin 1beta and caspase 1 in spinal cord inflammation. J. Neurochem. *99*, 868–880.

Clark, A.K., Yip, P.K., Grist, J., Gentry, C., Staniland, A.A., Marchand, F., Dehvari, M., Wotherspoon, G., Winter, J., Ullah, J., Bevan, S., and Malcangio, M. (2007). Inhibition of spinal microglial cathepsin S for the reversal of neuropathic pain. Proc. Natl. Acad. Sci. U. S.A. *104*, 10655–10660.

Constantin, C.E., Mair, N., Sailer, C.A., Andratsch, M., Xu, Z.Z., Blumer, M.J., Scherbakov, N., Davis, J.B., Bluethmann, H., Ji, R.R., and Kress, M. (2008). Endogenous tumor necrosis factor alpha (TNFalpha) requires TNF receptor type 2 to generate heat hyperalgesia in a mouse cancer model. J. Neurosci. *28*, 5072–5081.

Costigan, M., Befort, K., Karchewski, L., Griffin, R.S., D'Urso, D., Allchorne, A., Sitarski, J., Mannion, J.W., Pratt, R.E., and Woolf, C.J. (2002). Replicate high-density rat genome oligonucleotide microarrays reveal hundreds of regulated genes in the dorsal root ganglion after peripheral nerve injury. BMC. Neurosci. *3*, 16.

Coull, J.A., Beggs, S., Boudreau, D., Boivin, D., Tsuda, M., Inoue, K., Gravel, C., Salter, M.W., and De Koninck, Y. (2005). BDNF from microglia causes the shift in neuronal anion gradient underlying neuropathic pain. Nature *438*, 1017–1021.

Coull, J.A., Boudreau, D., Bachand, K., Prescott, S.A., Nault, F., Sik, A., De Koninck, P., and De Koninck, Y. (2003). Trans-synaptic shift in anion gradient in spinal lamina I neurons as a mechanism of neuropathic pain. Nature *424*, 938–942.

DeLeo, J.A. and Yezierski, R.P. (2001). The role of neuroinflammation and neuroimmune activation in persistent pain. Pain *90*, 1–6.

Devor, M. (1991). Neuropathic pain and injured nerve: peripheral mechanisms. Br. Med. Bull. 47, 619–630.

Devor, M., Wall, P.D., and Catalan, N. (1992). Systemic lidocaine silences ectopic neuroma and DRG discharge without blocking nerve conduction. Pain *48*, 261–268.

Djouhri, L., Koutsikou, S., Fang, X., McMullan, S., and Lawson, S.N. (2006). Spontaneous pain, both neuropathic and inflammatory, is related to frequency of spontaneous firing in intact C-fiber nociceptors. J. Neurosci. *26*, 1281–1292.

Fukuoka, T., Kondo, E., Dai, Y., Hashimoto, N., and Noguchi, K. (2001). Brain-derived neurotrophic factor increases in the uninjured dorsal root ganglion neurons in selective spinal nerve ligation model. J. Neurosci. *21*, 4891–4900.

Hains, B.C. and Waxman, S.G. (2006). Activated microglia contribute to the maintenance of chronic pain after spinal cord injury. J. Neurosci. *26*, 4308–4317.

Hokfelt, T., Zhang, X., and Wiesenfeld-Hallin, Z. (1994). Messenger plasticity in primary sensory neurons following axotomy and its functional implications. Trends Neurosci. *17*, 22–30.

Hudmon, A., Choi, J.S., Tyrrell, L., Black, J.A., Rush, A.M., Waxman, S.G., and Dib-Hajj, S.D. (2008). Phosphorylation of sodium channel Na(v)1.8 by p38 mitogen-activated protein kinase increases current density in dorsal root ganglion neurons. J. Neurosci. *28*, 3190–3201.

Ji, R.R., Kawasaki, Y., Zhuang, Z.Y., Wen, Y.R., and Zhang, Y.Q. (2007). Protein kinases as potential targets for the treatment of pathological pain. Handb. Exp. Pharmacol. 359–389.

Ji, R.R., Samad, T.A., Jin, S.X., Schmoll, R., and Woolf, C.J. (2002). p38 MAPK activation by NGF in primary sensory neurons after inflammation increases TRPV1 levels and maintains heat hyperalgesia. Neuron *36*, 57–68.

Ji, R.R. and Strichartz, G. (2004). Cell signaling and the genesis of neuropathic pain. Sci. STKE. *2004*, reE14.

Ji, R.R. and Suter, M.R. (2007). p38 MAPK, microglial signaling, and neuropathic pain. Mol. Pain *3*, 33.

Ji, R.R. and Woolf, C.J. (2001). Neuronal plasticity and signal transduction in nociceptive neurons: implications for the initiation and maintenance of pathological pain. Neurobiol. Dis. *8*, 1–10.

Ji, R.R., Zhang, Q., Zhang, X., Piehl, F., Reilly, T., Pettersson, R.F., and Hokfelt, T. (1995). Prominent expression of bFGF in dorsal root ganglia after axotomy. Eur. J. Neurosci. *7*, 2458–2468.

Jin, X. and Gereau, R.W. (2006). Acute p38-mediated modulation of tetrodotoxin-resistant sodium channels in mouse sensory neurons by tumor necrosis factor-alpha. J. Neurosci. *26*, 246–255.

Jin, S.X., Zhuang, Z.Y., Woolf, C.J., and Ji, R.R. (2003). p38 mitogen-activated protein kinase is activated after a spinal nerve ligation in spinal cord microglia and dorsal root ganglion neurons and contributes to the generation of neuropathic pain. J. Neurosci. *23*, 4017–4022.

Kawasaki, Y., Zhang, L., Cheng, J.K., and Ji, R.R. (2008). Cytokine mechanisms of central sensitization: distinct and overlapping role of interleukin-1beta, interleukin-6, and tumor necrosis factor-alpha in regulating synaptic and neuronal activity in the superficial spinal cord. J. Neurosci. *28*, 5189–5194.

Kehlet, H., Jensen, T.S., and Woolf, C.J. (2006). Persistent postsurgical pain: risk factors and prevention. Lancet *367*, 1618–1625.

Kobayashi, K., Yamanaka, H., Fukuoka, T., Dai, Y., Obata, K., and Noguchi, K. (2008). P2Y12 receptor upregulation in activated microglia is a gateway of p38 signaling and neuropathic pain. J. Neurosci. *28*, 2892–2902.

Kumar, S., Boehm, J., and Lee, J.C. (2003). p38 MAP kinases: key signalling molecules as therapeutic targets for inflammatory diseases. Nat. Rev. Drug Discov. *2*, 717–726.

Ma, C., Shu, Y., Zheng, Z., Chen, Y., Yao, H., Greenquist, K.W., White, F.A., and LaMotte, R.H. (2003). Similar electrophysiological changes in axotomized and neighboring intact dorsal root ganglion neurons. J. Neurophysiol. *89*, 1588–1602.

Ma, W., Zhang, Y., Bantel, C., and Eisenach, J.C. (2005). Medium and large injured dorsal root ganglion cells increase TRPV-1, accompanied by increased alpha2C-adrenoceptor co-expression and functional inhibition by clonidine. Pain *113*, 386–394.

Moore, K.A., Kohno, T., Karchewski, L.A., Scholz, J., Baba, H., and Woolf, C.J. (2002). Partial peripheral nerve injury promotes a selective loss of GABAergic inhibition in the superficial dorsal horn of the spinal cord. J. Neurosci. *22*, 6724–6731.

Obata, K., Katsura, H., Mizushima, T., Yamanaka, H., Kobayashi, K., Dai, Y., Fukuoka, T., Tokunaga, A., Tominaga, M., and Noguchi, K. (2005). TRPA1 induced in sensory neurons contributes to cold hyperalgesia after inflammation and nerve injury. J. Clin. Invest *115*, 2393–2401.

Obata, K., Yamanaka, H., Kobayashi, K., Dai, Y., Mizushima, T., Katsura, H., Fukuoka, T., Tokunaga, A., and Noguchi, K. (2004). Role of mitogen-activated protein kinase activation in injured and intact primary afferent neurons for mechanical and heat hypersensitivity after spinal nerve ligation. J. Neurosci. *24*, 10211–10222.

Pabbidi, R.M., Cao, D.S., Parihar, A., Pauza, M.E., and Premkumar, L.S. (2008). Direct role of streptozotocin in inducing thermal hyperalgesia by enhanced expression of transient receptor potential vanilloid 1 in sensory neurons. Mol. Pharmacol. *73*, 995–1004.

Porreca, F., Ossipov, M.H., and Gebhart, G.F. (2002). Chronic pain and medullary descending facilitation. Trends Neurosci. *25*, 319–325.

Raghavendra, V., Tanga, F., and DeLeo, J.A. (2003). Inhibition of microglial activation attenuates the development but not existing hypersensitivity in a rat model of neuropathy. J. Pharmacol. Exp. Ther. *306*, 624–630.

Rush, A.M., Dib-Hajj, S.D., Liu, S., Cummins, T.R., Black, J.A., and Waxman, S.G. (2006). A single sodium channel mutation produces hyper- or hypoexcitability in different types of neurons. Proc. Natl. Acad. Sci. U.S.A. *103*, 8245–8250.

Schafers, M., Lee, D.H., Brors, D., Yaksh, T.L., and Sorkin, L.S. (2003a). Increased sensitivity of injured and adjacent uninjured rat primary sensory neurons to exogenous tumor necrosis factor-alpha after spinal nerve ligation. J. Neurosci. *23*, 3028–3038.

Schafers, M., Svensson, C.I., Sommer, C., and Sorkin, L.S. (2003b). Tumor necrosis factor-alpha induces mechanical allodynia after spinal nerve ligation by activation of p38 MAPK in primary sensory neurons. J. Neurosci. *23*, 2517–2521.

Sung, C.S., Wen, Z.H., Chang, W.K., Chan, K.H., Ho, S.T., Tsai, S.K., Chang, Y.C., and Wong, C.S. (2005). Inhibition of p38 mitogen-activated protein kinase attenuates interleukin-1beta-induced thermal hyperalgesia and inducible nitric oxide synthase expression in the spinal cord. J. Neurochem. *94*, 742–752.

Suter, M.R., Wen, Y.R., Decosterd, I., and Ji, R.R. (2007). Do glial cells control pain? Neuron Glia Biol. *3*, 255–268.

Svensson, C.I., Fitzsimmons, B., Azizi, S., Powell, H.C., Hua, X.Y., and Yaksh, T.L. (2005a). Spinal p38beta isoform mediates tissue injury-induced hyperalgesia and spinal sensitization. J. Neurochem. *92*, 1508–1520.

Svensson, C.I., Schafers, M., Jones, T.L., Powell, H., and Sorkin, L.S. (2005b). Spinal blockade of TNF blocks spinal nerve ligation-induced increases in spinal P-p38. Neurosci. Lett. *379*, 209–213.

Sweitzer, S.M., Schubert, P., and DeLeo, J.A. (2001). Propentofylline, a glial modulating agent, exhibits antiallodynic properties in a rat model of neuropathic pain. J. Pharmacol. Exp. Ther. *297*, 1210–1217.

Tang, H.B., Li, Y.S., Arihiro, K., and Nakata, Y. (2007). Activation of the neurokinin-1 receptor by substance P triggers the release of substance P from cultured adult rat dorsal root ganglion neurons. Mol. Pain *3*, 42.

Trang, T., Beggs, S., Wan, X. and Salter, M.W., 2009. P2X4-receptor-mediated synthesis and release of brain-derived neurotrophic factor in microglia is dependent on calcium and p38-mitogen-activated protein kinase activation. J Neurosci. 29, 3518–3528.

Tsuda, M., Inoue, K., and Salter, M.W. (2005). Neuropathic pain and spinal microglia: a big problem from molecules in "small" glia. Trends Neurosci. *28*, 101–107.

Tsuda, M., Mizokoshi, A., Shigemoto-Mogami, Y., Koizumi, S., and Inoue, K. (2004). Activation of p38 mitogen-activated protein kinase in spinal hyperactive microglia contributes to pain hypersensitivity following peripheral nerve injury. Glia *45*, 89–95.

Tsuda, M., Shigemoto-Mogami, Y., Koizumi, S., Mizokoshi, A., Kohsaka, S., Salter, M.W., and Inoue, K. (2003). P2X4 receptors induced in spinal microglia gate tactile allodynia after nerve injury. Nature *424*, 778–783.

Watkins, L.R., Martin, D., Ulrich, P., Tracey, K.J., and Maier, S.F. (1997). Evidence for the involvement of spinal cord glia in subcutaneous formalin induced hyperalgesia in the rat. Pain *71*, 225–235.

Watkins, L.R., Milligan, E.D., and Maier, S.F. (2001). Glial activation: a driving force for pathological pain. Trends Neurosci. *24*, 450–455.

Wen, Y.R., Suter, M.R., Kawasaki, Y., Huang, J., Pertin, M., Kohno, T., Berde, C.B., Decosterd, I., and Ji, R.R. (2007). Nerve conduction blockade in the sciatic nerve prevents but does not reverse the activation of p38 mitogen-activated protein kinase in spinal microglia in the rat spared nerve injury model. Anesthesiology *107*, 312–321.

White, F.A., Jung, H., and Miller, R.J. (2007). Chemokines and the pathophysiology of neuropathic pain. Proc. Natl. Acad. Sci. U.S.A. *104*, 20151–20158.

Wilson-Gerwing, T.D., Dmyterko, M.V., Zochodne, D.W., Johnston, J.M., and Verge, V.M. (2005). Neurotrophin-3 suppresses thermal hyperalgesia associated with neuropathic pain and attenuates transient receptor potential vanilloid receptor-1 expression in adult sensory neurons. J. Neurosci. *25*, 758–767.

Woolf, C.J. and Mannion, R.J. (1999). Neuropathic pain: aetiology, symptoms, mechanisms, and management. Lancet *353*, 1959–1964.

Wu, G., Ringkamp, M., Hartke, T.V., Murinson, B.B., Campbell, J.N., Griffin, J.W., and Meyer, R.A. (2001). Early onset of spontaneous activity in uninjured C-fiber nociceptors after injury to neighboring nerve fibers. J. Neurosci. *21*, RC140.

Xiao, H.S., Huang, Q.H., Zhang, F.X., Bao, L., Lu, Y.J., Guo, C., Yang, L., Huang, W.J., Fu, G., Xu, S.H., Cheng, X.P., Yan, Q., Zhu, Z.D., Zhang, X., Chen, Z., Han, Z.G., and Zhang, X. (2002). Identification of gene expression profile of dorsal root ganglion in the rat peripheral axotomy model of neuropathic pain. Proc. Natl. Acad. Sci. U.S.A. *99*, 8360–8365.

Xu, J.T., Xin, W.J., Wei, X.H., Wu, C.Y., Ge, Y.X., Liu, Y.L., Zang, Y., Zhang, T., Li, Y.Y., and Liu, X.G. (2007). p38 activation in uninjured primary afferent neurons and in spinal microglia contributes to the development of neuropathic pain induced by selective motor fiber injury. Exp. Neurol. *204*, 355–365.

Yamanaka, H., Obata, K., Kobayashi, K., Dai, Y., Fukuoka, T., and Noguchi, K. (2007). Activation of fibroblast growth factor receptor by axotomy, through downstream p38 in dorsal root ganglion, contributes to neuropathic pain. Neuroscience *150*, 202–211.

Zhang, J., Shi, X.Q., Echeverry, S., Mogil, J.S., De Koninck, Y., and Rivest, S. (2007). Expression of CCR2 in both resident and bone marrow-derived microglia plays a critical role in neuropathic pain. J. Neurosci. *27*, 12396–12406.

Zhuang, Z.Y., Gerner, P., Woolf, C.J., and Ji, R.R. (2005). ERK is sequentially activated in neurons, microglia, and astrocytes by spinal nerve ligation and contributes to mechanical allodynia in this neuropathic pain model. Pain *114*, 149–159.

Zhuang, Z.Y., Kawasaki, Y., Tan, P.H., Wen, Y.R., Huang, J., and Ji, R.R. (2007). Role of the CX3CR1/p38 MAPK pathway in spinal microglia for the development of neuropathic pain following nerve injury-induced cleavage of fractalkine. Brain Behav Immun. *21*, 642–651.

Zhuang, Z.Y., Wen, Y.R., Zhang, D.R., Borsello, T., Bonny, C., Strichartz, G.R., Decosterd, I., and Ji, R.R. (2006). A peptide c-Jun N-terminal kinase (JNK) inhibitor blocks mechanical allodynia after spinal nerve ligation: respective roles of JNK activation in primary sensory neurons and spinal astrocytes for neuropathic pain development and maintenance. J. Neurosci. *26*, 3551–3560.

Chapter 21
Microglia and Trophic Factors in Neuropathic Pain States

Simon Beggs

Abstract Neuropathic pain is a complex phenomenon and abundant evidence suggests it involves molecular and cellular physiological, structural and pharmacological changes within the peripheral and central nervous systems. Traditionally considered to be mediated by neuronal changes, it is now becoming increasingly clear that neuro-immune interactions are key mediators of neuropathic pain states. Following peripheral nerve injury in adults, through an as yet not fully understood process, the peripherally injured neurons signal to spinal microglia and induce their activation and proliferation. A key change is the upregulation of microglial $P2X_4$ receptors; ATP acting on $P2X_4$ receptors cause microglia to synthesize and release BNDF which in turn signal through neuronal TrkB receptors to cause a downregulation of KCC2. This results in a disruption of the cellular chloride homeostasis, impairing inhibition and increasing neuronal excitability. Microglia can no longer be regarded as simply immune effectors of the CNS, clearing cellular debris. It is now evident that they are crucial mediators of neuro-immune signaling, controlling neuronal excitability and contributing to the pathology of post-injury pain states.

Abbreviations

AIF-1	allograft inflammatory factor-1
ATP	adenosine triphosphate
BDNF	brain-derived neurotrophic factor
CNS	central nervous system
ERK	extracellular signal-regulated kinase
GABA	gamma amino butyric acid
Iba1	ionized calcium binding adaptor protein-1
IPSC	inhibitory postsynaptic current

S. Beggs (✉)
Program in Neurosciences & Mental Health, Centre for the Study of Pain, Hospital for Sick Children, University of Toronto, Toronto, ON, Canada, M5G 1X8
e-mail: simon.beggs@utoronto.ca

M. Malcangio (ed.), *Synaptic Plasticity in Pain*,
DOI 10.1007/978-1-4419-0226-9_21, © Springer Science+Business Media, LLC 2009

KCC2   potassium-chloride cotransporter 2
MAPK   mitogen-activated protein kinase
MHC   major histocompatibility complex
NMDA N-methyl-D-aspartic acid
PNI   peripheral nerve injury
PPADS   pyridoxal-phosphate-6-azophenyl-2′,4′-disulfonate
SNI   spared nerve injury
TNP-ATP   2″,3″-O-(2,4,6-trinitrophenyl)adenosine 5″-triphosphate
TrkB   tropomyosin-related kinase B

21.1 Introduction

Peripheral nerve injury resulting in neuropathic pain has until comparatively recently been considered the direct consequence of alterations in peripheral and central neuronal function. It is without doubt that peripheral nerve injury activates a myriad changes at the molecular and cellular level capable of directly affecting neuronal plasticity, in turn initiating considerable physiological and anatomical alterations in pain pathways (Woolf and Salter, 2000; Scholz and Woolf, 2002; Zimmermann, 2001; Woolf, 2004). Furthermore, these effects are key to understanding the pathophysiology of neuropathic pain. However, it is becoming increasingly clear that an approach to understanding pain restricted to neuronal function alone is an over-simplification and that a more comprehensive approach to central responses to peripheral nerve injury is required that takes into account other cellular components within the CNS and their interactions.

The dorsal horn of the spinal cord contains a rich and complex cellular network responsible for transducing and modulating somatosensory (both innocuous and nociceptive) input from the periphery. Nociceptive processing within this network takes the form of local and descending inhibitory control that regulates the output to other areas of the CNS which in turn imbue the nociceptive signal with the necessary factors to produce pain; sensory, emotional, autonomic and motor processes. Under normal conditions nociceptive input from peripheral primary afferents is processed within the dorsal horn by a careful balance of inhibitory and excitatory influence such that the spinal cord output is appropriate to the peripheral damage. This has been described as 'nociceptive pain' (Woolf and Salter, 2000) and is associated with defensive behaviour from noxious stimuli. Chronic pain, such as can result from direct injury to peripheral nerves, is often associated with a disruption of the inhibitory/excitatory balance within the spinal dorsal horn leading to aberrant activity within dorsal horn nociceptive processing and a consequent increase in spinal output to higher areas of the CNS. Diverse mechanisms have been described that mediate changes in excitability and inhibitory control and it is now clear that neuronal-glial signaling is implicitly involved.

Historically, glial cells have been considered as having a supportive role within the central nervous system, their primary function being to maintain neuronal 'health'. The CNS contains three glial populations: astrocytes, oligodendrocytes and microglia. The latter, comprising up to 10% of the total glial population and therefore as numerous as neurons, have been implicated both in neuronal support and as a key component of the immune system. Conventionally regarded as reactive macrophages, microglia have far more diverse roles within the CNS, responding to conditions such as trauma, ischaemia, inflammation and infection. Considerable evidence now exists that implicate microglia in having a critical role in the development and maintenance of neuropathic pain (Inoue and Tsuda, 2006; Tsuda et al., 2005; Salter, 2005; Watkins et al., 2001; Watkins and Maier, 2003). Microglia have yet to be shown to be involved in non-pathological nociception, i.e., acute pain. This apparent restrictive role to pathological chronic neuropathic pain makes these cells of considerable interest as targets for future neuropathic pain therapies, particularly important given the often refractory nature of neuropathic pain to conventional treatments.

21.2 Microglial Activation Following Neuronal Injury

Under normal conditions, microglia appear dormant, with a small soma and long spindle-like radiating processes. Often described as in a 'resting' state it is now clear that far from being inactive, microglia continuously monitor their local microenvironment (Nedergaard et al., 2003; Davalos et al., 2005). In imaging experiments conducted *in vivo*, local damage to the parenchymal environment of the microglia by focal laser elicits a rapid response, the microglia extending processes via an ATP-dependent mechanism that converge on the site of injury within minutes (Davalos et al., 2005). Longer term changes to microglia occur in a process described as 'activation'. This consists of changes in morphology, gene/protein expression, migration and proliferation (Kreutzberg 1996; Stoll and Jander 1999; Nakajima and Kohsaka, 2001; Perry, 2004). The process of activation is likely a misnomer; often described in terms of a single phenotype or event, it is likely a continuum of events that occur on a time scale of seconds to several days. Morphologically, microglia undergo a distinct anatomical change, from the ramified, 'resting' state, retracting their secondary and tertiary processes and taking on an amoeboid 'activated' state. In this form microglia are capable of phagocytosis and have been described as the resident macrophages of the CNS. This 'activated' state is characterized by the expression of many cell surface molecules. Typical of these, and widely used as markers of microglia, are CD11b/CD18 (also known as complement receptor 3 – CR3, or Mac-1), recognized by the antibody ox-42 and ionized calcium binding adaptor protein1 – IBA-1 (also known as AIF1). In addition, activation is marked by the synthesis and secretion of immunomolecules such as major histocompatibility complex (MHC) class I and II and proinflammatory cytokines that also contribute to the initiation and maintenance of pain hypersensitivity. The correlation between an upregulation of many of these markers in the spinal cord and peripheral nerve injury has long been reported (Streit et al., 1988; Eriksson et al., 1993; Liu et al., 1995) (Fig. 21.1), and subsequently with the onset of allodynia (Colburn et al., 1997; Coyle, 1998; Colburn et al., 1999) but it is only comparatively recently that a causal role of microglial activation in nerve injury-induced pain behaviours has been established (Tsuda et al., 2003; Jin et al., 2003).

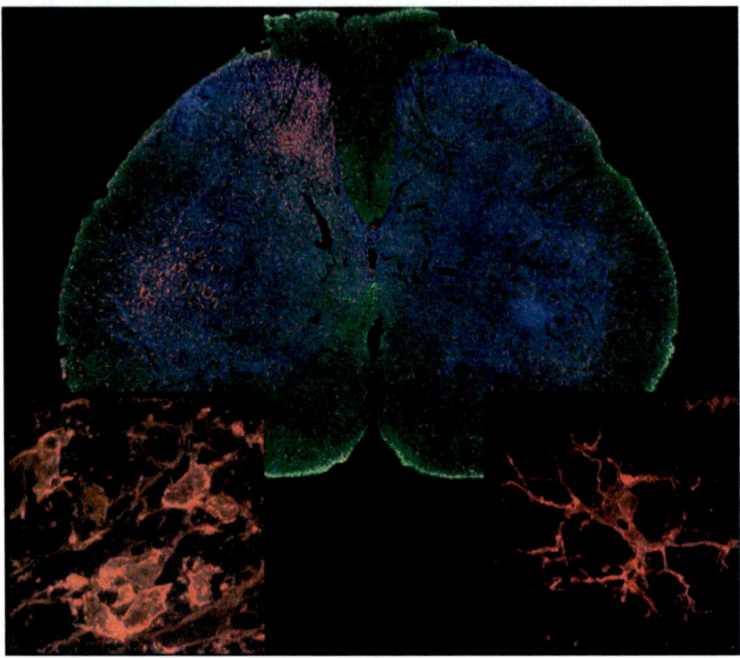

Fig. 21.1 Microglial proliferation in the dorsal horn of the spinal cord 2 weeks following peripheral nerve injury (spared nerve injury, SNI) as revealed by iba1 immunoreactivity (*red*). The proliferation is restricted to the ipsilateral *grey* matter (MAP2 labeling, *blue*. *White* matter shown by NF200, *green*). Insets are high power images and show the morphological change that individual microglia undergo following peripheral nerve injury. The *right* inset shows the normal ramified state, and on the *left* the proliferating microglia have taken on an amoeboid appearance

A stereotypical 'activation' response of spinal microglia has been reported in all experimental models of neuropathic pain involving peripheral nerve injury; ligation, compression or transection (Beggs and Salter, 2007). This response is associated with a morphological change as described above and a concurrent proliferation of microglia, increasing their number and density in the ipsilateral spinal cord. Following spared nerve injury (transaction of the tibial and common peroneal branches of the sciatic nerve) the pattern of microglial activation generally follows somatotopic boundaries, with the majority of microglia within the territory occupied by peripherally axotomised primary afferents with some spread into regions occupied by intact, 'spared' central projections of the sural nerve (Fig. 21.2).

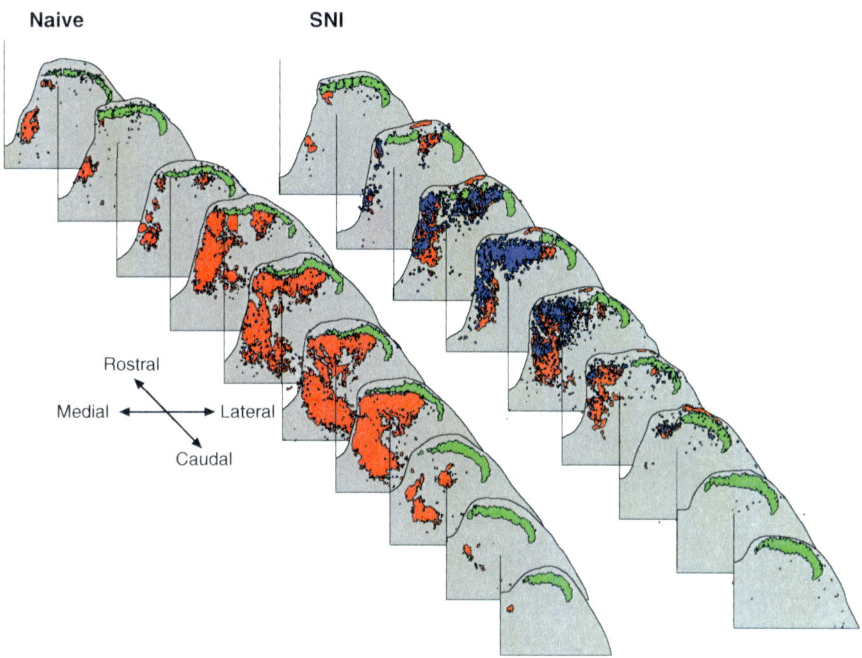

Fig. 21.2 Reconstructions of naïve and SNI lumbar dorsal horns showing the somatotopic nature of the microglial response to peripheral nerve injury. Microglia are indicated in *blue*, the central terminals of sciatic nerve primary afferents (as revealed by retrograde labelling of CTB) in *red* and *green* (IB4)

21.3 Role of ATP and P2X$_4$ Receptor in Microglial Activation

The microglial response to peripheral nerve injury is well documented as are their functions as support cells for neurons and as phagocytes, removing the debris of degenerating primary afferent central terminals (Rotshenker, 2003; Aldskogius, 2001). However, an explicit role as mediators of neuropathic pain represented a paradigm shift in microglial physiology (Coull et al., 2005; Tsuda et al., 2003). A critical causal role of microglia in nerve injury-induced pain was established in studies implicating the P2X$_4$ receptors, first demonstrated by Tsuda et al. (2003). Through pharmacological manipulation of the P2X receptor family they were able to identify the receptor subtypes involved in central responses to peripheral nerve injury. Previous studies have focused on neuronal

expression of P2X receptors and shown that P2X2 and P2X3 are expressed
specifically on nociceptive neurons (Liu and Salter, 2005; North, 2004).
However, pharmacological blockade of these two receptors by their antagonist PPADS had no effect on established post-nerve injury allodynia, whereas
treatment with an alternative P2X antagonist, the ATP analogue TNP-ATP,
produced a rapid and transient reversal of the nerve injury-induced pain
behaviour. The pharmacological profiles of these two antagonists indicated
$P2X_4$ receptors were being selectively targeted. Crucially, an increase in $P2X_4$
protein in the ipsilateral dorsal horn was detected following peripheral nerve
injury and treatment with $P2X_4$ anti-sense significantly attenuated the associated pain behaviour. Normally at extremely low constitutive levels, the
$P2X_4$ protein upregulation was restricted to microglia within the spinal
dorsal horn. This combination of pharmacological, molecular and behavioural findings indicated that $P2X_4$ receptor activation on spinal microglia
are necessary to induce the tactile allodynia associated with neuropathic
pain. Indeed, in $P2X_4$ null mice, nerve injury does not elicit tactile allodynia
(Ulmann et al., 2008).

The cellular and molecular machinery that drives $P2X_4$ upregulation in the
spinal cord following injury to a peripheral nerve is unclear. As described
above, the process of microglial 'activation' is best considered a cascade or
continuum of events. One facet of this process is a wide-ranging increase in
transcription (Perry, 1994; Wieseler-Frank et al., 2005) which may include
$P2X_4$ (Inoue and Tsuda, 2006), although transcription factors and specific
promoter regions involved in the regulation of the $P2X_4$ gene remain unidentified. However, key intracellular signaling molecules with known transcriptional activity are modulated by peripheral nerve injury. Pharmacological
inhibition of p38 and ERK MAP kinases has been shown to both reverse
and prevent the maintenance and induction of allodynia following experimental peripheral nerve lesions (Jin et al., 2003; Schafers et al., 2003; Tsuda
et al., 2004; Zhuang et al., 2005). Indeed, peripheral nerve injury leads to a
persistent, microglial specific, activation of p38 MAP kinase. The maintenance of behavioural pain states is therefore dependent upon both $P2X_4$ and
p38 protein expression (Fig. 21.3). Whether the pathways influenced by $P2X_4$
receptor activation and p38 MAP kinase phosphorylation are parallel or
convergent is not known but it is tempting to speculate a transcriptional role
for p38 on $P2X_4$ gene expression.

Recently a mechanism has been proposed for $P2X_4$ upregulation following
peripheral nerve injury involving the extracellular matrix molecule fibronectin
(Nasu-Tada et al., 2006). Fibronectin is upregulated in the ipsilateral dorsal
horn as a consequence of nerve injury and is therefore ideally placed to influence
$P2X_4$ receptor expression. Microglia grown in culture in the presence of fibronectin show a marked increase in $P2X_4$ expression and an enhanced calcium
response to ATP stimulation. Neuronal $P2X_4$ receptors are recycled; internalized and reinserted into the membrane (Bobanovic et al., 2002) and whether
this is also the case for microglial $P2X_4$ receptors is not known. Enhanced

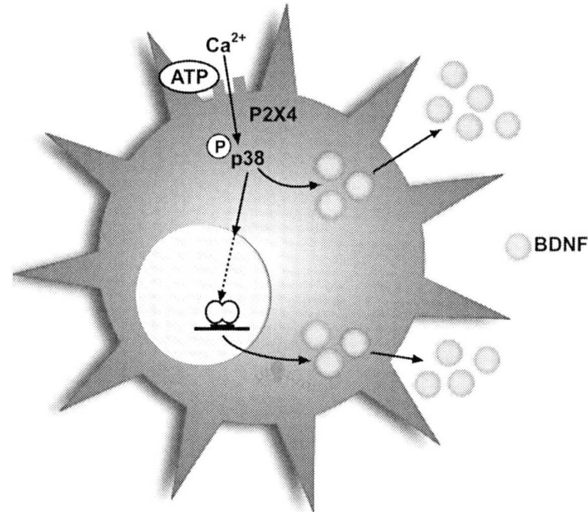

Fig. 21.3 ATP stimulated activation of microglial P2X$_4$ receptors causes p38-dependent BDNF synthesis and release

protein translation, such as via fibronectin (or other extracellular matrix molecules) interactions, or augmented trafficking could influence the activity of ATP on P2X$_4$-expressing cells. Multiple potential pathways exist through which P2X$_4$ expression can be increased following damage to a peripheral nerve and these mechanisms may have considerable therapeutic potential for future treatment of neuropathic disorders.

ATP is an endogenous ligand for the P2 purinergic receptor family. Expressed on many cell types, the P2 family consists of two sub-divisions, the G-protein-coupled P2Y receptors and the ATP-gated cation channels of the P2X receptors. ATP, acting through various neuronally-expressed P2 receptors, has been shown to be explicitly involved both in non-pathological pain transmission and pain hypersensitivity (Burnstock, 2006; Liu and Salter, 2005; Salter et al., 1993), specifically acute pain and inflammatory hyperalgesia.

Many P2Y receptors are expressed by microglia, (Boucsein et al., 2003; Sasaki et al., 2003; Inoue and Tsuda, 2006), yet no clear role has been established in pain processing, their effects being related to chemotaxis and migration toward areas of damage. P2X receptor expression on microglia is restricted to P2X$_4$ and P2X$_7$ (Tsuda et al., 2003; Collo et al., 1997; Ferrari et al., 1996).

ATP is released from damaged and/or inflamed tissue, from central terminals of nociceptive afferents within the spinal cord (Bardoni et al., 1997) and from astrocytes during calcium wave propagation. It has been proposed to mediate both neuronal and injury-induced activation of microglia (Zimmermann, 1994). Neuron-microglia communication is a bidirectional process and purinergic signaling is a key molecular component (Di Virgilio, 2006). Neuronal release of ATP directly modulates microglial functioning, stimulating the release of cytokines, chemokines and neurotrophic factors that in turn affect neuronal function

(Inoue and Tsuda, 2006; Tsuda et al., 2005; Watkins et al., 2001; DeLeo and Yezierski, 2001). It is likely, therefore, that damage to a peripheral nerve will result in the release of ATP (amongst other molecules) from the nociceptive primary afferent terminals. This local release in itself will result in activation and recruitment of microglia to the source of ATP (Davalos et al., 2005). Whether continuous ATP release from peripherally damaged primary afferents is in itself sufficient to maintain chronic pain states is unlikely.

Neuronal excitability within the spinal dorsal horn is controlled by a fine balance of excitatory and inhibitory influences. Disruption of neuronal, and indeed network, excitability can therefore be achieved by either increasing excitatory input, or decreasing inhibitory tone. Disinhibition has been of considerable interest as a key mediator of the transmission of augmented sensory input to higher CNS regions in neuropathic pain states e.g., polysynaptic low threshold inputs onto nociceptive pathways that are normally suppressed by local inhibitory activity (Torsney and McDermott, 2006). Furthermore, in extreme cases, disinhibition can reverse the action of GABA to net excitation, thereby mediating a direct excitatory effect on nociceptive pathways by low threshold input.

Inhibitory transmission, be it GABAergic or glycinergic, is mediated by the net flow of chloride ions. Chloride is driven through open channels on the neuronal membrane by an electrical gradient. During normal inhibitory transmission, chloride ions pass from the extracellular space, an area of relatively high chloride concentration, into neurons where the chloride concentration is maintained at a low level by the actions of cation-chloride cotransporters. These cotransporters establish and maintain the reversal potential for GABAergic and glycinergic receptor channels and therefore directly control chloride ion flux.

One mechanism of increasing dorsal horn excitability following peripheral nerve injury that has recently been demonstrated in the spinal dorsal horn is a depolarizing shift in the chloride reversal potential (Coull et al., 2003). Experiments to investigate the potential of altered KCC2 levels to influence neuronal excitability using antagonists or antisense targetted against the gene resulted in a much reduced nociceptive threshold, mimicking the behavioural consequences of peripheral nerve injury; with disrupted anion homeostasis of lamina I neurons mediating the effect. Subtle changes in the balance of inhibition may have compounding effects as a consequence of disinhibition, resulting in net excitation, mediated through voltage-sensitive calcium channels and NMDA receptors.

As already described, the previously neurocentric approach to research into the mechanisms underlying pain states is fundamentally limited and such a reductionist view of the complex cellular interactions that occur within the central nervous system fails to encompass all possible interactions. However, fundamentally, the perception of pain requires the involvement and active participation of higher CNS centres. Whatever form the central modulation takes at the spinal level, following damage to a peripheral nerve, some form of altered communication to the brain is inevitable and necessary. For this transduction of afferent input to higher CNS areas, neuronal activity is the key.

One proposed mechanism underlying neuropathic pain involves spinal hyperexcitability mediated by local disinhibition. The environment of the spinal dorsal horn consists of a rich network of multiple cellular populations. Within the neuronal population there is considerable diversity, with excitation driven by primary afferent input dampened by local and descending inhibitory circuitry. A large proportion of inhibitory control is via postsynaptic action on second order dorsal horn neurons. An imbalance of these two opposing systems will likely have profound effects on local excitability and subsequent ascending transduction of peripheral nociceptive input.

The question was then raised, if dorsal horn neuron chloride reversal potential and the upregulation of $P2X_4$ receptors on spinal microglia are implicated in the modulation of neuropathic pain states, is it possible that one may influence the other? In the neurocentric model of spinal networks processing and relaying nociceptive input from the periphery, microglia had not been considered as functionally relevant. However, much is known of the responses of microglia to peripheral nerve injury and that their subsequent 'activation' includes the release of many pro-inflammatory molecules. The microglial transfer experiments, first demonstrated by Tsuda et al. (2003), showed that ATP acting through $P2X_4$ receptors in microglia was necessary to maintain neuropathic pain states. This technique was again used to investigate whether microglia-derived factors have the potential to influence the chloride reversal potential of lamina I projection neurons and therefore directly influence nociceptive processing at the spinal level. In a series of experiments by Coull et al., electrophysiological recordings were made from lamina I neurons in slices taken from rats in which behavioural allodynia had been induced by the intrathecal application of ATP-stimulated microglia. The results showed that not only did the rats behaviour mimic that of rats that had previously received a peripheral nerve injury, but that the chloride reversal potential also followed that found in the nerve-injured rats, with a significant depolarizing shift (Coull et al., 2005). These experiments confirmed an instructive role for microglia in disrupting neuronal chloride homeostasis. However, It is clear that microglia are capable of releasing a complex inflammatory 'soup' of chemokines, cytokines and other molecules, and the question remained which was active in ion gradient modulation.

21.4 BDNF as Signaling Molecule Between Microglia and Neurons

The first demonstration of a potential intercellular signalling event modulating chloride homeostasis in adult tissue was the observation that brain-derived neurotrophic factor (BDNF), when applied to hippocampal neurons in culture caused a selective shift in reversal potential of inhibitory synaptic currents (IPSCs) in a subpopulation of cells (Wardle and Poo, 2003). It was subsequently shown that BDNF caused a rapid (less than one hour) down-regulation of KCC2 and consequent disruption in the cell's ability to extrude chloride (Rivera et al., 2002). Crucially, these changes could also be manifested by activity within the slice preparation: kindling-induced epileptogenesis caused both upregulation of BDNF and downregulation of KCC2 over a similar temporal time course. Consistent with these original findings, BDNF applied to adult spinal cord slices caused a depolarizing shift in the chloride reversal potential, decreasing hyperpolarizing inhibition. In a subset of neurons, BDNF application even caused a switching of GABAergic currents into net excitatory events (Coull et al., 2005). Intrathecal administration of BDNF into naïve rats was shown to induce an allodynic like state of behavioural sensitivity and subsequent electrophysiological studies on these rats showed a depolarized anion reversal potential (Coull et al., 2005). Whether endogenous BDNF was responsible for the shift in E_{GABA} after nerve injury was tested by either sequestering endogenous BDNF using a TrkB-Fc fusion protein (Mannion et al., 1999; Thompson et al., 1999) or blocking its binding to TrkB receptors using a function blocking anti-TrkB antibody (Balkowiec and Katz, 2000) in spinal slices taken from rats with peripheral nerve injury. Both methods produced an acute blockade of BDNF-TrkB signalling and reversed the depolarizing shift in E_{GABA} (Coull et al., 2005).

The evidence that BDNF is microglia-derived came from experiments in which cultured microglia were treated with siRNA targeted against BDNF before being stimulated with ATP and intrathecally injected into naïve rats. In vitro ATP-stimulated secretion of BDNF from purified microglial cultures was prevented by concurrent TNP-ATP application and prior siRNA transfection of the microglia. Whether BDNF release from endogenous spinal microglia acts directly on lamina I neurons following peripheral nerve injury to set off the cascade of physiological and biophysical events that leads to a behavioural mechanical allodynia, i.e., the neuropathic pain state, remains to be shown conclusively. BDNF has been implicated as a key mediator of central nociceptive processing, although the main focus of research has been neuronally derived BDNF and evidence suggests that inflammatory pain is indeed mediated by neuronally derived BDNF (Mannion et al., 1999; Kerr et al., 1999). Compelling evidence comes from the demonstration of conditional mouse knock-outs where BDNF was selectively removed from nociceptive sensory neurons (Zhao et al., 2006). In these animals pain-related behaviours were markedly reduced, both at basal levels and particularly in response to inflammation. However, in contrast to these effects, neuropathic pain in response to a peripheral nerve injury developed normally, suggesting a specific role for neuronally-derived BDNF in inflammatory pain and consistent with a role for microglia-derived BDNF in post nerve injury chronic pain states.

These findings indicated that endogenously released BDNF is responsible for the depolarizing shift in chloride reversal potential. Importantly, they also show that this shift is maintained at a depolarized level in dorsal horn neurons by tonically released BDNF at the spinal level after peripheral nerve injury. This release of BDNF in turn requires ongoing microglial $P2X_4$ receptor activation. Established neuropathic pain behaviours can be rapidly and transiently

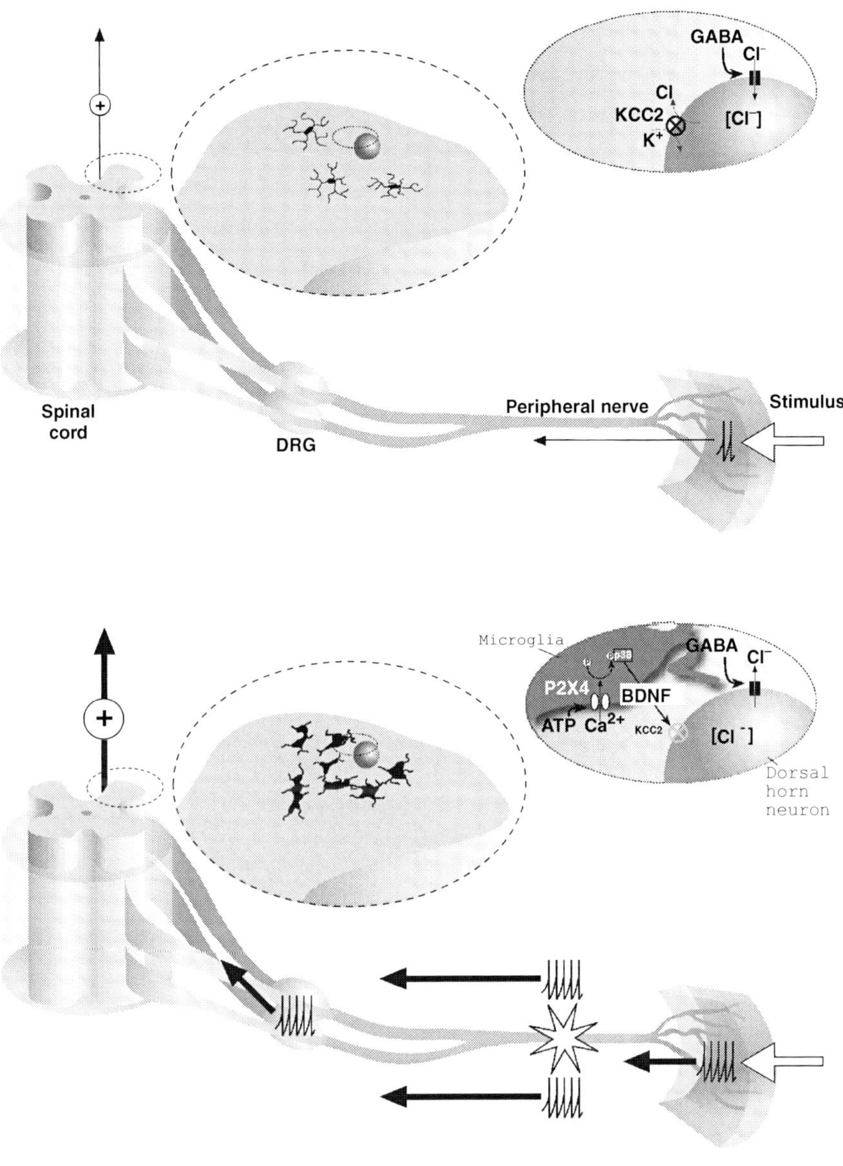

Fig. 21.4 *Top panel*: In the naïve spinal cord, microglia maintain a ramified morphology. Dorsal horn neurons express the potassium-chloride cotransporter KCC2. GABA activation allows calcium influx, resulting in net hyperpolarization, maintaining inhibition. *Lower panel*: Following peripheral nerve injury, microglia proliferate and take on an amoeboid morphology. The microglia upregulate the $P2X_4$ receptor which, following activation, induce the synthesis and secretion of BDNF. BDNF signalling to dorsal horn neurons (via neuronal trkB) causes KCC2 expression to downregulate, disrupting cellular chloride homeostasis and a consequent net disinhibition within the dorsal horn

reversed by intrathecal administration of the P2X$_4$ antagonist TNP-ATP (Tsuda et al., 2003). Furthermore, application of ATP onto microglial cells cultured from P2X$_4$ null mice failed to cause BDNF release (Ulmann et al., 2008)

21.5 Concluding Remarks

This series of experiments defines a novel signaling pathway from microglial P2X$_4$-activated, p38-mediated BDNF release to neuronal trkB activation and subsequent disruption of chloride homeostasis via downregulation of KCC2 (Fig. 21.4). This mechanism reveals a number of potential therapeutic strategies for neuropathic pain conditions. Drugs targetted at restoring chloride homeostasis would have the advantage of not directly affecting neuronal excitability, instead modulating the efficacy of endogenous inhibition. Furthermore, restoring endogenous inhibition rather than actively depressing excitability may also yield more specific therapeutic treatments with less detrimental side effects (De Koninck, 2007).

The finding that normal expression of KCC2 is restored upon blocking BDNF-TrkB or ATP- P2X$_4$ signalling *after* the pathogenesis has developed in the case of nerve-injury induced pain hypersensitivity (Coull et al., 2005) would suggest that restoring transporter function is a viable therapeutic target. While BDNF-TrkB signalling is itself too widespread a signaling mechanism throughout the CNS and PNS to be targetted, specifically preventing activated microglia from secreting BDNF by blocking microglial P2X$_4$ receptors or p38 MAP kinase would represent a more specific strategy. Current pain management strategies act generally by suppressing neuronal activity. The neuro-immune interactions that have now been shown to be key signaling mediators underlying neuropathic pain represent new targets for drug discovery.

References

Aldskogius, H. (2001) Microglia in neuroregeneration. Microsc. Res. Technol. 54(1):40–46.
Balkowiec, A. and Katz, D. M. (2000) Activity-dependent release of endogenous brain-derived neurotrophic factor from primary sensory neurons detected by ELISA in situ. J Neurosci. 20(19):7417–7423.
Bardoni, R., Goldstein, P. A., Lee, C. J., Gu, J. G., and Macdermott, A. B. (1997) ATP P2X receptors mediate fast synaptic transmission in the dorsal horn of the rat spinal cord. J. Neurosci. 17:5297–5304.
Beggs, S. and Salter, M. W. (2007) Stereological and somatotopic analysis of the spinal microglial response to peripheral nerve injury. Brain Behav. Immun. 21(5):624–633.
Bobanovic, L. K., Royle, S. J., and Murrell-Lagnado, R. D. (2002) P2X receptor trafficking in neurons is subunit specific. J. Neurosci. 22:4814–4824.
Boucsein, C., Zacharias, R., Farber, K., Pavlovic, S., Hanisch, U. K., and Kettenmann, H. (2003) Purinergic receptors on microglial cells: functional expression in acute brain slices and modulation of microglial activation in vitro. Eur. J. Neurosci. 17:2267–2276.

Burnstock, G. (2006) Purinergic P2 receptors as targets for novel analgesics. Pharmacol Ther. 110(3):433–454.

Colburn, R. W., DeLeo, J. A., Rickman, A. J., Yeager, M. P., Kwon, P., and Hickey, W. F. (1997) Dissociation of microglial activation and neuropathic pain behaviors following peripheral nerve injury in the rat. J. Neuroimmunol. 79(2):163–75.

Colburn, R. W., Rickman, A. J., and DeLeo, J. A. (1999) The effect of site and type of nerve injury on spinal glial activation and neuropathic pain behavior. Exp. Neurol. 157(2): 289–304.

Collo, G., Neidhart, S., Kawashima, E., Kosco-Vilbois, M., North, R. A., and Buell, G. (1997) Tissue distribution of the P2X7 receptor. Neuropharmacology. 36(9):1277–1283.

Coull, J. A., Boudreau, D., Bachand, K., Prescott, S. A., Nault, F., SÚk, A., De Koninck, P., and De Koninck, Y. (2003) Trans-synaptic shift in anion gradient in spinal lamina I neurons as a mechanism of neuropathic pain. Nature. 424(6951):938–942.

Coull, J. A., Beggs, S., Boudreau, D., Boivin, D., Tsuda, M., Inoue, K., Gravel, C., Salter, M. W., and De, Koninck Y. (2005) BDNF from microglia causes the shift in neuronal anion gradient underlying neuropathic pain. Nature 438:1017–1021.

Coyle, D. E. (1998) Partial peripheral nerve injury leads to activation of astroglia and microglia which parallels the development of allodynic behavior. Glia 23(1):75–83.

Davalos, D., Grutzendler, J., Yang, G., Kim, J. V., Zuo, Y., Jung, S., Littman, D. R., Dustin, M. L., and Gan, W. B. (2005) ATP mediates rapid microglial response to local brain injury in vivo. Nat. Neurosci. 8(6):752–758.

DeLeo, J. A. and Yezierski, R. P. (2001) The role of neuroinflammation and neuroimmune activation in persistent pain. Pain 90(1–2):1–6.

De Koninck, Y. (2007) Altered chloride homeostasis in neurological disorders: a new target. Curr Opin Pharmacol. 7(1):93–99.

Di Virgilio, F. (2006) Purinergic signalling between axons and microglia. Novartis Found Symp. 276:253–258.

Eriksson, N. P., Persson, J. K., Svensson, M., Arvidsson, J., Molander, C., and Aldskogius, H. (1993) A quantitative analysis of the microglial cell reaction in central primary sensory projection territories following peripheral nerve injury in the adult rat. Exp. Brain Res. 96(1):19–27.

Ferrari, D., Villalba, M., Chiozzi, P., Falzoni, S., Ricciardi-Castagnoli, P., and Di Virgilio, F. (1996) Mouse microglial cells express a plasma membrane pore gated by extracellular ATP. J Immunol. 156(4):1531–1539.

Inoue, K. and Tsuda, M. (2006) The role of microglia and ATP receptors in a mechanism of neuropathic pain. Nippon. Yakurigaku Zasshi 127:14–17.

Jin, S. X., Zhuang, Z. Y., Woolf, C. J., and Ji, R. R. (2003) p38 mitogen-activated protein kinase is activated after a spinal nerve ligation in spinal cord microglia and dorsal root ganglion neurons and contributes to the generation of neuropathic pain. J. Neurosci. 23(10):4017–4022.

Kerr, B. J., Bradbury, E. J., Bennett, D. L., Trivedi, P. M., Dassan, P., French, J., Shelton, D. B., McMahon, S. B., and Thompson, S. W. (1999) Brain-derived neurotrophic factor modulates nociceptive sensory inputs and NMDA-evoked responses in the rat spinal cord. J. Neurosci. 19(12):5138–5148.

Kreutzberg, G. W. (1996) Microglia: a sensor for pathological events in the CNS. Trends Neurosci. 19:312–318.

Liu, L., Tornqvist, E., Mattsson, P., Eriksson, N. P., Persson, J. K., Morgan, B. P., Aldskogius, H., and Svensson, M. (1995) Complement and clusterin in the spinal cord dorsal horn and gracile nucleus following sciatic nerve injury in the adult rat. Neuroscience 68(1):167–179.

Liu, X. J. and Salter, M. W. (2005) Purines and pain mechanisms: recent developments. Curr. Opin. Investig. Drugs 6(1):65–75.

Mannion, R. J., Costigan, M., Decosterd, I., Amaya, F., Ma, Q. P., Holstege, J. C., Ji, R. R., Acheson, A., Lindsay, R. M., Wilkinson, G. A., and Woolf, C. J. (1999) Neurotrophins:

peripherally and centrally acting modulators of tactile stimulus-induced inflammatory pain hypersensitivity. Proc. Natl. Acad. Sci. 96(16):9385–9390.

Nakajima, K. and Kohsaka, S. (2001) Microglia: activation and their significance in the central nervous system. J. Biochem. (Tokyo) 130:169–175.

Nasu-Tada, K., Koizumi, S., Tsuda, M., Kunifusa, E., and Inoue, K. (2006) Possible involvement of increase in spinal fibronectin following peripheral nerve injury in upregulation of microglial P2X$_4$, a key molecule for mechanical allodynia. Glia 53(7):769–775.

Nedergaard, M., Ransom, B., and Goldman, S. A. (2003) New roles for astrocytes: redefining the functional architecture of the brain. Trends Neurosci. 26:523–530.

North, R. A. (2004) P2X3 receptors and peripheral pain mechanisms. J. Physiol. 554(Pt 2): 301–308.

Perry, V. H. (1994) Modulation of microglia phenotype. Neuropathol. Appl. Neurobiol. 20:177.

Perry, V. H. (2004) The influence of systemic inflammation on inflammation in the brain: implications for chronic neurodegenerative disease. Brain Behav Immun. 18(5):407–413.

Rivera, C., Li, H., Thomas-Crusells, J., Lahtinen, H., Viitanen, T., Nanobashvili, A., Kokaia, Z., Airaksinen, M. S., Voipio, J., Kaila, K., and Saarma, M. (2002) BDNF-induced TrkB activation down-regulates the K + -Cl- cotransporter KCC2 and impairs neuronal Cl-extrusion. J. Cell Biol. 159:747–752.

Rotshenker, S. (2003) Microglia and macrophage activation and the regulation of complement-receptor-3 (CR3/MAC-1)-mediated myelin phagocytosis in injury and disease. J. Mol. Neurosci. 21(1):65–72.

Salter, M. W., De Koninck, Y., and Henry, J. L. (1993) Physiological roles for adenosine and ATP in synaptic transmission in the spinal dorsal horn. Prog Neurobiol. 41(2):125–156.

Salter, M. W. (2005) Cellular signalling pathways of spinal pain neuroplasticity as targets for analgesic development. Curr. Top. Med. Chem. 5:557–567.

Sasaki, Y., Hoshi, M., Akazawa, C., Nakamura, Y., Tsuzuki, H., Inoue, K., and Kohsaka, S. (2003) Selective expression of Gi/o-coupled ATP receptor P2Y12 in microglia in rat brain. Glia 44:242–250.

Schafers, M., Svensson, C. I., Sommer, C., and Sorkin, L. S. (2003) Tumor necrosis factor-alpha induces mechanical allodynia after spinal nerve ligation by activation of p38 MAPK in primary sensory neurons. J. Neurosci. 23:2517–2521.

Scholz, J. and Woolf, C. J. (2002) Can we conquer pain? Nat. Neurosci. 5(Suppl):1062–1067.

Stoll, G. and Jander, S. (1999) The role of microglia and macrophages in the pathophysiology of the CNS. Prog. Neurobiol. 58:233–247.

Streit, W. J., Graeber, M. B., Kreutzberg, G. W. (1988) Functional plasticity of microglia: a review. Glia 1(5):301–307.

Thompson, S. W., Bennett, D. L., Kerr, B. J., Bradbury, E. J., and McMahon, S. B. (1999) Brain-derived neurotrophic factor is an endogenous modulator of nociceptive responses in the spinal cord. Proc Natl Acad Sci U S A. 96(14):7714–7718.

Torsney, C. and MacDermott, A. B. (2006) Disinhibition opens the gate to pathological pain signaling in superficial neurokinin 1 receptor-expressing neurons in rat spinal cord. J Neurosci. 26(6):1833–1843.

Tsuda, M., Shigemoto-Mogami, Y., Koizumi, S., Mizokoshi, A., Kohsaka, S., Salter, M. W., Inoue, K. (2003) P2X$_4$ receptors induced in spinal microglia gate tactile allodynia after nerve injury. Nature 424(6950):778–783.

Tsuda, M., Mizokoshi, A., Shigemoto-Mogami, Y., Koizumi, S., and Inoue, K. (2004) Activation of p38 mitogen-activated protein kinase in spinal hyperactive microglia contributes to pain hypersensitivity following peripheral nerve injury. Glia 45:89–95.

Tsuda, M., Inoue, K., and Salter, M. W. (2005) Neuropathic pain and spinal microglia: a big problem from molecules in "small" glia. Trends Neurosci. 28:101–107.

Ulmann, L., Hatcher, J. P., Hughes, J. P., Chaumont, S., Green, P. J., Conquet, F., Buell, G. N., Reeve, A. J., Chessell, I. P., and Rassendren, F. (2008) Up-regulation of P2X$_4$ receptors in spinal microglia following peripheral nerve injury mediates BDNF release and neuropathic pain. J. Neurosci. 28(44):11263–11268.

Wardle, R. A. and Poo, M. M. (2003) Brain-derived neurotrophic factor modulation of GABAergic synapses by postsynaptic regulation of chloride transport. J Neurosci. 23(25):8722–8732.

Watkins, L. R., Milligan, E. D., and Maier, S. F. (2001) Glial activation: a driving force for pathological pain. Trends Neurosci. 24:450–455.

Watkins, L. R. and Maier, S. F. (2003) Glia: a novel drug discovery target for clinical pain. Nat. Rev. Drug Discov. 2:973–985.

Wieseler-Frank, J., Maier, S. F., and Watkins, L. R. (2005) Central proinflammatory cytokines and pain enhancement. Neurosignals 14:166–174.

Woolf, C. J. (2004) Dissecting out mechanisms responsible for peripheral neuropathic pain: implications for diagnosis and therapy. Life Sci. 74:2605–2610.

Woolf, C. J. and Salter, M. W. (2000) Neuronal plasticity: increasing the gain in pain. Science 288:1765–1769.

Zhao, J., Seereeram, A., Nassar, M. A., Levato, A., Pezet, S., Hathaway, G., Morenilla-Palao, C., Stirling, C., Fitzgerald, M., McMahon, S. B., Rios, M., and Wood, J. N. (2006) Nociceptor-derived brain-derived neurotrophic factor regulates acute and inflammatory but not neuropathic pain. Mol. Cell. Neurosci. 31(3):539–548.

Zhuang, Z. Y., Gerner, P., Woolf, C. J., and Ji, R. R. (2005) ERK is sequentially activated in neurons, microglia, and astrocytes by spinal nerve ligation and contributes to mechanical allodynia in this neuropathic pain model. Pain 114:149–159.

Zimmermann, H. (1994) Signaling via ATP in the nervous system. Trends Neurosci. 17:420–426.

Zimmermann, M. (2001) Pathobiology of neuropathic pain. Eur. J. Pharmacol. 429:23–37.

Chapter 22
The Cathepsin S/Fractalkine Pair: New Players in Spinal Cord Neuropathic Pain Mechanisms

Anna K. Clark and Marzia Malcangio

Abstract A recent major conceptual advance has been the recognition of the importance of immune system-neuronal interactions in the modulation of brain function. One example of which is spinal pain processing in neuropathic states. Mounting evidence supports the hypothesis that pro-inflammatory mediators secreted by microglia modulate nociceptive function in the injured CNS and following peripheral nerve damage. Here we examine the evidence that one such mediator, the lysosomal cysteine protease cathepsin S (CatS), is critical for the maintenance of neuropathic pain and spinal microglia activation in neuropathic pain states. CatS exerts its pro-nociceptive effects via cleavage of the transmembrane chemokine fractalkine (FKN). Under conditions of increased nociception, microglial CatS is responsible for the liberation of neuronal FKN which stimulates p38 MAPK phosphorylation in microglia thereby activating neurons via the release of pro-nociceptive mediators, thus establishing a new role for the CatS/FKN pair in the maintenance of neuropathic hypersensitivity and suggest that CatS inhibition constitutes a novel therapeutic approach for the treatment of chronic pain.

Abbreviations

CatS	cathepsin S
FKN	fractalkine
APCs	antigen presenting cells
DCs	dendritic cells
MHC	major histocompatibility complex
Ii	invariant chain
CLIP	class II-associated Ii peptide
HLA-DM	human leukocyte antigen DM

A.K. Clark (✉)
Wolfson Centre for Age Related Diseases, King's College London, Guy's Campus, London, SE1 1UL, UK
e-mail: anna.clark@kcl.ac.uk

M. Malcangio (ed.), *Synaptic Plasticity in Pain*,
DOI 10.1007/978-1-4419-0226-9_22, © Springer Science+Business Media, LLC 2009

LHVS   morpholinurea-leucine-homophenylalanine-vinyl sulfone-phenyl
PNL   partial sciatic nerve ligation
CCI   chronic constriction injury
SNL   spinal nerve ligation
ADAM   a disintegrin and metalloprotease domain
TACE   tumour necrosis factor-α converting enzyme

22.1 Introduction

22.1.1 Biochemical Characteristics of Cathepsin S

Cathepsin S (CatS) is a lysosomal enzyme belonging to the papain family of cysteine proteases and was initially purified from bovine lymph nodes in 1975 (Turnsek et al., 1975). Almost all cells express some level of papain-like lysosomal cysteine proteases. Of these, CatB is the most abundant member of the cathepsin family. However, several of the cathepsins, including CatS, have been found to display tissue specific distribution. CatS is expressed by antigen presenting cells (APCs) and this distribution underlies the enzymes major physiological roles. CatS expression has been observed in both professional, bone marrow derived, APCs including macrophages (Liuzzo et al., 1999b; Shi et al., 1992), B-lymphocytes (Riese et al., 1996), dendritic cells (DCs) (Shi et al., 1999) and microglia (Liuzzo et al., 1999b; Petanceska et al., 1994; Liuzzo et al., 1999a), as well as in non-professional APCs, such as epithelial cells (Beers et al., 2005).

The regulation of cysteine protease activity is achieved via a number of factors. One major regulator of protease over activity is the narrow pH range at which the lysosomal proteases are able to function. Cysteine proteases are unstable at neutral pH and are optimised to function at the acidic pH found in intracellular compartments such as the lysosomal compartment. Such instability acts as a mechanism for limiting proteolytic damage if an enzyme escapes from its intracellular compartment. The exception is CatS, which remains extremely stable, and functionally active, at neutral pH (Vasiljeva et al., 2005; Liuzzo et al., 1999a; Petanceska and Devi, 1992; Shi et al., 1992). This unique characteristic may account for specific roles of CatS outside of lysosomes. Indeed, the activity of CatS is not restricted to intracellular compartments, as the secretion of CatS has been observed from a number of cell types, including macrophages and microglia (Liuzzo et al., 1999b,a; Petanceska et al., 1996). The secretion of CatS from both macrophages and microglia is enhanced by a number of inflammatory mediators including Lipopolysaccharide, pro-inflammatory cytokines and neurotrophins (Liuzzo et al., 1999a,b; Petanceska et al., 1996; Riese et al., 1996).

A second major means of regulation of cathepsins is by their endogenous protein inhibitors, the cystatins. The cystatins are a large superfamily of naturally occurring molecules that inhibit cysteine proteases by forming a tight, reversible complex with the active site of the enzyme. Of particular interest with regard to CatS is cystatin C, the most extensively studied human cystatin. Cystatin C was originally isolated from the serum of patients with autoimmune diseases (Brzin et al., 1984), and has since been isolated from a number of other species. Cystatin C expression modulates the enzymatic activity of CatS. CatS plays a vital role in antigen presentation and the intracellular expression of cystatin C appears to tightly regulate CatS activity (Pierre and Mellman, 1998).

22.2 Established Physiological Functions of Cathepsin S

Lysosomal proteases have long been viewed as mediators of terminal protein degradation. Although the majority of cathepsins are widely expressed throughout the body, CatS has been demonstrated to have tissue specific localisation, suggesting that the protease is more than just a mediator of protein fate and function, and implies tissue specific physiological roles.

22.2.1 Antigen Presentation

Over recent years a number of studies have identified a critical role for CatS in antigen presentation (see Fig. 22.1). The process of antigen presentation, by which short peptide fragments of pathogen-derived proteins are displayed on the cell surface of specialised APCs, is vital for effective $CD4^+$ T-lymphocyte immune responses. Class II major histocompatibility complex (MHC) molecules are expressed by bone marrow derived APCs including B-lymphocytes, DCs and macrophages. These molecules sample antigenic peptides from lysosomal compartments and display them for T-lymphocyte recognition. The transport of MHC class II molecules from their site of synthesis to lysosomal compartments is heavily regulated by a chaperone molecule, the invariant chain (Ii). During transport a specific region of the Ii known as CLIP (class II-associated Ii peptide) binds to the peptide groove of the αβ heterodimer (Ghosh et al., 1995), preventing premature peptide loading into the MHC class II peptide groove before transport to the lysosome (Roche and Cresswell, 1991; Roche and Cresswell, 1990). This molecule is also required for correct folding and assembly of MHC class II and for transport of the MHC class II complex into the lysosomal pathway. Once inside the lysosomal compartment degradation of the Ii is vital for antigenic peptide fragments to access the MHC class II binding groove.

The early stages of degradation are carried out by aspartyl proteases. Cathepsin S then further cleaves Ii until only the CLIP region remains associated with the αβ heterodimer. An additional class II-like molecule, human leukocyte

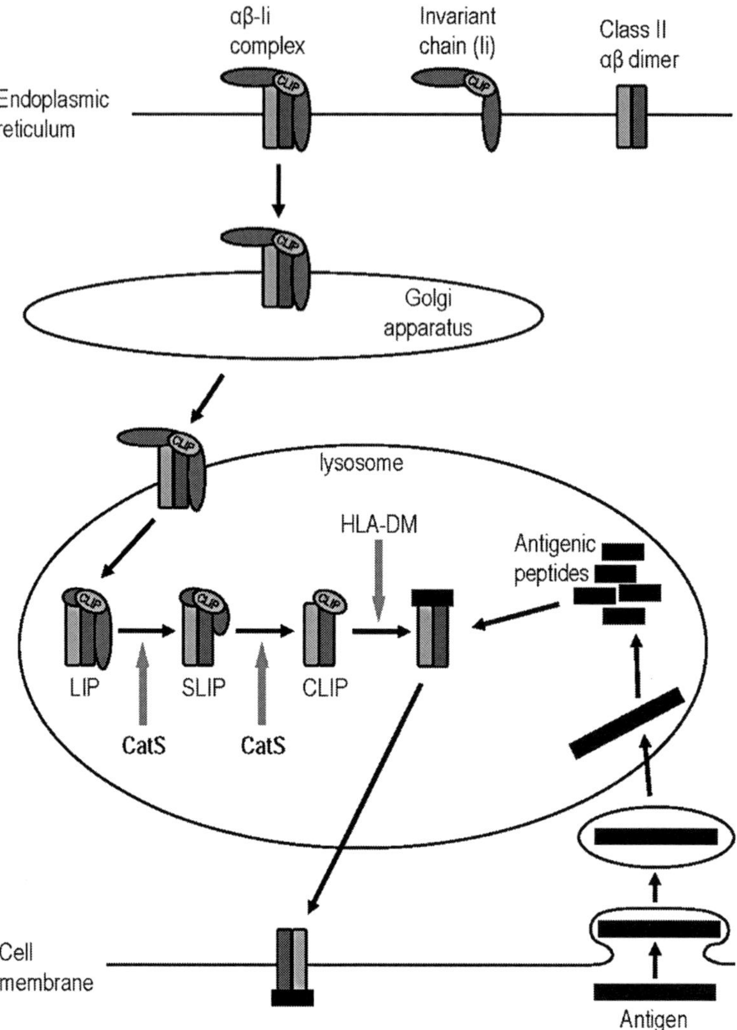

Fig. 22.1 Schematic of class II MHC molecule antigen presentation. αβ-Ii complexes are formed in the endoplasmic reticulum, transported through the Golgi apparatus, and then diverted to the lysosomal pathway. Ii is then degraded to CLIP in a series of stages by lysosomal proteases, including CatS. HLA-DM then catalyses the removal of CLIP from the αβ dimer, facilitating the loading of pathogenic antigen fragments previously endocytosed by the APC

antigen DM (HLA-DM), is then required to catalyse the liberation of CLIP from the αβ heterodimer, and facilitate loading of the antigenic peptide (Denzin and Cresswell, 1995; Sherman et al., 1995; Sloan et al., 1995). The MHC class II molecule is then transported to the cell surface where the antigenic peptide fragment is presented for recognition by T-lymphocytes.

Cysteine proteases were initially implicated in the late stages of Ii degradation following observed accumulation of the Ii fragments in APCs following incubation with the cysteine protease inhibitor leupeptin (Blum and Cresswell, 1988). In vitro biochemical studies by Chapman and colleagues proposed CatS to have major involvement in proteolysis of Ii (Riese et al., 1996; Riese et al., 1998; Villadangos et al., 1997). Incubation of various APCs with the specific CatS inhibitor LHVS (Morpholinurea-leucine-homophenylalanine-vinyl sulfone-phenyl) resulted in the accumulation of Ii fragments, and addition of recombinant CatS protein to these cells restored digestion of Ii (Riese et al., 1996; Villadangos et al., 1997). In addition, in vivo administration of LHVS to mice resulted in deficits in Ii degradation, antigen presentation and immune responses following immunisation with exogenous antigen (Riese et al., 1998). More recently the use knockout mice has determined that CatS is vital for degrading Ii fragments to CLIP in vivo. B-lymphocytes and DCs from mice deficient in CatS are unable to degrade Ii to CLIP, resulting in the accumulation of Ii fragments (Shi et al., 1999; Nakagawa et al., 1999), and impaired antigen presentation (Shi et al., 1999; Nakagawa et al., 1999; Driessen et al., 1999).

22.2.2 *Tissue Remodelling and Extracellular Matrix Degradation*

In addition to the intracellular role of CatS in antigen presentation, CatS can be secreted by activated macrophages and microglia in response to inflammatory mediators (Liuzzo et al., 1999a,b; Petanceska et al., 1996). This secretion as well as the remarkable stability of the enzyme at neutral, extracellular pH (Liuzzo et al., 1999a; Vasiljeva et al., 2005; Petanceska and Devi, 1992; Shi et al., 1992) led to investigation of the extracellular properties of CatS. Indeed, CatS has been observed to cleave some extracellular matrix proteins and may therefore play a role in tissue remodelling. In vitro CatS is able to degrade extracellular matrix proteins including laminin, fibronectin, and several collagens (Petanceska et al., 1996), chondroitin sulphate (Petanceska et al., 1996), heparan sulfate and basement membrane proteoglycans (Liuzzo et al., 1999b; Petanceska et al., 1996), myelin basic protein and amyloid beta peptide (Liuzzo et al., 1999a).

22.3 A New Role for Cathepsin S in Nociception

It is now well established that the contribution of immune cells, in both the peripheral and central nervous systems, are critical for the full expression of neuropathic pain behaviours in a vast range of rodent models (Thacker et al., 2007; Scholz and Woolf, 2007). The high stability of CatS at neutral pH, in conjunction with an established role in APCs led us to hypothesis that one possible extracellular function of CatS may be an involvement in nociceptive transmission. Using Affymetrix GeneChip^{TM} technology to screen for changes in gene expression in rat lumbar DRG following two models of peripheral nerve injury CatS was identified as a possible component that may underlie neuropathic pain and a potential target for therapeutic intervention. The data from these microarray experiments has been deposited in the Gene Expression Omnibus (Accession: GSE2636) at NCBI. Increased CatS mRNA levels in the lumbar DRG were confirmed in both the partial sciatic nerve ligation (PNL) and chronic constriction injury (CCI) models of neuropathy by RT-PCR (Barclay et al., 2007). The CatS gene had been previously found to be up-regulated in the distal stump of the sciatic nerve 7 days after axotomy (Kubo et al., 2002), in the DRG 3 days after axotomy (Costigan et al., 2002) and 7 days following spinal nerve ligation (SNL) (Valder et al., 2003), however, the role of CatS in nociceptive processing had not previously been examined. Immunohistochemical studies revealed that following PNL, CatS expression is observed in cells surrounding the neuronal cell bodies in the ipsilateral DRG and in cells located close to the ligation site. CatS expressing cells at both sites were identified as macrophages based on co-expression of CatS with several macrophage markers, suggesting that increased CatS expression in the DRG and at the injury site is due to an infiltration of CatS expressing macrophages (Barclay et al., 2007).

22.3.1 CatS is Expressed by Spinal Microglia

Several lines of evidence indicate that activated microglia in the spinal cord are an important source of CatS. As the resident macrophages of the CNS, CatS expression has been observed in microglial cell lines (Liuzzo et al., 1999a,b) and in brain microglia (Petanceska et al., 1994). In the dorsal horn of the spinal cord punctate CatS expression is observed in spinal microglia under naïve conditions, and is extensively upregulated following PNL in the ipsilateral dorsal horn. This upregulation occurs in the area where damaged fibres terminate and occurs exclusively in microglial cells, as observed by co-localisation of the microglial markers OX42 (Clark et al., 2007a) and Iba-1 (see Fig. 22.2). Microglial cells are established secretory cells that release CatS upon activation by inflammatory mediators (Liuzzo et al., 1999a,b). Therefore, following a peripheral nerve injury it is likely that the activation of spinal microglia will result in secretion of a number of inflammatory mediators, one of which may be CatS.

Following PNL, the temporal profile of CatS expression in microglia appears to correlate with the maintenance of neuropathy, rather than its development. A few CatS positive cells are present in the ipsilateral dorsal horn of the spinal cord 1 day after injury when mechanical hyperalgesia and allodynia are

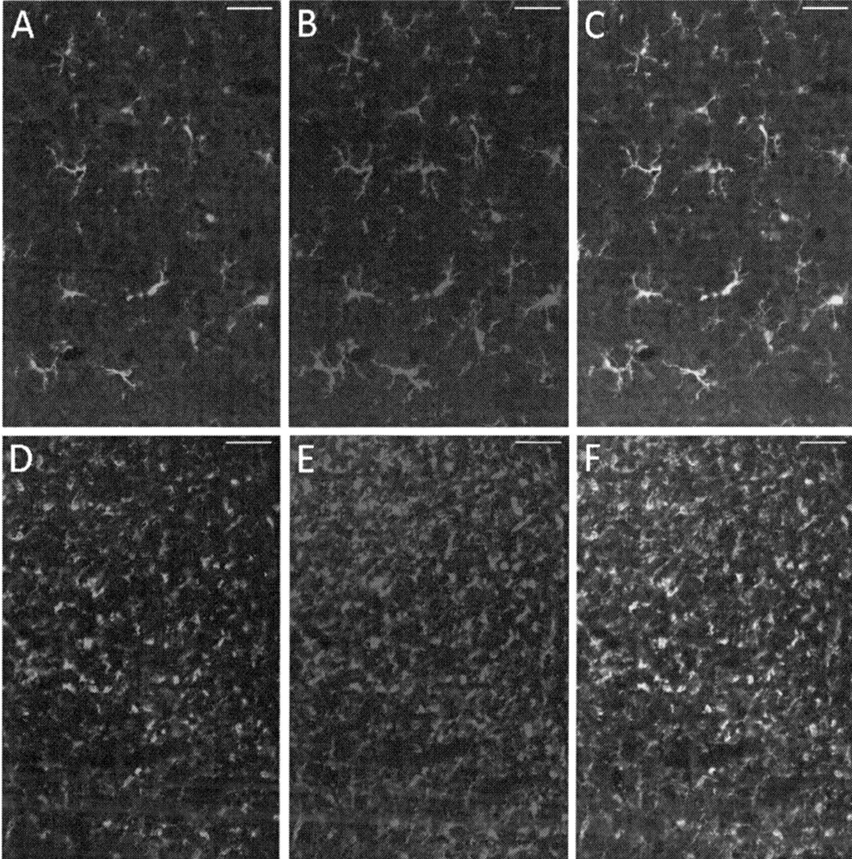

Fig. 22.2 CatS IR co-localises with the microglia marker Iba-1. (A–C) Photomicrographs show that in the naïve dorsal horn CatS IR (**A**) co-localises with Iba-1 (**B**; *merge*, **C**). (**D–F**) Photomicrographs show that in the ipsilateral dorsal horn, 7 days following peripheral nerve injury, CatS IR (**D**) co-localises with Iba-1 (**E**; *merge*, **F**). Scale bars = 50 μm

known to be already well established. Over the next 2 days, CatS expression increases, peaking at 7 days and remains elevated for a total of 21 days. The time course of CatS expression is consistent with reported microglial activation in the ipsilateral dorsal horn following PNL (Coyle, 1998; Clark et al., 2007b). Microglial activation is evident by postoperative day 2 within the somatotopic localization of sciatic nerve primary afferents. This activation is maximal by day 14 and remains high for several weeks in association with neuropathic pain (Coyle, 1998; Clark et al., 2007b). Indeed, the temporal profile of CatS expression in the dorsal horn correlates with the activation of microglial cells following PNL (see Fig. 22.3).

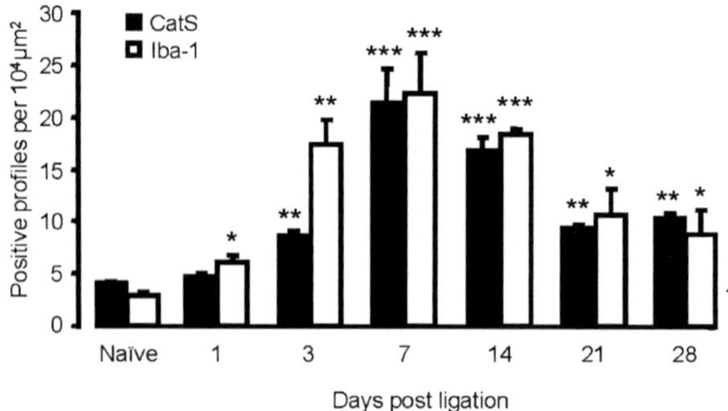

Fig. 22.3 Timecourse of CatS and Iba-1 expression in the dorsal horn following peripheral nerve injury. Quantification of number of CatS or Iba-1 positive cells in the dorsal horn of the spinal cord, per 10^4 μm^2. *$P<0.05$, **$P<0.01$, ***$P<0.001$, compared to naïve tissue, One-way ANOVA, post-hoc Tukey's test

22.3.2 CatS Inhibition Attenuates Neuropathic Pain Behaviour

Spinal administration of LHVS, an irreversible, synthetic CatS inhibitor (Riese et al., 1998), suggests that microglial CatS may contribute to mechanical allodynia. Intrathecal administration of the CatS inhibitor LHVS transiently reverses allodynia in 7-day and 14-day neuropathic rats, when CatS expression in the dorsal horn is high (Clark et al., 2007a). However, 3 days following injury, when CatS expression is sub-maximal, LHVS does not alter mechanical allodynia. Interestingly, in 14-day neuropathic rats LHVS dose-dependently reverses established mechanical hyperalgesia suggesting that CatS contained in spinal microglial cells is a major contributor to the maintenance of neuropathic pain rather than to the initial development of mechanical hypersensitivity. LHVS preferentially reverses hyperalgesia in comparison to allodynia when administered acutely. Chronic inhibition of CatS appears to be more efficacious in the reversal of mechanical allodynia. This conclusion is reinforced by the observation that the chronic administration of LHVS from day 0–7 post-PNL fails to prevent the development of mechanical allodynia, whereas treatment from day 7–14 post-PNL attenuates already established allodynia (Clark et al., 2007a). Importantly, the central microglial source of the enzyme is essential for the persistence of neuropathic allodynia.

In addition, the extent of ipsilateral microglial activation following PNL is significantly reduced following chronic delivery of LHVS 7–14 days post-PNL suggesting that the anti-allodynic effects of LHVS following chronic delivery may be partially due to an attenuation of microglial activation in the dorsal horn of the spinal cord. Many other mediators and receptors have been suggested to play a role in microglial activation after peripheral nerve injury. Of these, evidence suggests that ATP results in the release of BDNF via activation of microglial P2X$_4$ receptors (Tsuda et al., 2003; Coull et al., 2005). Tsuda and colleagues observe extensive upregulation of the purinergic $P2X_4$ receptor on spinal microglial cells following a peripheral nerve injury, and an attenuation of mechanical allodynia following impairment of $P2X_4$ signaling (Tsuda et al., 2003). However, as opposed to $P2X_4$ inhibition which did not modify microglia activation (Tsuda et al., 2003), inhibition of CatS reduces OX42 labeling (Clark et al., 2007a), suggesting that this enzyme is essential for the microglial cells to remain in their hyperactive phenotype.

Due to the role of CatS in antigen presentation CatS inhibitors have long been seen as an attractive therapeutic target for autoimmune disorders. As a result many inhibitors have already been developed for this purpose (Ward et al., 2002; Wei et al., 2007; Bekkali et al., 2007; Gauthier et al., 2007; Inagaki et al., 2007; Chatterjee et al., 2007; Irie et al., 2008a). Importantly, two highly selective CatS inhibitors have recently been shown to be efficacious in attenuating neuropathic pain behaviours following PNL (Irie et al., 2008b), suggesting that these newly developed compounds may be beneficial in the treatment of neuropathic pain.

22.3.3 Exogenous CatS Induces Pain Behaviours

Several observations support the view that CatS is pro-nociceptive. Spinal delivery of activated recombinant CatS evokes dose-dependent mechanical hypersensitivity that develops within 30 min of administration (Clark et al., 2007a). This hyperalgesia requires active enzyme and is specific to CatS since the related enzymes Cathepsin B and Cathepsin L have no pro-nociceptive actions. The effects of spinally administered CatS are completely prevented by prior LHVS administration, indicating a local, spinal site of action for the enzyme (Clark et al., 2007a). The inhibitory effects of LHVS are specific for CatS-induced pain behaviours, since LHVS has no effect on the hyperalgesia induced by injections of either Substance P or NMDA. The finding that exogenous CatS is pro-nociceptive clearly points to an extracellular site of action and is consistent with the observations that the behavioural effects are rapid and maximal at the first time point studied, 30 min after enzyme injection. Indeed, unlike other cathepsins, CatS is extremely stable and functionally active at neutral extracellular pH (Vasiljeva et al., 2005; Liuzzo et al., 1999a; Petanceska and Devi, 1992; Shi et al., 1992).

22.3.4 CatS Pro-Nociceptive Effects Are Mediated Via Fractalkine Cleavage

Chemokines comprise a large family of structurally homologous cytokines that at present number over 50 individual chemokines, signalling via 19 receptors (reviewed in Bajetto et al., 2002; Murphy et al., 2000). Chemokines are subdivided into four groups based on the relative position of their first N-terminal cysteine residues. FKN (CX3CL1) is at present the only member of the CX3C class of chemokines, existing as both a membrane bound and a soluble form (Bazan et al., 1997; Pan et al., 1997). The binding of chemokines to their receptors is highly promiscuous. However, several monogamous relationships exist, one of which is the binding of FKN to the CX3CR1 receptor. The classical view of chemokine function is in the control of immune cell trafficking, a process important in host immune surveillance as well as acute and chronic inflammation. However, in recent years a large number of chemokines and their receptors have been identified in the CNS under both normal and pathological conditions, and evidence suggests that chemokines play a role in the regulation of neuronal activity and synaptic plasticity.

The structure of FKN means that in contrast to most other chemokines FKN can exist as both transmembrane and soluble forms (Bazan et al., 1997). These two forms of FKN have differing functions, with transmembrane FKN acting as a unique adhesion molecule, whereas as soluble FKN (sFKN) acts as a chemoattractant for CX3CR1 expressing cells. The expression of FKN appears to be highly regulated. FKN is synthesised as a 50–75 kDa precursor, that following glycosylation yields a mature 100 kDa transmembrane protein (Garton et al., 2001). In both endothelial cells (Liu et al., 2005), and DRG neurons (Clark et al., 2007a) mature FKN is observed in two separate compartments. FKN is located on the plasma membrane as well as in an intracellular compartment. In endothelial cells, the recycling of FKN between these two compartments is observed in unstimulated cells (Liu et al., 2005). The regulation of this recycling may be key in altering the expression of membrane bound FKN in response to cell activation. In addition, the shedding of membrane bound FKN may represent a key regulatory mechanism for FKN signaling. As an adhesion molecule expressed by endothelium, membrane bound FKN is able to capture cells expressing the CX3CR1 receptor (Imai et al., 1997), however sFKN acts as a chemoattractant. Shedding of FKN from the cell membrane of endothelial cells occurs constitutively from unstimulated cells but can also be induced. Constitutive expression has been demonstrated to involve the disintegrin-like metalloproteinase ADAM (*a d*isintegrin *a*nd *m*etalloprotease domain) 10 (Hundhausen et al., 2003). Following stimulation of cells with phorbol esters (e.g. PMA) shedding of FKN is markedly enhanced. This inducible shedding is regulated by a second protease of the ADAM family, tumour necrosis factor-α converting enzyme (TACE; or ADAM 17) (Garton et al., 2001; Tsou et al., 2001). Stimulation of FKN shedding by PMA reduces the adhesion of monocytes to FKN expressing endothelium (Hundhausen et al., 2003), suggesting that increased FKN shedding may function as a mechanism to decrease adhesion whilst enhancing sFKN-induced chemotaxis. However, not all of the shedding of FKN observed in these studies can be accounted for by cleavage of ADAM 10 and TACE, as following metalloproteinase inhibition formation of sFKN is observed (Hundhausen et al., 2003).

In particular, a role for FKN in nociceptive transmission has been proposed. FKN is expressed by neurons in the spinal cord (Lindia et al., 2005; Verge et al., 2004), and in a number of brain areas (Harrison et al., 1998; Nishiyori et al., 1998; Schwaeble et al., 1998; Hughes et al., 2002). In contrast, its receptor,

CX3CR1, is expressed by microglia in both the spinal cord (Lindia et al., 2005; Verge et al., 2004) and brain (Harrison et al., 1998; Nishiyori et al., 1998; Hughes et al., 2002). In addition, both FKN mRNA and protein are expressed in cell bodies of DRG neurons (Verge et al., 2004). Following a peripheral nerve injury extensive up-regulation of CX3CR1 is observed in microglia (Verge et al., 2004; Lindia et al., 2005). Interestingly, de novo expression of FKN is observed in astrocytes following SNL (Lindia et al., 2005), however these changes are absent in other models of neuropathy (Verge et al., 2004). Intrathecal administration of FKN results in thermal (Milligan et al., 2005; Milligan et al., 2004) and mechanical (Milligan et al., 2005; Milligan et al., 2004; Clark et al., 2007a) hyperalgesia that is prevented by pre-treatment with a neutralising agent against CX3CR1 (Milligan et al., 2005; Milligan et al., 2004) or FKN itself (Clark et al., 2007a), and is absent in CX3CR1 knock-out mice (Clark et al., 2007a). Importantly, FKN induced pain behaviours result following microglia activation, as observed by phosphorylation of p38 MAPK (Zhuang et al., 2007; Clark et al., 2007a), and appear to be mediated in part by microglial derived IL-1β, IL-6 and NO (Milligan et al., 2005). In addition, impairment of FKN/CX3CR1 signalling following nerve injury either by administration of a CX3CR1 (Milligan et al., 2004) or FKN antibody (Clark et al., 2007a) attenuates neuropathic pain behaviours. Interestingly, in naïve rats FKN is able to cause hyper-responsiveness of wide dynamic range neurons the lumbar spinal cord (Owolabi and Saab, 2006), suggesting that FKN is able to modulate neuronal activity to peripheral stimuli.

In the CNS membrane associated FKN protein may constitutively signal via direct interaction with CX3CR1 expressing microglia in the vicinity. It is still unclear as to whether constitutive shedding of FKN occurs in the CNS. However, metalloproteinase mediated cleavage of FKN has been shown from cortical neurons in culture following glutamate-induced excitotoxicity (Chapman et al., 2000). To date neither ADAM 10 nor TACE have been implicated in the cleavage of neuronal FKN in the spinal cord. We suggest that microglial CatS is responsible for the shedding of FKN in the spinal cord. The increase in sFKN in response to brain injury promotes chemotaxis of both microglia and monocytes (Chapman et al., 2000), suggesting that sFKN may signal via 'volume transmission' and induce the chemotaxis of microglia to the injury site. The proposed role of microglia in the modulation of nociceptive transmission following injury suggests that neuronal FKN may regulate microglial function in chronic pain states.

A bioinformatics approach identified FKN as a possible cleavage target for CatS, and the predicted cleavage site for FKN would liberate the soluble, biologically active extracellular domain of this chemokine. Several observations provide evidence that the molecular substrate for CatS is neuronal FKN and that the mechanism responsible for the pro-nociceptive effects of CatS is FKN shedding from neuronal membranes. The intrathecal injection of both CatS (Clark et al., 2007a) and FKN (Milligan et al., 2005; Milligan et al., 2004; Clark et al., 2007a) is pro-nociceptive. In addition, pre-administration of a FKN neutralising antibody is able to prevent the development of mechanical hyperalgesia following either FKN or CatS, but does not modify either NMDA or Substance P-induced pain behaviours (Clark et al., 2007a). Extracellular CatS appears to indirectly activate microglial p38 MAPK by proteolytically liberating the soluble form of neuronal FKN. Intrathecal administration of both FKN (Zhuang et al., 2007; Clark et al., 2007a) and CatS (Clark et al., 2007a) result in extensive phosphorylation of p38 MAPK, a marker of rapid microglial cell activation, which is attenuated following pre-administration of FKN neutralising antibody (Clark et al., 2007a). In addition, Cats-induced mechanical allodynia is absent in mice lacking the CX3CR1 receptor, indicating that CatS pro-nociceptive effects are mediated via cleavage of FKN and activation of the CX3CR1 receptor (Clark et al., 2007a).

22.4 Concluding Remarks

The following scheme (Fig. 22.4) is proposed. Following peripheral nerve injury, CatS expressing spinal cord microglia in the area of the dorsal horn innervated by damaged fibres release CatS. Extracellular CatS then liberates sFKN from the membranes of spinal neurons. The released sFKN then feeds back onto the microglial cells via the CX3CR1 receptor to activate the p38 MAPK pathway. Activation of this intracellular pathway is thought to contribute to neuropathic pain (Jin et al., 2003; Tsuda et al., 2004) by modulating the synthesis and release of pro-nociceptive mediators such as cytokines, including IL-1β (Baldassare et al., 1999; Kim et al., 2004; Caivano and Cohen,

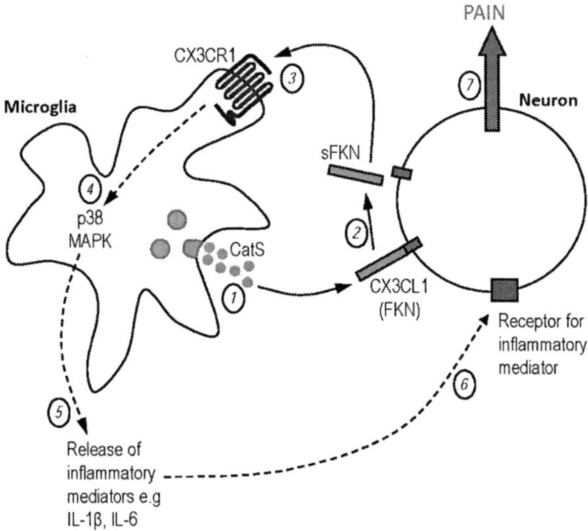

Fig. 22.4 Schematic of proposed pro-nociceptive mechanism of CatS. In the dorsal horn area innervated by damaged fibres activated microglia release CatS (1) which then liberates soluble FKN from neurons (2). FKN feeds back onto the microglial cells via the CX3CR1 receptor (3) to activate the p38 MAPK pathway (4) and release inflammatory mediators (5) that activate neurons (6) and result in neuropathic pain (7). Pathways not elucidated in this study are shown by *dotted lines*

2000; Clark et al., 2006). In summary, these data establish a new role for the CatS/FKN pair in the maintenance of neuropathic hypersensitivity and suggest that CatS inhibition constitutes a novel therapeutic approach for the treatment of chronic pain.

References

Bajetto A, Bonavia R, Barbero S, Schettini G (2002) Characterization of chemokines and their receptors in the central nervous system: physiopathological implications. J Neurochem 82:1311–1329.

Baldassare JJ, Bi Y, Bellone CJ (1999) The role of p38 mitogen-activated protein kinase in IL-1{beta} transcription. J Immunol 162:5367–5373.

Barclay J, Clark AK, Ganju P, Gentry C, Patel S, Wotherspoon G, Buxton F, Song C, Ullah J, Winter J, Fox A, Bevan S, Malcangio M (2007) Role of the cysteine protease cathepsin S in neuropathic hyperalgesia. Pain 130:225–234.

Bazan JF, Bacon KB, Hardiman G, Wang W, Soo K, Rossi D, Greaves DR, Zlotnik A, Schall TJ (1997) A new class of membrane-bound chemokine with a CX3C motif. Nature 385:640–644.

Beers C, Burich A, Kleijmeer MJ, Griffith JM, Wong P, Rudensky AY (2005) Cathepsin S controls MHC class II-mediated antigen presentation by epithelial cells in vivo. J Immunol 174:1205–1212.

Bekkali Y, Thomson DS, Betageri R, Emmanuel MJ, Hao MH, Hickey E, Liu W, Patel U, Ward YD, Young ERR, Nelson R, Kukulka A, Brown ML, Crane K, White D, Freeman DM, Labadia ME, Wildeson J, Spero DM (2007) Identification of a novel class of succinyl-nitrile-based cathepsin S inhibitors. Bioorgan Med Chem Lett 17:2465–2469.

Blum JS, Cresswell P (1988) Role for intracellular proteases in the processing and transport of class II HLA antigens. PNAS 85:3975–3979.

Brzin J, Popovic T, Turk V, Borchart U, Machleidt W (1984) Human cystatin, a new protein inhibitor of cysteine proteinases. Biochem Biophys Res Commun 118:103–109.

Caivano M, Cohen P (2000) Role of mitogen-activated protein kinase cascades in mediating lipopolysaccharide-stimulated induction of cyclooxygenase-2 and IL-1{beta} in RAW264 macrophages. J Immunol 164:3018–3025.

Chapman GA, Moores K, Harrison D, Campbell CA, Stewart BR, Strijbos PJLM (2000) Fractalkine cleavage from neuronal membranes represents an acute event in the inflammatory response to excitotoxic brain damage. J Neurosci 20:87RC.

Chatterjee AK, Liu H, Tully DC, Guo J, Epple R, Russo R, Williams J, Roberts M, Tuntland T, Chang J, Gordon P, Hollenbeck T, Tumanut C, Li J, Harris JL (2007) Synthesis and SAR of succinamide peptidomimetic inhibitors of cathepsin S. Bioorgan Med Chem Lett 17:2899–2903.

Clark AK, D'Aquisto F, Gentry C, Marchand F, McMahon SB, Malcangio M (2006) Rapid co-release of interleukin 1beta and caspase 1 in spinal cord inflammation. J Neurochem 99:868–880.

Clark AK, Yip PK, Grist J, Gentry C, Staniland AA, Marchand F, Dehvari M, Wotherspoon G, Winter J, Ullah J, Bevan S, Malcangio M (2007a) Inhibition of spinal microglial cathepsin S for the reversal of neuropathic pain. PNAS 104:10655–10660.

Clark AK, Gentry C, Bradbury EJ, McMahon SB, Malcangio M (2007b) Role of spinal microglia in rat models of peripheral nerve injury and inflammation. Eur J Pain 11:223–230.

Costigan M, Befort K, Karchewski L, Griffin RS, D'Urso D, Allchorne A, Sitarski J, Mannion JW, Pratt RE, Woolf CJ (2002) Replicate high-density rat genome oligonucleotide microarrays reveal hundreds of regulated genes in the dorsal root ganglion after peripheral nerve injury. BMC Neurosci 3:16.

Coull JA, Beggs S, Boudreau D, Boivin D, Tsuda M, Inoue K, Gravel C, Salter MW, De Koninck Y (2005) BDNF from microglia causes the shift in neuronal anion gradient underlying neuropathic pain. Nature 438:1017–1021.

Coyle DE (1998) Partial peripheral nerve injury leads to activation of astroglia and microglia which parallels the development of allodynic behavior. Glia 23:75–83.

Denzin LK, Cresswell P (1995) HLA-DM induces clip dissociation from MHC class II [alpha][beta] dimers and facilitates peptide loading. Cell 82:155–165.

Driessen C, Bryant RAR, Lennon-Dumenil AM, Villadangos JA, Bryant PW, Shi GP, Chapman HA, Ploegh HL (1999) Cathepsin S controls the trafficking and maturation of MHC class II molecules in dendritic cells. J Cell Biol 147:775–790.

Garton KJ, Gough PJ, Blobel CP, Murphy G, Greaves DR, Dempsey PJ, Raines EW (2001) Tumor necrosis factor-alpha-converting enzyme (ADAM17) mediates the cleavage and shedding of fractalkine (CX3CL1). J Biol Chem 276:37993–38001.

Gauthier JY, Black WC, Courchesne I, Cromlish W, Desmarais S, Houle R, Lamontagne S, Li CS, Massq F, McKay DJ, Ouellet M, Robichaud J, Truchon JF, Truong VL, Wang Q, Percival MD (2007) The identification of potent, selective, and bioavailable cathepsin S inhibitors. Bioorgan Med Chem Lett 17:4929–4933.

Ghosh P, Amaya M, Mellins E, Wiley DC (1995) The structure of an intermediate in class II MHC maturation: CLIP bound to HLA-DR3. Nature 378:457–462.

Harrison JK, Jiang Y, Chen S, Xia Y, Maciejewski D, McNamara RK, Streit WJ, Salafranca MN, Adhikari S, Thompson DA, Botti P, Bacon KB, Feng L (1998) Role for neuronally derived fractalkine in mediating interactions between neurons and CX3CR1-expressing microglia. PNAS 95:10896–10901.

Hughes PM, Botham MS, Frentzel S, Mir A, Perry VH (2002) Expression of fractalkine (CX3CL1) and its receptor, CX3CR1, during acute and chronic inflammation in the rodent CNS. Glia 37:314–327.

Hundhausen C, Misztela D, Berkhout TA, Broadway N, Saftig P, Reiss K, Hartmann D, Fahrenholz F, Postina R, Matthews V, Kallen KJ, Rose-John S, Ludwig A (2003) The disintegrin-like metalloproteinase ADAM10 is involved in constitutive cleavage of CX3CL1 (fractalkine) and regulates CX3CL1-mediated cell-cell adhesion. Blood 102:1186–1195.

Imai T, Hieshima K, Haskell C, Baba M, Nagira M, Nishimura M, Kakizaki M, Takagi S, Nomiyama H, Schall TJ, Yoshie O (1997) Identification and molecular characterization of fractalkine receptor CX3CR1, which mediates both leukocyte migration and adhesion. Cell 91:521–530.

Inagaki H, Tsuruoka H, Hornsby M, Lesley SA, Spraggon G, Ellman JA (2007) Character-ization and optimization of selective, nonpeptidic inhibitors of cathepsin S with an unprecedented binding mode. J Med Chem 50:2693–2699.

Irie O, Kosaka T, Ehara T, Yokokawa F, Kanazawa T, Hirao H, Iwasaki A, Sakaki J, Teno N, Hitomi Y, Iwasaki G, Fukaya H, Nonomura K, Tanabe K, Koizumi S, Uchiyama N, Bevan SJ, Malcangio M, Gentry C, Fox AJ, Yaqoob M, Culshaw A J, Hallett A (2008a) Discovery of orally bioavailable cathepsin S inhibitors for the reversal of neuropathic pain. J Med Chem.

Irie O, Ehara T, Iwasaki A, Yokokawa F, Sakaki J, Hirao H, Kanazawa T, Teno N, Horiuchi M, Umemura I, Gunji H, Masuya K, Hitomi Y, Iwasaki G, Nonomura K, Tanabe K, Fukaya H, Kosaka T, Snell CR, Hallett A (2008b) Discovery of selective and nonpeptidic cathepsin S inhibitors. Bioorgan Med Chem Lett 51:5502–5505.

Jin SX, Zhuang ZY, Woolf CJ, Ji RR (2003) p38 mitogen-activated protein kinase is activated after a spinal nerve ligation in spinal cord microglia and dorsal root ganglion neurons and contributes to the generation of neuropathic pain. J Neurosci 23: 4017–4022.

Kim SH, Smith CJ, Van Eldik LJ (2004) Importance of MAPK pathways for microglial pro-inflammatory cytokine IL-1[beta] production. Neurobiol Aging 25:431–439.

Kubo T, Yamashita T, Yamaguchi A, Hosokawa K, Tohyama M (2002) Analysis of genes induced in peripheral nerve after axotomy using cDNA microarrays. J Neurochem 82:1129–1136.

Lindia JA, McGowan E, Jochnowitz N, Abbadie C (2005) Induction of CX3CL1 expression in astrocytes and CX3CR1 in microglia in the spinal cord of a rat model of neuropathic pain. J Pain 6:434–438.

Liu GY, Kulasingam V, Alexander RT, Touret N, Fong AM, Patel DD, Robinson LA (2005) Recycling of the membrane-anchored chemokine, CX3CL1. J Biol Chem 280:19858–19866.

Liuzzo JP, Petanceska SS, Devi LA (1999a) Neurotrophic factors regulate cathepsin S in macrophages and microglia: a role in the degradation of myelin basic protein and amyloid beta peptide. Mol Med 5:334–343.

Liuzzo JP, Petanceska SS, Moscatelli D, Devi LA (1999b) Inflammatory mediators regulate cathepsin S in macrophages and microglia: a role in attenuating heparan sulfate interactions. Mol Med 5:320–333.

Milligan ED, Zapata V, Chacur M, Schoeniger D, Biedenkapp J, O'Connor KA, Verge GM, Chapman G, Green P, Foster AC, Naeve GS, Maier SF, Watkins LR (2004) Evidence that exogenous and endogenous fractalkine can induce spinal nociceptive facilitation in rats. Eur J Neurosci 20:2294–2302.

Milligan E, Zapata V, Schoeniger D, Chacur M, Green P, Poole S, Martin D, Maier SF, Watkins LR (2005) An initial investigation of spinal mechanisms underlying pain enhancement induced by fractalkine, a neuronally released chemokine. Eur J Neurosci 22:2775–2782.

Murphy PM, Baggiolini M, Charo IF, Hebert CA, Horuk R, Matsushima K, Miller LH, Oppenheim JJ, Power CA (2000) International union of pharmacology. XXII. Nomenclature for chemokine receptors. Pharmacol Rev 52:145–176.

Nakagawa TY, Brissette WH, Lira PD, Griffiths RJ, Petrushova N, Stock J, McNeish JD, Eastman SE, Howard ED, Clarke SRM (1999) Impaired invariant chain degradation and antigen presentation and diminished collagen-induced arthritis in cathepsin S null mice. Immunity 10:207–217.

Nishiyori A, Minami M, Ohtani Y, Takami S, Yamamoto J, Kawaguchi N, Kume T, Akaike A, Satoh M (1998) Localization of fractalkine and CX3CR1 mRNAs in rat brain: does fractalkine play a role in signaling from neuron to microglia? FEBS Lett 429:167–172.

Owolabi SA, Saab CY (2006) Fractalkine and minocycline alter neuronal activity in the spinal cord dorsal horn. FEBS Lett 580:4306–4310.

Pan Y, Lloyd C, Zhou H, Dolich S, Deeds J, Gonzalo JA, Vath J, Gosselin M, Ma J, Dussault B, Woolf E, Alperin G, Culpepper J, Gutierrez-Ramos JC, Gearing D (1997) Neurotactin, a membrane-anchored chemokine upregulated in brain inflammation. Nature 387:611–617.

Petanceska S, Devi L (1992) Sequence analysis, tissue distribution, and expression of rat cathepsin S. J Biol Chem 267:26038–26043.

Petanceska S, Burke S, Watson SJ, Devi L (1994) Differential distribution of messenger RNAs for cathepsins B, L and S in adult rat brain: an in situ hybridization study. Neuroscience 59:729–738.

Petanceska S, Canoll P, Devi LA (1996) Expression of rat cathepsin S in phagocytic cells. J Biol Chem 271:4403–4409.

Pierre P, Mellman I (1998) Developmental regulation of invariant chain proteolysis controls MHC class II trafficking in mouse dendritic cells. Cell 93:1135–1145.

Riese RJ, Mitchell RN, Villadangos JA, Shi GP, Palmer JT, Karp ER, De Sanctis GT, Ploegh HL, Chapman HA (1998) Cathepsin S activity regulates antigen presentation and immunity. J Clin Invest 101:2351–2363.

Riese RJ, Wolf PR, Bromme D, Natkin LR, Villadangos JA, Ploegh HL, Chapman HA (1996) Essential role for cathepsin S in MHC class II-associated invariant chain processing and peptide loading. Immunity 4:357–366.

Roche PA, Cresswell P (1990) Invariant chain association with HLA-DR molecules inhibits immunogenic peptide binding. Nature 345:615–618.

Roche PA, Cresswell P (1991) Proteolysis of the class II-associated invariant chain generates a peptide binding site in intracellular HLA-DR molecules. PNAS 88:3150–3154.

Scholz J, Woolf CJ (2007) The neuropathic pain triad: neurons, immune cells and glia. Nat Neurosci 10:1361–1368.

Schwaeble WJ, Stover CM, Schall TJ, Dairaghi DJ, Trinder PKE, Linington C, Iglesias A, Schubart A, Lynch NJ, Weihe E, Schafer MK (1998) Neuronal expression of fractalkine in the presence and absence of inflammation. FEBS Lett 439:203–207.

Sherman MA, Weber DA, Jensen PE (1995) DM enhances peptide binding to class II MHC by release of invariant chain-derived peptide. Immunity 3:197–205.

Shi GP, Munger JS, Meara JP, Rich DH, Chapman HA (1992) Molecular cloning and expression of human alveolar macrophage cathepsin S, an elastinolytic cysteine protease. J Biol Chem 267:7258–7262.

Shi GP, Villadangos JA, Dranoff G, Small C, Gu L, Haley KJ, Riese R, Ploegh HL, Chapman HA (1999) Cathepsin S required for normal MHC class II peptide loading and germinal center development. Immunity 10:197–206.

Sloan VS, Cameron P, Porter G, Gammon M, Amaya M, Mellins E, Zaller DM (1995) Mediation by HLA-DM of dissociation of peptides from HLA-DR. Nature 375: 802–806.

Thacker MA, Clark AK, Marchand F, McMahon SB (2007) Pathophysiology of peripheral neuropathic pain: immune cells and molecules. Anesth Analg 105:838–847.

Tsou CL, Haskell CA, Charo IF (2001) Tumor necrosis factor-alpha-converting enzyme mediates the inducible cleavage of fractalkine. J Biol Chem 276:44622–44626.

Tsuda M, Shigemoto-Mogami Y, Koizumi S, Mizokoshi A, Kohsaka S, Salter MW, Inoue K (2003) P2X4 receptors induced in spinal microglia gate tactile allodynia after nerve injury. Nature 424:778–783.

Tsuda M, Mizokoshi A, Shigemoto-Mogami Y, Koizumi S, Inoue K (2004) Activation of p38 mitogen-activated protein kinase in spinal hyperactive microglia contributes to pain hypersensitivity following peripheral nerve injury. Glia 45:89–95.

Turnsek T, Kregar I, Lebez D (1975) Acid sulphydryl protease from calf lymph nodes. Biochimica et Biophysica Acta (BBA) – Enzymology 403:514–520.

Valder CR, Liu JJ, Song YH, Luo ZD (2003) Coupling gene chip analyses and rat genetic variances in identifying potential target genes that may contribute to neuropathic allodynia development. J Neurochem 87:560–573.

Vasiljeva O, Dolinar M, Pungercar JR, Turk V, Turk B (2005) Recombinant human pro-cathepsin S is capable of autocatalytic processing at neutral pH in the presence of glycosaminoglycans. FEBS Lett 579:1285–1290.

Verge GM, Milligan ED, Maier SF, Watkins LR, Naeve GS, Foster AC (2004) Fractalkine (CX3CL1) and fractalkine receptor (CX3CR1) distribution in spinal cord and dorsal root ganglia under basal and neuropathic pain conditions. Eur J Neurosci 20:1150–1160.

Villadangos JA, Riese RJ, Peters C, Chapman HA, Ploegh HL (1997) Degradation of mouse invariant chain: roles of cathepsins S and D and the influence of major histocompatibility complex polymorphism. J Exp Med 186:549–560.

Ward YD, Thomson DS, Frye LL, Cywin CL, Morwick T, Emmanuel MJ, Zindell R, McNeil D, Bekkali Y, Girardot M, Hrapchak M, DeTuri M, Crane K, White D, Pav S, Wang Y, Hao MH, Grygon CA, Labadia ME, Freeman DM, Davidson W, Hopkins JL, Brown ML, Spero DM (2002) Design and synthesis of dipeptide nitriles as reversible and potent cathepsin S inhibitors. J Med Chem 45:5471–5482.

Wei J, Pio BA, Cai H, Meduna SP, Sun S, Gu Y, Jiang W, Thurmond RL, Karlsson L, Edwards JP (2007) Pyrazole-based cathepsin S inhibitors with improved cellular potency. Bioorgan Med Chem Lett 17:5525–5528.

Zhuang ZY, Kawasaki Y, Tan PH, Wen YR, Huang J, Ji RR (2007) Role of the CX3CR1/p38 MAPK pathway in spinal microglia for the development of neuropathic pain following nerve injury-induced cleavage of fractalkine. Brain Behav Immun 21:642–651.

Index